Advances in Intelligent Systems and Computing 190

Editor-in-Chief

Prof. Janusz Kacprzyk
Systems Research Institute
Polish Academy of Sciences
ul. Newelska 6
01-447 Warsaw
Poland
E-mail: kacprzyk@ibspan.waw.pl

For further volumes:
http://www.springer.com/series/11156

Rudolf Kruse, Michael R. Berthold,
Christian Moewes, María Ángeles Gil,
Przemysław Grzegorzewski,
and Olgierd Hryniewicz (Eds.)

Synergies of Soft Computing and Statistics for Intelligent Data Analysis

Editors
Prof. Dr. Rudolf Kruse
Otto-von-Guericke University of Magdeburg
Faculty of Computer Science
Magdeburg
Germany

Prof. Dr. María Ángeles Gil
Department of Statistics and OR
University of Oviedo
Oviedo
Spain

Prof. Dr. Michael R. Berthold
University of Konstanz
FB Informatik & Informationswissenschaft
Konstanz
Germany

Prof. Dr. Przemysław Grzegorzewski
Polish Academy of Sciences
Systems Research Institute
Warsaw
Poland

Dipl.-Inform. Christian Moewes
Otto-von-Guericke University of Magdeburg
Faculty of Computer Science
Magdeburg
Germany

Prof. Dr. Olgierd Hryniewicz
Polish Academy of Sciences
Systems Research Institute
Warsaw
Poland

ISSN 2194-5357
e-ISSN 2194-5365
ISBN 978-3-642-33041-4
e-ISBN 978-3-642-33042-1
DOI 10.1007/978-3-642-33042-1
Springer Heidelberg New York Dordrecht London

Library of Congress Control Number: 2012945296

© Springer-Verlag Berlin Heidelberg 2013
This work is subject to copyright. All rights are reserved by the Publisher, whether the whole or part of the material is concerned, specifically the rights of translation, reprinting, reuse of illustrations, recitation, broadcasting, reproduction on microfilms or in any other physical way, and transmission or information storage and retrieval, electronic adaptation, computer software, or by similar or dissimilar methodology now known or hereafter developed. Exempted from this legal reservation are brief excerpts in connection with reviews or scholarly analysis or material supplied specifically for the purpose of being entered and executed on a computer system, for exclusive use by the purchaser of the work. Duplication of this publication or parts thereof is permitted only under the provisions of the Copyright Law of the Publisher's location, in its current version, and permission for use must always be obtained from Springer. Permissions for use may be obtained through RightsLink at the Copyright Clearance Center. Violations are liable to prosecution under the respective Copyright Law.
The use of general descriptive names, registered names, trademarks, service marks, etc. in this publication does not imply, even in the absence of a specific statement, that such names are exempt from the relevant protective laws and regulations and therefore free for general use.
While the advice and information in this book are believed to be true and accurate at the date of publication, neither the authors nor the editors nor the publisher can accept any legal responsibility for any errors or omissions that may be made. The publisher makes no warranty, express or implied, with respect to the material contained herein.

Printed on acid-free paper

Springer is part of Springer Science+Business Media (www.springer.com)

Preface

We are proud to present the proceedings of the 6th International Conference on Soft Methods in Probability and Statistics. The conference took place in Konstanz, Germany, October 4–6, 2012. The SMPS conferences series started in 2002 in Warsaw and moved from Oviedo (2004), Bristol (2006), Toulouse (2008) to Mieres-Oviedo (2010). SMPS'2012 was organized by the Computational Intelligence group in Magdeburg and the Intelligent Data Analysis group in Konstanz. The theme of SMPS2012 was "Synergies of Soft Computing and Statistics for Intelligent Data Analysis".

The main objective of the SMPS conference series is to strengthen the dialogue between various research communities in the field of data analysis in order to cross-fertilize the fields and generate mutually beneficial activities. In SMPS'2012 we were especially interested in bringing experts from the areas of Soft Computing and Statistics together. Both branches have different intentions on data analysis as they stem from computer science and mathematics, respectively. Soft computing is able to quickly produce low-cost solutions using nature-inspired problem-solving strategies. Its ability to adapt to different problems and models led to its success in real-world applications. Also, its inherent necessity to construct understandable and interpretable solutions made soft computing very popular in economical fields. The field of statistics aims at much less subjective goals. It focuses on the need for mathematical methods that validate models and ensure their applicability based on observations maximizing some kind of likelihood. It is our hope that the synergies of both fields improve intelligent data analysis methods in terms of robustness to noise and applicability to larger datasets, while being able to efficiently obtain understandable solutions of real-world problems.

SMPS'2012 provided an attractive interdisciplinary forum for discussions and mutual exchange of knowledge in the field of intelligent data analysis. The 58 papers in this volume were carefully selected by an extensive reviewing process. Every paper has been reviewed by three of 113 international experts. We are delighted that Christian Borgelt, principal researcher at the European Centre for Soft Computing, Lawrence O'Higgins Hall, professor of computer science at the University of South Florida, and Hannu T. Toivonen, professor of computer science at the University of Helsinki, accepted our invitation to present keynote lectures. Part I of the volume contains these invited papers. Part II encloses contributions to the foundations of uncertainty theories including imprecise probability, probability theory and fuzzy set theory. Part III consists of a variety of papers dealing with soft statistical methods ranging from statistical inference to statistical tests. Part IV focuses on mathematical aspects of soft methods applied to probability and statistics, e.g. copulas, decision making, partial knowledge and conditional uncertainty measures. And Part V finally comprises application-orientated papers devoted to engineering. The methods described here exploit information mining, machine learning techniques and computational intelligence. Most applications stem from bioinformatics, human sciences and automobile industry.

The editors are very grateful to all contributing authors, invited speakers, program committee members, and additional referees who made it possible to put together an attractive program for the conference. This conference has benefited from the financial support of the Spanish bank CajAstur, which covered all the production and distribution costs of the proceedings. We thank the editor of the Springer series Advances in Soft Computing, Janusz Kacprzyk, and Springer-Verlag for the dedication to the production of this volume. We are particularly grateful to the universities of Konstanz and Magdeburg as well as the German Society of Computer Science and the European Society for Fuzzy Logic and Technology for their continuous support. Finally, we thank Heather Fyson who did an outstanding job putting the conference on the ground in Konstanz. Months before this conference, she already started working on everything that made this conference run smoothly and so enjoyable for all attendees.

Konstanz and Magdeburg
July 2012

Rudolf Kruse
Michael R. Berthold
Christian Moewes
María Ángeles Gil
Przemysław Grzegorzewski
Olgierd Hryniewicz

Organization

General Chairs

Michael R. Berthold (University of Konstanz, Germany)
Rudolf Kruse (University of Magdeburg, Germany)

Advisory Committee (Core SMPS Group)

María Ángeles Gil (University of Oviedo, Spain)
Przemysław Grzegorzewski (Polish Academy of Sciences, Warsaw, Poland)
Olgierd Hryniewicz (Polish Academy of Sciences, Warsaw, Poland)

Program Committee

Bernard de Baets (Ghent University, Belgium)
Michael Beer (University of Liverpool, United Kindom)
Radim Belohlavek (Palacký University, Olomouc, Czech Republic)
Piero Bonissone (GE Global Research, Niskayuna, NY, USA)
Christian Borgelt (European Centre for Soft Computing, Mieres, Spain)
Bernadette Bouchon-Meunier (Université Pierre et Marie Curie, Paris, France)
Giulianella Coletti (University of Perugia, Italy)
Ana Colubi (University of Oviedo, Spain)
Gert de Cooman (Ghent University, Belgium)
Ines Couso (University of Oviedo, Spain)
Thierry Denœux (Université de Technologie de Compiègne, France)
Sébastien Destercke (Université de Technologie de Compiègne, France)
Didier Dubois (Université Paul Sabatier, Toulouse, France)
Fabrizio Durante (Free University of Bozen, Bolzano, Italy)
Peter Filzmoser (TU Vienna, Austria)
Jörg Gebhardt (Volkswagen AG, Wolfsburg, Germany)
Lluís Godo Lacasa (Universitat Autònoma de Barcelona, Spain)
Gil González-Rodríguez (University of Oviedo, Spain)

Michel Grabisch (Université Panthéon-Sorbonne, France)
Mario Guarracino (Second University of Naples, Italy)
Frank Hoffmann (TU Dortmund, Germany)
Eyke Hüllermeier (University of Marburg, Germany)
Janusz Kacprzyk (Polish Academy of Sciences, Warsaw, Poland)
Uzay Kaymak (Eindhoven University of Technology, The Netherlands)
Etienne Kerre (Ghent University, Belgium)
Vladik Kreinovich (University of Texas at El Paso, TX, USA)
Jonathan Lawry (University of Bristol, United Kingdom)
Marie-Jeanne Lesot (Université Pierre et Marie Curie, Paris, France)
Shoumei Li (Beijing University Of Technology, China)
Edwin Lughofer (Johannes Kepler University, Linz, Austria)
Ralf Mikut (Karlsruhe Institute of Technology, Germany)
Serafin Moral (University of Granada, Spain)
Detlef D. Nauck (British Telecom, Ipswich, United Kindom)
Mirko Navara (Czech Technical University, Praha, Czech Republic)
Vilem Novak (University of Ostrava, Czech Republic)
Henri Prade (Université Paul Sabatier, Toulouse, France)
Dan Ralescu (University of Cincinnati, USA)
Thomas A. Runkler (Siemens AG, Munich, Germany)
Enrique Ruspini (European Centre for Soft Computing, Mieres, Spain)
João Miguel da Costa Sousa (Technical University of Lisbon, Portugal)
Wolfgang Trutschnig (European Centre for Soft Computing, Mieres, Spain)
Bülent Tütmez (İnönü University, Malatya, Turkey)
Stefan Van Aelst (Ghent University, Belgium)
Michel Verleysen (Université catholique de Lovain, Louvain-la-Neuve, Belgium)
Peter Winker (Justus-Liebig-Universität Gießen, Germany)
Marco Zaffalon (University of Lugano, Switzerland)

Additional Referees

Andreas Alfons, Andrés David Báez Sánchez, Pablo Bermejo, Holger Billhardt, Angela Blanco-Fernández, Christian Braune, Andrea Capotorti, María Rosa Casals Varela, Andrea Cerioli, Yun-Shiow Chen, Norberto Corral, Pierpaolo D'Urso, Serena Doria, Luca Fanelli, Juan Fernández Sánchez, Maria Brigida Ferraro, Rachele Foschi, Enrico Foscolo, Marta García-Bárzana, Luis Angel García Escudero, Roberto Ghiselli Ricci, Angelo Gilio, Paolo Giordani, Pascal Held, Mehran Hojati, Karel Hron, Mia Hubert, Jui-Chung Hung, Piotr Jaworski, Huidong Jin, Radim Jirousek, Mehdi Khashei, Gernot D. Kleiter, Tomáš Kroupa, Alexander Lepskiy, Baoding Liu, María Teresa Lopez García, Kevin Loquin, María Asunción Lubiano, Marek T. Malinowski, Peter McCullagh, Deyu Meng, Radko Mesiar, Mariusz Michta, Enrique Miranda, Stephan Morgenthaler, Zoran Ognjanović, Roberta Pappadà, Franco Pellerey, Domenico Perrotta, Georg Peters, Davide Petturiti, Niki Pfeifer,

Florin Popescu, Erik Quaeghebeur, Ana Belén Ramos-Guajardo, Giuliana Regoli, Marco Riani, José Antonio Rodriguez Lallena, Peter Sarkoci, Carlo Sempi, Fabio Spizzichino, Olivier Strauss, Pedro Terán, Valentin Todorov, Matthias Troffaes, Lev Utkin, Alain C. Vandal, Barbara Vantaggi, Jinping Zhang

Local Organization

Heather Fyson (University of Konstanz, Germany)

Publication Chair

Christian Moewes (University of Magdeburg, Germany)

Special Sessions Chair

Frank Klawonn (Ostfalia University of Applied Sciences, Wolfenbüttel, Germany)

Publicity Chair

Andreas Nürnberger (University of Magdeburg, Germany)

Contents

Part I: Invited Papers

Part II: Foundations

Imprecise Probabilities

Probability and Fuzzy Sets

Part III: Statistical Methods

Statistics with Imprecise Data

Fuzzy Random Variables

Computational Intelligence

Industrial Applications

Machine Learning

Part I
Invited Papers

Soft Pattern Mining in Neuroscience

Christian Borgelt

Abstract. While the lower-level mechanisms of neural information processing (in biological neural networks) are fairly well understood, the principles of higher-level processing remain a topic of intense debate in the neuroscience community. With many theories competing to explain how stimuli are encoded in nerve signal (spike) patterns, data analysis tools are desired by which proper tests can be carried out on recorded parallel spike trains. This paper surveys how pattern mining methods, especially soft methods that tackle the core problems of *temporal imprecision* and *selective participation*, can help to test the *temporal coincidence coding hypothesis*. Future challenges consist in extending these methods, in particular to the case of *spatio-temporal coding*.

1 Introduction

Basically all information transmission and processing in humans and animals is carried out by the *nervous system*, which is a network of special cells called *neurons* or *nerve cells*. These cells communicate with each other by electrical and chemical signals. While the lower-level mechanisms are fairly well understood (see Section 2) and it is widely accepted in the neuroscience community that stimuli are encoded and processed by cell assemblies rather than single cells [17, 23], it is still a topic of intense ongoing debate how exactly information is encoded and processed on such a higher level: there are many competing theories, each of which has its domain of validity. Due to modern multi-electrode arrays, which allow to record the electrical signals emitted by hundreds of neurons in parallel [8], more and more data becomes available in the form of (massively) *parallel spike trains* that can help to tackle the challenge of understanding higher-level neural information processing.

Christian Borgelt
European Centre for Soft Computing, Edificio de Investigación, 33600 Mieres, Asturias, Spain
e-mail: `christian@borgelt.net`

R. Kruse et al. (Eds.): Synergies of Soft Computing and Statistics, AISC 190, pp. 3–10.
springerlink.com © Springer-Verlag Berlin Heidelberg 2013

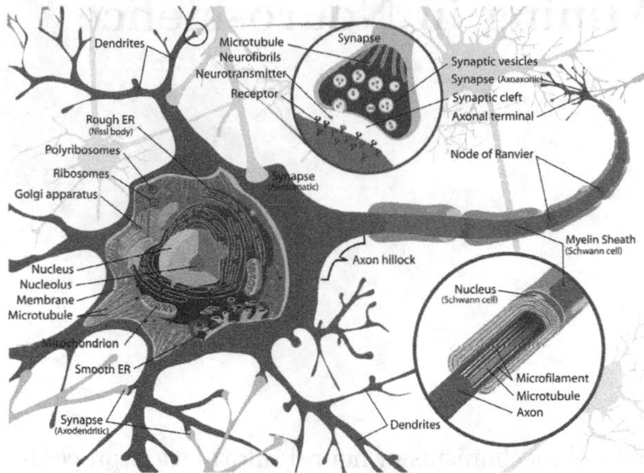

Fig. 1 Diagram of a typical myelinated vertebrate motoneuron (source: Wikipedia [27]), showing the main parts involved in its signaling activity like the *dendrites*, the *axon*, and the *synapses*

After reviewing some of the main competing models of neural information coding (Section 2), this paper focuses on the *temporal coincidence coding hypothesis*. It explores how pattern mining methods can help in the search for synchronous spike patterns in parallel spike trains (Section 3) and considers, in particular, soft methods that can handle the core problems of *temporal imprecision* and *selective participation* (Section 4). The paper closes with an outlook on future work, especially tackling the challenge of identifying *spatio-temporal patterns* under such conditions (Section 5).

2 Neural Information Processing

Essentially, neurons are electrically excitable cells that send signals to each other. The mechanisms are well understood on a physiological and chemical level, but how several neurons coordinate their activity is not quite clear yet.

Physiology and Signaling Activity. Neurons are special types of cells that can be found in most animals. They connect to each other, thus forming complex networks. Attached to the *cell body* (or *soma*) are several arborescent branches that are called *dendrites* and one longer cellular extension called the *axon*. The axon terminals form junctions, so-called *synapses*, with the dendrites or the cell bodies of other neurons (see Figure 1) [14].

The most typical form of communication between neurons (this is a *very* simplified description!) is that the axon terminals of a neuron release chemical substances, called *neurotransmitters*, which act on the membrane of the connected neuron and change its polarization (its electrical potential). Synapses that reduce the potential difference between the inside and the outside of the membrane are called *excitatory*, those that increase it, *inhibitory*. Although the change caused by a single synapse is comparatively small, the effects of multiple synapses accumulate. If the total excitatory input is large enough, the start of the axon becomes, for a short period of time (around 1ms), depolarized (i.e. the potential difference is inverted). This sudden change of the electrical potential, called *action potential*, travels along the axon, with the speed depending on the amount of *myelin* present. When this nerve impulse reaches the end of the axon, it triggers the release of neurotransmitters. Thus the signal is passed on to the next neuron [14]. The electrical signals can be recorded with electrodes, yielding so-called *spike trains.*

Neural Information Coding. It is widely accepted in the neuroscience community that stimuli and other pieces of information are not represented by individual neurons and their action potentials, but that multiple neurons work together, forming so-called *cell assemblies.* However, there are several competing theories about how exactly the information is encoded. The main models that are considered include, but are not limited to the following [23]:

- **Frequency Coding** [29, 12]
  Neurons generate spikes trains with varying frequency as a response to different stimulus intensities: the stronger the stimulus, the higher the spike frequency. Frequency coding is used in the motor system, which directly or indirectly controls muscles, because the rate at which a muscle contracts is correlated with the number of spikes it receives. Frequency coding has also been shown to be present in the sensory system.
- **Temporal Coincidence Coding** [21, 30, 19, 25]
  Tighter coincidence of spikes recorded from different neurons represent higher stimulus intensity, with spike occurrences being modulated by local field oscillation [23]. A temporal coincidence code has the advantage that it leads to shorter "switching times," because it avoids the need to measure a frequency, which requires to observe multiple spikes. Therefore it appears to be a better model for neural processing in the cerebral cortex.
- **Delay Coding** [18, 9]
  The input stimulus is converted into a spike delay (possibly relative to some reference signal). A neuron that is stimulated more strongly reaches the depolarization threshold earlier and thus initiates a spike (action potential) sooner than neurons that are stimulated less strongly.
- **Spatio-Temporal Coding** [2, 1]
  Neurons emit a causal sequence of spikes in response to a stimulus configuration. A stronger stimulus induces spikes earlier and initiates spikes in other, connected cells. The sequence of spike propagation is determined by the spatio-temporal configuration of the stimulus as well as the connectivity of the network [23]. This coding model can be seen as integrating the temporal coincidence and the delay coding principles.

Among other models a spatio-temporal scheme based on a frequency code [28] is noteworthy. In this model the increased spike frequencies form specific spatio-temporal patterns over the involved neurons. Thus it can be seen as combining spatio-temporal coding with frequency coding.

3 Detecting Synchronous Activity

This paper focuses on the temporal coincidence coding hypothesis and thus on the task to detect unusual synchronous spiking activity in recorded parallel spike trains, where "unusual" means that it cannot be explained as a chance event. In addition, we do not merely consider whether a parallel spike train contains synchronous spiking activity (e.g. [31]) or whether a given neuron participates in the synchronous spiking activity of a cell assembly (of otherwise unknown composition) (e.g. [4]). Rather we concentrate on the most complex task of identifying specific assemblies that exhibit(s) (significant) synchronous spiking activity (e.g. [13, 3]). Tackling this task is computationally expensive for (massively) parallel spike trains due to a combinatorial explosion of possible neuron groups that have to be examined.

Other core problems are *temporal imprecision* and *selective participation*. The former means that it cannot be expected that spikes are temporally perfectly aligned, while the latter means that only a subset of the neurons in an assembly may participate in any given synchronous spiking event, with the subset varying between different such events. Note that both may be the effect of deficiencies of the spike recording process (the spike time or even whether a spike occurred is not correctly extracted from the measured profile of the electrical potential) or may be due to the underlying biological process (delays or even failures to produce a spike due to lower total synaptic input, as neurons may receive signals coding different information in parallel).

The most common (or even: the almost exclusively applied) method of handling temporal imprecision is time binning: given a user-specified bin width, a spike train, which is originally a (continuous) point process of spike times, is turned into a binary sequence: a 1 means that the corresponding neuron produced a spike and a 0 that there is no spike in the corresponding time bin. In this way the problem is essentially transformed into a frequent item set mining problem [3]. The translation of the relevant notions to market basket analysis (for which frequent item set mining was originally developed) and to spike train analysis is shown in Table 1. Clearly, the problems are structurally equivalent and thus can be attacked with the same means.

The standard problem of frequent item set mining—namely that a huge number of frequent item sets may be found, most of them *false discoveries*—is best addressed by randomization methods [22, 15]. In spike train analysis, these methods take the form of surrogate data generation schemes, since one tries to preserve as many properties (that are deemed biologically relevant, e.g. inter-spike intervals) as possible, while destroying the coincidences. A survey of such surrogate data generation methods can be found in [20].

In essence, an assembly detection method then works as follows: a sufficient number of surrogate data sets (say, 1000 or 10,000) are created and mined for frequent item sets, which are identified by their size (number of neurons) and support (number of coincidences). Then the original data set is mined and if patterns of a size and support (but ignoring the exact composition by

Table 1 Translation of basic notions of frequent item set mining to market basket analysis (for which it was originally developed) and to spike train analysis

mathematical problem	market basket analysis	spike train analysis
item	product	neuron
item base	set of all products	set of all neurons
— (transaction id)	customer	time bin
transaction	set of products bought by a customer	set of neurons firing in a time bin
frequent item set	set of products frequently bought together	set of neurons frequently firing together

neurons) can be found that do not show up in any of the surrogate data sets, these patterns can be considered significant results.

4 Soft Pattern Mining

Accepting time binning for now as a simple (though deficient, see below) method for handling temporal imprecision, let us turn to the problem of selective participation. In the framework of frequent item set mining this is a well-known problem for which many approaches exist (see, e.g., [5]). The core idea is this: in standard frequent item set mining a transaction (time bin) supports an item set (neuron set) only if all items in the set are present. By relaxing the support definition, allowing for some items of a given set to be missing from a transaction, we arrive at *fault-tolerant item set mining*. The various algorithms for this task can be roughly categorized into (1) error-based approaches, which allow for a maximum number of missing items, (2) density-based approaches, which allow for a maximum fraction of missing items, and (3) cost-based approaches, which reduce the support contribution of a transaction depending on the number of missing items (and may, in addition, restrict the number of missing items) [5].

However, such approaches suffer from the even larger search space (as more item sets need to be examined) and thus can increase the computational costs considerably. An alternative approach that avoids an exhaustive enumeration relies on distance measures for binary vectors [10] and uses multi-dimensional scaling [11] to a single dimension to group neurons together that exhibit similar spiking activity [7]. The actual assemblies are then discovered by traversing the neurons according to their image location and testing for dependence. The approach of computing distances of time-binned spike trains has been extended to various well-known clustering methods in [6].

All of the mentioned methods work on time binned data. However, the time binning approach has several severe drawbacks. In the first place, the induced concept of synchrony is two-valued, that is, spikes are either synchronous

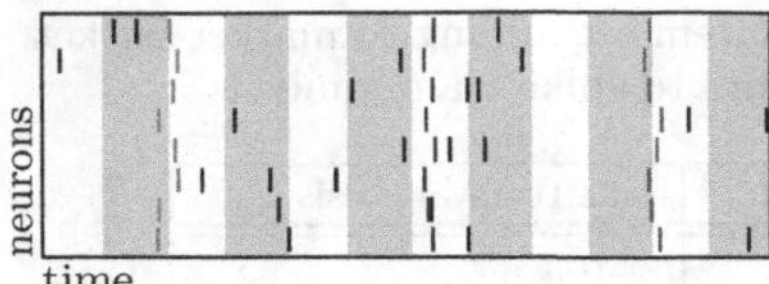

Fig. 2 Eight parallel spike trains with three coincident spiking events (shown in color), two of which are disrupted by time bin boundaries (time bins indicated by gray and white stripes)

(namely if they lie in the same time bin) or not. We have no means to express that the spikes of some coincident event are better aligned than those of another. Secondly, time binning leads to anomalies: two spikes that are (very) close together in time, but happen to be on different sides of a time bin boundary are seen as not synchronous, while two spikes that are almost as far apart as the length of a time bin, but happen to fall into the same time bin, are seen as synchronous. Generally, the location of the time bin boundaries can have a disruptive effect. This is illustrated in Figure 2, where two of the three coincidences of the eight neurons (shown in color) cannot be detected, because they are split by badly placed time bin boundaries.

These problems have been addressed with the influence map approach (see [24, 6]), which bears some resemblance to the definition of a distance measure for continuous spike trains suggested in [26]. The core idea is to surround each spike time with an influence region, which specifies how imprecisely another spike may be placed, which is still to be considered as synchronous. Thus one can define a graded notion of synchrony based on the (relative) overlap of such influence regions. Unfortunately, a direct generalization of binary distance measures to this case (using properly scaled durations instead of time bin counts) seems to lose too much information due to the fact that full synchrony can only be achieved with perfectly aligned spikes [6].

As a solution one may consider specific groups of spikes, one from each neuron, rather than intersecting, over a set of neurons, the union of the influence regions of the spikes of each neuron. This allows to define ϵ-tolerant synchrony, which is 1 as long as the temporal imprecision is less than a user-specified ϵ and becomes graded only beyond that. In addition, extensions to the fault-tolerant case are possible by allowing some spikes to be missing.

5 Future Challenges

The methods reviewed in this paper were devised to detect synchronous activity. However, attention in the neuroscience community shifts increasingly towards spatio-temporal spike patterns as the more general concept, which contains synchronous spiking as a special case. If the time binning approach is accepted, frequent pattern mining offers readily available solutions, for example, in the form of the Spade [33] and cSpade algorithms [32]. However, these approaches require discretized time. Similarly, approaches developed in the neuroscience community (e.g. [2]) are based on time bins, and thus suffer from the mentioned anomalies. In addition, these methods cannot handle faults, in the sense of individual missing spikes: they only count full occurrences of the potential patterns. It is a challenging, but very fruitful problem to extend these approaches (possibly with influence maps) to continuous time or find alternative methods that can handle both faults and continuous time.

Acknowledgements. I am grateful to Denise Berger, Christian Braune, Iván Castro-León, George Gerstein, Sonja Grün, Sebastien Louis, and David Picado-Muiño (listed alphabetically by last name, no precedence expressed), with whom I have cooperated on several of the topics described in this paper.

References

1. Abeles, M., Bergman, H., Margalit, E., Vaadia, E.: Spatiotemporal firing patterns in the frontal cortex of behaving monkeys. J. Neurophysiol. 70(4), 1629–1638 (1993)
2. Abeles, M., Gerstein, G.L.: Detecting spatiotemporal firing patterns among simultaneously recorded single neurons. J. Neurophysiol. 60(3), 909–924 (1988)
3. Berger, D., Borgelt, C., Diesmann, M., Gerstein, G., Grün, S.: An accretion based data mining algorithm for identification of sets of correlated neurons. BMC Neurosci. 10, 1–2 (2009)
4. Berger, D., Borgelt, C., Louis, S., Morrison, A., Grün, S.: Efficient identification of assembly neurons within massively parallel spike trains. Comput. Intell. Neurosci. 2010, 1–18 (2010)
5. Borgelt, C., Braune, C., Kötter, T., Grün, S.: New algorithms for finding approximate frequent item sets. Soft Comput 16(5), 903–917 (2012)
6. Braune, C.: Analysis of parallel spike trains with clustering methods. Master's thesis, Faculty of Computer Science, Otto-von-Guericke University of Magdeburg, Germany (2012)
7. Braune, C., Borgelt, C., Grün, S.: Finding Ensembles of Neurons in Spike Trains by Non-linear Mapping and Statistical Testing. In: Gama, J., Bradley, E., Hollmén, J. (eds.) IDA 2011. LNCS, vol. 7014, pp. 55–66. Springer, Heidelberg (2011)
8. Buzsáki, G.: Large-scale recording of neuronal ensembles. Nat. Neurosci. 7(5), 446–451 (2004)
9. Buzsáki, G., Chrobak, J.J.: Temporal structure in spatially organized neuronal ensembles: a role for interneuronal networks. Curr. Opin. Neurobiol. 5(4), 504–510 (1995)
10. Choi, S., Cha, S., Tappert, C.C.: A survey of binary similarity and distance measures. J. Systemics Cybern. Inform. 8(1), 43–48 (2010)
11. Cox, T.F., Cox, M.A.A.: Multidimensional Scaling, 2nd edn. Chapman & Hall, London (2001)
12. Eccles, J.C.: The Physiology of Nerve Cells. Johns Hopkins University Press, Baltimore (1957)
13. Gerstein, G.L., Perkel, D.H., Subramanian, K.N.: Identification of functionally related neural assemblies. Brain Res. 140(1), 43–62 (1978)
14. Gilman, S., Newman, S.: Essentials of Clinical Neuroanatomy and Neurophysiology, 10th edn. F.A. Davis Company, Philadelphia (2002)

15. Gionis, A., Mannila, H., Mielikäinen, T., Tsaparas, P.: Assessing data mining results via swap randomization. ACM Trans. Knowl. Discov. Data 1(3) (2007)
16. Grün, S., Rotter, S. (eds.): Analysis of Parallel Spike Trains. Springer Series in Computational Neuroscience, vol. 7. Springer US, New York (2010)
17. Hebb, D.O.: The Organization of Behavior: A Neuropsychological Theory. John Wiley & Sons, Ltd., New York (1949)
18. Hopfield, J.J.: Pattern recognition computation using action potential timing for stimulus representation. Nature 376(6535), 33–36 (1995)
19. König, P., Engel, A.K., Singer, W.: Integrator or coincidence detector? the role of the cortical neuron revisited. Trends Neurosci. 19(4), 130–137 (1996)
20. Louis, S., Borgelt, C., Grün, S.: Generation and selection of surrogate methods for correlation analysis. In: Grün and Rotter [16], pp. 359–382
21. von der Malsburg, C., Bienenstock, E.: A neural network for the retrieval of superimposed connection patterns. Europhys. Lett. 3(11), 1243–1249 (1987)
22. Megiddo, N., Srikant, R.: Discovering predictive association rules. In: Proc. 4th Int. Conf. on Knowledge Discovery and Data Mining (KDD 1998), New York, NY, pp. 274–278 (1998)
23. Nádasy, Z.: Spatio-temporal patterns in the extracellular recording of hippocampal pyramidal cells: From single spikes to spike sequences. Ph.D. thesis, Rutgers University, Newark, NJ, USA (1998)
24. Picado-Muiño, D., Castro-León, I., Borgelt, C.: Continuous-time characterization of spike synchrony and joint spiking activity in parallel spike trains (submitted, 2012)
25. Riehle, A., Grün, S., Diesmann, M., Aertsen, A.: Spike synchronization and rate modulation differentially involved in motor cortical function. Science 278(5345), 1950–1953 (1997)
26. van Rossum, M.C.: A novel spike distance. Neural Comput. 13(4), 751–763 (2001)
27. Ruiz Villarreal, M.: Complete neuron cell diagram (2007), `http://commons.wikimedia.org/wiki/File:Complete_neuron_cell_diagram_en.svg`
28. Seidemann, E., Meilijson, I., Abeles, M., Bergman, H., Vaadia, E.: Simultaneously recorded single units in the frontal cortex go through sequences of discrete and stable states in monkeys performing a delayed localization task. J. Neurosci. 16(2), 752–768 (1996)
29. Sherrington, C.S.: The Integrative Action of the Nervous System. Yale University Press, New Haven (1906)
30. Singer, W.: Synchronization of cortical activity and its putative role in information processing and learning. Annu. Rev. Physiol. 55, 349–374 (1993) PMID: 8466179
31. Staude, B., Rotter, S., Grün, S.: CuBIC: cumulant based inference of higher-order correlations in massively parallel spike trains. J. Comput. Neurosci. 29(1-2), 327–350 (2010)
32. Zaki, M.J.: Sequence mining in categorical domains: incorporating constraints. In: Proc. 9th Int. Conf. on Information and Knowledge Management, CIKM 2000, pp. 422–429. ACM, New York (2000)
33. Zaki, M.J.: SPADE: an efficient algorithm for mining frequent sequences. Mach. Learn. 42(1-2), 31–60 (2001)

Exploring Big Data with Scalable Soft Clustering

Lawrence O. Hall

Abstract. Sky surveys for Astronomy are expected to generate 2.5 petabytes a year. Electronic medical records hold the promise of treatment comparisons, grouping patients by outcomes but will be contained in petabyte data storage. We can store lots of data and much of it wont have labels. How can we analyze or explore the data? Unsupervised clustering, fuzzy, possibilistic or probabilistic will allow us to group data. However, the algorithms scale poorly in terms of computation time as the data gets large and are impractical without modification when the data exceeds the size of memory. We will explore distributed clustering, stream data clustering and subsampling approaches to enable scalable clustering. Examples will show that one can scale to build good models of the data without necessarily seeing all the data and, if needed, modified algorithms can be applied to terabytes and more of data.

1 Introduction

There is a deluge of electronic data currently available and more coming available. This data comes from diverse sources such as Astronomy where PAN-Starrs is expected to generate 2.5 petabytes of data per year, daily collections of text from newspapers, blogs, etc., medical data that is being collected digitally (including medical images), images of underwater plankton, and more. There will be no class labels for most of this data. For some data sets all of the data will be unlabeled. To explore and make sense of the data, we need approaches such as clustering which groups data into clusters of like data without requiring any class labels [8, 1].

Lawrence O. Hall
Department of Computer Science and Engineering, ENB118,
University of South Florida, Tampa, FL 33620-9951, USA
e-mail: hall@cse.usf.edu

R. Kruse et al. (Eds.): Synergies of Soft Computing and Statistics, AISC 190, pp. 11–15.
springerlink.com © Springer-Verlag Berlin Heidelberg 2013

Very large data sets can be practically defined as any which exceed the size of a computers memory. Once data exceeds the size of memory, algorithms to cluster or group the data will run very, very slowly. There will be lots of accesses to secondary storage, typically a disk drive, that will cause the computation time to be bounded by the transfer speed of the disk. Unfortunately, this time is orders of magnitude slower than the speed at which the CPU operates. Hence, we need to have clustering algorithms that are designed for large data and minimize the number of disk accesses. Ideally, all data is loaded into memory only one time. This will mean data is discarded and summaries of it must be retained if all the data is accessed.

In this paper, we discuss variations of fuzzy c-means [7] (FCM) which is an iterative clustering algorithm based on the venerable k-means clustering algorithm [8]. The FCM algorithm is known to converge to a local minima or saddle point [4] as do the variants of it discussed here [3]. In addition to modified clustering algorithms to handle very large data sets, subsampling the data can be explored. The subsample can be random or use intelligent selection. The subsample needs to be smaller than the size of memory to avoid significant slowdowns. Intelligently selecting samples from a very large data set is often infeasible because one likely needs to examine all of the examples, potentially accessing some multiple times.

In this paper, we will discuss the advantages, disadvantages and performance of two large-scale fuzzy clustering algorithms and a subsampling approach that can be used with fuzzy c-means.

2 Single Pass and Online Fuzzy C-Means

One way to process all the data of a very large data set is to apply divide and conquer principles. Load as much data as will fit in memory, cluster it, and keep some type of model of the data for future use. We present two extensions to FCM which work on subsets of the data in different ways. They both rely on weighted examples. We can view each example $\mathbf{x}$ as a feature vector of dimension $\mathbf{s}$. The default weight for an example w_i is 1. FCM variants produce cluster centers or centroids which are representative of the cluster. These centroids can be assigned weights based on the fuzzy memberships of examples in the clusters. So, a cluster would have the weight calculated as shown in Equation 1:

$$w_{c_i} = \sum_{j=1}^{n} u_{ij}, \tag{1}$$

where u_{ij} is the membership of example $\mathbf{x}_j$ in the ith cluster c_i. We can then use the weighted examples representing cluster centers in the clustering process [2, 10]. The clustering is done with a weighted version of the classic fuzzy c-means algorithm.

The single pass fuzzy c-means clustering (SPFCM) [6] algorithm clusters a chunk of data, creates weighted cluster centers and then processes another chunk of data together with the weighted cluster centers from the previous chunk. The process continues until all the data has been processed. It makes one pass through the data and outputs a set of cluster centers. To determine the cluster each example belongs to a separate pass through the data may be necessary. For newly encountered examples, just a comparison to the final cluster centers is needed.

The online fuzzy c-means (OFCM) clustering algorithm [5] clusters a chunk of data that will fit in memory and obtains weighted cluster centers. It stores the weighted cluster centers. At the end of processing (or intermediate steps) it clusters all of the weighted cluster centers to get a final set of cluster centers. It is designed for streaming (i.e. never ending) data. It can also be applied to existing data sets, where it has some potential advantages.

The SPFCM algorithm will perform poorly if the data is ordered by class, for example. In that case, some chunks will likely have examples from only one, or certainly less than all, class(es) in the data. Since the number of clusters is fixed, this will likely cause a poor set of cluster centers to be created. When there is more than one heavily weighted example from the same cluster it tends to result in multiple final clusters from the same class. Alternatively, the weighted example is assigned to a cluster that represents another class where it will have a strong negative effect on the cluster center location. The problem is exacerbated when clusters are close together.

On the other hand the OFCM clustering algorithm will simply have multiple clusters that can later be combined into one cluster, as long as we choose a fixed number of clusters greater than or equal to the true number. So, in theory it should have less problems when the data for a chunk does not contain a set of examples reflective of the true class distribution.

When applied to existing data, OFCM can be run in parallel on as many processors as necessary (if available). Then the resulting weighted clusters can be clustered. So, it can be completed with just 2 sequential applications of the clustering algorithm, thus allowing for it to be fast in a parallel processing environment.

3 Subsampling

One way of doing subsampling is to select random examples until the subset passes a test [10]. Extensible Fast Fuzzy c-means (eFFCM) randomly samples the dataset (with replacement) in an effort to obtain a statistically significant sample. Statistical significance is tested for with the Chi-square (χ^2) statistic or divergence. If the initial sample fails testing, additional data is added to the sample until the statistical test is passed [9]. There is time required to do the necessary statistical tests on the random data. However, there is only one data set needed for fuzzy c-means clustering (assuming the size is less than available memory, if not SPFCM or OFCM can be applied).

4 Experiments and Results

One way to evaluate the large data clustering approaches discussed here is to evaluate how close their final cluster centers are to FCM applied to all the data. If they are close or the same, then the speed-ups become important. The algorithms have mostly been evaluated on large volumes of magnetic resonance images of the human brain [6]. The images have 3 features and between 3.8 and 4.3 million examples. The fast algorithm results generally have very good fidelity to FCM. The fastest algorithm is usually SPFCM (with a between 3 and 8 times speed-up depending on chunk size) and eFCM also has a good speed-up and sometimes the best fidelity.

5 Conclusions

Mofications to fuzzy c-means can be used to effectively cluster very large data sets. They have the advantage of convergence and inheriting the well understood properties of FCM. OFCM needs to run in parallel to get the most speed advantages on existing data. SPFCM and eFCM are effective on large data sets with perhaps a little more speed-up using SPFCM. There is a need for scalable clustering algorithms that can find very small clusters and that is a challenge the FCM variants may not be up to.

References

1. Bezdek, J.C.: Pattern Recognition with Fuzzy Objective Function Algorithms. Plenum Press, New York City (1981)
2. Gu, Y., Hall, L.O., Goldgof, D.B.: Evaluating scalable fuzzy clustering. In: Proc. 2010 IEEE Int. Conf. on Systems Man and Cybernetics (SMC), Istanbul, Turkey, October 10-13, pp. 873–880. IEEE Press (2010)
3. Hall, L., Goldgof, D.: Convergence of the single-pass and online fuzzy c-means algorithms. IEEE Trans. Fuzzy Syst. 19(4), 792–794 (2011)
4. Hathaway, R.J., Bezdek, J.C., Tucker, W.T.: An improved convergence theory for the fuzzy c-means clustering algorithms. In: Bezdek, J.C. (ed.) Analysis of Fuzzy Information: Applications in Engineering and Science, vol. 3, pp. 123–131. CRC Press, Boca Raton (1987)

5. Hore, P., Hall, L., Goldgof, D., Cheng, W.: Online fuzzy c means. In: Ann. Meeting of the North American Fuzzy Information Processing Society (NAFIPS 2008), pp. 1–5 (2008)
6. Hore, P., Hall, L.O., Goldgof, D.B., Gu, Y., Maudsley, A.A., Darkazanli, A.: A scalable framework for segmenting magnetic resonance images. J. Sign. Process. Syst. 54, 183–203 (2009)
7. Hung, M.C., Yang, D.L.: An efficient fuzzy c-means clustering algorithm. In: Proc. 2001 IEEE Int. Conf. on Data Mining (ICDM 2001), pp. 225–232. IEEE Press (2001)
8. Jain, A.K.: Data clustering: 50 years beyond k-means. Pattern Recogn. Lett. 31(8), 651–666 (2010)
9. Pal, N.R., Bezdek, J.C.: Complexity reduction for "large image" processing. IEEE Trans. Syst. Man Cybern. 32(5), 598–611 (2002)
10. Parker, J.K., Hall, L.O., Bezdek, J.C.: Comparison of scalable fuzzy clustering methods. In: Proc. IEEE Int. Conf. on Fuzzy Systems (FUZZ-IEEE 2012), Brisbane, Australia, June 10-15, pp. 359–367. IEEE Press (2012)

On Creative Uses of Word Associations

Hannu Toivonen, Oskar Gross, Jukka M. Toivanen, and Alessandro Valitutti

Abstract. The ability to associate concepts is an important factor of creativity. We investigate the power of simple word co-occurrence analysis in tasks requiring verbal creativity. We first consider the Remote Associates Test, a psychometric measure of creativity. It turns out to be very easy for computers with access to statistics from a large corpus. Next, we address generation of poetry, an act with much more complex creative aspects. We outline methods that can produce surprisingly good poems based on existing linguistic corpora but otherwise minimal amounts of knowledge about language or poetry. The success of these simple methods suggests that corpus-based approaches can be powerful tools for computational support of creativity.

1 Introduction

The ability to associate concepts, ideas, and problems is an important factor of creativity. Creative people often are able to see or establish connections and analogies where others could not, and this ability may lead to better solutions to problems or new pieces of art.

We are interested in using computers to support or even accomplish tasks involving verbal creativity. In this paper, we will more specifically look at methods that use word associations derived from word co-occurrences in large corpora. For instance, words 'hand' and 'fist' occur relatively often together, indicating that they are semantically related.

More specifically, our goal is to explore the power of word co-occurrences on tasks that require lexical creativity. We keep all other linguistic and world knowledge at a minimum to test how far plain word associations can take

Hannu Toivonen · Oskar Gross · Jukka M. Toivanen · Alessandro Valitutti
Department of Computer Science and HIIT, University of Helsinki, Finland
e-mail: **firstname.lastname@cs.helsinki.fi**

R. Kruse et al. (Eds.): Synergies of Soft Computing and Statistics, AISC 190, pp. 17–24.
springerlink.com © Springer-Verlag Berlin Heidelberg 2013

us. On the other hand, the methods are less dependent on any particular language and resources.

We address two specific tasks. The first one is the taking the Remote Associates Test [10], a psychometric test of creativity. It directly measures the ability to associate words. The second task is generation of poetry, an act with much more complex creative aspects. Also in this case, word associations can be used as a key component of a poetry generation system.

This paper is structured as follows. We first review some background in Section 2. We then address the Remote Associates Test of creativity in Section 3 and generation of poems in Section 4. We conclude in Section 5.

2 Background

We next provide a brief background for word associations: first the RAT creativity test and then word co-occurrence measures.

Remote Associates Test. The Remote Associates Test (RAT) measures the test subject's ability to find associations between words. In the test, three unrelated *cue words* are presented to the subject, e.g., 'thread', 'pine', and 'pain'. The person then tries to identify a fourth word, the *answer word*, which is related to each of the cue words. In this example, the solution is 'needle'.

The Remote Associates Test was developed by Mednick [10] in the 1960s to test creativity defined as "*the forming of associative elements into new combinations, which either meet specified requirements or are in some way useful*". The test is frequently used by psychologists even if some argue that it is not a good measure of creativity.

In practice, RAT measures the ability to discover new associations between concepts that are not typically connected. Performance on RAT also relates to how well one can generate original ideas [5].

Log-likelihood Ratio. We now describe how we use log-likelihood ratio (LLR) to measure how strongly two words are related in a give corpus. We assume a corpus of unstructured documents, and we treat documents as bags of sentences and sentences as bags of words. Instead of sentences, we can consider all n-grams, i.e., sequences of n consecutive words.

The LLR as applied here is based on a multinomial model of co-occurrences of words (see, e.g., Dunning [4]). The multinomial model of any pair $\{x, y\}$ of words has four parameters $p_{11}, p_{12}, p_{21}, p_{22}$, corresponding to the probabilities of events $\{x, y\}$, $\{\neg x, y\}$ $\{x, \neg y\}$ $\{\neg x, \neg y\}$. The ratio of likelihoods of two multinomial models is computed, a null model and an alternative model. The null model assumes independence of words x and y. Their probabilities are estimated as their frequencies in the data, and the probabilities of their different combinations $(p_{11}, \ldots, p_{22})$ are obtained by simple multiplication (assuming independence). The alternative model, in turn, is the maximum likelihood model which assigns all four parameters from their observed frequencies.

The log-likelihood ratio test is then defined as

$$LLR(x,y) = -2\sum_{i=1}^{2}\sum_{j=1}^{2} k_{ij}\log(p_{ij}^{null}/p_{ij}), \quad (1)$$

where k_{ij} are the respective counts. LLR measures how much the observed joint distribution of words x and y differs from their distribution under the null hypothesis of independence, i.e., how strong the association between them is in the given corpus.

Related work. Literature on measuring co-occurrences or collocations of words is abundant. Standard techniques include the following.

Log-likelihood ratio is a non-parametric statistical test often used for co-occurrence analysis [4]. Unlike some other measures, log-likelihood ratio does not overestimate the importance of very frequent words.

Latent Semantic Analysis [3] aims to find a set of concepts (instead of terms) in a corpus using singular value decomposition. The semantic similarity (relatedness) of two words can then be estimated by comparing them in the concept space. Latent semantic analysis has then evolved to *Probabilistic Latent Semantic Analysis* [8] and later to *Latent Dirichlet Allocation* [1].

We are also interested in building *networks of word associations.* Concepts maps, mind maps, and mental maps are some well-known examples of specific types of networks designed to help learning and creativity or to model subjective information processing. As an example of work in this area, Tseng et al. [15] proposed a two-phase concept map construction algorithm which uses fuzzy sets and multiple types of rules to generate concept maps.

3 Solving the Remote Associates Test of Creativity

We now illustrate the power of simple word co-occurrence analysis for the RAT test of creativity [7]. This is, admittedly, a narrow and specific context. However, if the human capability to perform well in RAT is related to creativity, then certainly the capability of a computer performing well is an encouraging indication of its ability to potentially perform creative tasks, or at least to help humans in tasks requiring creativity. The more complex task of creating poetry will be addressed in the next section.

Data. We used 212 RAT items of Bowers *et al.* [2] and Mednick & Mednick [11], divided to a training set of 140 items and a test set of 72 items. As a corpus, we use Google 2-grams [12]. We removed stopwords, i.e., common and therefore uninformative English words, using the NLTK stopword list.

3.1 Modeling RAT Items Computationally

Let quadruple $r = (c_1, c_2, c_3, a)$ denote a RAT item, where c_i is the ith cue word and a is the answer word. In a probabilistic formulation, the task is to predict the most likely answer word a given cue words c_1, c_2, c_3. Assuming independence between the cue words, i.e., using the Naïve Bayes model, we obtain

$$P(a|c_1, c_2, c_3) \propto P(a, c_1, c_2, c_3) = P(a) \prod_{i=1}^{3} P(c_i|a). \qquad (2)$$

We estimate the (conditional) probabilities from the relative frequencies of the words in the Google 2-grams, and find the word a that maximizes Eq. 2. For more details, see Gross *et al.* [7].

The problem is challenging. There are millions of words to choose from, and even when only considering words that co-occur with each of the cue words, there are thousands of possibilities.

3.2 Experiments

When tested on the RAT items from psychometric literature, the above model provided the correct answer in 66% of cases both in the training and the test sets. Clearly, computers can perform well in such limited tests of creativity by simple co-occurrence analysis even if the search space is very large.

Looking at the 33% of unsuccessful cases, the system often answered with a plural form when the correct answer was singular. Additionally, in some of the test items, a cue word does not occur in the 2-grams at all as an individual word, but only as part of a compound word (with the answer word, for instance). Obviously, one could engineer the method to deal with such issues with plurals and compound words, but the main point is already clear: the performance of the system is better than that of an average person. Item-wise solution rates are typically 30–70%, so the performance of 66% correct solutions can actually be considered very good. This is a clear indication that computers can solve some tasks that are considered to require creativity.

4 Creation of Poetry

We now move on to a much more demanding creative task, writing of poems. We outline a corpus-based approach for this task; more details are given by Toivanen *et al.* [14].

In the literature, several different methods and systems have been proposed for poetry generation (e.g., [9, 6, 16, 13]). They use, among others, statistical approaches, case-based reasoning, and evolutionary algorithms. Many of the best performing systems are based on explicitly coded knowledge about the world (e.g., using formal logic) as well as rich linguistic knowledge (e.g., a generative grammar or a tagged corpus of poetical text fragments). A different family of approaches is based on Markov chains or n-grams. They learn a model of word sequences from a given corpus and use this model to produce new poetry. The typical shortcoming of such approaches is that longer sequences of text make no sense grammatically or semantically.

Our goal is to minimize all explicit knowledge about the world or the language, and instead rely on given corpora for implicit knowledge about them. Additionally, some off-the-shelf linguistic analysis tools are needed (lemmatizer, part-of-speech tagger, morphological analyzer and synthesizer). We take corpora as input, just like Markov models, but the method is completely different.

We use two corpora. The first one, called *background corpus*, is used to analyze word co-occurrences and to construct a word association network. This network is used to control the topic and semantic coherence of poetry. The second corpus, called *grammar corpus*, is used as a set of grammatical examples or templates in an instance-based manner.

Data. We currently generate poetry in Finnish. The background corpus is Finnish Wikipedia, and the grammar corpus consists of older Finnish poetry.

4.1 Method

The input to the method essentially consists of three items: the background corpus, the grammar corpus, and a topic word.

The contents and coherence of the poem are controlled by using words that are related to the given topic word in the background corpus, as measured by LLR. The grammatical correctness, in turn, is partially guaranteed by taking a random fragment (e.g., a sentence or a poem) from the grammar corpus, and using its grammatical structure in the generated poem.

More specifically, an example fragment of the desired length is chosen from the grammar corpus. It is then analyzed morphologically for the part of speech, case, verb tense, clitics, etc. of each word.

Then, words (especially verbs, nouns, adjectives and adverbs) in the fragment are substituted independently, one by one, by words associated with the given topic. The substitutes are of the same type with the original words and are transformed to similar morphological forms. The original word is left intact, however, if there are no words associated with the topic that can be transformed to the correct morphological form. This can happen, e.g., if the morphological form is rare or complex.

4.2 Results

We next give some example poems generated by the method, translated from Finnish originals. The first poem is about children's play (in the left column). The original text on which it is based (in the right column) is a fragment of a poem by Uuno Kailas.

Computer-generated poem	Text used as a template
How she played then	how she played once
in a daring, daring whispering	in a big green park
under the pale trees.	under the lovely trees.
She had heard for fun	She had watched for fun
how her whispering	how her smile
drifted as jingle to the wind.	fell down as flowers,

The following poem is about hand. The poem fragment used as a template is by Edith Södergran.

Computer-generated poem	Text used as a template
In a pale fist	In a gloomy forest
in a well-balanced fist,	In a dim forest
the buds are so pale	flowers are so pale
in your image lies a dear child god.	In the shadow lies a sick god

The last example is about snow. The text used as a template is by Eino Leino.

Computer-generated poem	Text used as a template
Lives got the frolic ways,	Waves fared the wind's ways,
snow the home of time,	sun the track of time,
softly chimed abandoned homes,	slowly skied for long days,
softly got frolics beloved –	slowly crept for long nights –
ripening crop got the snows' joys.	day wove the deeds of moons

We evaluated the poetry using a panel of twenty random subjects. Each of them evaluated 22 poems, of which 11 were computer-generated and 11 human-written. The poems were presented in a random order and the subjects were not informed that some of the poems are computer-generated. Each poem was evaluated qualitatively along six dimensions: (1) How typical is the text as a poem? (2) How understandable is it? (3) How good is the language? (4) Does the text evoke mental images? (5) Does the text evoke emotions? (6) How much does the subject like the text? These dimensions were evaluated on the scale from one (very poor) to five (very good).

On each of the dimensions, the 67% confidence intervals of the answers for computer-generated vs. human-written poetry overlap a lot (Figure 1), indicating that a large fraction of computer-generated poetry is as good as human-written poetry, even if on average human-written poetry is better.

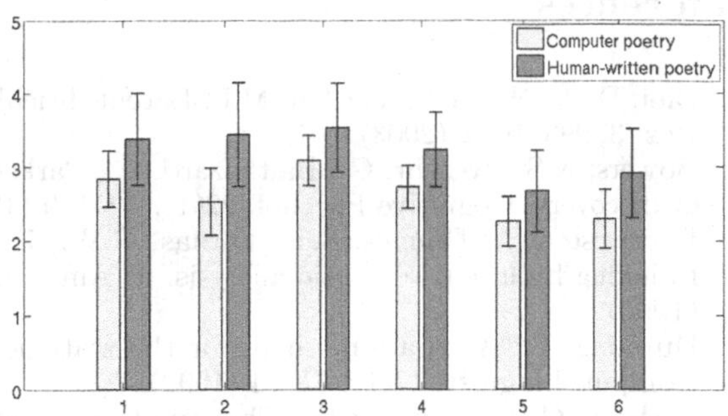

Fig. 1 Subjective evaluation of computer-generated and human-written poetry along six dimensions (see text). Results are averaged over all subjects and poems; whiskers include one standard deviation above and below the mean.

This is a striking result given the simplicity of the methods, and again indicates that simple text analysis methods can be powerful components of verbally creative systems.

5 Conclusions

We have shown how word co-occurrence analysis can be used to perform acts requiring verbal creativity. The Remote Associates Test directly measures the capability to associate words, which is a relatively easy task for a computer when it is given a large corpus. Generation of poetry is a much more complex problem, but word associations together with existing poetry as templates can give surprisingly good results.

The results indicate that word co-occurrence analysis can be a powerful building block of creative systems or systems that support human creativity. While 2-grams were sufficient for achieving a high score on RAT, more relaxed co-occurrences are likely to provide more interesting semantic associations to support or inspire creativity, as suggested by Gross *et al.* [7].

We have used statistical, co-occurrence-based associations of words. The benefit is that their coverage is large, but at the same time they lack explicit semantics. Our results on computational generation of poetry [14] show that this does not prevent them from being used in tasks that demand higher verbal creativity.

In this paper, we have only touched on some specific problems in verbal creativity. We believe that corpus-based approaches can be powerful for many other creative problems, too: they are adaptive and the methods are largely independent of language and resources such as lexicons or knowledge-bases.

Acknowledgements. This work has been supported by the Algorithmic Data Analysis (Algodan) Centre of Excellence of the Academy of Finland.

References

1. Blei, D.M., Ng, A.Y., Jordan, M.I.: Latent dirichlet allocation. J. Mach. Learn. Res. 3, 993–1022 (2003)
2. Bowers, K.S., Regehr, G., Balthazard, C., Parker, K.: Intuition in the context of discovery. Cognitive Psychol. 22(1), 72–110 (1990)
3. Deerwester, S., Dumais, S.T., Furnas, G.W., Landauer, T.K., Harshman, R.: Indexing by latent semantic analysis. J. Am. Soc. Inform. Sci. 41(6), 391–407 (1990)
4. Dunning, T.: Accurate methods for the statistics of surprise and coincidence. Comput. Linguist. 19(1), 61–74 (1993)
5. Forbach, G.B., Evans, R.G.: The remote associates test as a predictor of productivity in brainstorming groups. Applied Psychological Measurement 5(3), 333–339 (1981)
6. Gervás, P.: An expert system for the composition of formal spanish poetry. Knowl-Based Syst. 14(3-4), 181–188 (2001)
7. Gross, O., Toivonen, H., Toivanen, J.M., Valitutti, A.: Lexical creativity from word associations. In: Proc. 7th Int. Conf. on Knowledge, Information and Creativity Support Systems (KICSS 2012), Melbourne, Australia, November 8-10. IEEE Press (accepted for publication, 2012)
8. Hofmann, T.: Probabilistic latent semantic indexing. In: ACM SIGIR Conf. on Research and Development in Information Retrieval, SIGIR 1999, pp. 50–57. ACM, New York (1999)
9. Manurung, H., Ritchie, G., Thompson, H.: Towards a computational model of poetry generation. In: Proc. AISB 2000 Symposium on Creative and Cultural Aspects and Applications of AI, pp. 79–86 (2000)
10. Mednick, S.: The associative basis of the creative process. Psychol. Rev. 69(3), 220–232 (1962)
11. Mednick, S., Mednick, M.: Remote associates test, examiner's manual: college and adult forms 1 and 2, Boston, MA, USA (1967)
12. Michel, J., Shen, Y.K., Aiden, A.P., Veres, A., Gray, M.K., Pickett, J.P., Hoiberg, D., Clancy, D., Norvig, P., Orwant, J., Pinker, S., Nowak, M.A., Aiden, E.L.: Quantitative analysis of culture using millions of digitized books. Science 331(6014), 176–182 (2011)
13. Netzer, Y., Gabay, D., Goldberg, Y., Elhadad, M.: Gaiku: generating haiku with word associations norms. In: Proc. Workshop on Computational Approaches to Linguistic Creativity (CALC 2009), pp. 32–39. Association for Computational Linguistics, Stroudsburg (2009)
14. Toivanen, J.M., Toivonen, H., Valitutti, A., Gross, O.: Corpus-based generation of content and form in poetry. In: Proc. Int. Conf. on Computational Creativity (ICCC 2012), Dublin, Ireland, May 30-June 1, pp. 175–179 (2012)
15. Tseng, S., Sue, P., Su, J., Weng, J., Tsai, W.: A new approach for constructing the concept map. Comput. Educ. 49(3), 691–707 (2007)
16. Wong, W.T., Chun, A.H.W.: Automatic haiku generation using VSM. In: Proc. Int. Conf. on Advances on Applied Computer and Applied Computational Science (ACACOS 2008), pp. 318–323 (2008)

Part II
Foundations

Combining Imprecise Probability Masses with Maximal Coherent Subsets: Application to Ensemble Classification

Sébastien Destercke and Violaine Antoine

Abstract. When working with sets of probabilities, basic information fusion operators quickly reach their limits: intersection becomes empty, while union results in a poorly informative model. An attractive means to overcome these limitations is to use maximal coherent subsets (MCS). However, identifying the maximal coherent subsets is generally NP-hard. Previous proposals advocating the use of MCS to merge probability sets have not provided efficient ways to perform this task. In this paper, we propose an efficient approach to do such a merging between imprecise probability masses, a popular model of probability sets, and test it on an ensemble classification problem.

Keywords: Ensemble, inconsistency, information fusion, maximal coherent subsets.

1 Introduction

When multiple sources provide information about the ill-known value of some variable X it is necessary to aggregate these pieces of information into a single model. In the case where the initial uncertainty models are precise probabilities and where the aggregated model is constrained to be precise as well, there are only a few options to combine the information (see [3] for a complete review).

The situation changes when one considers imprecision-tolerant uncertainty theories, such as possibility theory, evidence theory or imprecise

Sébastien Destercke
UMR Heudiasyc, Université de Technologie de Compiègne,
60200 Compiègne, France
e-mail: sebastien.destercke@hds.utc.fr

Violaine Antoine
LISTIC, Polytech Annecy-Chambéry, 74944 Annecy le Vieux, France
e-mail: violaine.antoine@univ-savoie.fr

R. Kruse et al. (Eds.): Synergies of Soft Computing and Statistics, AISC 190, pp. 27–35.
springerlink.com © Springer-Verlag Berlin Heidelberg 2013

probability theory (see [7]). As they extend both set-theoretic and probabilistic approaches[1], these theories can use aggregation operators coming from both frameworks, i.e., they can generalise intersections and unions of sets as well as averaging methods.

When there is (strong) conflict between information pieces, both conjunctive (intersection) and disjunctive (union) aggregation face some problems: conjunction results are often empty and disjunction results are often too imprecise to be really useful. A theoretically attractive solution to these problems is to use maximal coherent subsets (MCS) [11], that is to consider subsets of sources who are consistent and that are maximal with this property. Aggregation can then be done by combining conjunction within maximal coherent subsets with other aggregation operators, e.g., disjunction. Practically, the main difficulty that faces this approach is to identify MCS, a NP-hard problem in the general case.

Different solutions have been proposed to combine inconsistent pieces of information within the framework of imprecise probability theory. In [10] and [12], hierarchical models are considered. In [1] and [8], Bayesian-like methods (i.e., using conditional probabilities) of aggregation are proposed. In [9] and [14], non-Bayesian methods are studied (although [14] considers that combination methods should commute with Bayesian updating). In the two latter references, MCS are proposed as a solution to combine information pieces that are partially inconsistent, but no practical methods are given to identify MCS.

In this paper, we concentrate on imprecise probability masses and propose a practical approach to apply MCS inspired combination methods to such models. We work in a non-Bayesian framework. Section 2 recalls the necessary background on imprecise probabilities and information fusion. Section 3 describes our approach, of which the most important part is the algorithm to identify MCS. Finally, Section 4 presents an application to ensemble classification, in which resulting classification models are combined using MCS.

2 Preliminaries

The theory of imprecise probabilities [15] is a highly expressive framework to represent uncertainty. This section presents the basics of imprecise probabilities.

2.1 Imprecise Probabilities

Consider a variable X taking values in a finite domaine D_x of n elements $\{x_1, x_2, \ldots, x_n\}$. Basically, imprecise probabilities characterize uncertainty about X by a closed convex set $\mathcal{P}$ of probabilities defined on D_x. To this

[1] Except possibility theory, that does not encompass probabilities as special cases.

set $\mathcal{P}$ can be associated Lower and upper probabilities that are mappings from the power set 2^{D_x} to $[0,1]$. They are respectively denoted $\underline{P}$ and $\overline{P}$ and are defined, for an event $A \subseteq D_x$, as $\underline{P}(A) = \inf_{p \in \mathcal{P}} P(A)$ and $\overline{P}(A) = \sup_{p \in \mathcal{P}} P(A)$. These two measures are dual, in the sense that $\underline{P}(A) = 1 - \overline{P}(A^c)$, with A^c the complement of A. Hence, all the information is contained in only one of them.

Alternatively, one can start from a lower measure $\underline{P}$ and compute the convex set $\mathcal{P}_{\underline{P}} = \{P \in \mathbb{P}(D_x) | P(A) \geq \underline{P}(A),\ \forall A \subseteq D_x\}$ of dominating probability measures ($\mathbb{P}(D_x)$ is the set of all probabilities on D_x). Note that the lower value $P_*(A) = \inf_{P \in \mathcal{P}_{\underline{P}}} P(A)$ need not coincide with $\underline{P}(A)$ in general. If the equality $P_* = \underline{P}$ holds, then $\underline{P}$ is said to be *coherent*. In this paper, we will deal exclusively with coherent lower probabilities. Note that lower probabilities are not sufficient to represent every possible convex sets of probabilities.To represent any convex set $\mathcal{P}$, one actually needs to consider bounds on expectations (see [15]).

2.2 Imprecise Probability Masses (IPM)

Usually, the handling of generic sets $\mathcal{P}$ (and sets represented by lower probabilities) represent a heavy computational burden. In practice using simpler models alleviate this computational burden to the cost of a lower expressivity. Imprecise probability masses [4] (IPM) are such simpler models.

IPM can be represented as a family of intervals $L = \{[l_i, u_i], i = 1, \ldots, n\}$ verifying $0 \leq l_i \leq u_i \leq 1 \forall i$. The interval bounds are interpreted as probability bounds over singletons. They induce a set $\mathcal{P}_L = \{p \in \mathbb{P}(D_x) | l_i \geq p(x_i) \geq u_i,\ \forall x_i \in D_x\}$. An extensive study of IPM and their properties can be found in [4].

A set L of IPM is said to be proper if the condition $\sum_{i=1}^n l_i \geq 1 \geq \sum_{i=1}^n u_i$ holds, and $\mathcal{P}_L \neq \emptyset$ if and only if L is proper. Considered sets are always proper, other types having no interest. To guarantee that lower and upper bounds are reachable for each singleton x_i by at least one probability in $\mathcal{P}_L$, the intervals must verify:

$$\sum_{i \neq j} l_j + u_i \leq 1 \text{ and } \sum_{i \neq j} u_j + l_i \geq 1 \quad \forall i. \tag{1}$$

If L is reachable, lower and upper probabilities of $\mathcal{P}_L$ can be computed as follows:

$$\underline{P}(A) = \max(\sum_{x_i \in A} l_i, 1 - \sum_{x_i \notin A} u_i), \quad \overline{P}(A) = \min(\sum_{x_i \in A} u_i, 1 - \sum_{x_i \notin A} l_i). \tag{2}$$

If L is not reachable, a reachable set L' is obtained by applying Eq. (2) to singletons.

2.3 Basic Combinations of Imprecise Probabilities

When M sources provide information, there are three basic ways to combine this information: through a conjunction, a disjunction or a weighted mean. When information is given by credal sets $\mathcal{P}_i$, $i = 1, \dots, M$, computing these basic combination results present some computational difficulties [9]. Computations become much easier if we consider a set $L_1, \dots, L_M$ of IPM. In this case, if $l_{i,j}, u_{i,j}$ denote respectively the lower and upper probability bounds on element x_i given by source j, (approximated) combinations are as follows:

- Weighted mean ($L_{\sum}$): $l_{i,\sum} = \sum_{j=1,M} w_j l_{i,j}, \quad u_{i,\sum} = \sum_{j=1,M} w_j u_{i,j}$
- Disjunction ($L_{\cup}$): $l_{i,\cup} = \min_{j=1,M} l_{i,j}, \quad u_{i,\cup} = \max_{j=1,M} u_{i,j}$
- Conjunction ($L_{\cap}$):

$$l_{i,\cap} = \max(\max_{j=1,M} l_{i,j}, 1 - \sum_{k \neq i} \min_{j=1,M} u_{i,j}), \tag{3}$$

$$u_{i,\cap} = \min(\min_{j=1,M} u_{i,j}, 1 - \sum_{k \neq i} \max_{i=1,M} l_{i,j})$$

In general, the bounds obtained by conjunction (3) may be non-proper, i.e. may result in an empty $\mathcal{P}_{L_{\cap}}$. $L_1, \dots, L_M$ have a non-empty intersection iff the following conditions [4] hold:

$$\max_{j=1,M} l_{i,j} \leq \min_{j=1,M} u_{i,j} \text{ for every } i \in [1, n] \tag{4}$$

$$\sum_{i=1,n} \max_{j=1,M} l_{i,j} \leq 1 \leq \sum_{i=1,n} \min_{j=1,M} u_{i,j} \tag{5}$$

The first condition ensures that intervals have a non-empty intersection for every singleton, while the second makes sure that the result is a proper probability interval.

3 Maximal Coherent Subsets (MCS) and IPM

This section describes the methods to identify and combine MCS.

3.1 Identifying MCS

When sources provide sets $\mathcal{P}_1, \dots, \mathcal{P}_M$, finding MCS comes down to find every subset $K \subseteq [1, M]$ such that $\bigcap_{i \in K} \mathcal{P}_i \neq \emptyset$ and such that K is maximum with this property (i.e., adding a new set would make the intersection empty).

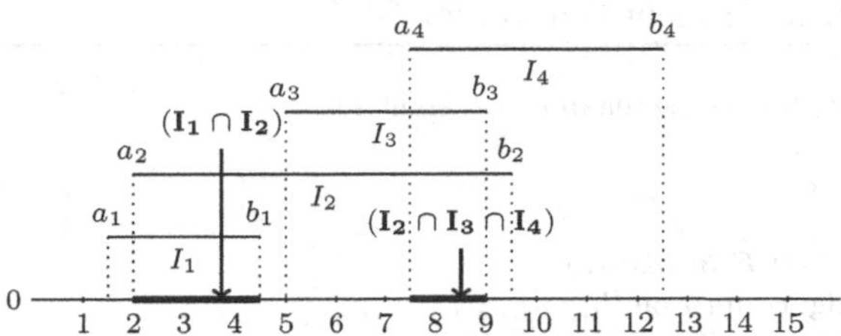

Fig. 1 Maximal coherent subsets on Intervals

Usually, identifying every possible coherent subset among $\mathcal{P}_1, \ldots, \mathcal{P}_M$ is NP-hard, making it a difficult problem to solve in practice.

A particularly interesting case where MCS can be found easily is when each sources provide intervals $[a_i, b_i]$, $i = 1, \ldots, M$. In this case, Algorithm 1 given in [6] finds MCS. It requires to sort values $\{a_i, b_i | i = 1, \ldots, M\}$ (complexity in $\mathcal{O}(M \log M)$), and is then linear in the number of sources.

Algorithm 1. Maximal coherent subsets of intervals

Input: M intervals
Output: List of S maximal coherent subsets K_j

List = $\emptyset$, j=1, $K = \emptyset$;
Order in an increasing order $\{a_i | i = 1, \ldots, M\} \cup \{b_i | i = 1, \ldots, M\}$;
Rename them $\{c_i | i = 1, \ldots, 2M\}$ with *type*$(i) = a$ if $c_i = a_k$ and *type*$(i) = b$ if $c_i = b_k$;
for $i = 1, \ldots, 2M - 1$ **do**
  if *type*$(i) = a$ **then**
    Add Source k to K s.t. $c_i = a_k$;
    if *type*$(i + 1) = b$ **then**
      Add K to List ($K_j = K$) ;
      $j = j + 1$;
  else
    Remove Source k from K s.t. $c_i = b_k$;

This algorithm can be applied directly to IPM intervals to check MCS satisfying Condition (4) (which is necessary for a subset of IPM to have a non-empty intersection). Indeed, consider a singleton x_i and the set of intervals $L^i = [l_{i,j}, u_{i,j}]$, $j = 1, M$: if $K \subseteq [1, M]$ is not a MCS of L^i, then the credal sets $\{\mathcal{P}_j | j \in K\}$ do not form a MCS. Hence, iteratively applying Algorithm 1 as exposed in Algorithm 2 allows to easily identify possible MCS among sets $\mathcal{P}_{L_1}, \ldots, \mathcal{P}_{L_M}$. In each iteration (Line 2), Algorithm 2 refines the MCS found in the previous one (stored in *List*) by finding MCS for probaiblity intervals of singleton x_i (Line 5).

Algorithm 2. MCS identification for IPM

Input: M IPM
Output: List of S possible maximal coherent subsets K_j
1 List = $\{\{1, M\}\}$;
2 **for** $i = 1, \ldots, n$ **do**
3   K=$\emptyset$;
4   **foreach** *subset E in List* **do**
5   Run Algorithm 1 on $[l_{i,j}, u_{i,j}]$, $j \in E$;
6   Add resulting list of MCS to K;
7   List= K;

Example 1. Consider the IPM defined on $D_x = \{x_1, x_2, x_3\}$ and summarised in Table 1. Running Algorithm 2 then provides successively the following MCS: $K = \{\{1,2,3\},\{2,3,4\}\}$ after the first iteration ($i = 1$ in Line 2 of Algorihm 2); $K = \{\{1,2\},\{3\},\{2,4\}\}$ after the second iteration, and K is not changed during the third iteration.

Table 1 Examples of IPM

*Source*1				*Source*2				*Source*3				*Source*4			
	x_1	x_2	x_3		x_1	x_2	x_3		x_1	x_2	x_3		x_1	x_2	x_3
$u_{i,1}$	0.6	0.5	0.2	$u_{i,2}$	0.55	0.55	0.2	$u_{i,3}$	0.5	0.2	0.6	$u_{i,4}$	0.35	0.6	0.35
$l_{i,1}$	0.4	0.3	0.	$l_{i,2}$	0.35	0.35	0.	$l_{i,3}$	0.3	0.	0.4	$l_{i,4}$	0.15	0.4	0.15

Some subsets $K_1, \ldots, K_S$ of sources resulting from Algorithm 2 do not satisfy Condition (5). If K_ℓ is such a set, then one can either make an (exponential) exhaustive search of MCS within K_ℓ, or correct IPM in K_ℓ in a minimal way, so that they satisfy (5). In this last case we can transform [9] bounds $l_{i,j}$ and $u_{i,j}$, $j \in K_\ell, i \in [1,n]$ into $l'_{i,j} = \epsilon l_{i,j}$ and $u'_{i,j} = \epsilon u_{i,j} + (1-\epsilon)$ with ϵ the minimal value such that

$$\sum_{i=1,n} \max_{j \in K_\ell} l'_{i,j} \leq 1 \leq \sum_{i=1,n} \min_{j \in K_\ell} u'_{i,j}. \tag{6}$$

This strategy makes the identification of MCS easy. Roughly speaking, it applies Algorithm 1 to probabilistic (expectation) bounds coming from different sources but bearing on common events (functions). Note that the same strategy can be applied to models based on peculiar families of events (functions), such as p-boxes [5].

3.2 Combination with MCS

Once MCS $K_1, \ldots, K_S$ of sources have been identified, they can be used to combine inconsistent information. Without loss of generality, consider the indexing such that $|K_1| \geq \ldots \geq |K_S|$ where $|K_i|$ is the cardinality of K_i (i.e., the number of sources within it).

We then propose two ways of combining the probability sets $\mathcal{P}_{L_1}, \ldots, \mathcal{P}_{L_M}$. In both of them, we consider the IPM L_{K_ℓ}, $\ell = 1, \ldots, S$ obtained by combining IPM in K_ℓ according to the conjunctive rule (3).

The first rule combines disjunctively the first n IPM L_{K_i}, that is

$$l_{i,\cup\cap_n} = \min_{\ell=1,n} l_{i,K_\ell}, \quad u_{i,\cup\cap_n} = \max_{\ell=1,n} u_{i,K_\ell} \tag{7}$$

where $l_{i,K_\ell}, u_{i,K_\ell}$ are the probability bounds given by L_{K_ℓ} on x_i.

The second rule combines by a weighted mean the first n, L_{K_i}, that is

$$l_{i,\cup\cap_n} = \sum_{\ell=1,n} w_{\ell,n} l_{i,K_\ell}, \quad u_{i,\cup\cap_n} = \sum_{\ell=1,n} w_{\ell,n} u_{i,K_\ell} \tag{8}$$

where $w_{\ell,n} = {|K_\ell|}/{\sum_{i=1,n} |K_i|}$ is the importance of K_ℓ in number of sources (a similar strategy is used in [9]). If $n = S$ the rules simply combine every MCS.

4 Application to Ensemble Classification

Combination is an essential feature of ensemble classification. As classifiers often disagree together, using a MCS based approach to combine the different sources appears sensible. We have therefore tested our approach in the following way: we have trained forest of decision trees; for a given instance and for each decision trees, we have built an IPM model using the Imprecise Dirichlet model (IDM) with an hyperparameter $s = 4$ (see [2] for details) and taking the samples in the tree leaves as observations. We then combined the different IPM with the two rules (7) and (8) (n=5) and selected the final class according to the maximin and maximality criterion (see [13] for details). The former results in a unique decision while the latter results in a set of possible optimal decisions.

Classifier performances are estimated using discounted accuracy: assume we have T observations whose classes $x^i, i = 1, \ldots, T$ are known and for which T (possibly imprecise) predictions $\widehat{X}^1, \ldots, \widehat{X}^T$ have been made. The discounted accuracy $d - acc$ of the classifier is then

$$d - acc = \frac{1}{T} \sum_{i=1}^{T} \frac{\Delta_i}{f(|\widehat{X}_i|)}, \tag{9}$$

with $\Delta_i = 1$ if $x^i \in \widehat{X}^i$, zero otherwise and f an increasing function such that $f(1) = 1$. Set accuracy ($s - acc$) is obtained with $f(|\widehat{X}_i|) = 1$.

Results are summarized in Table 2. Numbers of trees in the forest are $\{10, 20, 50\}$ and the data sets are Zoo, Segment and Satimage (taken from UCI), all of them with 7 classes. Results were compared to a classical voting strategy. We have also indicated the average CPU time needed to apply the different combination rules. From the results, it appears that using a conjunctive rule between provided imprecise probabilistic models does not improve much the results of classical voting. This is not surprising as we use precise decision trees and IDM to build our models, and it would be worthwhile to check whether these conclusions still hold when using credal classifiers. The interest of using imprecise probabilistic models appears when we allow for some imprecision, that is when we adopt a partially disjunctive rule (Rule (7) with n=5). In this latter case, allowing for imprecise classification increases the percentage of well-recognized instances while not decreasing too much the precision. Finally, we can notice that the average computational time does not increase much when the number of sources increases.

Table 2 Results summary. d-acc: discounted accuracy, s-acc: set accuracy, acc: standard accuracy (with maximin)

data set	Tree nb	Single tree	Votes	Rule (7) (n=1) d-acc	s-acc	acc	Rule (7) (n=5) d-acc	s-acc	acc	Rule (8) (n=5) d-acc	s-acc	acc	avg CPU time
Sat	10	0.81	0.88	0.87	0.88	0.87	0.64	0.98	0.81	0.87	0.89	0.87	15.16
	20	0.81	0.89	0.88	0.89	0.88	0.61	0.98	0.82	0.88	0.90	0.88	23.80
	50	0.81	0.89	0.89	0.90	0.89	0.63	0.97	0.85	0.89	0.90	0.89	61.28
Zoo	10	0.91	0.92	0.86	0.86	0.86	0.69	0.96	0.80	0.88	0.92	0.92	0.28
	20	0.91	0.93	0.79	0.79	0.79	0.60	0.99	0.68	0.85	0.91	0.88	0.50
	50	0.91	0.93	0.90	0.92	0.91	0.69	0.96	0.87	0.86	0.88	0.88	2.34
Seg	10	0.93	0.96	0.96	0.96	0.96	0.72	1.00	0.84	0.95	0.97	0.96	5.59
	20	0.93	0.95	0.95	0.95	0.95	0.66	0.99	0.81	0.95	0.96	0.95	8.03
	50	0.93	0.96	0.96	0.96	0.96	0.64	0.98	0.84	0.96	0.96	0.96	17.64

5 Conclusion

In this paper, we have proposed an efficient way to find MCS with imprecise probability masses, and have applied it to the combination of multiple classifiers. First results indicate that using a disjunctive approach to combine conjunctively merged MCS may quickly result in poorly informative models, hence it may be safer in general to adopt other strategies (e.g., combining only a limited number of MCS or using a weighted mean).

Note that the algorithms presented here can be applied to other imprecise probabilistic models as well, as long as they are defined by probability bounds bearing on the same events (or by expectation bounds bearing on the same function).

References

1. Benavoli, A., Antonucci, A.: An aggregation framework based on coherent lower previsions: Application to zadeh's paradox and sensor networks. Int. J. Approx. Reason. 51(9), 1014–1028 (2010)
2. Bernard, J.: An introduction to the imprecise Dirichlet model for multinomial data. Int. J. Approx. Reason. 39(2-3), 123–150 (2005)
3. Cooke, R.: Experts in Uncertainty. Oxford University Press, Oxford (1991)
4. de Campos, L., Huete, J., Moral, S.: Probability intervals: a tool for uncertain reasoning. Int. J. Uncertain. Fuzz. 2, 167–196 (1994)
5. Destercke, S., Dubois, D.: The role of generalised p-boxes in imprecise probability models. In: Proc. of the 6th Int. Symp. on Imprecise Probability: Theories and Applications, pp. 179–188 (2009)
6. Dubois, D., Fargier, H., Prade, H.: Multi-source information fusion: a way to cope with incoherences. In: Proc. of French Days on Fuzzy Logic and Applications (LFA), La Rochelle, France, pp. 123–130 (2000)
7. Dubois, D., Prade, H.: Formal representations of uncertainty. In: Decision-Making Process, ch. 3, pp. 85–156. ISTE, London (2010)
8. Karlsson, A., Johansson, R., Andler, S.F.: On the behavior of the robust Bayesian combination operator and the significance of discounting. In: Proc. of the 6th Int. Symp. on Imprecise Probability: Theories and Applications, pp. 259–268 (2009)
9. Moral, S., Sagrado, J.: Aggregation of imprecise probabilities. In: Aggregation and Fusion of Imperfect Information, pp. 162–188. Physica, Heidelberg (1997)
10. Nau, R.: The aggregation of imprecise probabilities. J. Stat. Plan. Infer. 105, 265–282 (2002)
11. Rescher, N., Manor, R.: On inference from inconsistent premises. Theor. Decis. 1, 179–219 (1970)
12. Troffaes, M.: Generalising the conjunction rule for aggregating conflicting expert opinions. Int. J. Intell. Syst. 21(3), 361–380 (2006)
13. Troffaes, M.: Decision making under uncertainty using imprecise probabilities. Int. J. Approx. Reas. 45, 17–29 (2007)
14. Walley, P.: The elicitation and aggregation of beliefs. Technical report, University of Warwick (1982)
15. Walley, P.: Statistical reasoning with imprecise Probabilities. Chapman and Hall, New York City (1991)

The Goodman-Nguyen Relation in Uncertainty Measurement

Renato Pelessoni and Paolo Vicig

Abstract. The Goodman-Nguyen relation generalises the implication (inclusion) relation to conditional events. As such, it induces inequality constraints relevant in extension problems with precise probabilities. We extend this framework to imprecise probability judgements, highlighting the role of this relation in determining the natural extension of lower/upper probabilities defined on certain sets of conditional events. Further, a generalisation of the Goodman–Nguyen relation to conditional random numbers is proposed.

Keywords: Goodman-Nguyen relation, imprecise probabilities.

1 Introduction

It is well known that probability constraints depend essentially on relations among events. In particular, the *monotonicity* requirement that

$$(E \Rightarrow F) \to \mu(E) \leq \mu(F), \tag{1}$$

μ being a probability or a more general uncertainty measure, is a very minimal one. In fact, $E \Rightarrow F$ means that F is certainly true whenever E is true, but might possibly be true even in cases when E is false: then obviously F must be at least as likely as E. In fact, (1) holds also when μ is a coherent lower/upper probability, or a capacity. In the latter case, it is generally taken as one of the defining properties of capacities.

The implication relation '$\Rightarrow$' also plays a role in the following problem, a special case of de Finetti's Fundamental Theorem [3]: given a coherent

Renato Pelessoni · Paolo Vicig
Università di Trieste, 34127, Trieste, Italy
e-mail: {renato.pelessoni,paolo.vicig}@econ.units.it

R. Kruse et al. (Eds.): Synergies of Soft Computing and Statistics, AISC 190, pp. 37–44.
springerlink.com © Springer-Verlag Berlin Heidelberg 2013

probability P on the set $\mathcal{A}(\mathbb{P})$ of all events (logically) dependent on a given partition $\mathbb{P}$, which are its coherent extensions to an additional event $E \notin \mathcal{A}(\mathbb{P})$? The well known answer is that $P(E)$ must be chosen in a closed interval, $P(E_*) \leq P(E) \leq P(E^*)$, with $E_* = \vee\{e \in \mathbb{P} : e \Rightarrow E\}$, $E^* = \vee\{e \in \mathbb{P} : e \wedge E \neq \varnothing\}$, E_*, $E^* \in \mathcal{A}(\mathbb{P})$.

A generalisation of the implication relation to *conditional* events, the Goodman-Nguyen (in short: GN) relation $\leq_{GN}$ was apparently first introduced in [4], and some of its implications for precise conditional probabilities were studied in [1, 2, 5].

The main purpose of this paper is to further explore the relevance of the GN relation in more general cases. Precisely, we consider imprecise conditional probabilities which are either coherent or C-convex. Some preliminary material is recalled in Section 2, including a survey of known facts about the GN relation. We then discuss the generalisation of the basic result (1) to

$$A|B \leq_{GN} C|D \rightarrow \mu(A|B) \leq \mu(C|D) \tag{2}$$

in Section 3, and the role of the GN relation in extension problems in Section 3.1. The main results here are Propositions 3 and 4; in particular, Proposition 4 determines the natural extension of a coherent lower probability (alternatively the convex natural extension of a C-convex lower probability) assessed on a structured set $\mathcal{A}_C$ of conditional events. In Section 4 we explore the possibility of further extending the GN relation to conditional random numbers, and hence to employ it with coherent lower previsions. To the best of our knowledge, these questions have not been tackled yet in the relevant literature. By Proposition 5, the generalisation we propose ensures an analogue of eq. (2), whilst it is less clear if and how it may be employed in extension problems. Some final considerations are included in Section 5. Due to space limitations, most proofs have been omitted.

2 Preliminaries

In the sequel, following [3, 4] and others, we employ the logical rather than the set theoretical notation for operations with events. In terms of a truth table, a conditional event $A|B$ can be thought of as true, when A and B are true, false when A is false and B true, undefined when B is false. It ensues that $A|B$ and $A \wedge B|B$ have the same logical values, i.e. $A|B = A \wedge B|B$.

Given a *partition* $\mathbb{P}$, i.e. a set of pairwise disjoint events whose logical sum (union) is the sure event Ω, an event E is *logically dependent* on $\mathbb{P}$ iff E is a logical sum of events of $\mathbb{P}$, $E = \vee\{\omega \in \mathbb{P} : \omega \Rightarrow E\}$. The set $\mathcal{A}(\mathbb{P})$ of all events logically dependent on $\mathbb{P}$ is a field (also called the powerset of $\mathbb{P}$).

The precise or imprecise *conditional* probabilities considered in the sequel will often be defined on $\mathcal{A}_C = \mathcal{A}_C(\mathbb{P}) = \{A|B : A, B \in \mathcal{A}(\mathbb{P}), B \neq \varnothing\}$.

Given $\mathbb{P}$, a conditional random number $X|B$, $B \in \mathcal{A}(\mathbb{P}) - \{\varnothing\}$, takes up the values $X(\omega)$, for $\omega \in \mathbb{P}$, $\omega \Rightarrow B$, is undefined for $\omega \Rightarrow B^c$. When $B = \Omega$, $X|\Omega = X$ is a(n unconditional) random number. The *indicator* I_A of an event A is the simplest non-trivial random number; we shall often denote A and its indicator I_A with the same letter A. Note that $A \Rightarrow B$ is equivalent to $I_A \leq I_B$. An arbitrary set of events $S = \{E_i : i \in I\}$ does generally not constitute a partition, but originates the *partition* $\mathbb{P}_g$ *generated* by S, whose elements are the logical products $\wedge_{i \in I} E'_i$, where for each $i \in I$ the symbol E'_i can be replaced by either event E_i or its negation E_i^c. By specifying E'_i for each i in all possible ways, we get the elements (some of them, in general, may be impossible) of $\mathbb{P}_g$. The events in S belong to $\mathcal{A}(\mathbb{P}_g)$.

A *lower prevision* $\underline{P}$ on a set S of (*bounded*, in what follows) conditional random numbers is a map $\underline{P} : S \longmapsto \mathbb{R}$. If S has the property $X|B \in S \rightarrow -X|B \in S$, its conjugate *upper prevision* is defined as $\overline{P}(X|B) = -\underline{P}(-X|B)$. Because of conjugacy, we may employ lower or alternatively upper previsions only.

Definition 1. A lower prevision $\underline{P} : S \longmapsto \mathbb{R}$ is *W-coherent* iff, for all $n \in \mathbb{N}$, $\forall X_0|B_0, \ldots, X_n|B_n \in S$, $\forall\ s_0, s_1, \ldots, s_n$ real and *non-negative*, defining $B^* = \bigvee_{i=0}^n B_i$ and $\underline{G} = \sum_{i=1}^n s_i B_i(X_i - \underline{P}(X_i|B_i)) - s_0 B_0(X_0 - \underline{P}(X_0|B_0))$, the following condition holds: $\sup(\underline{G}|B^*) \geq 0$.

This is essentially Williams' definition of coherence [12], as restated in [8], and is equivalent to Walley's definition 7.1.4(b) in [11], if S is made up of a finite number of conditional random numbers, each with finitely many values. If $X|B = X|\Omega = X$, $\forall X|B \in S$, it reduces to Walley's (unconditional) coherence ([11], Sec. 2.5.4 (a)).

A weaker concept than W-coherence is that of *C–convex* conditional lower prevision, obtained from Definition 1 by introducing just the extra *convexity constraint* $\sum_{i=1}^n s_i = s_0\ (> 0)$ and requiring that $\underline{P}(0) = 0$. These previsions were studied in [7] and correspond to certain kinds of risk measures.

Coherent conditional previsions may be defined similarly:

Definition 2. $P : S \longmapsto \mathbb{R}$ is a coherent conditional prevision iff, for all $n \in \mathbb{N}$, $\forall\ X_1|B_1, \ldots, X_n|B_n \in S$, $\forall\ s_i \in \mathbb{R}$ $(i = 1, \ldots, n)$, defining $G = \sum_{i=1}^n s_i B_i(X_i - P(X_i|B_i))$, $B^* = \bigvee_{i=1}^n B_i$, it holds that $\sup(G|B^*) \geq 0$.

In the consistency concepts above, we may speak of (lower, upper or precise) *probability* μ instead of prevision if, for any $X|B \in S$, X is (the indicator of) an event. In all such cases, the following are necessary consistency conditions:

$$\mu(A|B) \in [0; 1], \mu(\varnothing|B) = 0, \mu(B|B) = 1. \tag{3}$$

In general, results for upper probabilities follow from those for lower probabilities by the conjugacy equality $\overline{P}(A) = 1 - \underline{P}(A^c)$.

Definition 3. *(Goodman–Nguyen relation.)* We say that $A|B \leq_{GN} C|D$ iff

$$A \wedge B \Rightarrow C \wedge D \text{ and } C^c \wedge D \Rightarrow A^c \wedge B. \tag{4}$$

The GN relation was introduced in an equivalent form in [4], while investigating *conditional event algebras*. As observed in [1, 5, 6], the intuition behind Definition 3 can be explained easily resorting to betting arguments, much in the style of de Finetti [3]. In fact, (4) states that whenever we bet both on $A|B$ and on $C|D$ (iff $B \wedge D$ is true), if we win the bet on $A|B$, we also win the bet on $C|D$ (because $A \wedge B \Rightarrow C \wedge D$), and losing the bet on $C|D$ implies also our losing the bet on $A|B$ (because of $C^c \wedge D \Rightarrow A^c \wedge B$). When $B = D = \Omega$, just one of the implications in (4) is needed, because of the tautology $A \Rightarrow C \leftrightarrow C^c \Rightarrow A^c$.

Not surprisingly then, (2) should hold. In the case that μ is a conditional probability P, (2) was stated without proof in [4] (assuming P defined on $\mathcal{A}_C$), and proved in [1] (under general assumptions) and independently (in a less general case) in [5]. The result ensues also from Proposition 1 in Section 3, supplying a unique proof for either precise or imprecise probabilities. The GN relation in extension problems is explored in [1], showing that, given a coherent probability P on a *finite* set of conditional events, the bounds on its coherent extensions on one additional event $C|D$ depend on the values of P on two events, $(C|D)_*$ and $(C|D)^*$, determined by the GN relation. We consider extension problems for *arbitrary* sets of events and precise or imprecise probability assessments in Section 3.

3 The GN Relation with Imprecise Probabilities

Remark 1. If $A|B \leq_{GN} C|D$, then necessarily the partition $\mathbb{P}_g$ generated by A, B, C, D allows for at most 7 non-impossible events that imply $B \vee D$: $\omega_1 = ABCD$, $\omega_2 = A^cBCD$, $\omega_3 = AB^cCD$, $\omega_4 = A^cB^cCD$, $\omega_5 = A^cBC^cD$, $\omega_6 = A^cBCD^c$, $\omega_7 = A^cBC^cD^c$. This is easily seen from (4), using $A \Rightarrow B \leftrightarrow A \wedge B^c = \varnothing$.

Example 1. If $A \Rightarrow C \Rightarrow D \Rightarrow B$, then $A|B \leq_{GN} C|D$. Of the 7 events in Remark 1, only 4 (at most, iff $A \neq \varnothing$) are non-impossible: ω_1, ω_2, ω_5, ω_7.

We prove now equation (2) for C-convex probabilities.[1]

Proposition 1. *Let μ be a C-convex lower (or upper) probability defined on $S = \{A|B, C|D\}$. Then, $A|B \leq_{GN} C|D$ implies $\mu(A|B) \leq \mu(C|D)$.*

Proof. Consider, in the lower probability case,

$$\underline{G}|B^* = \underline{G}|B \vee D = B(A - \mu(A|B)) - D(C - \mu(C|D))|B \vee D. \qquad (5)$$

Assuming $A|B \leq_{GN} C|D$, by Remark 1 $\underline{G}|B \vee D$ can take up at most 7 values, actually fewer distinct ones: $\underline{G}(\omega_1) = \underline{G}(\omega_5) = \mu(C|D) - \mu(A|B)$,

[1] Since any W-coherent or coherent probability is C-convex, Proposition 1 applies to these probabilities too.

$\underline{G}(\omega_2) = \mu(C|D) - \mu(A|B) - 1 \leq 0$, $\underline{G}(\omega_3) = \underline{G}(\omega_4) = \mu(C|D) - 1 \leq 0$, $\underline{G}(\omega_6) = \underline{G}(\omega_7) = -\mu(A|B) \leq 0$. If either $\mu(A|B) = 0$ or $\mu(C|D) = 1$, (2) is trivial. If not, $\max \underline{G}|B^* = \underline{G}(\omega_1) = \mu(C|D) - \mu(A|B) \geq 0$, i.e. (2) holds. A similar line of reasoning applies to upper probabilities. □

Example 2. If $A \Rightarrow B_1 \Rightarrow B_0$, it is easy to check that $A|B_0 \leq_{GN} A|B_1$. This is in fact a special case of Example 1. By Proposition 1,

$$\mu(A|B_0) \leq \mu(A|B_1) \tag{6}$$

if μ is either a probability (well known from the product rule $P(A|B_0) = P(A|B_1)P(B_1|B_0)$) or an upper/lower probability which is W-coherent (established in a different way in [10], Proposition 13) or C-convex. If $B_1 \Rightarrow B_0$, the GN relation and hence (6) still hold if $A \Rightarrow B_1 \vee B_0^c$, while, when A is arbitrary, (6) is replaced by

$$\mu(A \wedge B_1|B_0) \leq \mu(A|B_1). \tag{7}$$

To see this, use (6) and Proposition 1: $A \wedge B_1|B_0 \leq_{GN} A \wedge B_1|B_1 = A|B_1$.

3.1 The GN Relation in Extension Problems

An interesting feature of the GN relation is that it allows comparing conditional events whose conditioning events are possibly different. This fact is useful in the generalisations of the extension problem presented in the introduction that we are going to discuss now.

Let $\mathbb{P}$ be any partition. We wish to extend an uncertainty measure μ, assessed on $\mathcal{A}_C(\mathbb{P})$, to an arbitrary event $C|D$. We assume in what follows $C|D \neq \varnothing|D$, $C|D \neq D|D$, ruling out limiting cases where the extension is already known by (3).

Definition 4. Define $m(C|D) = \{A|B \in \mathcal{A}_C(\mathbb{P}) : A|B \leq_{GN} C|D\}$, $M(C|D) = \{A|B \in \mathcal{A}_C(\mathbb{P}) : C|D \leq_{GN} A|B\}$.

It is easy to see that

Proposition 2. *The sets m, M are non-empty and have, respectively, a maximum $(C|D)_*$ and a minimum $(C|D)^*$ conditional event w.r.t. $\leq_{GN}$,*

$$\begin{aligned} (C|D)_* &= (C \wedge D)_*|[(C \wedge D)_* \vee (C^c \wedge D)^*], \\ (C|D)^* &= (C \wedge D)^*|[(C \wedge D)^* \vee (C^c \wedge D)_*]. \end{aligned} \tag{8}$$

where $(C \wedge D)_* = \vee\{e \in \mathbb{P} : e \Rightarrow C \wedge D\}$, $(C^c \wedge D)^* = \vee\{e \in \mathbb{P} : e \wedge C^c \wedge D \neq \varnothing\}$, $(C \wedge D)^* = \vee\{e \in \mathbb{P} : e \wedge C \wedge D \neq \varnothing\}$, $(C^c \wedge D)_* = \vee\{e \in \mathbb{P} : e \Rightarrow C^c \wedge D\}$.

It holds that

Proposition 3. *Let $P(\cdot|\cdot)$ be a coherent precise probability on $\mathcal{A}_C$. Any of its extensions on $\mathcal{A}_C \cup \{C|D\}$ is a coherent precise probability iff $P(C|D) \in [P((C|D)_*); P((C|D)^*)]$.*

For extensions on an *arbitrary* set of conditional events, we have:

Proposition 4. *Let $\underline{P}$ ($\overline{P}$) be a W-coherent, respectively C-convex lower (upper) probability defined on $\mathcal{A}_C$ and $\mathcal{E}$ be an arbitrary set of conditional events. Then, the extension of $\underline{P}$ ($\overline{P}$) on $\mathcal{A}_C \cup \mathcal{E}$, such that $\underline{P}(C|D) = \underline{P}((C|D)_*)$ ($\overline{P}(C|D) = \overline{P}((C|D)^*)$), $\forall C|D \in \mathcal{E}$, is a W-coherent, respectively C-convex lower (upper) probability.*

In the special case of a coherent (precise) probability P, Propositions 3 and 4 show that both its extension $P(C|D) = P((C|D)_*)$, $\forall C|D \in \mathcal{E}$, and its extension $P(C|D) = P((C|D)^*)$, $\forall C|D \in \mathcal{E}$, are surely again coherent precise probabilities on $\mathcal{A}_C \cup \mathcal{E}$ only when $\mathcal{E}$ is a singleton. In general, they are either a lower (the former) or an upper (the latter) W-coherent probability on $\mathcal{A}_C \cup \mathcal{E}$. Like its unconditional counterpart, this extension problem naturally generates imprecise uncertainty measures. When the initial measure is a lower (upper) W-coherent probability $\underline{P}$ ($\overline{P}$), so is its extension by Proposition 4. Let us term it *GN-extension.* Interestingly, as a straightforward consequence of Propositions 1 and 2, the GN-extension is the *least committal* extension of $\underline{P}$ (or $\overline{P}$): in fact, taking for instance $\underline{P}$, for any other W-coherent extension $\underline{Q}$ of $\underline{P}$, $\underline{Q}(C|D) \geq \underline{Q}((C|D)_*) = \underline{P}((C|D)_*) = \underline{P}(C|D)$ (for the first equality, note that $\underline{Q}$ and $\underline{P}$ coincide on $\mathcal{A}_C$). As well known, this property identifies the *natural extension* (cf. [11]) of $\underline{P}$ on $\mathcal{A}_C(\mathbb{P}) \cup \mathcal{E}$. Similarly, when the initial $\underline{P}$ (or $\overline{P}$) is C-convex the GN-extension is what is termed *C-convex natural extension* in [7]. In both instances, computing the natural extensions is straightforward using the GN-relation: we just have to detect $(C|D)_*$ (or $(C|D)^*$).

The procedure requires that the starting P, $\underline{P}$ or $\overline{P}$ are defined on $\mathcal{A}_C$. If they are assessed on an arbitrary set S of conditional events, we should first extend P, $\underline{P}$ or $\overline{P}$ on some $\mathcal{A}_C(\mathbb{P}) \supset S$ to apply Proposition 4; a convenient $\mathbb{P}$ is the partition generated by S. Clearly, Proposition 4 does not add much, *operationally*, in this case: we would still need an operational procedure for the extension on $\mathcal{A}_C(\mathbb{P})$. It is meaningful at a *theoretical* level, as an explanation of how logical constraints determine our inferences on additional events.

4 A GN Type Relation with Imprecise Previsions

How could the GN relation $\leq_{GN}$ be defined and employed to compare conditional random numbers? We next propose a possible generalisation.

Definition 5. $X|B \leq_{GN} Y|D$ iff

$$BX(\omega) + \sup_B X \cdot B^c D(\omega) \leq DY(\omega) + \inf_D Y \cdot BD^c(\omega), \forall\omega \Rightarrow B \vee D. \quad (9)$$

The motivation for this definition is very similar to the betting argument in [5, 6], recalled in Section 2 to justify the GN relation. In fact:

a) if $\omega \Rightarrow B \wedge D$, (9) reduces to $X(\omega) \leq Y(\omega)$. This means: whenever we bet both on $X|B$ and on $Y|D$, we gain at least as much with the bet on $Y|D$;
b) for $\omega \Rightarrow B^c \wedge D$, it reduces to $\sup_B X = \sup\{X|B\} \leq Y(\omega)$. For such ω, we bet on $Y|D$ but not on $X|B$. By the last inequality the gain from our bet on $Y|D$ is not less than our (potential) gain on $X|B$, had we bet on it;
c) if $\omega \Rightarrow B \wedge D^c$, (9) reduces to $X(\omega) \leq \inf_D Y = \inf\{Y|D\}$, whose interpretation is specular to that in b) above.

When $X|B \neq \varnothing|B$ and $Y|D \neq D|D$, Definition 5 generalises Definition 3: if X, Y are (indicators of) events, say $X = A$, $Y = C$, (9) becomes $AB + \max\{A|B\} \cdot B^c D \leq CD + \min\{C|D\} \cdot BD^c$, and it can be shown that $A|B \leq_{GN} C|D$ (by Definition 3) iff this inequality holds.

The partial ordering *among conditional random numbers* of the generalised GN relation induces an agreeing ordering on their uncertainty measures in the cases stated by the next result:

Proposition 5. *Let* $S = \{X|B, Y|D\}$. *Then*

$$X|B \leq_{GN} Y|D \Rightarrow \mu(X|B) \leq \mu(Y|D) \quad (10)$$

whenever μ *is either a coherent precise prevision* P *or a W-coherent lower (upper) prevision* $\underline{P}$ $(\overline{P})$, *defined on* S.

Proposition 5 generalises Proposition 1, except for C-convex probabilities. It is not clear at present whether the result applies to such previsions too.

As for the inferential use of the GN relation with conditional random numbers, the problem is considerably more complex than with events and largely to be investigated yet. Consider for this the following example:

Example 3. How does (7) in Example 2 generalise with random numbers? Intuitively, we should first check whether, replacing A with X, $B_1X|B_0 \leq_{GN} X|B_1$ is still true, if $B_1 \Rightarrow B_0$. Ignoring the trivial case $B_0 = B_1$, by Definition 5 this inequality is equivalent to $B_1B_0X + \sup_{B_0} X \cdot (B_0^c B_1) \leq B_1X + \inf_{B_1} X \cdot (B_0 B_1^c)$, which is equivalent ($B_1B_0 = B_1$, $B_0^c B_1 = 0$, $B_0 B_1^c \geq 0$) to $\inf(X|B_1) \geq 0$, a condition always fulfilled if $X|B$ is a conditional event. On the contrary, it is easy to check that $X|B_1 \leq_{GN} B_1X|B_0$ iff $\sup(X|B_1) \leq 0$. Hence, $X|B_1$ and $B_1X|B_0$ are GN-incomparable iff $\inf(X|B_1) \cdot \sup(X|B_1) < 0$. In this example, the GN relation is *sign sensitive*.

5 Conclusions

With imprecise probabilities, the GN relation allows computing easily the natural extension of a W-coherent (or C-convex) assessment on the special set $\mathcal{A}_C$, a problem which may be viewed as a generalisation of de Finetti's Fundamental Theorem. While the exact relevance of the GN relation with imprecise previsions remains an open question, it may be asserted that it supplies bounds on uncertainty measures, in certain specific instances. One such case is considered in Example 3. However, the bound it achieves is rather rough. For a stricter bound under looser hypotheses, see Proposition 3.1 in [9]. For another example, let $X|B \geq 0$ take up a finite number of values $x_1, x_2, \ldots, x_n$. Then W-coherence requires that $\underline{P}(X|B) = \underline{P}(\sum_{i=1}^n x_i e_i|B) \geq \sum_{i=1}^n x_i \underline{P}(e_i|B)$, where e_i is the event '$X = x_i$', $i = 1, \ldots, n$. Applying Proposition 1 to each $\underline{P}(e_i|B)$, we get the lower bound $\underline{P}(X|B) \geq \sum_{i=1}^n x_i \underline{P}((e_i|B)_*)$.

References

1. Coletti, G., Gilio, A., Scozzafava, R.: Comparative probability for conditional events: A new look through coherence. Theory and Decision 35(3), 237–258 (1993)
2. Coletti, G., Scozzafava, R.: Characterization of coherent conditional probabilities as a tool for their assessments and extensions. Int. J. Uncertain. Fuzz. 4(2), 103–127 (1993)
3. de Finetti, B.: Theory of Probability. John Wiley & Sons, Ltd., New York City (1974)
4. Goodman, I.R., Nguyen, H.T.: Conditional objects and the modeling of uncertainties. In: Gupta, M., Yamakawa, T. (eds.) Fuzzy Computing, pp. 119–138. North Holland, Amsterdam (1988)
5. Milne, P.: Bruno de Finetti and the logic of conditional events. Br. J. Philos. Sci. 48(2), 195–232 (1997)
6. Milne, P.: Bets and boundaries: assigning probabilities to imprecisely specified events. Studia Logica 90, 425–453 (2008)
7. Pelessoni, R., Vicig, P.: Uncertainty modelling and conditioning with convex imprecise previsions. Int. J. Approx. Reason. 39(2-3), 297–319 (2005)
8. Pelessoni, R., Vicig, P.: Williams coherence and beyond. Int. J. Approx. Reason. 50(4), 612–626 (2009)
9. Pelessoni, R., Vicig, P.: Bayes' Theorem Bounds for Convex Lower Previsions. J. Stat. Theor. Pract. 3(1), 85–101 (2009)
10. Pelessoni, R., Vicig, P., Zaffalon, M.: Inference and risk measurement with the parimutuel model. Int. J. Approx. Reason. 51(9), 1145–1158 (2010)
11. Walley, P.: Statistical Reasoning with Imprecise Probabilities. Chapman and Hall, London (1991)
12. Williams, P.M.: Notes on conditional previsions. Int. J. Approx. Reason. 44(3), 366–383 (2007)

The CONEstrip Algorithm

Erik Quaeghebeur

Abstract. Uncertainty models such as sets of desirable gambles and (conditional) lower previsions can be represented as convex cones. Checking the consistency of and drawing inferences from such models requires solving feasibility and optimization problems. We consider finitely generated such models. For closed cones, we can use linear programming; for conditional lower prevision-based cones, there is an efficient algorithm using an iteration of linear programs. We present an efficient algorithm for general cones that also uses an iteration of linear programs.

Keywords: Consistency, convex cones, feasibility, inference, linear programming.

1 Introduction

Mathematically speaking, frameworks for modeling uncertainty consist of rules that specify what constitutes a within the framework valid model and rules to perform computations with such models. For a number of frameworks under the imprecise probability umbrella [8, 9], checking validity—i.e., the consistency criteria of avoiding sure & partial loss and coherence—and calculating an inference—i.e., natural extension—involves solving feasibility and optimization problems.

We illustrate in Section 2 that the feasibility aspect of these problems essentially boils down to checking whether some vector lies in a *general convex cone*, called *general cone* from now on, a cone that may be closed, open, or *ajar*, i.e., neither open nor closed. For models specified in a finitary way, algorithms to do this for closed and specific general cases can be found in

Erik Quaeghebeur
SYSTeMS Research Group, Ghent University, Belgium
e-mail: erik.quaeghebeur@ugent.be

R. Kruse et al. (Eds.): Synergies of Soft Computing and Statistics, AISC 190, pp. 45–54.
springerlink.com © Springer-Verlag Berlin Heidelberg 2013

the literature. In Section 4 we present an efficient algorithm for all general finitary instances. But to do this, we first need to make a small detour with Section 3 to discuss how we can represent finitary general cones—and therefore the feasibility and optimization problems that interest us—in a way that is conducive to algorithm formulation.

Concepts & Notation. We assume that the *possibility space* Ω is non-empty and finite. *Gambles* are real-valued functions on Ω, i.e., elements of $\mathcal{L} := [\Omega \to \mathbb{R}]$. The *indicator* 1_B of an event $B \subseteq \Omega$ is 1 on B and 0 elsewhere; $1_\omega := 1_{\{\omega\}}$ for $\omega \in \Omega$. The finite subset relation is $\Subset$. A set superscripted with '$*$' denotes the set of all its finite subsets. Vector inequalities: $\gg$ ($\geq$) is pointwise (non-)strict; $>$ means "$\geq$ but not $=$".

2 Problems Solved in the Literature

In this section, we present a number of problems from and solved in the literature. On the one hand, they are meant to make a link with the literature, and on the other hand, we use them to illustrate that these problems essentially boil down to checking whether some vector lies in a cone.

Lower Previsions and Sets of Almost-Desirable Gambles. The most basic consistency criterion for lower previsions and sets of almost desirable gambles is *avoiding sure loss* [8, §2.4 & §3.7.1]. Checking whether a lower prevision $\underline{P} \in [\mathcal{K} \to \mathbb{R}]$, with $\mathcal{K} \Subset \mathcal{L}$, or set of almost desirable gambles $\mathcal{A} \Subset \mathcal{L}$ *incurs sure loss* amounts to solving the feasibility problem below, where in the former case $\mathcal{A} := \{h - \underline{P}h : h \in \mathcal{K}\}$:

$$\begin{aligned} \text{find} \quad & \lambda \in \mathbb{R}^{\mathcal{A}}, \\ \text{subject to} \quad & \textstyle\sum_{g \in \mathcal{A}} \lambda_g \cdot g \lessdot 0 \quad \text{and} \quad \lambda \geq 0. \end{aligned}$$

We can get an equivalent feasibility problem that can however be solved using linear programming [10, §2.4] by replacing the constraints by

$$\textstyle\sum_{g \in \mathcal{A}} \lambda_g \cdot g \leq -1 \quad \text{and} \quad \lambda \geq 0.$$

By introducing (slack) variables $\mu \in \mathbb{R}^{\Omega}$, these can also be written as

$$\textstyle\sum_{g \in \mathcal{A}} \lambda_g \cdot g + \sum_{\omega \in \Omega} \mu_\omega \cdot 1_\omega = 0 \quad \text{and} \quad \lambda \geq 0 \quad \text{and} \quad \mu \geq 1,$$

which express that the origin must lie in a closed cone spanned by the elements of $\mathcal{A}$ and $\{1_\omega : \omega \in \Omega\}$.

A typical inference drawn from a set of almost desirable gambles $\mathcal{A} \Subset \mathcal{L}$ (or the lower prevision it may be derived from) is the lower prevision for a gamble $f \in \mathcal{L}$. This is calculated using *natural extension* [8, §3.1], i.e., the linear program below:

$$\begin{aligned} &\text{maximize} \quad \alpha \in \mathbb{R}, \\ &\text{subject to} \quad f - \alpha \geq \textstyle\sum_{g \in \mathcal{A}} \lambda_g \cdot g \quad \text{and} \quad \lambda \geq 0. \end{aligned}$$

By introducing variables $\mu \in \mathbb{R}^\Omega$, the constraints can be written as

$$\textstyle\sum_{g \in \mathcal{A}} \lambda_g \cdot g + \sum_{\omega \in \Omega} \mu_\omega \cdot 1_\omega + \alpha = f \quad \text{and} \quad \lambda \geq 0 \quad \text{and} \quad \mu \geq 0,$$

which express that f must lie in the closed cone spanned by the elements of $\mathcal{A}$, $\{1_\omega : \omega \in \Omega\}$, and $\{1_\Omega, -1_\Omega\}$. This problem is always feasible: take, for example, $\lambda = 0$, $\alpha = \min f$, and $\mu = f - \min f$; it will be unbounded if $\mathcal{A}$ incurs sure loss.

On top of avoiding sure loss, a lower prevision $\underline{P} \in [\mathcal{K} \to \mathbb{R}]$, with $\mathcal{K} \Subset \mathcal{L}$, may be required to be *coherent* [8, §2.5]. This can be checked by verifying that for all $f \in \mathcal{L}$ the natural extension coincides with $\underline{P}f$, so we do not need to investigate this further.

Conditional Lower Previsions. Moving from unconditional to conditional lower previsions, the basic ideas stay the same, but become more involved because we need to take the conditioning events into account.

Avoiding partial loss [8, §7.1.2–3 & Notes 7.1(7.)] replaces avoiding sure loss. Checking whether a conditional lower prevision $\underline{P} \in [\mathcal{N} \to \mathbb{R}]$, with $\mathcal{N} \Subset \mathcal{L} \times \Omega^*$ non-empty, *incurs partial loss* amounts to solving the feasibility problem below [3, (17)], in which $\mathcal{B} := \{([h - \underline{P}(h|B)] \cdot 1_B, B) : (h, B) \in \mathcal{N}\}$:

$$\begin{aligned} &\text{find} \quad (\lambda, \varepsilon) \in \mathbb{R}^\mathcal{B} \times \mathbb{R}^\mathcal{B}, \\ &\text{subject to} \quad \textstyle\sum_{(g,B) \in \mathcal{B}} \lambda_{g,B} \cdot [g + \varepsilon_{g,B} \cdot 1_B] \leq 0 \quad \text{and} \quad \lambda > 0 \quad \text{and} \quad \varepsilon \gg 0. \end{aligned}$$

An algorithm for solving this *bilinearly* constrained feasibility problem using a sequence of linear programs has been first presented by Walley et al. [10, Alg. 2] for a subclass and made explicit for the general case by Couso and Moral [3, Alg. 1]. By introducing variables $\nu \in \mathbb{R}^\mathcal{B} \times \mathbb{R}^\mathcal{B}$ and $\mu \in \mathbb{R}^\Omega$, the constraints can be written as

$$\begin{aligned} &\textstyle\sum_{(g,B) \in \mathcal{B}} \lambda_{g,B} \cdot [\nu_{g,B,g} \cdot g + \nu_{g,B,B} \cdot 1_B] + \sum_{\omega \in \Omega} \mu_\omega \cdot 1_\omega = 0 \\ &\text{and} \quad \lambda > 0 \quad \text{and} \quad \nu \gg 0 \quad \text{and} \quad \mu \geq 0, \end{aligned}$$

which express that the origin must lie in a general cone spanned by elements of $\bigcup_{(g,B) \in \mathcal{B}} \{(1 - \delta) \cdot g + \delta \cdot 1_B : 0 < \delta < 1\}$ and $\{1_\omega : \omega \in \Omega\}$.

Inferring the lower prevision of a gamble $f \in \mathcal{L}$ conditional on an event $C \subseteq \Omega$ from a given conditional lower prevision $\underline{P} \in [\mathcal{N} \to \mathbb{R}]$ as above is calculated using a generalization of the natural extension procedure seen before [3, (21)]:

$$\begin{aligned}&\text{maximize} && \alpha \in \mathbb{R},\\ &\text{subject to} && f \cdot 1_C - \alpha \cdot 1_C \geq \textstyle\sum_{(g,B)\in\mathcal{B}} \lambda_{g,B} \cdot [g + \varepsilon_{g,B} \cdot 1_B]\\ &&& \text{and} \quad \lambda \geq 0 \quad \text{and} \quad \varepsilon > 0.\end{aligned}$$

Again, this problem can be solved using linear program iteration [10, Alg. 4]-[3, Alg. 2]. By introducing variables $\nu \in \mathbb{R}^{\mathcal{B}} \times \mathbb{R}^{\mathcal{B}}$ and $\mu \in \mathbb{R}^{\Omega}$, the constraints can be written as

$$\textstyle\sum_{(g,B)\in\mathcal{B}} \lambda_{g,B} \cdot [\nu_{g,B,g} \cdot g + \nu_{g,B,B} \cdot 1_B] + \sum_{\omega\in\Omega} \mu_\omega \cdot 1_\omega + \alpha \cdot 1_C = f \cdot 1_C$$
$$\text{and} \quad \lambda \geq 0 \quad \text{and} \quad \nu > 0 \quad \text{and} \quad \mu \geq 0,$$

which express that $f \cdot 1_C$ must lie in a general cone spanned by elements of $\bigcup_{(g,B)\in\mathcal{B}}\{(1-\delta)\cdot g + \delta \cdot 1_B : 0 < \delta < 1\}$, $\{1_\omega : \omega \in \Omega\}$, and $\{1_C, -1_C\}$. This is always feasible: take, e.g., $\lambda = 0$, $\alpha = \min(f \cdot 1_C)$, and $\mu = [f - \min(f \cdot 1_C)] \cdot 1_C$.

3 Representation and Problem Formulation

In the previous section, we presented four problems from the literature that can essentially be formulated in terms of (conditional) lower previsions and that result in *specific* general cones. General sets of desirable gambles [8, App. F]-[9, §6]-[4] take the form of far more *general* general cones, and in generalizations of desirability [6] essentially any general cone could appear. So we must be able to deal with the feasibility and optimization problems that arise out of working with such models.

In this section, we will first discuss a representation for finitary general cones. Then we use that representation to formulate the general problem we want to tackle.

Representation of General Cones. We use the idea of Couso and Moral [3, Thm. 13; the need for such a representation is illustrated in Examples 3 & 4] to represent a finitary general cone as a convex hull of a finite number of finitary *open* cones. Formally, given a finite set of finite sets of gambles $\mathcal{R} \Subset \mathcal{L}^*$, the corresponding cone is

$$\underline{\mathcal{R}} := \{\textstyle\sum_{\mathcal{D}\in\mathcal{R}} \lambda_{\mathcal{D}} \cdot \sum_{g\in\mathcal{D}} \nu_{\mathcal{D},g} \cdot g : \lambda > 0, \nu > 0\}.$$

Whereas 'finitary' has a well-known meaning for open and closed cones—pointwise strict and non-strict convex hulls of finite sets of rays—, it is a concept we have fixed for ajar and therefore general cones by choosing a representation. To justify this choice, we employ facets [11]—the closed cones that are a closed cone's maximally dimensional non-trivial faces—to give an appealing polytope-theoretically flavored definition and then show that the chosen representation satisfies it:

Definition. *An ajar cone $\mathcal{C}$ is* finitary *iff its closure* $\operatorname{cl}\mathcal{C}$ *is finitary and the intersection of $\mathcal{C}$ with each of* $\operatorname{cl}\mathcal{C}$*'s facets is a finitary (open, closed, or ajar) cone.*

Facet recursion is bound to stop in a finite number of steps: open and closed cones are terminal, with each step the dimension strictly decreases, and $\emptyset$ is clopen.

Theorem. *$\underline{\mathcal{R}}$ is a finitary general cone for every $\mathcal{R} \Subset \mathcal{L}^*$.*

Proof. Because $\mathcal{R}$ and its elements have finite cardinality, $\operatorname{cl}\underline{\mathcal{R}}$ is finitary; also $\operatorname{cl}\underline{\mathcal{R}} = \operatorname{cl}\underline{\mathcal{E}}$, with $\mathcal{E} := \bigcup\mathcal{R}$. Let $\mathcal{F}$ be one of its (finite number of) facets, let $\mathcal{S} := \{\mathcal{D} \in \mathcal{R} : \mathcal{D} \subset \mathcal{F}\}$, and let $\underline{\mathcal{D}} := \underline{\{\mathcal{D}\}}$. Then $\underline{\mathcal{S}} = \underline{\mathcal{R}} \cap \mathcal{F}$, because $\underline{\mathcal{D}} \setminus \mathcal{F} \neq \emptyset$ implies $\underline{\mathcal{D}} \cap \mathcal{F} = \emptyset$ by definition of $\underline{\mathcal{D}}$ and the fact that $\mathcal{F}$ is a facet. The proof is complete by facet recursion (replacing $\mathcal{R}$ by $\mathcal{S}$ in the first sentence). □

The approach of the proof also allows us to construct a canonical *cone-in-facet* representation (cf. concept of *zero-layers* [2, Ch. 12]). We illustrate this in Figure 1.

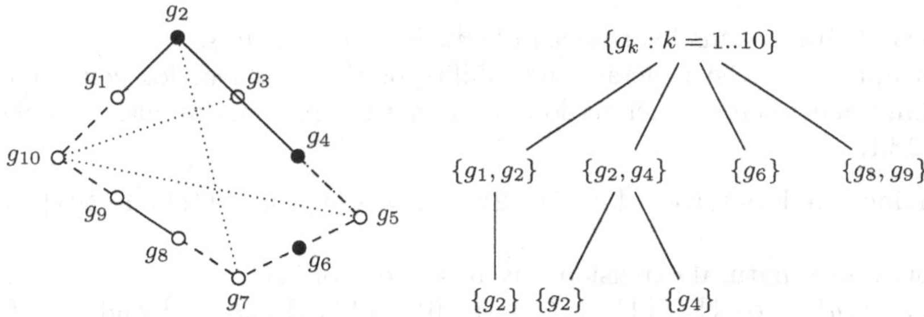

Fig. 1 On the left, we show the intersection of a cone $\underline{\mathcal{R}}$ with a plane. It can be represented by $\mathcal{R} := \big\{\{g_3, g_5, g_{10}\}, \{g_1, g_2\}, \{g_2, g_7\}, \{g_8, g_9\}, \{g_2\}, \{g_4\}, \{g_6\}\big\}$. On the right, we order the sets $\mathcal{E}$—as defined in the proof of the Theorem—for each facet encountered in the facet recursion of $\mathcal{R}$ according to facet inclusion. Then $\big\{\{g_k : k = 1..10\}, \{g_1, g_2\}, \{g_2, g_4\}, \{g_6\}, \{g_8, g_9\}, \{g_2\}, \{g_4\}\big\}$ is the resulting cone-in-facet representation for $\underline{\mathcal{R}}$. Note that in terms of set sizes, this representation is not minimal, as $\big\{\{g_5, g_7, g_{10}\}, \{g_1, g_2\}, \{g_8, g_9\}, \{g_2\}, \{g_4\}, \{g_6\}\big\}$ also represents $\underline{\mathcal{R}}$.

We use the CONEstrip algorithm to check whether g_3 and g_1 lie in the cone $\underline{\mathcal{R}}$ of Figure 1. First g_3: in the first iteration, $\tau_{\{g_2\}} = \tau_{\{g_4\}} = \tau_{\{-g_3\}} = 1$ and τ is zero for other indices; because possibly, e.g., $\mu_{\{g_3,g_5,g_{10}\},g_3} > 0$, we might need a second iteration—with $\mathcal{S} = \big\{\{g_2\}, \{g_4\}, \{-g_3\}\big\}$—in which $\tau = 1$, so $g_3 \in \underline{\mathcal{R}}$. Next g_1: in the first iteration, $\tau_{\{g_2\}} = \tau_{\{g_1,g_2\}} = \tau_{\{-g_1\}} = 1$ and τ is zero for other indices; because necessarily $\mu_{\{g_3,g_5,g_{10}\},g_{10}} > 0$, we need a second iteration—with $\mathcal{S} = \big\{\{g_2\}, \{g_1, g_2\}, \{-g_1\}\big\}$—which is infeasible, so $g_1 \notin \underline{\mathcal{R}}$.

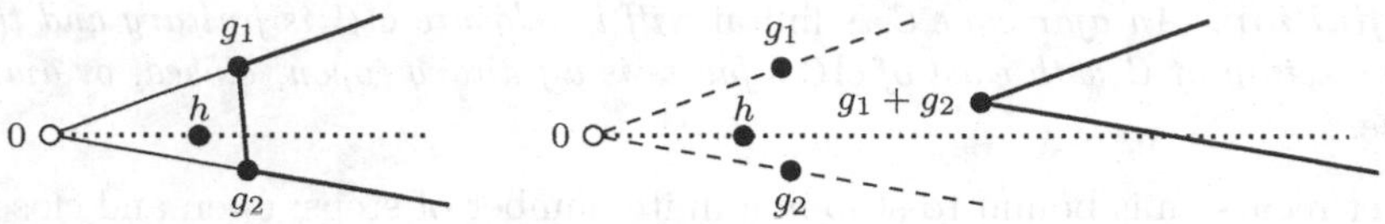

Fig. 2 On the left, we show the blunt closed cone $\mathcal{C} := \{\lambda_1 \cdot g_1 + \lambda_2 \cdot g_2 : \lambda > 0\}$ (thin lines) and its proxy $\mathcal{C}' := \{\mu_1 \cdot g_1 + \mu_2 \cdot g_2 : \mu \geq 0, \mu_1 + \mu_2 \geq 1\}$ (thick lines): $\{\kappa \cdot h : \kappa \geq 1\} \cap \mathcal{C}' \neq \emptyset$ is equivalent to $h \in \mathcal{C}$. On the right, we show the open cone $\mathcal{C} := \{\nu_1 \cdot g_1 + \nu_2 \cdot g_2 : \nu \gg 0\}$ (dashed lines) and its proxy $\mathcal{C}' := \{\mu_1 \cdot g_1 + \mu_2 \cdot g_2 : \mu \geq 1\}$ (thick lines): $\{\kappa \cdot h : \kappa \geq 1\} \cap \mathcal{C}' \neq \emptyset$ is equivalent to $h \in \mathcal{C}$.

Formulation of the General Problem. Given a general cone represented by $\mathcal{R} \Subset \mathcal{L}^*$ and a gamble $h \in \mathcal{L}$, we wish to

$$\begin{aligned} \text{find} \quad & (\lambda, \nu) \in \mathbb{R}^{\mathcal{R}} \times \mathsf{X}_{\mathcal{D} \in \mathcal{R}} \mathbb{R}^{\mathcal{D}} \\ \text{or maximize an affine function of} \quad & \mu := (\lambda_{\mathcal{D}} \cdot \nu_{\mathcal{D},g} : \mathcal{D} \in \mathcal{R}, g \in \mathcal{D}), \\ \text{subject to} \quad & \textstyle\sum_{\mathcal{D} \in \mathcal{R}} \lambda_{\mathcal{D}} \cdot \sum_{g \in \mathcal{D}} \nu_{\mathcal{D},g} \cdot g = h \\ \text{and} \quad & \lambda > 0 \quad \text{and} \quad \nu \gg 0 \\ \text{and} \quad & \text{POLC on } \mu \end{aligned}$$

whereas POLC stands for *possibly other, linear constraints.*

To appreciate the general applicability of this problem for consistency checking and inference, let us look at how the problems discussed in Section 2 fit:

Incurring sure loss (given $\mathcal{A}$ of Sec. 2): $\mathcal{R} := \{\{1_\omega : \omega \in \Omega\} \cup \mathcal{A} \setminus \{0\}\}$ and $h := 0$.
Unconditional natural extension (given $\mathcal{A}$ and f of Sec. 2):
$\mathcal{R} := \{\{g\} : g \in \mathcal{A}\} \cup \{\{1_\Omega\}, \{-1_\Omega\}, \{0\}\} \cup \{\{1_\omega\} : \omega \in \Omega\}$ and $h := f$;
objective function expression $\mu_{\{1_\Omega\},1_\Omega} - \mu_{\{-1_\Omega\},-1_\Omega}$.
Incurring partial loss (given $\mathcal{B}$ of Sec. 2):
$\mathcal{R} := \{\{g, 1_B\} : (g, B) \in \mathcal{B} \setminus [\{0\} \times \Omega^*]\} \cup \{\{1_\omega\} : \omega \in \Omega\}$ and $h := 0$.
Conditional natural extension (given $\mathcal{B}$, f, and C of Sec. 2):
$\mathcal{R} := \{\{g, 1_B\} : (g, B) \in \mathcal{B}\} \cup \{\{1_C\}, \{-1_C\}, \{0\}\} \cup \{\{1_\omega\} : \omega \in \Omega\}$ and $h := f \cdot 1_C$;
objective function expression $\mu_{\{1_C\},1_C} - \mu_{\{-1_C\},-1_C}$.

The inclusion and exclusion of $\{0\}$ is essentially a way of selecting between constraints $\lambda \geq 0$ and $\lambda > 0$.

4 The CONEstrip Algorithm

Now that we have formulated the general problem and have a feel for the general cone representation used in this formulation, we are ready to work towards the actual algorithm that will allow us to solve this general problem. Before presenting the algorithm itself, we perform a supporting analysis of the general problem.

Analysis of the General Problem. In the general problem formulation we gave in Section 3, the (non-additional) constraints express that h must lie in the general cone represented by $\mathcal{R}$. For the feasibility problem, we can actually assume in all generality that $h = 0$, because we allow additional linear constraints on μ. To see this, consider the original feasibility problem and write it as

$$\begin{aligned} \text{find} \quad & (\lambda, \nu) \in \mathbb{R}^{\mathcal{R}\cup\{\{-h\}\}} \times \mathsf{X}_{\mathcal{D}\in\mathcal{R}\cup\{\{-h\}\}} \mathbb{R}^{\mathcal{D}}, \\ \text{subject to} \quad & \textstyle\sum_{\mathcal{D}\in\mathcal{R}\cup\{\{-h\}\}} \lambda_{\mathcal{D}} \cdot \sum_{g\in\mathcal{D}} \nu_{\mathcal{D},g} \cdot g = 0 \\ \text{and} \quad & \lambda > 0 \quad \text{and} \quad \nu \gg 0 \quad \text{and} \quad \mu_{\{-h\},-h} = \lambda_{\{-h\}} \cdot \nu_{\{-h\},-h} \geq 1 \\ \text{and} \quad & \text{POLC on } \mu := (\lambda_{\mathcal{D}} \cdot \nu_{\mathcal{D},g} : \mathcal{D} \in \mathcal{R}, g \in \mathcal{D}), \end{aligned}$$

whose feasible solutions μ can be related to those of the original problem by dividing them by $\mu_{\{-h\},-h}$.

For this feasibility problem, checking whether or not 0 lies in the *blunt* closure of $\underline{\mathcal{R}}$ can be done by solving the following linear programming feasibility problem:

$$\begin{aligned} \text{find} \quad & \mu \in \mathsf{X}_{\mathcal{D}\in\mathcal{R}} \mathbb{R}^{\mathcal{D}}, \\ \text{subject to} \quad & \textstyle\sum_{\mathcal{D}\in\mathcal{R}} \sum_{g\in\mathcal{D}} \mu_{\mathcal{D},g} \cdot g = 0 \quad \text{and} \quad \mu \geq 0 \quad \text{and} \quad \text{POLC on } \mu \\ \text{and} \quad & \textstyle\sum_{\mathcal{D}\in\mathcal{R}} \sum_{g\in\mathcal{D}} \mu_{\mathcal{D},g} \geq 1. \end{aligned}$$

where the last constraint is a proxy for the blunting—i.e., using $>$ instead of $\geq$—implied by the constraint $\lambda > 0$ (cf. Figure 2). If this problem is feasible, then 0 lies either in the interior of $\underline{\mathcal{R}}$ or in a facet. The interior case can be checked by solving another linear programming feasibility problem:

$$\begin{aligned} \text{find} \quad & \mu \in \mathsf{X}_{\mathcal{D}\in\mathcal{R}} \mathbb{R}^{\mathcal{D}}, \\ \text{subject to} \quad & \textstyle\sum_{\mathcal{D}\in\mathcal{R}} \sum_{g\in\mathcal{D}} \mu_{\mathcal{D},g} \cdot g = 0 \quad \text{and} \quad \mu \geq 1 \quad \text{and} \quad \text{POLC on } \mu, \end{aligned}$$

where $\mu \geq 1$ is a proxy for the constraint $\nu \gg 0$ (cf. Figure 2).

Now, if h lies in the closure, but not the interior, it must lie on a facet. From the proof of the Theorem we know that we are then actually faced with the same type of feasibility problem we started out with, but now with $\mathcal{R}$ replaced by $\mathcal{S} \subset \mathcal{R}$. Based on this insight, we could construct a recursion algorithm. However, this would involve facet enumeration, which is a computationally expensive operation [1]. Nevertheless, the key insight is that, to solve the feasibility problem, we must identify the facet containing 0. Put differently, we must eliminate those elements from $\mathcal{R}$ that preclude 0 from lying in the (relative) interior of $\underline{\mathcal{R}}$.

The Algorithm for the Feasibility Problem. The idea is to relax the general feasibility problem to a linear programming problem and detect which elements $\mathcal{D}$ of $\mathcal{R}$ make the problems infeasible if $\lambda_{\mathcal{D}} > 0$. This is done by transforming it into a blunted closure-case optimization problem with an objective that essentially rewards solutions that are close to interior-case solutions. So, given the general feasibility problem of Section 3 with arbitrary $\mathcal{R} \Subset \mathcal{L}^*$ and $h := 0$, the algorithm is:

1.
$$\begin{aligned} &\text{maximize} && \textstyle\sum_{\mathcal{D}\in\mathcal{R}} \tau_{\mathcal{D}}, \\ &\text{subject to} && \textstyle\sum_{\mathcal{D}\in\mathcal{R}} \sum_{g\in\mathcal{D}} \mu_{\mathcal{D},g} \cdot g = 0 \quad \text{and} \quad \mu \geq 0 \quad \text{and} \quad \text{POLC on } \mu \\ &&& \text{and} \quad 0 \leq \tau \leq 1 \quad \text{and} \quad \forall \mathcal{D} \in \mathcal{R} : \tau_{\mathcal{D}} \leq \mu_{\mathcal{D}} \quad \text{and} \quad \textstyle\sum_{\mathcal{D}\in\mathcal{R}} \tau_{\mathcal{D}} \geq 1. \end{aligned}$$

2. a. If there is no feasible solution, then the general problem is infeasible.
   b. Otherwise set $\mathcal{S} := \{\mathcal{D} \in \mathcal{R} : \tau_{\mathcal{D}} > 0\}$; τ is equal to 1 on $\mathcal{S}$:
      i. If $\forall \mathcal{D} \in \mathcal{R} \setminus \mathcal{S} : \mu_{\mathcal{D}} = 0$, then the general problem is feasible.
      ii. Otherwise, return to step 1 with $\mathcal{R}$ replaced by $\mathcal{S}$.

We call this the *CONEstrip algorithm* because of step 2(b)ii, in which the irrelevant parts of the cone (representation) are stripped away. The description of the algorithm on the cone of Figure 1 can be found there as well. It is implemented in and tested with murasyp [5].

Proposition. *The claims made in the CONEstrip algorithm are veracious and it terminates after at most* $|\mathcal{R}| - 1$ *iterations.*

Proof. First the claim in step 2a: the feasibility requirements of the problem in step 1 are weaker than those of the general problem for the *current* representation $\mathcal{R}$ ($\sum_{\mathcal{D}\in\mathcal{R}} \tau_{\mathcal{D}} \geq 1$ is a proxy for $\lambda > 0$). Next the claim in step 2b: if μ is a solution, then $\mu / \min\{\tau_{\mathcal{D}} : \mathcal{D} \in \mathcal{S}\}$ is a solution for which the claim can be satisfied, which will be the case, because it increases the objective. Finally the claim in step 2(b)i: if the condition of the claim is verified, then μ is a solution to the general problem for the *current* representation $\mathcal{R}$ (take λ equal to 1 and ν equal to μ for indices in $\mathcal{S}$, and λ equal to 0 and ν arbitrary—e.g., 1—for other indices).

Now let $\mathcal{E} := \{g \in \bigcup\mathcal{R} : (\exists \mathcal{D} \in \mathcal{R} : \mu_{\mathcal{D},g} > 0)\}$; by step 2(b)ii we know that $\underline{\mathcal{E}}$ contains 0. Moreover, $\underline{\mathcal{S}} = \underline{\mathcal{R}} \cap \underline{\mathcal{E}}$, so reiterating with $\mathcal{R}$ replaced by $\mathcal{S}$ leads to an equivalent problem feasibility-wise. Each iteration, by step 2(b)ii, we know that $|\mathcal{S}| < |\mathcal{R}|$, so at most $|\mathcal{R}| - 1$ iterations are necessary to decide feasibility of the original problem. □

The Algorithm for the Optimization Problem. The idea is to split off the non-linear aspect of the feasibility part of the general optimization problem and deal with it using the CONEstrip algorithm. The optimization itself is then reduced to a linear programming problem. So, given the general optimization problem of Section 3 with arbitrary $\mathcal{R} \Subset \mathcal{L}^*$ and $h \in \mathcal{L}$, the proposed algorithm is:

1. Apply the CONEstrip algorithm to $\mathcal{R} \cup \{-h\}$ with $\mu_{\{-h\},-h} \geq 1$ as an additional constraint; if feasible, continue to the next step with the terminal set $\mathcal{S}$.
2.
$$\begin{aligned} \text{maximize an affine function of} \quad & \mu \in \times_{\mathcal{D}\in\mathcal{R}} \mathbb{R}^{\mathcal{D}}, \\ \text{subject to} \quad & \textstyle\sum_{\mathcal{D}\in\mathcal{S}} \sum_{g\in\mathcal{D}} \mu_{\mathcal{D},g} \cdot g = h \\ \text{and} \quad & \mu \geq 0 \quad \text{and} \quad \text{POLC on } \mu, \end{aligned}$$

   where $\mu \geq 0$ is a proxy for $\mu \gg 0$ by continuity of the linear objective.

5 Conclusion

We now have an efficient, polynomial time algorithm for consistency checking and inference in uncertainty modeling frameworks using general cones: the number of linear programs to solve has worst-case complexity linear in the cardinality of the cone representation. The work of Walley et al. [10] made me believe such an efficient algorithm was possible, the representation of Couso and Moral [3] provided useful structure, and a variable-bounding technique spotted in the 'zero norm'-minimization literature [7, (1) to (2)] made everything come together.

The CONEstrip algorithm—formulated in terms of linear programs—is rather high-level. Integrating it with a specific linear programming solver might allow for a practical increase in efficiency: e.g., the stripping step can be seen as a form of column elimination. Also, heuristics could be found to reduce the representation size.

The question may arise whether the representation and the algorithm are also applicable when modeling uncertainty using general bounded polytopes, such as non-closed credal sets (arising, e.g., when strict bounds on expectations are allowed). Yes: such polytopes can be seen as intersections of a general cone and a hyperplane.

Acknowledgements. For useful discussion, I thank Dirk Aeyels, Gert de Cooman, Nathan Huntley, and especially Filip Hermans, who also provided very helpful pointers and feedback. I thank the reviewers for their effort and a much appreciated critical reading.

References

1. Avis, D., Bremner, D., Seidel, R.: How good are convex hull algorithms? Comput. Geom-Theor. Appl. 7(5-6), 265–301 (1997)
2. Coletti, G., Scozzafava, R.: Probabilistic logic in a coherent setting. Kluwer Academic Publishers, Dordrecht (2002)

3. Couso, I., Moral, S.: Sets of desirable gambles: conditioning, representation, and precise probabilities. Int. J. Approx. Reason. 52(7), 1034–1055 (2011)
4. Quaeghebeur, E.: Desirability. In: Coolen, F.P.A., Augustin, T., De Cooman, G., Troffaes, M.C.M. (eds.) Introduction to Imprecise Probabilities. John Wiley & Sons, Ltd., New York City (at the editor)
5. Quaeghebeur, E.: murasyp: Python software for accept/reject statement-based uncertainty modeling (in progress), `http://equaeghe.github.com/murasyp`
6. Quaeghebeur, E., De Cooman, G., Hermans, F.: Accept & reject statement-based uncertainty models (in preparation)
7. Rinaldi, F., Schoen, F., Sciandrone, M.: Concave programming for minimizing the zero-norm over polyhedral sets. Comput. Optim. Appl. 46(3), 467–486 (2010)
8. Walley, P.: Statistical Reasoning with Imprecise Probabilities. Chapman & Hall, London (1991)
9. Walley, P.: Towards a unified theory of imprecise probability. Int. J. Approx. Reason. 24(2-3), 125–148 (2000)
10. Walley, P., Pelessoni, R., Vicig, P.: Direct algorithms for checking consistency and making inferences from conditional probability assessments. J. Stat. Plan. Infer. 126(1), 119–151 (2004)
11. Ziegler, G.M.: Lectures on Polytopes. Springer, Berlin (1995)

Towards a Robust Imprecise Linear Deconvolution

Oliver Strauss and Agnes Rico

Abstract. Deconvolution consists of reconstructing a signal from blurred (and usually noisy) sensory observations. It requires perfect knowledge of the impulse response of the sensor. Relevant works in the litterature propose methods with improved precision and robustness. But those methods are not able to account for a partial knowledge of the impulse response of the sensor. In this article, we experimentally show that inverting a Choquet capacity-based model of an imprecise knowledge of this impulse response allows to robustly recover the measured signal. The method we use is an interval valued extension of the well known Schultz procedure.

Keywords: Choquet capacities, deconvolution, inverse problem, robustness.

1 Introduction

Discrete signals used in engineering applications are recorded through sensors. The measured signal is usually a smeared version of the signal under consideration that highly affects its resolution. Deconvolution consists in improving the resolution of the signal by removing the smoothing effect of the measuring instrument by using its resolution (or point spread) function.

The measured signal being discrete, deconvolution consists in inverting a system of N linear discrete convolution equations:

Oliver Strauss · Agnes Rico
LIRMM, Université Montpellier 2, France
e-mail: strauss@lirmm.fr

Agnes Rico
ERIC, Université Lyon 1, France
e-mail: agnes.rico@univ-lyon1.fr

R. Kruse et al. (Eds.): Synergies of Soft Computing and Statistics, AISC 190, pp. 55–62.
springerlink.com © Springer-Verlag Berlin Heidelberg 2013

$$\forall n \in \{0, \ldots, N-1\}, m_n = \sum_{k=0}^{K-1} h_{n-k} s_k + \eta_n, \tag{1}$$

m being the measured signal, h the impulse response of the sensor, s the signal under consideration, η a measurement noise, N the number of samples of the measured signal and K the number of samples of the signal to be reconstructed.

Equation (1) can be rewritten in matrix form:

$$M = AS + E, \tag{2}$$

M being the measurement vector of length N, S being the signal vector of length K, A being a $N \times K$ matrix, and E being an error vector of length N. Deconvolution thus consists of inverting the matrix eq. (2) by minimizing a risk function [10]. A very common risk function is the Euclidian distance: $J_A(S, M) = ||M - AS||^2$. Minimizing this risk function leads to computing the following solution:

$$\hat{S} = A^+ M, \tag{3}$$

where A^+ is the pseudo inverse matrix of A. Without any noise nor uncertainty, the least square solution leads to recovering S from M. It should also be the case if the error vector E can be considered as a white gaussian additive noise, and if the matrix A can be properly pseudo-inverted. However, deconvolution is known as an ill-posed problem, i.e. small divergence in the model can cause high divergence in the solution.

In the relevant literature, most work focusses on considering the measurement noise properties and designing appropriate inversion algorithms [13, 7]. Regularizing the obtained algorithms to provide solutions that have a relevant physical interpretation and stable behavior is the second track that gathers many research effort [8]. Those work try and find methods with improved precision and robustness. Improving the precision consists in trying to reduce the distance between the sought after signal and the reconstructed signal. Improving the robustness consists in keeping this distance as low as possible even if there is deviation in the system and/or noise modeling.

However, few attention has been payed to considering an imprecise knowledge in the sensor's impulse response. The traditional approach, called blind deconvolution, consider the shape of the impulse response to be known and estimate its parametric description during the reconstruction process [9]. Therefore, within this kind of approach, the resolution of the sensor is supposed to be completely unknown, while it is usually only imprecisely known.

In this paper, we consider an alternative technique that enables using an imprecise knowledge of the impulse response of the considered sensor. This technique is based on modeling this imprecise knowledge by means of a concave Choquet capacity. Compared to blind deconvolution, this technique can be can be thought as myopic, since it supposes the impulse response to be imprecisely known. For seek of simplicity, this work is restricted to positive centered symmetric impulse responses.

The paper is structured as follows. The next section presents how filtering can be seen as an expectation operation. Section 3 is devoted to the description of interval-valued extension of linear filtering. Section 4 focusses on the extension of the Schultz iterative procedure. Before the conclusion, section 5 is devoted to illustrating how this method improves the robustness of a least square inversion based signal deblurring.

2 Filtering Can Be Seen as an Expectation Operation

For seek of simplicity, we will consider that the number of measurement samples equals the number of sought after signal values (i.e. $N = K$) which is usually the case when considering the measured discrete signal as being a blurred version of the original discrete signal. Let $\Omega = \{0, \cdots, N-1\}$. In classical representation, h being the impulse response of the measurement device, each value m_n ($n \in \Omega$) is a linear combination of the signal values (see Eq. (1)).

Let $\zeta = \sum_{k \in \mathbb{Z}} h_k$ with $h_k \geq 0$ for all k and $\rho_k = \zeta^{-1} h_k$. So $\rho = (\rho_k)_{k \in \mathbb{Z}}$ can be considered as a discrete probability distribution on $\mathbb{Z}$. Let $\rho^n = (\rho^n_k)_{k \in \mathbb{Z}}$ be the probability distribution defined by translating ρ in n: $\rho^n_k = \rho_{n-k}$. Hence, a noise free version of Eq. (1) can be re-written as follows:

$$m_n = \zeta \sum_{k \in \Omega} s_k \rho^n_k = \zeta \mathbb{E}_{P_n}(S), \tag{4}$$

P_n being the probability measure induced by ρ^n and $\mathbb{E}_{P_n}$ being the expectation operator induced by P_n on $\mathbb{Z}$.

Thus, filtering a signal with a linear filter whose impulse response is positive can be seen as an expectation operation multiplied by a constant real value. This operation can be presented in matrix form:

$$M = \zeta A_P S, \tag{5}$$

where M and S respectively denote the measurement and the input signal vector, and with A_P defined by:

$$A_P = \begin{bmatrix} \rho^0_0 & \rho^0_1 & \cdots & \rho^0_{N-1} \\ \rho^1_0 & \rho^1_1 & \cdots & \rho^1_{N-1} \\ \cdots & \cdots & \cdots & \cdots \\ \rho^{N-1}_0 & \rho^{N-1}_1 & \cdots & \rho^{N-1}_{N-1} \end{bmatrix} \tag{6}$$

3 Interval-Valued Extension of Linear Filtering

In [14], we have proposed to extend the measurement Eq. (4) to represent an imprecise knowledge on the impulse response h. This extension is based on replacing the probability measure in Eq. (4) by a more general confidence measure called a concave capacity (see e.g. [1]). The use of a capacity to represent a confidence measure entails using a more general expectation operator called the Choquet integral (see [2]).

Let $\mathcal{P}(\Omega)$ be the set of all subsets of Ω and V be the set all the real functions defined on Ω. A capacity (or non-additive or monotone or fuzzy measure) ν is a set function $\nu : \mathcal{P}(\Omega) \to [0,1]$ such that $\nu(\varnothing) = 0$, $\nu(\Omega) = 1$, and $\forall A \subseteq B \Rightarrow \nu(A) \leq \nu(B)$. The conjugate ν^c of a capacity ν is defined as: $\nu^c(A) = 1 - \nu(A^c)$, for any subset A of Ω, with A^c being the complementary subset of A in Ω.

A capacity ν such that $\forall A, B \in \mathcal{P}(\Omega)$, $\nu(A \cup B) + \nu(A \cap B) \leq \nu(A) + \nu(B)$ is said to be concave or submodular or 2-alternating measure. The core of a concave capacity ν, denoted *core*(ν), is the set of probability measures P defined on Ω such that $\forall A \in \mathcal{P}(\Omega)$, $\nu(A) \geq P(A)$.

Let ν be a capacity on $\mathcal{P}(\Omega)$, and $X = \{x_0, \ldots, x_{N-1}\} \in V$ be a finite positive real function (or vector), then the Choquet integral [5] of X with respect to ν is defined by: $\check{\mathbb{C}}_\nu(X) = \sum_{n \in \Omega} x_{(n)}(\nu(A_{(n)}) - \nu(A_{(n+1)}))$, where $(.)$ indicates a permutation that sorts the x_n in non-decreasing order: $x_{(0)} \leq \ldots \leq x_{(N-1)}$, and subsets $A_{(i)}$ being defined by: $A_{(i)} = \{(i), \ldots, (N-1)\}$, and $A_{(N)} = \varnothing$.

The expectation operator can easily be extended to concave capacities (see [14]). Let ν be a concave capacity on $\mathcal{P}(\Omega)$, then $\forall X \in V$, $\overline{\underline{\mathbb{E}}}_\nu(X) = [\check{\mathbb{C}}_{\nu^c}(X), \check{\mathbb{C}}_\nu(X)]$. By using the results proved by [3], it can be easily proved that: $\forall X \in V$, $\forall P \in$ *core*(ν), $\mathbb{E}_P(X) \in \overline{\underline{\mathbb{E}}}_\nu(X)$, and $\forall X \in V$, $\forall Y \in \overline{\underline{\mathbb{E}}}_\nu(X)$, $\exists P \in$ *core*(ν) such that $Y = \mathbb{E}_P(X)$.

Thus a concave capacity can represent a set of normalized impulse responses. Let ν be a concave capacity representing a convex set of normalized impulse responses, the interval-valued extension of Eq. (4) is:

$$[m_n] = \zeta \overline{\underline{\mathbb{E}}}_{\nu_n}(S), \tag{7}$$

$[m_n]$ being an interval-valued measure and ν_n being the capacity ν translated in n.

Equation (7) can be rewritten in a simple way:

$$[M] = \zeta \mathcal{A}_\nu(S), \tag{8}$$

where $[M] = [\underline{M}, \overline{M}]$ is an interval-valued vector and $\mathcal{A}_\nu$ is an interval valued function that sums up the N imprecise measurement equations. $[M]$ is the convex hull of all the values that could have been obtained by using Eq. (4) and a probability measure $P \in$ *core*(ν).

Imprecise expectation can be extended to interval-valued inputs due to the monotony of the Choquet integral. Let $[X] = ([\underline{x_1}, \overline{x_1}], \cdots, [\underline{x_N}, \overline{x_N}])$ be an interval-valued vector then $\overline{\mathbb{E}}_\nu([X]) = [\check{\mathbb{C}}_{\nu^c}(\underline{X}), \check{\mathbb{C}}_\nu(\overline{X})]$ where $\underline{X}$ is the vector $(\underline{x_1}, \cdots, \underline{x_N})$ and $\overline{X}$ is the vector $(\overline{x_1}, \cdots, \overline{x_N})$. Hence Eq. (7) can be extended to interval-valued inputs:

$$[m_n] = \zeta \overline{\mathbb{E}}_{\nu_n}([S]), \tag{9}$$

and this equation can be rewritten:

$$[M] = \zeta \mathcal{A}_\nu([S]), \tag{10}$$

In this context, a useful concave capacity is the possibility measure induced by a centered symmetric triangular possibility distribution whose support is $[-\Delta, \Delta]$ ($\Delta \in \mathbb{R}^+$). As shown in [11], the core of such a capacity includes any finite positive normalized centered symmetric impulse response whose support is included in $[-\Delta, \Delta]$.

4 Inverting the Imprecise Filtering by Extending the Schultz Iterative Procedure

Inverting Eq. (2) can be obtained by computing A^+ the pseudo-inverse of matrix A. The solution $\hat{S} = A^+ M$ is the standard solution of the regularized equation $(A^T A)S = A^T M$. However, due to the ill-posedeness, the matrix $(A^T A)$ is ill-conditioned. Thus direct estimation of $\hat{S}$ has to be replaced by other procedures, like the Schultz iterative procedure (often called the Hotelling iterative procedure, see [6]). Starting from a wrong solution (e.g. $S^0 = 0$), the Schultz procedure iteratively corrects this value and converges towards the least squares solution. A simplified version of the Schultz procedure is given by:

$$S^{i+1} = S^i + \lambda A^T (M - AS^i), \tag{11}$$

where S^i is the estimation of S at the ith iteration and λ is a positive real that ensures the convergence [4] (e.g. $\lambda = 1/\sum_{k \in \mathbb{Z}} h_k^2$). Note that, in our case, the impulse response being centered and symmetric, $A = A_P = A_P^T$.

In [16], we have proposed to extend the Schultz procedure in order to invert Eq. (10). This extension can be written in that way:

$$[S^{i+1}] = [S^i] \boxplus \mathcal{A}_\nu(M \ominus \mathcal{A}_\nu[S^i]), \tag{12}$$

where $[S^{i+1}]$ is the interval valued estimation of S at the ith iteration, $\ominus$ is the term by term Minkowski interval extension of the subtraction defined by: $[\underline{a}, \overline{a}] \ominus [\underline{b}, \overline{b}] = [\underline{a} - \overline{b}, \overline{a} - \underline{b}]$, and $\boxplus$ is the term by term dual Minkowski interval extension of the addition defined by: $[\underline{a},\overline{a}] \boxplus [\underline{b},\overline{b}] = [\underline{a}+\overline{b},\overline{a}+\underline{b}]$. Note that the result of a $\boxplus$ operation can be an improper interval (i.e. the lower value is upper than the upper value). This situation has to be considered as an intermediate calculus steps (see e.g. [15]).

5 Experiments

This experimental section aims at showing that such a method can help to improve the robustness of a least square inversion when knowledge in the impulse response of the sensor is imprecise. This experiment consists in filtering an electrocardiogram (ECG) signal[1] of 10000 samples [12] by using different randomly chosen impulse responses (see Fig. 1) and computing the Euclidian distance between the original signal and the reconstructed signals.

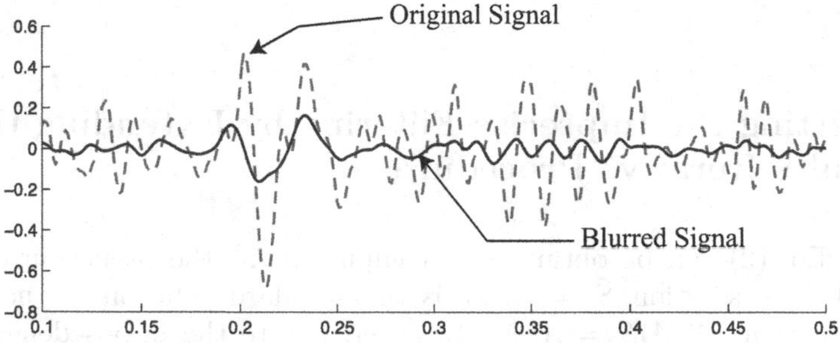

Fig. 1 Original and blurred ECG signals

In this experiment, we suppose that the only thing that is known about the filter is that its impulse response is positive normalized centered symmetric with a bounded support included in $[-15, 15]$. We aim at comparing a traditional deconvolution method using a wrong impulse response with our method based on imprecise knowledge about the *true* impulse response. For this experiment, 200 pairs of impulse responses with a bounded support included in $[-15, 15]$ have been randomly chosen. For each pair of impulse responses, one is used to blur the signal and the other one is used to reconstruct the signal by using the traditional least squares method. The blurred signal is also reconstructed with our method by considering a capacity whose core contains all the envisaged impulse responses. This capacity is a possibility measure based on the triangular possibility distribution whose support is $[-15, 15]$.

For both methods, the iterative reconstructing procedure is stopped when the distance between the reconstructed signal and the original signal is

[1] ECG data were courteously provided by the LTSI laboratory, Rennes.

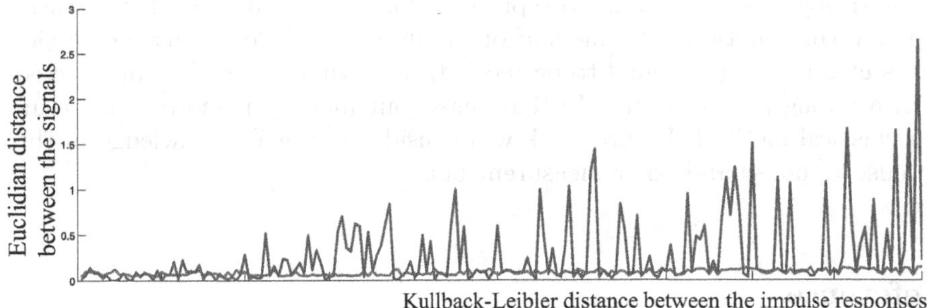

Fig. 2 Distances between the original and the reconstructed signal for the classical method (in blue) and for the new interval valued method (in red)

minimal. Concerning our method, the distance between the interval valued reconstructed signal and the original signal is computed by considering the median signal, i.e. the signal $\tilde{S}$ such that $\forall k \in \{1, \ldots, N\}$, $\tilde{s}_k = \frac{1}{2}(\underline{s}_k + \overline{s}_k)$.

Figure 2) shows the distances between the reconstructed signal and the original signal for increasing Kullback-Leibler distances between the impulse response that has been used for smoothing the signal and the impulse response that has been used to reconstruct the signal. Distances for the classical method are plotted in blue, while distances for the new method are plotted in red.

When the two impulse responses are close one from one-other, the classical method provides a reconstructed signal that is close from the original signal (left part of Fig. 2). Though, when the two impulse responses are different (right part of Fig. 2), then the classical solution is rather unstable, i.e. the distance can be low or high, and this fact seems not to only depend on the distance between the real and the considered impulse response. On the other hand, the interval-valued method seems to be more robust, i.e. the distance between the median of the interval valued reconstructed signal and the original signal seems to be less affected by the choice of the smoothing filter.

6 Conclusion

In this paper, we have presented a reconstruction method that is able to deal with imprecise knowledge in the impulse response of the sensor that has been used to measure the signal. One of the main feature of this method is that it leads to reconstruct an interval-valued signal.

What is claimed and illustrated here is that, when knowledge about the impulse response used to blur a signal is high enough for the deconvolution kernel to be close from the blurring kernel, a better result is obtained by using a traditional deconvolution approach. But, when this knowledge is poor, using an imprecise kernel based representation leads to improved robustness in the reconstruction : the median of the interval-valued reconstructed signal is closer from the signal to be reconstructed than a signal reconstructed with a wrong precise kernel. In that sense, our method is more stable than the classical method. Future work will consider imprecise knowledge in the impulse response and noisy measurements.

References

1. Campos, L., Huete, J., Moral, S.: Probabilities intervals: a tool for uncertain reasoning. Int. J. Uncertain. Fuzz. 2, 167–196 (1994)
2. Choquet, G.: Theory of capacities. Ann. Inst. Fourier 5, 131–295 (1953)
3. Denneberg, D.: Non-Additive Measure and Integral. Kluwer Academic Publishers, Dordrecht (1994)
4. Eggermont, P., Herman, G.: Iterative algorithms for large partitioned linear systems with applications to image reconstruction. Linear Algebra Appl. 40, 37–67 (1981)
5. Grabisch, M., Labreuche, C.: The symmetric and asymmetric Choquet integral on finite spaces for decision making. Stat. Pap. 43, 37–52 (2002)
6. Herzberger, J., Petkovi, L.: Efficient iterative algorithms for bounding the inverse of a matrix. Computing 44, 237–244 (1990)
7. Hudson, M., Lee, T.: Maximum likelihood restoration and choice of smoothing parameter in deconvolution of image data subject to poisson noise. Comput. Stat. Data Anal. 26(4), 393–410 (1998)
8. Jalobeanu, A., Blanc-Ferraud, L., Zerubia, J.: Hyperparameter estimation for satellite image restoration using a MCMC maximum-likelihood method. Pattern Recogn. 35(2), 341–352 (2002)
9. Kaaresen, K.F.: Deconvolution of sparse spike trains by iterated window maximization. IEEE Trans. Signal Process. 45(5), 1173–1183 (1997)
10. Kalifa, J., Mallat, S.: Thresholding estimators for linear inverse problems and deconvolutions. Ann. Stat. 1, 58–109 (2003)
11. Loquin, K., Strauss, O.: On the granularity of summative kernels. Fuzzy Set. Syst. 159(15), 1952–1972 (2008)
12. Poree, F., Bansard, J.-Y., Kervio, G., Carrault, G.: Stability analysis of the 12-lead ECG morphology in different physiological conditions of interest for biometric applications. In: Proc. of the Int. Conf. on Computers in Cardiology 2009, Park City, UT, USA, vol. 36, pp. 285–288 (2009)
13. Rice, J.: Choice of smoothing parameter in deconvolution problems. Contemp Math. 59, 137–151 (1986)
14. Rico, A., Strauss, O.: Imprecise expectations for imprecise linear filtering. Int. J. Approx. Reason. 51, 933–947 (2010)
15. Sainz, M., Herrero, P., Armengol, J., Vehí, J.: Continuous minimax optimization using modal intervals. J. Math. Anal. Appl. 339, 18–30 (2008)
16. Strauss, O., Rico, A.: Towards interval-based non-additive deconvolution in signal processing. Soft Comput. 16(5), 809–820 (2012)

2D Probability-Possibility Transformations

Alessandro Ferrero, Marco Prioli,
Simona Salicone, and Barbara Vantaggi

Abstract. Probability-possibility transformations are useful whenever probabilistic information must be dealt with in the possibility theory. In this paper, two-dimensional probability-possibility transformations of joint probability densities are considered, to build joint possibilities such that the marginals preserve the same information content as the marginals of the joint probability densities.

Keywords: Joint possibility distributions, metrology, probability-possibility transformations, uncertainty.

1 Introduction

The probability-possibility transformations [5, 10, 13] allow one to build a possibility distribution starting from a probability density function. These transformations are useful whenever probabilistic information and statistical data must be dealt with in the possibility theory. The interest in these transformations is not only from the mathematical and theoretical point of view. An increasing interest comes also from the engineering field, and, in particular, from the field of measurements, as here motivated.

In fact, the best practice in measurement requires that measurement uncertainty is always associated to a measurement result. The mathematical

Alessandro Ferrero · Marco Prioli · Simona Salicone
Politecnico di Milano
e-mail: {alessandro.ferrero,simona.salicone}@polimi.it,
marco.prioli@mail.polimi.it

Barbara Vantaggi
Universita di Roma La Sapienza
e-mail: barbara.vantaggi@dmmm.uniroma1.it

R. Kruse et al. (Eds.): Synergies of Soft Computing and Statistics, AISC 190, pp. 63–72.
springerlink.com © Springer-Verlag Berlin Heidelberg 2013

theory that is actually referred to for evaluating and expressing measurement uncertainty is the probability theory, as recommended by the Guide to the Expression of Uncertainty in Measurement (GUM) [2], that represents the present reference Standard document. Within this approach, a measurement result, together with its associated measurement uncertainty, is represented by a probability density function (pdf).

However, in the recent years, the limitations of this approach to uncertainty evaluation have been outlined, and several Authors have proposed the possibility theory as an alternative, more general method [7, 8, 9, 11, 12, 15] to uncertainty evaluation and representation.

Within this theory, a measurement result, together with its associated measurement uncertainty, is represented by a possibility distribution [11, 15].

This approach has similar limitation as the traditional one. As a matter of fact, in the measurement field, two contributions to uncertainty of different nature are recognized: the random contributions and the systematic ones. The random contributions to uncertainty are associated with the fact that, when a measurement is repeated several times, it will generally provide different measured values. A random contribution to uncertainty *"arises from unpredictable or stochastic temporal and spatial variations of influence quantities. The effects of such variations, ..., give rise to variations in repeated observations of the measurand"* [2]. On the other hand, a systematic contribution to uncertainty *"arises from a recognized effect of an influence quantity on a measurement result"* [2]. Hence, in repeated observations of the measurand, these contributions always show the same value and sign. In other words, the measured value contains a *bias*, whose value is often unknown. Generally, the effect of the unknown systematic uncertainty contribution on the measurement result is given (i. e. by the calibration certificate or by the instrument datasheet) in terms of an interval about the measured value, within which the measurand will surely lie, but it is not known whereabout[1].

According to the above definitions, the random contributions to uncertainty distribute *randomly* and can be effectively mathematically represented by random variables, that is, by pdfs. On the contrary, the unknown systematic contributions to uncertainty, whose behavior is *not random*, can be effectively mathematically represented by fuzzy variables, that is by possibility distributions (PDs).

In general, a measurement result is affected by several contributions to uncertainty: some of these contributions are random, others are systematic.

[1] The best practice in measurement, reflected by the GUM [2], requires that *"all recognized significant systematic effects"* have been corrected for, so that only random contributions remain and can be handled by the probability theory [2]. However, the *significance* of an effect can be assessed, from a metrological point of view, by stating whether its contribution to the final uncertainty leads to exceed the given *target uncertainty* [3]; and this can be done only if a suitable mathematical approach is defined to evaluate and propagate all different kinds of uncertainty contributions.

The systematic contributions are naturally mathematically represented by PDs, and the probability-possibility transformations allow one to represent also the random ones in terms of PDs.

Hence, also in the field of measurements, any time the available information is given in terms of a pdf, it could be necessary to translate this information within the possibility theory, in terms of a PD.

Many probability-possibility (p-p) transformations have been defined, which transform a monovariate pdf into a monovariate PD. The aim of this paper is to extend these transformations also to bivariate distributions. In particular, this paper proposes a new 2D probability-possibility transformation satisfying the principle of maximum specificity for the marginal distributions, which appears to be a suitable choice for metrological applications.

The paper is organized as follows. In Sec. 2, the definition of the 1D probability-possibility transformations is briefly recalled, while in Sec. 3 possible extensions to two-dimensional ones are discussed.

2 One-Dimensional Probability-Possibility Transformations

Let p_X be a continuous probability density of X around x^*. For a given confidence interval I_γ^* around x^*, the associated confidence level γ is given by:

$$\gamma = P(X \in I_\gamma^*) = \int_{x \in I_\gamma^*} p_X(x)dx.$$

Varying the confidence level γ, different nested confidence intervals are obtained (i.e. $I_\gamma^* \subseteq I_\beta^*$ for $\gamma \leq \beta$). The possibility distribution r_X which encodes the whole set of confidence intervals I_γ^* is defined as [5, 10]:

$$r_X(x) = 1 - P(X \in I_\gamma^*).$$

Moreover, for any real interval I, the possibility measure on I, is defined as $z_X(I) = \sup_{x \in I} r_X(x)$.

Let us now consider unimodal pdfs p_X, with mode x^m, such that $p_X(x) \leq p_X(x')$ for any $x \leq x' \leq x^m$, and $p_X(x) \geq p_X(x')$ for any $x^m \leq x \leq x'$. For such pdfs, it is possible to choose a family of intervals I_γ^* in order to find the corresponding maximally specific PDs.

The maximally specific PD r_X^{ms} associated to a given pdf is the PD that preserves the maximum amount of information contained in the pdf [5, 10], that is, the minimal possibility distribution dominating the pdf.

From the mathematical point of view, a PD $\pi(x)$ dominates a given pdf if, for any real interval I, its associated possibility measure $z_X(I)$ satisfies to:

$$z_X(I) \geq P(X \in I)$$

The maximally specific PD r_X^{ms} is the smallest PD among the family of the dominating possibilities $\pi(x)$ ($r_X^{ms}(x) \leq \pi(x)$).

Given a pdf, the corresponding PD r_X^{ms} is obtained choosing, as the confidence intervals I_γ^*, the cuts I_x of the pdf itself [5, 10]:

$$\begin{aligned} I_x &= \{\psi : p_X(\psi) \geq p_X(x)\} \\ r_X^{ms}(x) &= 1 - P(X \in I_x) \end{aligned} \tag{1}$$

Fig. 1 shows an example of the maximally specific p-p transformation applied to a triangular pdf.

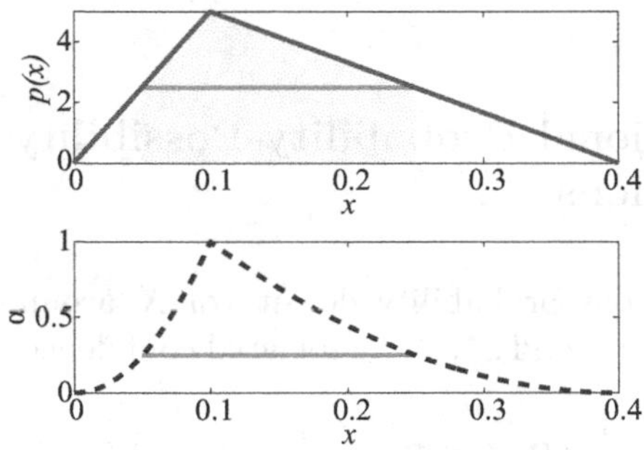

Fig. 1 Example of application of (1). p_X: red line; r_X^m: dashed blue line; a particular I_x and its mapping on the α-cut of r_X^{ms}: green line

3 Two-Dimensional Probability-Possibility Transformations

In order to define a 2D p-p transformation, let us now consider a convex joint pdf $p_{X,Y}$. The aim is to find the transformation that gives the joint PD $r_{X,Y}$ whose marginal distributions coincide with the PDs obtained through the 1D p-p transformation of the corresponding marginals of $p_{X,Y}$. This is an open problem since the joint PD obtained through the product or minimum t-norm by two PDs maximally specific dominating the associated pdfs could be not maximally specific and dominating the relevant joint pdf [14, 1].

Two possible 2D p-p transformations are considered in the following.

3.1 A Natural Extension of the 1D p-p Transformation

As recalled in Sec. 2, for monovariate distributions, the maximally specific PD obtained from a given pdf is the one whose α-cuts are exactly the cuts of the pdf. A possible two-dimensional p-p transformation can be defined by naturally extending to two dimensions (1), which mean choosing, as the α-cuts of the joint PD, the cuts of the joint pdf:

$$\begin{aligned} A_{xy} &= \{(\chi, \varphi) \mid p_{X,Y}(\chi, \varphi) \geq p_{X,Y}(x, y)\} \\ r_{X,Y}(x, y) &= 1 - P(X, Y \in A_{xy}) \end{aligned} \tag{2}$$

The great advantage of this transformation is that the confidence areas (and the associated levels of confidence) are strictly maintained, by definition, in their shape and dimension. Moreover, the obtained joint PD $r_{X,Y}$ dominates the joint pdf $p_{X,Y}$ (the proof is similar to that followed in [5]).

However, this transformation does not satisfy our requirement, that is, the marginals of the obtained joint PD are not maximally specific with respect to the marginals of the joint pdf.

Example 1. Let us consider a bivariate normal pdf of X and Y, with X and Y independent and having standard normal distribution: $p_{X,Y} = \mathcal{N}_2(\mathbf{0}, \mathbf{I})$. The cuts A_{xy} are given by: $A_{xy} = \{(\chi, \varphi) : \chi^2 + \varphi^2 \leq x^2 + y^2\}$.
From (2), it follows:

$$r_{X,Y}(x, y) = 1 - P(X, Y \in A_{xy}) = e^{-\frac{x^2+y^2}{2}}.$$

Then, the marginal possibility distributions are:

$$r_X(z) = r_Y(z) = e^{-\frac{z^2}{2}}.$$

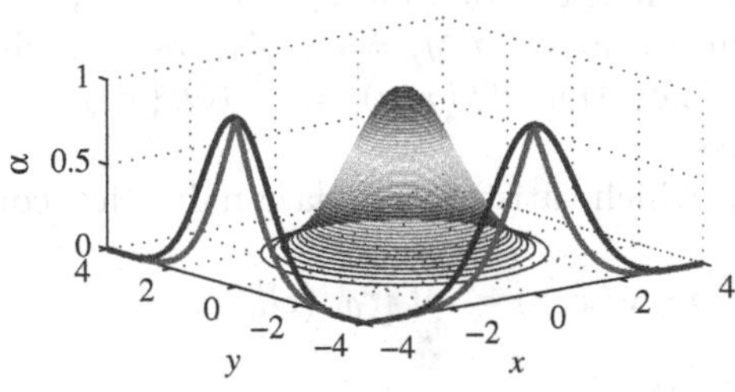

Fig. 2 Joint PD $r_{X,Y}$ obtained by applying (2) to $p_{X,Y}$ of example 1; marginal PDs r_X and r_Y: blue lines; maximally specific PDs $\hat{r}_X$ and $\hat{r}_Y$: red lines.

On the other hand, from the marginal pdfs of $p_{X,Y}$, the following maximally specific marginal PDs are obtained applying (1). For any positive real z:

$$\hat{r}_X(-z) = \hat{r}_X(z) = 1 - P(-z \leq X \leq z) = 1 - (2\phi(z) - 1) = 2 - 2\phi(z),$$

where ϕ is the cumulative distribution function. Then, the marginal PDs, obtained by the joint PD $r_{X,Y}$ induced by the joint pdf through the natural extension of the 1D p-p transformation, do not coincide with the PDs induced by the marginal pdfs. This example is reported in Fig. (2).

3.2 An Ad-Hoc 2D p-p Transformation

3.2.1 The Transformation

A different transformation than (2) shall be defined to preserve the maximal specificity for the marginal distributions [6]. Since the choice of a Cartesian coordinate system is arbitrary, this condition must be satisfied for every possible choice of the Cartesian axes.

Let Θ_k be the 1 to 1 map on $\Re^2$ related to a rotation of an angle ϑ_k around the mean vector $[\mu_1, \mu_2]$ of $p_{X,Y}$, and $\Theta_k(X,Y) = \left(\Theta_k^1(X,Y), \Theta_k^2(X,Y)\right)$ be the random vector associated to the rotation with joint pdf $p_{X',Y'}^{\vartheta_k}$.

From $p_{X',Y'}^{\vartheta_k}$ it is possible to compute the marginal pdf $p_{X'}^{\vartheta_k}$ of X'. Note that $p_{X'}^{\vartheta_k}$ coincides with $p_{Y'}^{\vartheta_j}$ for $\vartheta_k = \vartheta_j + \frac{\pi}{2}$. Then, it is enough to consider only the pdf $p_{X'}^{\vartheta_k}$ for ϑ_k varying in $[0°, 180°]$.

By applying (1) to $p_{X'}^{\vartheta_k}$, it is possible to determine the corresponding maximally specific PD $r_{X'}^{\vartheta_k}$.

Let us define the surface $r_{X',Y'}^{\vartheta_k}(x', y') = r_{X'}^{\vartheta_k}(x')$ for any real y'. Since the marginal PD of X' of the desired joint PD $r_{X,Y}$ must be $r_{X'}^{\vartheta_k}$, it follows that $r_{X,Y}$ must be contained within the surface $r_{X',Y'}^{\vartheta_k}$.

As for the pdfs, also for PDs the dependence on the original coordinates derives from the transformation Θ_k. Since, given ϑ_k, for any (x', y') there is a unique point (x, y) such that $\Theta_k(x, y) = (x', y')$, we use, when it is possible, the following abuse of notation $r^{\vartheta_k}(x, y) := r^{\vartheta_k}(\Theta_k^1(x, y))$. Then, we refer to the original (x, y)-axes.

The joint PD $r_{X,Y}$ which satisfies the marginalization condition for every ϑ_k is given by:

$$r_{X,Y}(x, y) = \inf_{\theta_k} \left[r_{\theta_k} \left(\Theta_k^1(x, y) \right) \right] \tag{3}$$

with ϑ_k varying in $[0°, 180°]$.

From an operational point of view, a finite number of angles ϑ_k is taken, so that a finite number n of surfaces $r_{X',Y'}^{\vartheta_k}$ is identified.

The approximated joint PD is given by the intersection of the n considered surfaces:

$$r_{X,Y}(x,y) = \min\left[r^{\theta_1}_{X',Y'}(x,y), r^{\theta_2}_{X',Y'}(x,y), ..., r^{\theta_n}_{X',Y'}(x,y)\right] \quad (4)$$

Fig. 3 shows, as an example, four surfaces $r^{\vartheta_k}_{X',Y'}$ and the final joint PD, obtained by applying (4) to the joint pdf $\mathcal{N}_2(\mathbf{0},\mathbf{I})$. The marginal PDs along (x,y)-axes are also reported, which coincide, by definition, with the maximally specific PDs induced by the marginal pdfs.

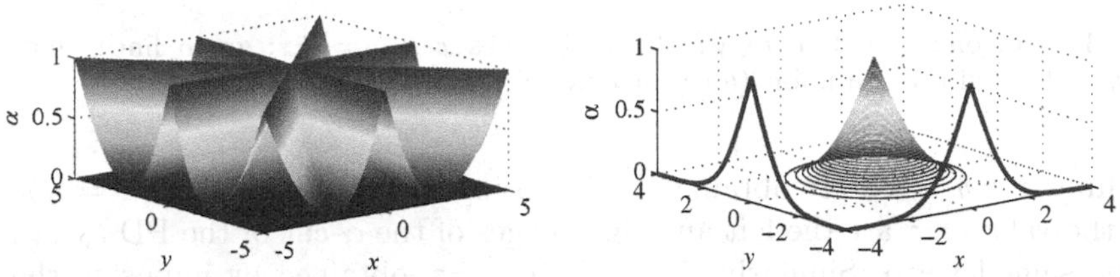

Fig. 3 Example of four surfaces $r^{\vartheta_k}_{X',Y'}$ (for $\vartheta_k \in \{0°, 45°, 90°, 135°\}$) and the final joint PD $r_{X,Y}$ (obtained, with $n = 40$), induced by the joint pdf $\mathcal{N}_2(\mathbf{0},\mathbf{I})$

3.2.2 Some Further Properties

In the previous Section, a 2D probability-possibility transformation is defined. An operative definition is also given, which allows a practical application and provides an approximated joint PD. It is obvious that, in (4), the greater is n, the better approximation is obtained. However, next theorem shows that, when only the information about the sum and difference of two variables needs to be preserved, it is sufficient to take $n = 4$.

Theorem 1. *Let $p_{X,Y}$ be a pdf of (X,Y) with convex cuts and $r_{X,Y}$ be the corresponding PD defined as in (4) with n=4 and $\vartheta_k \in \{0°, 45°, 90°, 135°\}$. Let r_{X+Y} and r_{X-Y} be the the maximally specific PDs associated to the pdfs p_{X+Y} and p_{X-Y} of $X+Y$ and $X-Y$, respectively. Then, for any real z:*

$$\sup_{(x,y):x+y=z} r_{X,Y}(x,y) = r_{X+Y}(z), \quad (5)$$

$$\sup_{(x,y):x-y=z} r_{X,Y}(x,y) = r_{X-Y}(z). \quad (6)$$

In Thm. 1, $r_{X,Y}$ is obtained imposing four marginal PDs, on axes rotated of angles $\vartheta_k \in \{0°, 45°, 90°, 135°\}$ with respect to the original x-axis. Under this hypothesis, since the cuts of the joint pdf are convex, the 2D α-cuts of the joint PD are octagons, as shown in Fig. 4.

As it will be shown in the following proof, the information about the sum $X+Y$ is contained in the marginal PD at $\vartheta = 45°$, while the information about the difference $X-Y$ is contained in the marginal PD at $\vartheta = 135°$. Fig. 4 shows an example, in case of variables $X+Y$ and $X-Y$ centered in the origin and symmetric. Under this assumption, for a generic value α, the

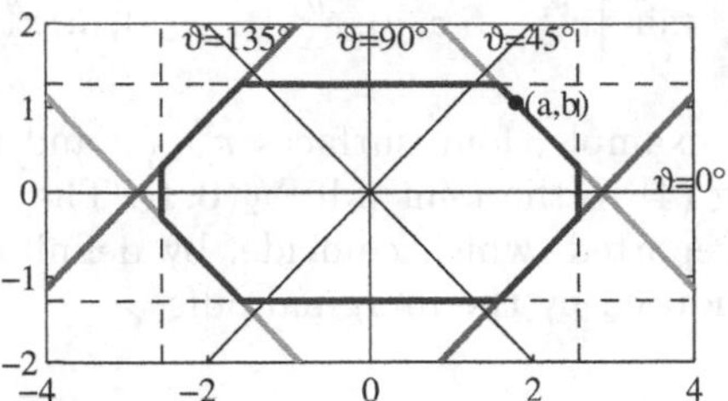

Fig. 4 Octagonal α-cut of $r_{X,Y}$: blue line; lines $x + y = \pm z$: green lines; lines $x - y = \pm z$: red lines; a point (a,b): black circle

two lines $x + y = \pm z$ are obtained by imposing the marginal PD at $\vartheta = 45°$ and the values $\pm z$ are the left and right edges of the α-cut of the PD r_{X+Y}, at the same level α. Similarly, lines $x - y = \pm z$, obtained by imposing the marginal PD at $\vartheta = 135°$, define the left and right edges of the α-cut of the PD r_{X-Y}. The marginal PDs at angles $\vartheta_k \in \{0°, 90°\}$ are considered as well to maintain also the information about the marginal PDs on the main axes.

Proof. Let us consider Eq. (5) first.

Without loss of generality, let us suppose that X and Y have null means, since the axes are centered in the means.

For any real z, the value $r_{X+Y}(z)$ corresponds to the probability of $P(X+Y \in I_c)$ where $I_c = \{\omega : p_{X+Y}(\omega) \leq p_{X+Y}(z)\}$, which is equal to $r_{X'}^{\vartheta=45°}(\frac{z}{2})$ (note that the point $(\frac{z}{2}, \frac{z}{2})$ is on the line $x + y = z$).

For any point (x, y) such that $x + y = z$, $r_{X'}^{\vartheta=45°}(\frac{z}{2}) = r_{X',Y'}^{\vartheta=45°}(x, y)$ by construction, since the line $x + y = z$ is orthogonal to the line rotated of an angle $\vartheta = 45°$ with respect to the original x-axis.

Let us consider the minimum octagonal α-cut $\hat{C}$ of $r_{X,Y}$ that has non empty intersection with the line $z = x+y$. Since $\hat{C}$ has at least one point (a, b) on the line $x + y = z$, for such point $r_{X,Y}(a, b) = r_{X',Y'}^{\vartheta=45°}(a, b) = r_{X'}^{\vartheta=45°}(\frac{z}{2}) = r_{X+Y}(z)$. Since the point (a, b) belongs to the minimum α-cut $\hat{C}$, the *sup* value of (5) is given by the α-level of $\hat{C}$: $\sup_{(x,y):x+y=z} r_{X,Y}(x, y) = r_{X,Y}(a, b)$.

Thus, Eq. (5) follows. Equation (6) is proved in a similar way, by considering the marginal PD $r_{X'}^{\vartheta=135°}$ and the line $x - y = z$.

4 Conclusions

Two different 2D p-p transformations to build joint PDs starting from joint pdfs are discussed.

The first transformation provides joint PDs dominating the original joint pdfs, but their marginal distributions do not coincide with the PDs obtained through the 1D p-p transformation of the corresponding marginal pdfs.

On the other hand, the second transformation provides joint PDs which generally do not dominate the original joint pdfs, but their marginal distributions coincide with the PDs obtained through the 1D p-p transformation of the corresponding marginal pdfs. This second transformation appears more suitable for metrological applications, even if it is an open problem to check whether it dominates the joint pdf.

Moreover, a theorem has been proved, showing that a simplified p-p transformation can be applied in the case it is not required the knowledge of the whole joint PD, but its construction is only an intermediate step to compute the sum and the difference of the considered variables. It would be interesting to link our result with various results of imprecise probability dealing with arithmetic operations.

References

1. Baudrit, C., Dubois, D.: Comparing methods for joint objective and subjective uncertainty propagation with an example in a risk assessment. In: Proc. 4th Int. Symposium on Imprecise Probabilities and Their Applications, Pittsburgh, PA, USA (2005)
2. BIPM, IEC, IFCC, ILAC, ISO, IUPAC, IUPAP, OIML: Evaluation of measurement data – Guide to the Expression of Uncertainty in Measurement. Joint Committee for Guides in Metrology, vol. 100 (2008)
3. BIPM, IEC, IFCC, ILAC, ISO, IUPAC, IUPAP, OIML: International vocabulary of metrology – Basic and general concepts and associated terms (VIM). Joint Committee for Guides in Metrology, vol. 200 (2008)
4. Carlsson, C., Fullér, R., Majlender, P.: On possibilistic correlation. Fuzzy Set. Syst. 155(3), 425–445 (2005)
5. Dubois, D., Foulloy, L., Mauris, G., Prade, H.: Probability-Possibility Transformations, Triangular Fuzzy Sets, and Probabilistic Inequalities. Reliab. Comput. 10(4), 273–297 (2004)
6. Ferrero, A., Prioli, M., Salicone, S.: A metrology-sound probability-possibility transformation for joint distributions. In: Proc. I2MTC 2012, Graz, Austria, May 14-16 (2012)
7. Ferrero, A., Salicone, S.: The random-fuzzy variables: a new approach for the expression of uncertainty in measurement. IEEE Trans. Instrum. Meas. 53(5), 1370–1377 (2004)
8. Ferrero, A., Salicone, S.: A comparative analysis of the statistical and random-fuzzy approaches in the expression of uncertainty in measurement. IEEE Trans. Instrum. Meas. 54(4), 1475–1481 (2005)
9. Ferrero, A., Salicone, S.: Fully-comprehensive mathematical approach to the expression of uncertainty in measurement. IEEE Trans. Instrum. Meas. 55(3), 706–712 (2006)
10. Mauris, G.: Representing and Approximating Symmetric and Asymmetric Probability Coverage Intervals by Possibility Distributions. IEEE Trans. Instrum. Meas. 58(1), 41–45 (2009)
11. Mauris, G., Lasserre, V., Foulloy, L.: A fuzzy approach for the expression of uncertainty in measurement. Measurement 29(3), 165–177 (2001)

12. Salicone, S.: Measurement Uncertainty: An Approach Via the Mathematical Theory of Evidence. Springer Series in Reliability Engineering. Springer US, New York (2007)
13. Shafer, G.: A Mathematical Theory of Evidence. Princeton University Press, Princeton (1976)
14. Tonon, F., Chen, S.: Inclusion properties for random relations under the hypotheses of stochastic independence and non-interactivity. Int. J. Gen. Syst. 34(5), 615–624 (2005)
15. Urbanski, M., Wasowsky, J.: Fuzzy approach to the theory of measurement inexactness. Measurement 34(1), 67–74 (2003)

A Note on the Convex Structure of Uncertainty Measures on MV-algebras

Tommaso Flaminio and Lluís Godo Lacasa

Abstract. In this paper we address the issue of providing a geometrical characterization for the decision problem of asking whether a partial assignment $\beta : f_i \mapsto \alpha_i$ mapping *fuzzy events* f_i into real numbers α_i $(i = 1, \ldots, n)$ extends to a generalized belief function on fuzzy sets, according to a suitable definition. We will characterize this problem in a way that allows to treat it as the membership problem of a point to a specific convex set.

Keywords: Belief functions, extendability problem, fuzzy sets, MV-algebras.

1 Introduction

The problem of deciding whether a partial assignment $v : f_i \mapsto \alpha_i$ mapping each *event* f_i into a real number α_i (for $i = 1, \ldots, s$) extends to a probability measure, is well known in the literature, and it is closely related with *de Finetti's no-Dutch Book coherence criterion* [5]. This criterion can be generalized in mainly two ways: by moving from classical to non-classical events (cf. [14, 16] for instance), or framing the problem out of the probabilistic setting, by taking into account alternative theories of uncertainty.

In [14, 16] the authors extend de Finetti's criterion for finitely-additive measures on non-classical events, and in particular in the case of those fuzzy events being representable as formulas of Łukasiewicz calculus. Following the proof of de Finetti's no-Dutch Book theorem (cf. [16, Theorem 2]), in [14, 16] *extendible* (i.e. *coherent* in de Finetti's terminology) assignments were

Tommaso Flaminio · Lluís Godo Lacasa
IIIA Artificial Intelligence Research Institute, CSIC Spanish
National Research Council, Campus UAB s/n, Bellaterra 08193, Spain
e-mail: {tommaso,godo}@iiia.csic.es

R. Kruse et al. (Eds.): Synergies of Soft Computing and Statistics, AISC 190, pp. 73–81.
springerlink.com © Springer-Verlag Berlin Heidelberg 2013

characterized as Euclidean convex subsets of the finite dimensional space $\mathbb{R}^s$, s being the number of events the assignment is defined over.

A similar approach to extendible assignments, but framed into an idempotent, rather than additive, setting has been developed in [8] in the frame of possibility theory. In particular, a geometrical characterization for the extendability problem of possibility and necessity assignments for fuzzy events over a finite domain has been presented. It is worth recalling that in the frame of possibility theory, extendible possibility and necessity assignments are again characterized as convex sets, but the Euclidean geometry has to be replaced by the min-plus geometry (see for instance [3]).

Belief functions are the measures used within Dempster-Shafer evidence theory [17] to quantify the amount of uncertainty associated to events. Among all the classical theories of uncertainty, Dempster-Shafer plays a pivotal role since both probability theory, and possibility theory can be obtained as particular cases. Belief functions have been recently extended to the MV-algebraic setting to cope with spaces of events that can be organized as MV-algebras of fuzzy subsets of a finite domain $X = \{x_1, \ldots, x_n\}$ [6, 11].

In this paper we will provide a geometrical characterization for the extendability problem for events being (not necessarily normalized) fuzzy sets over a finite domain in a generalized framework of belief function theory. In particular, we will show how a mixture of min-plus convex geometry and Euclidean convex geometry can be applied to translate the extendability problem for belief functions, into the membership problem of a point to a convex set.[1]

2 Preliminaries

In this section we will introduce the necessary preliminaries about the min-plus convex geometry and MV-algebras. We will assume the reader to be familiar with Euclidean convex geometry, reminding that in what follows, given any subset S of $\mathbb{R}^s$, $\mathrm{co}(S)$, and $\overline{\mathrm{co}}(S)$ will respectively stand for the convex hull of S and its topological closure (with respect to the Euclidean metric). We invite the reader to consult [3] and [4] for all the unexplained notions.

2.1 Preliminaries on Min-Plus Convexity

Let $\mathbf{x}_1, \ldots, \mathbf{x}_n \in \mathbb{R}^s$, and for every $i = 1, \ldots, n$ and for every $t = 1, \ldots, s$, let us denote by $\mathbf{x}_i(t)$ the t-th projection of $\mathbf{x}_i$. We then define the *min-plus convex hull* generated by $\mathbf{x}_1, \ldots, \mathbf{x}_n$ as the set

[1] Due to space reasons we will omit the proof of all the presented results, with the exception of Thm. 4 whose proof will be given below its statement. An extended version of this paper that contains all the needed proof, can be downloaded from `http://www.iiia.csic.es/files/pdfs/Belief.pdf`

$$\text{mp-co}(\mathbf{x}_1, \ldots, \mathbf{x}_n) = \{\mathbf{y} \in \mathbb{R}^s : \exists \lambda_1, \ldots, \lambda_n \in \mathbb{R}, \mathbf{y} = \min_{i \leq n}(\lambda_i + \mathbf{x}_i)\},$$

where $\mathbf{y}$ is a vector in $\mathbb{R}^s$ such that for every t, $\mathbf{y}(t) = \min_{i \leq n}(\lambda_i + \mathbf{x}_i(t))$. In the particular case of $\mathbf{x}_1, \ldots, \mathbf{x}_n \in [0,1]^s$, and $\lambda_1, \ldots, \lambda_n \in [0,1]$, we call *bounded* the combination $\min_{i \leq n}(\lambda_i \oplus \mathbf{x}_i)$ where the usual sum $+$ is replaced by the bounded sum $\oplus$, where for every $a, b \in [0,1]$, $a \oplus b = \min\{1, a+b\}$, and on $[0,1]^s$ is point-wise defined. Therefore, for $\mathbf{x}_1, \ldots, \mathbf{x}_n \in [0,1]^s$, we define the *bounded min-plus convex hull* generated by $\mathbf{x}_1, \ldots, \mathbf{x}_n$ as the set

$$\text{bmp-co}(\mathbf{x}_1, \ldots, \mathbf{x}_n) = \{\mathbf{y} \in [0,1]^s : \exists \lambda_1, \ldots, \lambda_n \in [0,1], \mathbf{y} = \min_{i \leq n}(\lambda_i \oplus \mathbf{x}_i)\}.$$

A bounded min-plus convex combination is said to be *normalized*, if the parameters $\lambda_1, \ldots, \lambda_n$ satisfy $\max_{i \leq n} \lambda_i = 1$, and a bounded min-plus convex hull is said to be *normalized* accordingly. We will denote by nmp-co(S) the normalized bounded min-plus convex hull generated by a set S.

2.2 MV-algebras of Fuzzy Sets

An *MV-algebra* [1, 15] is an algebra $(A, \oplus, \neg, \bot, \top)$ of type $(2,1,0,0)$ such that its reduct $(A, \oplus, \bot)$ is an abelian monoid, and the following equations hold for every $a, b \in A$: $\neg\neg a = a$, $a \oplus \top = \top$, and $\neg(\neg a \oplus b) \oplus b = \neg(\neg b \oplus a) \oplus a$.

Let A and B be MV-algebras. A *MV-homomorphism* is a map $h : A \to B$ sending $\bot$ and $\top$ of A in $\bot$ and $\top$ of B respectively, and such that, for every $a, a' \in A$, $h(a \oplus a') = h(a) \oplus h(a')$, and $h(\neg a) = \neg h(a)$. We will denote by $\mathfrak{H}(A, B)$ the class of homomorphisms between A and B.

Let $X = \{x_1, \ldots, x_n\}$ be a finite set of cardinality n, and consider the class $[0,1]^X$ of all fuzzy subsets of X, i.e. all functions from X into the real unit interval $[0,1]$. This set can obviously be identified with the direct product $[0,1]^n$.[2] The algebra obtained endowing $[0,1]^n$ with the point-wise operations $a \oplus b = \min\{1, a+b\}$, $\neg a = 1 - a$, and the two functions constantly equal to 0 and 1, also denoted $\bot$ and $\top$ respectively, is the typical example of MV-algebra that we will consider in this paper as domain for uncertainty measures.

We will equivalently denote by $[0,1]^n$ or $[0,1]^X$ both the domain, and the MV-algebra above defined without danger of confusion, moreover, we will always assume X to be finite. Notice that, whenever X consists of just one element, $[0,1]^1$ is the (linearly ordered) MV-algebra over the real unit interval. This algebra, that is usually named the *standard* MV-algebra, will be denoted by $[0,1]_{MV}$.

[2] The set $X = \{x_1, \ldots, x_n\}$ can be equivalently identified with the set of its indices $\{1, \ldots, n\}$. This allows in turn to identify each function $f \in [0,1]^X$ as a point $\langle f(1), \ldots, f(n) \rangle \in [0,1]^n$ and vice-versa.

For every $f = \langle f(1), \ldots, f(n)\rangle \in [0,1]^n$, we will henceforth consider the function $\rho_f : [0,1]^n \to [0,1]$ mapping every $b \in [0,1]^n$ into

$$\rho_f(b) = \min_{i \leq n}(\neg b(i) \oplus f(i)), \tag{1}$$

Those mappings ρ_f can be regarded as *generalized inclusion operators* between fuzzy sets (cf. [6] for further details). For every $f \in \{0,1\}^n$ (i.e. whenever f is identified with a vector in $[0,1]^n$ with *integer* components), the map $\rho_f : [0,1]^n \to [0,1]$ is a pointwise minimum of finitely many linear functions with integer coefficients, and hence ρ_f is a non-increasing McNaughton function [1, 15]. Letting $\mathcal{R}$ to be the MV-algebra generated by the set $\{\rho_f : f \in [0,1]^n\}$, [1, Theorem 3.4.3] and [13, Theorem 2.5], allow to prove the following result.

Theorem 1. *There exists a one-to-one correspondence between the points of* $[0,1]^n$ *and the class* $\mathfrak{H}(\mathcal{R}, [0,1]_{MV})$ *of homomorphisms of* $\mathcal{R}$ *into the standard MV-algebra* $[0,1]_{MV}$.

Thanks to the above Thm. 1 we will henceforth identify points in $[0,1]^n$ (hence fuzzy subsets of X) with homomorphisms of $\mathcal{R}$ in the MV-algebra $[0,1]_{MV}$ without loss of generality. Moreover, the following holds:

Corollary 1. *Let* $\{\tau_1, \ldots, \tau_s\}$ *be a finite subset of* $\mathcal{R}$. *Then* $\{\langle h(\tau_1), \ldots, h(\tau_s)\rangle \in [0,1]^s : h \in \mathfrak{H}(\mathcal{R}, [0,1]_{MV})\} = \{\langle \tau_1(a), \ldots, \tau_s(a)\rangle : a \in [0,1]^n\}$.

3 Uncertainty Measures on MV-algebras of Fuzzy Sets

In this section we are going to recall how *states* [13] and *necessity measures* [7], can be defined on MV-algebras (of fuzzy sets). In particular we will also recall how the problem of extending a partial assessment can be geometrically characterized in these frameworks.

A *state* on an MV-algebra A is a map $s : A \to [0,1]$ satisfying: (1) $s(\top) = 1$; (2) for every $a, b \in A$ such that $\neg(\neg a \oplus \neg b) = \bot$, $s(a \oplus b) = s(a) + s(b)$. A state s is said to be *faithful* provided that $s(a) = 0$, implies $a = \bot$.

States play the same role on MV-algebras as probability measures do on Boolean algebras. In particular the well known de Finetti's extension theorem was generalized to the case of states and MV-algebras by Mundici [14]. Below we recall it in the particular case of the MV-algebra being $[0,1]^X$.

Theorem 2. *Let* $f_1, \ldots, f_s$ *be elements in* $[0,1]^X$. *Then a map* $\sigma : f_i \mapsto \alpha_i \in [0,1]$ *extends to a state on* $[0,1]^X$ *iff* $\langle \alpha_1, \ldots, \alpha_s\rangle \in \overline{\mathrm{co}}\{\langle h(f_1), \ldots, h(f_s)\rangle : h \in \mathfrak{H}([0,1]^X, [0,1]_{MV})\}$.

As in Thm. 1, the class $\mathfrak{H}([0,1]^X, [0,1]_{MV})$ is in one-to-one correspondence with the set $X = \{x_1, \ldots, x_n\}$ and hence from Thm. 2 an assignment $\sigma : f_i \mapsto \alpha_i$ extends to a state on $[0,1]^X$ iff $\langle \alpha_1, \ldots, \alpha_s \rangle \in \mathrm{co}\{\langle f_1(x), \ldots, f_s(x) \rangle : x \in X \rangle\}$[3]. The result is shown in Fig. 1.

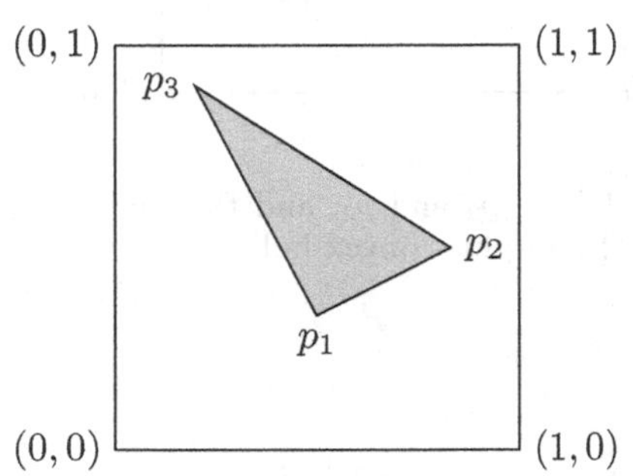

Fig. 1 Let $X = \{1,2,3\}$, consider two elements $f_1, f_2 \in [0,1]^3$, and the assignment $\sigma(f_1) = \alpha_1$, and $\sigma(f_2) = \alpha_2$. The two functions $f_1, f_2 : X \to [0,1]$ defines three points in the unit square $[0,1]^2$, namely:
$p_1 = (f_1(1), f_2(1))$;
$p_2 = (f_1(2), f_2(2))$;
$p_3 = (f_1(3), f_2(3))$.
Therefore the assignment σ extends to a state on $[0,1]^3$ iff the point $\langle \alpha_1, \alpha_2 \rangle$ belongs to the triangle with vertices p_1, p_2, and p_3.

Necessity measures on MV-algebras like $[0,1]^X$ have been introduced in [7]: a map $N : [0,1]^X \to [0,1]$ is a *necessity measure*, provided that $N(\top) = 1$, for every $f, f' \in [0,1]^X$, $N(f \wedge f') = \min\{N(f), N(f')\}$, and for every $f, \overline{r} \in [0,1]^X$, if $\overline{r}$ is the function constantly equal to r, then $N(\overline{r} \oplus f) = r \oplus N(f)$.[4]

Necessity measures on $[0,1]^X$ can be equivalently defined as follows: if $\pi : X \to [0,1]$ is a map called a *possibility distribution*, we define $N_\pi : f \in [0,1]^X \mapsto \min_{x \in X}\{(1 - \pi(x)) \oplus f(x)\} \in [0,1]$, and in this case we say that N_π is *defined by* π. In [7, Theorem 3.3] it is shown that for every necessity measure N on $[0,1]^X$, there exists a (unique) possibility distribution π defining N.

A necessity measure N is said to be *normalized* provided that the possibility distribution π defining N satisfies $\max_{x \in X} \pi(x) = 1$. Notice that, if N is not normalized, then $N(\bot) > 0$. On the other hand, normalized necessities always satisfy $N(\bot) = 0$. We will denote by $\mathcal{N}([0,1]^X)$ the class of necessity measures on $[0,1]^X$. Normalization will always be clear by the context.

The following can be proved going through the lines of [8, Theorem 4].

Theorem 3. *Let $f_1, \ldots, f_s \in [0,1]^X$, and let $\eta : f_i \mapsto \alpha_i$ be an assessment. Then the following hold:*

[3] Notice that since X is finite, so is $F = \{\langle f_1(x), \ldots, f_s(x) : x \in X \rangle\}$, and hence its convex hull is the polytope generated by F which is already closed. In other words, $\overline{\mathrm{co}}(F) = \mathrm{co}(F)$.

[4] In [7, 8] necessity measures also satisfying the last condition on constant functions were called *homogeneous* necessity measures. In this paper, since we will not distinguish between homogeneous and non-homogeneous mappings, we will use to call them *necessity measures* without danger of confusion.

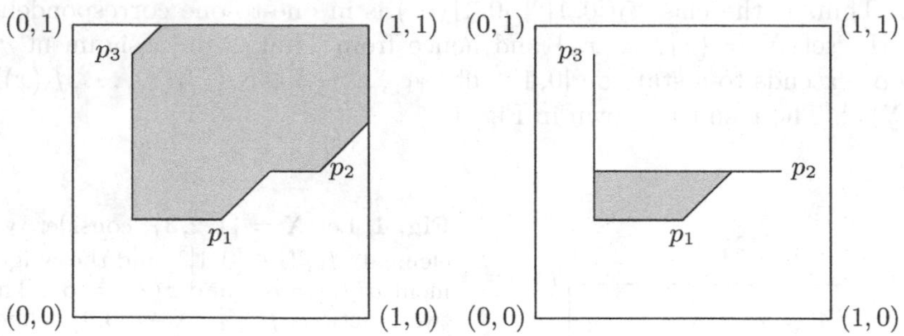

Fig. 2 Three points in the unit square $[0,1]^2$, p_1, p_2 and p_3, and their min-plus convex hull (left) and their normalized (right) min-plus convex hull

1. *η extends to a necessity measure iff*

$$\langle \alpha_1, \ldots, \alpha_s \rangle \in \text{bmp-co}(\{\langle f_1(x_i), \ldots, f_s(x_i) \rangle : 1 \leq i \leq n\}).$$

2. *η extends to a normalized necessity measure iff*

$$\langle \alpha_1, \ldots, \alpha_s \rangle \in \text{nmp-co}(\{\langle f_1(x_i), \ldots, f_s(x_i) \rangle : 1 \leq i \leq n\}).$$

Example 1. Let $X = \{1, 2, 3\}$, consider two elements $f_1, f_2 \in [0,1]^3$, and an assessment $\sigma(f_1) = \alpha_1$, and $\sigma(f_2) = \alpha_2$. The two functions $f_1, f_2 : X \to [0,1]$ define three points in the unit square $[0,1]^2$, namely: $p_1 = (f_1(1), f_2(1))$; $p_2 = (f_1(2), f_2(2))$; $p_3 = (f_1(3), f_2(3))$. Therefore the assignment σ respectively extends to a normalized (resp. non-normalized) necessity measure on $[0,1]^3$ iff the point $\langle \alpha_1, \alpha_2 \rangle$ belongs to the normalized (resp. non-normalized) min-plus convex polygon of vertices p_1, p_2, and p_3, depicted respectively in Fig. 2 on the left and on the right.

Next result shows that the inclusion operators $\rho_{(\cdot)}(b)$, with varying $b \in [0,1]^X$, generate the whole class of necessity measures on $[0,1]^X$, and will be useful in the next section.

Lemma 1. *(1) The class of all necessity measures on $[0,1]^X$ coincides with the class $\{\rho_{(\cdot)}(b) : f \in [0,1]^X \mapsto \rho_f(b) \mid b \in [0,1]^X\}$.*

(2) The class of all normalized necessity measures on $[0,1]^X$ coincides with the class $\{\rho_{(\cdot)}(b) : f \in [0,1]^X \mapsto \rho_f(b) \mid b \in [0,1]^X, \max_{x \in X} b(x) = 1\}$.

Corollary 2. *Let $f_1, \ldots, f_s \in [0,1]^X$, and let $\eta : f_i \mapsto \alpha_i$ be an assessment. Then:*

1. *η extends to a necessity in $\mathcal{N}([0,1]^X)$ iff*

$$\langle \alpha_1, \ldots, \alpha_s \rangle \in \{\langle \rho_{f_1}(b), \ldots, \rho_{f_s}(b) \rangle : b \in [0,1]^X\}.$$

2. *η extends to a normalized necessity in $\mathcal{N}([0,1]^X)$ iff*

$$\langle \alpha_1, \ldots, \alpha_s \rangle \in \{\langle \rho_{f_1}(b), \ldots, \rho_{f_s}(b) \rangle : b \in [0,1]^X, \max_{x \in X} b(x) = 1\}.$$

4 Belief Functions and the Extendability Problem

Let us recall from Sec. 2.2 that, the algebra $\mathcal{R}$ is defined as the MV-algebra generated by all functions ρ_f defined as in (1). Then *belief functions* on $[0,1]^X$ can be defined along the proposals in [6, 11, 12].

Definition 1. A map $Bel : [0,1]^X \to [0,1]$ is a *belief function* if there exists a state $s : \mathcal{R} \to [0,1]$ such that for every $a \in [0,1]^X$, $Bel(a) = s(\rho_a)$.

Notice that, in general, belief functions on $[0,1]^X$ fail to satisfy $Bel(\bot) = 0$. In fact, for every $b \in [0,1]^X$ such that $\max_{x \in X} b(x) < 1$, $\rho_\bot(b) > 0$, and hence $\rho_\bot$ does not coincide with the function constantly equal to 0. Therefore, whenever s is faithful, $Bel(\bot) = s(\rho_\bot) > 0$. We will henceforth call *normalized* any belief function satisfying $Bel(\bot) = 0$. The following result provides a geometrical characterization for the extendability problem for belief functions on MV-algebras.

Theorem 4. *Let $f_1, \ldots, f_s \in [0,1]^X$, and let $\beta : f_i \mapsto \alpha_i$ be a $[0,1]$-valued mapping. Then the following hold:*

1. *β extends to a belief function Bel on $[0,1]^X$ iff*

$$\langle \alpha_1, \ldots, \alpha_s \rangle \in \overline{\text{co}}(\text{bmp-co}(\{\langle f_1(x_i), \ldots, f_s(x_i) \rangle : i \leq n\})).$$

2. *β extends to a normalized belief function Bel on $[0,1]^n$ iff*

$$\langle \alpha_1, \ldots, \alpha_s \rangle \in \overline{\text{co}}(\text{nmp-co}(\{\langle f_1(x_i), \ldots, f_s(x_i) \rangle : i \leq n\})).$$

Proof. 1. The assignment β extends to a belief function $Bel : [0,1]^X \to [0,1]$ iff there exists a state $s : \mathcal{R} \to [0,1]$ such that, for every $t = 1, \ldots, s$, $\alpha_t = s(\rho_{f_t}) = Bel(f_t)$. From Thm. 2, this means that $\langle \alpha_1, \ldots, \alpha_s \rangle \in \overline{\text{co}}\{\langle h(\rho_{f_1}), \ldots, h(\rho_{f_s}) \rangle \in [0,1]^s : h \in \mathfrak{H}(\mathcal{R}, [0,1]_{MV})\}$. From Corollary 1, this is equivalent to:

$$\langle \alpha_1, \ldots, \alpha_s \rangle \in \overline{\text{co}}\{\langle \rho_{f_1}(b), \ldots, \rho_{f_s}(b) \rangle \in [0,1]^s : b \in [0,1]^X\}.$$

From (1) of Corollary 2, the set $\{\langle \rho_{f_1}(b), \ldots, \rho_{f_s}(b) \rangle \in [0,1]^s : b \in [0,1]^X\}$ coincides with the set of all the coherent necessity assignments over $\{f_1, \ldots, f_s\}$, in other words $y \in \{\langle \rho_{f_1}(b), \ldots, \rho_{f_s}(b) \rangle \in [0,1]^s : b \in [0,1]^X\}$ iff there exists a necessity measure N such that for every t, $N(f_t) = y(t)$. Finally Thm. 3 implies that the set of all coherent necessity assessments over $f_1, \ldots, f_s$ coincides with bmp-co$(\{\langle f_1(x_i), \ldots, f_s(x_i) \rangle : i \leq n\})$, and hence our the claim is settled.

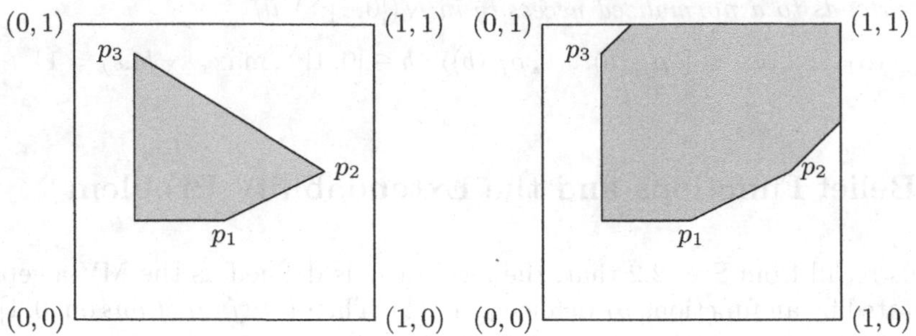

Fig. 3 Two convex polytopes in the unit square $[0,1]^2$ defined by vertices p_1, p_2 and p_3.

2. The proof runs completely parallel to the previous one and is omitted. □

Example 2. (Example 1 continued) Let $X = \{1,2,3\}$, consider two elements $f_1, f_2 \in [0,1]^3$, and an assessment $\sigma(f_1) = \alpha_1$, and $\sigma(f_2) = \alpha_2$. Therefore the assignment σ extends either to a normalized belief function, or to a belief function in general, on $[0,1]^3$ iff the point $\langle \alpha_1, \alpha_2 \rangle$ belongs respectively to the convex polytope on the left in Fig. 3, or to the convex polytope on the right of the same Fig. 3.

Acknowledgements. The authors acknowledge partial support from the Spanish projects TASSAT (TIN2010-20967-C04-01), Agreement Technologies (CONSOLIDER CSD2007-0022, INGENIO 2010) and ARINF (TIN2009-14704-C03-03). Flaminio acknowledges partial support from the Juan de la Cierva Program of the Spanish MICINN.

References

1. Cignoli, R., D'Ottaviano, I.M.L., Mundici, D.: Algebraic Foundations of Many-valued Reasoning. Kluwer, Dordrecht (2000)
2. Cohn, P.M.: Universal Algebra, rev. edn. D. Reidel Pub. Co., Dordrecht (1981)
3. Develin, M., Sturmfels, B.: Tropical convexity. Doc. Math. 9, 1–27 (2004)
4. Edwald, G.: Combinatorial Convexity and Algebraic Geometry. Springer US (1996)
5. de Finetti, B.: Theory of Probability, vol. 1. Wiley, New York (1974)
6. Flaminio, T., Godo, L., Marchioni, E.: Belief Functions on MV-Algebras of Fuzzy Events Based on Fuzzy Evidence. In: Liu, W. (ed.) ECSQARU 2011. LNCS (LNAI), vol. 6717, pp. 628–639. Springer, Heidelberg (2011)
7. Flaminio, T., Godo, L., Marchioni, E.: On the logical formalization of possibilistic counterpart of states over n-valued Łukasiewicz events. J. Logic Comput. 21, 447–464 (2011)

8. Flaminio, T., Godo, L., Marchioni, E.: Geometrical aspects of possibility measures on finite domain MV-clans. Soft Comput. (2012), doi:10.1007/s00500-012-0838-0
9. Goodearl, K.R.: Partially Ordered Abelian Group with Interpolation. Mathematical Surveys and Monographs, vol. 20. American Mathematical Society, Providence (1986)
10. Kroupa, T.: Every state on semisimple MV-algebra is integral. Fuzzy Set. Syst. 157(20), 2771–2782 (2006)
11. Kroupa, T.: From Probabilities to Belief Functions on MV-Algebras. In: Borgelt, C., González-Rodríguez, G., Trutschnig, W., Lubiano, M.A., Gil, M.Á., Grzegorzewski, P., Hryniewicz, O. (eds.) Combining Soft Computing and Statistical Methods in Data Analysis. AISC, vol. 77, pp. 387–394. Springer, Heidelberg (2010)
12. Kroupa, T.: Extension of Belief Functions to Infinite-valued Events. Soft Comput. (2012), doi:10.1007/s00500-012-0836-2
13. Mundici, D.: Averaging the truth-value in Łukasiewicz logic. Studia Logica 55, 113–127 (1995)
14. Mundici, D.: Bookmaking over infinite-valued events. Int. J. Approx. Reason. 43, 223–240 (2006)
15. Mundici, D.: Advanced Łukasiewicz calculus and MV-algebras. Trends in Logic, vol. 35. Springer, Heidelberg (2011)
16. Paris, J.B.: A note on the Dutch Book method. In: Proc. of the 2nd Int. Symp. on Imprecise Probabilities and their Applications, ISIPTA 2001, pp. 301–306. Shaker Publishing Company (2001)
17. Shafer, G.: A Mathematical Theory of Evidence. Princeton University Press (1976)

Cumulative Distribution Function Estimation with Fuzzy Data: Some Estimators and Further Problems

Xuecheng Liu and Shoumei Li

Abstract. In this paper, we discuss two types of estimators for cumulative distribution function (CDF) based on fuzzy data: substituting estimators and nonparametric maximum likelihood (NPML) based estimators, both of them are extensions of empirical distribution functions (EDF) of real-valued (non-fuzzy) data. We also list some further problems.

Keywords: Cumulative distribution function, empirical distribution functions, fuzzy data, nonparametric maximum likelihood, substituting function.

1 Introduction

Background: EDF and Its Properties. The CDF F for a real-valued random variable X is defined as, for $x \in \mathbb{R} := (-\infty, \infty)$,

$$F(x) := \Pr(X \leq x). \tag{1}$$

Given independent and identically distributed random variables $X_1, ..., X_n$ with CDF F, the EDF, as an estimator of F, is defined as, for $x \in \mathbb{R}$,

$$\hat{F}_n(x) := \frac{1}{n} \sum_{j=1}^{n} \mathbf{1}_{(-\infty,x]}(X_j). \tag{2}$$

Xuecheng Liu
Research Unit on Children's Psychosocial Maladjustment,
University of Montreal, Montreal, Quebec, Canada, H3T 1J7
e-mail: xuecheng.liu@umontreal.ca

Shoumei Li
Department of Applied Mathematics, Beijing University of Technology,
Beijing, 100124, China
e-mail: lisma@bjut.edu.cn

R. Kruse et al. (Eds.): Synergies of Soft Computing and Statistics, AISC 190, pp. 83–91.
springerlink.com © Springer-Verlag Berlin Heidelberg 2013

By the strong law of large numbers (SLLN), for each $x \in \mathbb{R}$, $\hat{F}_n(x)$ converges to $F(x)$ almost surely. Glivenko and Cantelli in 1933 (see, e.g., [18]) gave a theorem on uniform-type convergence of $\hat{F}_n$, often called Glivenko-Cantelli theorem:

$$\sup_{x\in\mathbb{R}} |\hat{F}_n(x) - F(x)| \to_{a.s.} 0.$$

For each $x \in \mathbb{R}$, define

$$Z_n(x) := \sqrt{n}(\hat{F}_n(x) - F(x)),$$

by the central limit theorem (CLT), for each $x \in \mathbb{R}$, $Z_n(x)$ converges weakly to $N(0, F(x)(1 - F(x))$. Donsker in 1952 proved a theorem of uniform-type convergence of Z_n, often called Donsker theorem,

$$Z_n \to_w U(F) \quad \text{in} \quad D(\mathbb{R}),$$

where $D(\mathbb{R})$ is the Skorokhod space and U a standard Brownian bridge in the unit interval $[0,1]$, see [18] for details.

Aim of the Paper. In many applications, observed data are imprecise rather than precise (i.e., real-valued). In this paper, we consider fuzzy number valued (imprecise) data, often called fuzzy data. With fuzzy data, two uncertainties, fuzziness and randomness, exist simultaneously. Fuzzy data are to fuzzy random variables in the sense of [9, 10, 8], or equivalently, [16], see [6] for overview discussion.

This paper is on CDF estimation with fuzzy data. We limit the estimated CDF $\hat{F}$ in the classical sense, i.e., $\hat{F}(x)$ is real-valued for each $x \in \mathbb{R}$. We will discuss two types of estimators: substituting estimators and nonparametric maximal likelihood (NPML) estimators. Another important component of this paper is to list some further (open) problems, among which, the important and very challengeable ones include point-wise or uniform-type convergence theorems, i.e., the extensions of SLLN, Glivenko-Cantelli theorem, CLT and Donsker theorems discussed above to estimators based on fuzzy data.

Notation of Fuzzy Numbers. A fuzzy number $\tilde{A}$ with a membership function $A : \mathbb{R} \to [0,1]$ is a fuzzy set [20] on $\mathbb{R}$ such that its α-level sets

$$A^\alpha := \{x \in \mathbb{R};\ A(x) \geq \alpha\} \quad \text{for} \quad \alpha \in [0,1]$$

are bounded closed intervals, where A^0 is, by convention, the closure of the set $\{x \in \mathbb{R};\ A(x) > 0\}$. We denote $\tilde{A}$'s α-level set by $[\underline{A}^\alpha, \overline{A}^\alpha]$ and call A^0 and A^1 the support and core of $\tilde{A}$.

A fuzzy set $\tilde{A}$ on $\mathbb{R}$ with support $[a, d]$ and core $[b, c]$ is a fuzzy number if and only if the two functions L and R of $\tilde{A}$'s membership function restricting on $[a, b)$ and $(c, d]$ respectively satisfy that L is non-decreasing and right-continuous and R is non-increasing and left-continuous.

We denote by $\mathcal{F}(\mathbb{R})$ the class of all fuzzy numbers.

2 CDF Estimation with Fuzzy Data: Substituting Estimators

In this section, we propose so called substituting estimators to the CDF F based on fuzzy data $\tilde{X}_1, ..., \tilde{X}_n$. For $X, x \in \mathbb{R}$ in Eqn. (1), denote

$$I(X,x) := \mathbf{1}_{(-\infty,x]}(X), \tag{3}$$

or, equivalently, with logical (true) indicator function,

$$I(X,x) = \mathbf{1}[X \leq x], \tag{4}$$

the EDF with real-valued data defined in Eqn. (2) can be rewritten as

$$\hat{F}_n(x) = \frac{1}{n}\sum_{j=1}^{n} I(X_j, x). \tag{5}$$

The idea of substituting estimators based on the fuzzy data is that we substitute $I(X_j, x)$ in Eqn. (5) by its fuzzy counterparts, denoted by $I^{(\mathrm{f})}(\tilde{X}_j, x)$, where $I^{(\mathrm{f})}$ is a function from $\mathcal{F}(\mathbb{R}) \times \mathbb{R}$ to $[0, 1]$. With $I^{(\mathrm{f})}$, we can estimate the CDF F as,

$$\hat{F}_n^{(\mathrm{f})}(x) := \frac{1}{n}\sum_{j=1}^{n} I^{(\mathrm{f})}(\tilde{X}_j, x). \tag{6}$$

Substituting Function. In order for such an $I^{(\mathrm{f})}$ to be used in estimating the CDF F in Eqn. (6), it is much reasonable that it meets the following 4 requirements, for $\tilde{A}, \tilde{B} \in \mathcal{F}(\mathbb{R})$ and $x \in \mathbb{R}$,

- **R1:** $I^{(\mathrm{f})}(\tilde{A}, x)$, as a function of x, is non-decreasing and right continuous with left limit;
- **R2:** If $\tilde{A} \subseteq (x, \infty)$ (i.e., $A_0 \subseteq (x, \infty)$), then $I^{(\mathrm{f})}(\tilde{A}, x) = 0$;
- **R3:** If $\tilde{A} \subseteq (-\infty, x]$ (i.e., $A_0 \subseteq (-\infty, x]$), then $I^{(\mathrm{f})}(\tilde{A}, x) = 1$;
- **R4:** If $\tilde{A} \leq \tilde{B}$, then $I^{(\mathrm{f})}(\tilde{A}, x) \geq I^{(\mathrm{f})}(\tilde{B}, x)$, where the partial order $\tilde{A} \leq \tilde{B}$ is defined as $\underline{A}^{\alpha} \leq \underline{B}^{\alpha}$, $\overline{A}^{\alpha} \leq \overline{B}^{\alpha}$ for all $\alpha \in [0, 1]$.

A function from $\mathcal{F}(\mathbb{R}) \times \mathbb{R}$ to $[0, 1]$ satisfying R1 to R4 is called a *substituting function.* R4 says that, for two fuzzy random variables $\tilde{X}, \tilde{Y}$ such that $\tilde{X} \leq \tilde{Y}$, $\Pr(\tilde{X} \leq x)$ should be at least $\Pr(\tilde{Y} \leq x)$. Functions satisfying all of R1, R2 and R3 may not satisfy R4, see below the counterexample in the discussion of substituting functions based on fuzzy number ranks with center of gravity defuzzification. Further, it is easy to see that R2 and R3 together imply $I^{(\mathrm{f})}(a, x) = \mathbf{1}[a \leq x]$ for all $a, x \in \mathbb{R}$.

In the following, we propose 4 substituting functions from different viewpoints of $\mathbf{1}_{(-\infty,x]}(X_j)$ and $\mathbf{1}[X_j \leq x]$ for real-valued random variables X_j and

$x \in \mathbb{R}$ in Eqns. (3) and (4) respectively. Each of the substituting functions corresponds to a substituting estimator in CDF estimation with fuzzy data.

Substituting Functions Based on Inclusion Measures of Fuzzy Sets. With the notation in [1], an inclusion measure of a fuzzy set $\tilde{A}$ in another fuzzy set $\tilde{B}$ is

$$I(\tilde{A}, \tilde{B}) := \inf_u i(A(u), B(u)),$$

where i is any fuzzy implication operator, and usually we choose

$$i(y_1, y_2) = \max(1 - y_1, y_2), \quad y_1, y_2 \in [0, 1]. \tag{7}$$

Inclusion measure based substituting functions focus on the inclusion relationship of X (regarding as a singleton) in $(-\infty, x]$ as in Eqn. (3). We interpret the value $I^{(\mathrm{f})}(\tilde{X}, x)$ as the degree (valued in [0, 1]) of $\tilde{X}$ being a subset of $(-\infty, x]$. As an example, with the implication operator in Eqn. (7), for $\tilde{A} \in \mathcal{F}(\mathbb{R})$ (with support $[a, d]$ and core $[b, c]$) and $x \in \mathbb{R}$,

$$\begin{aligned} I^{(\mathrm{f}),\mathrm{IM}}(\tilde{A}, x) &:= I(\tilde{A}, (-\infty, x]) \\ &= \inf_{u \in \mathbb{R}} \max(1 - A(u), \mathbf{1}_{(-\infty,x]}(u)) \\ &= \begin{cases} 0, & \text{if } x < c, \\ 1 - A(x), & \text{if } c \le x < d, \\ 1, & \text{if } x \ge d. \end{cases} \end{aligned}$$

We can show that $I^{(\mathrm{f}),\mathrm{IM}}$ is a substituting function under the condition that the functions R of the fuzzy data are continuous (see the notation of fuzzy numbers in Section 1).

Note that we could choose different inclusion measures (i.e., different implication operators) to get different substituting functions.

Substituting Functions Based on Fuzzy Number Ranks. Fuzzy number rank based substituting functions focus on the order relationship between X and x in Eqn. (4) for real-valued random variable X and $x \in \mathbb{R}$. We view the value of $I^{(\mathrm{f})}(\tilde{X}, x)$ as a degree (valued in [0, 1]) of a fuzzy random variable $\tilde{X}$ is smaller than x.

With any defuzzification method of fuzzy numbers, it is possible to construct a substituting function as

$$I^{(\mathrm{f}),\mathrm{defuz}}(\tilde{A}, x) := \mathbf{1}[\tilde{A}^{\mathrm{d}} \le x], \tag{8}$$

where $\tilde{A}^{\mathrm{d}}$ is the defuzzification value of $\tilde{A}$.

For most of defuzzification methods discussed in [11] such as BOA (bisector of area), MOM (mean of maximum, i.e., mean of the core), SOM (smallest of the maximum), LOM (largest of maximum), the functions $I^{(\mathrm{f}),\mathrm{defuz}}$ are substituting functions. However, for COG (center of gravity), the function

$I^{(\mathrm{f}),\mathrm{defuz}}$ satisfies R1, R2 and R3, but not R4, and hence, it is not a substituting function, since, with COG, for $\tilde{A}, \tilde{B} \in \mathcal{F}(\mathbb{R})$, $\tilde{A} \le \tilde{B}$ does not imply $\tilde{A}^{\mathrm{d}} \le \tilde{B}^{\mathrm{d}}$. For example, considering $\tilde{A}, \tilde{B} \in \mathcal{F}(\mathbb{R})$ with membership functions $A(x) := x\mathbf{1}_{[0,\frac{2}{3})}(x) + \mathbf{1}_{[\frac{2}{3},1]}(x)$ and $B(x) := x\mathbf{1}_{[0,1]}(x)$, we have $\tilde{A} \le \tilde{B}$ and $\tilde{A}^{\mathrm{d}} = \frac{61}{90} > \frac{2}{3} = \tilde{B}^{\mathrm{d}}$.

Substituting Functions Based on Extension Principles. Substituting functions based on extension principles are to apply extension principles to the function $I(X, x)$ in (3) or (4). We extend X to a fuzzy number $\tilde{X}$, but keep $x \in \mathbb{R}$. Hence, for any $\tilde{A} \in \mathcal{F}(\mathbb{R})$ and $x \in \mathbb{R}$, with any extension principle, we get the fuzzy set $I(\tilde{A}, x)$ on the universal set $\{0, 1\}$. We only use the membership grade of $I(\tilde{A}, x)$ at 1 in defining substituting functions.

As an example, with Zadeh's extension principle, for $\tilde{A} \in \mathcal{F}(\mathbb{R})$ (with support $[a, d]$ and core $[b, c]$) and $x \in \mathbb{R}$, we have

$$\begin{aligned} I^{(\mathrm{f}),\mathrm{Zadeh}}(\tilde{A}, x) &:= I(\tilde{A}, x)(1) \\ &= \begin{cases} 0 & \text{if } x < a, \\ A(x) & \text{if } a \le x < b, \\ 1 & \text{if } x \ge b. \end{cases} \end{aligned}$$

It is easy to show that $I^{(\mathrm{f}),\mathrm{Zadeh}}$ is a substituting function.

Note that Zadeh's extension principle is to T-norm "min" and T-conorm "max". We could study the possibilities of using extension principles with other T-norms and their T-conorms.

Substituting Functions Based on Possibility-Probability Transformations. There are many literatures on possibility-probability transformation. See detailed discussion in this topic in, e.g., [3, 7]. We require here that the support of the probability distribution is a subset of the support of the possibility distribution.

Suppose that we have chosen a method of transformation from possibility distributions to probability distributions, where the possibility distributions are determined by the fuzzy numbers of the fuzzy data [22]. For a fuzzy number $\tilde{A}$, denote by $F^{\tilde{A}}$ the CDF of the transformed probability distribution from $\tilde{A}$, and define

$$I^{(\mathrm{f}),\mathrm{PP}}(\tilde{A}, x) := F^{\tilde{A}}(x).$$

Then we can show that $I^{(\mathrm{f}),\mathrm{PP}}$ is a substituting function.

As examples, we consider two types of transformations with *interval* (fuzzy) data. With the re-scaled transform (i.e., the transformed probability distribution from an observation $[a, b]$ is the uniform distribution on the interval $[a, b]$), $I^{(\mathrm{f}),\mathrm{PP}}$ is just the smoothed empirical distribution function discussed in Chapter 6 of [19]; With the transformations of putting mass 1 on the left-end (respectively, right-end) point of each observation, $I^{(\mathrm{f}),\mathrm{PP}}$ is just the lower (respectively, upper) end-point of fuzzy histogram to the class $(-\infty, x]$ discussed in Chapter 5 of [19].

3 CDF Estimation with Fuzzy Data: NPML Estimators

In this section, assuming that the CDF F is to a continuous variable, we discuss NPML estimation with fuzzy data, adapted from the unpublished work of [13].

Likelihood. Let $\tilde{X}_1, ..., \tilde{X}_n$ be fuzzy data as above. By treating each fuzzy observation as a fuzzy event [21], under the condition that the joint membership function of $\tilde{X}_1, ..., \tilde{X}_n$ is decomposable in the sense of [2], the log-likelihood of F given the fuzzy data is

$$\ell(F|\tilde{X}_1, ..., \tilde{X}_n) := \sum_{j=1}^{n} \log \left[\int_{\mathbb{R}} X_j(x) \mathrm{d}F(x) \right].$$

When all $\tilde{X}_j$s are intervals $[\underline{X}_j, \overline{X}_j]$,

$$\ell(F|\tilde{X}_1, ..., \tilde{X}_n) = \sum_{j=1}^{n} \log \left[F(\overline{X}_j) - F(\underline{X}_j-) \right], \tag{9}$$

where $F(\underline{X}_j-)$ is the left limit of F at $\underline{X}_j$.

The NPML Estimation with Interval (Fuzzy) Data. In survival analysis, we model the time of an event of interest only known to have occurred within an interval, bounded or not. Under the assumption that the time of the event of interest is independent from the inspection process, the NPMLE for interval censored data and the NPMLE for interval (fuzzy) data in this paper share the same likelihood in Eqn. (9). Therefore, the estimator applied to the NPMLE for interval censored data can be used to the NPMLE for interval (fuzzy) data.

Applying NPML estimation with interval censored data to interval (fuzzy) data, we know that the support of the NPMLE for interval (fuzzy) data is the union of all maximal intersections (MIs) of the data, where an MI is the non-empty intersection of some observations and the other observations do not intersect it. (If an observation intersects none of the other observations, this observation itself is also an MI.) When moving (probability) mass from no MIs to MIs, intersected observations share mass as much as possible, hence the likelihood improves. The number of MIs, m, is finite and no more than n. Therefore, the NPML estimation is equivalent to the NPML estimation of probability vector $[p_1, ..., p_m]'$, where p_i's are the probability masses to MIs. Note that when a non-singleton MI receives positive mass, whatever the way the mass distributes inside the MI, the likelihood remains the same, we need to make convention to remove such nonuniqueness. For detailed discussion, see, e.g., [15, 17, 4].

An Approximate NPML Estimation with General Fuzzy Data. The notion of MI for interval (fuzzy) data cannot apply to general fuzzy data. For general fuzzy data, we propose an approximate estimator called synthesized estimator to alpha-level sets (SEALS) in the following 4 steps:

1. (Selection of alpha-levels.) Select several α levels $\alpha_1 = 0, ..., \alpha_K = 1$. (For example, with $K = 6$, we select $\alpha_1 = 0, \alpha_2 = 0.2, \alpha_3 = 0.4, ..., \alpha_6 = 1$);
2. (Extraction of alpha-level sets.) At each level of α_k, $k = 1, ..., K$, find the interval (fuzzy) data $\tilde{X}_1^{\alpha_k}, ..., \tilde{X}_n^{\alpha_k}$;
3. (Estimation with each alpha-level set.) For each of $\alpha_k, k = 1, ..., K$, find the NPMLE of the interval (fuzzy) data $\tilde{X}_1^{\alpha_k}, ..., \tilde{X}_n^{\alpha_k}$, denoted by $\hat{\mathbf{p}}^{\alpha_k}$, and let $\hat{F}_n^{\alpha_k}$ denote the CDF to $\hat{\mathbf{p}}^{\alpha_k}$ with some convention of mass distribution in each MI;
4. (Synthesization.) Synthesize $\hat{F}_n^{\alpha_k}$ $(k = 1, ..., K)$ into the final CDF estimate as follows. We introduce a new CDF family

$$\{w_1 \hat{F}_n^{\alpha_1} + \cdots + w_K \hat{F}_n^{\alpha_k}; \text{ all } w_k \in [0, 1] \text{ and } w_1 + \cdots + w_K = 1\},$$

then, by treating $w_1, ..., w_K$ as model parameters, search the CDF in this new CDF family instead of the family of all CDFs) in maximizing the likelihood. The parameters w_k for $k = 1, ..., K$ are certainly interpreted as weights.

4 Remarks and Further Problems

As already used in Section 3, for any finite CDF estimators $\hat{F}_n^1 ...., \hat{F}_n^K$, the convex combination $w_1 \hat{F}_n^1 + \cdots + w_K \hat{F}_n^K$ $(w_1, ..., w_K \in [0, 1]$, $w_1 + \cdots + w_K = 1)$ is also a CDF estimator. This is a powerful approach to construct new CDF estimators.

We did not discuss how to construct a confidence interval of CDF at any given $x \in \mathbb{R}$. When establishing a (possibly asymptotic) distribution of a CDF estimator at x, we could construct a confidence interval based on it. With unknown (asymptotic) distribution, (usually it is the case,) bootstrap confidence intervals (by random sampling with replacement from the fuzzy data set) could be used.

We list some further (open) problems.

- Search other CDF estimators with fuzzy data which extends EDF of real-valued data;
- Find conditions for a CDF estimator with fuzzy data to satisfy SLLN and CLT. For CLT, we also need to check the convergence rate of the estimator;

- Extend Glivenko-Cantelli and Donsker theorems to CDF estimators with fuzzy data;
- In this paper, the CDF estimators are to the set-class $\mathcal{C}_0 := \{(-\infty, x];\ x \in \mathbb{R}\}$, (i.e., we estimate $\Pr(X \in C)$ for $C \in \mathcal{C}_0$). Generation to more general classes, e.g., $\mathcal{C}_0 \cup \{(x_1, x_2];\ x_1, x_2 \in \mathbb{R},\ x_1 \leq x_2\}$ (a VC-class, see [18]) is natural and useful.
- Work on more general sample spaces such as $\mathbb{R}^2$. To bivariate interval (fuzzy) data, we can borrow the NPML estimation with bivariate interval censored data (see, for example, [4, 5, 12, 14]). To general bivariate fuzzy data, copula modeling could help.
- Fuzzify CDF estimators with fuzzy data. Although in principle, Zadeh's extension principle could be applied for such fuzzification, when the sample size gets large, it could be hard to manage the complexity of the resulting estimator. It is worth for us to explore other fuzzification possibilities.

Acknowledgements. The authors would like to thank the two anonymous reviewers for their valuable comments.

References

1. Denoeux, T., Masson, M.-H., Hébert, P.-A.: Nonparametric rank-based statistics and significance tests for fuzzy data. Fuzzy Set Syst. 153, 1–28 (2005)
2. Denoeux, T.: Maximum likelihood estimation from fuzzy data using the EM algorithm. Fuzzy Set Syst. 183, 72–91 (2011)
3. Dubois, D.: Possibility theory and statistical reasoning. Comput. Stat. Data An. 51, 47–69 (2006)
4. Gentleman, R., Vandal, A.C.: Computational Algorithms for Censored-Data Problems Using Intersection Graphs. J. Comput. Graph. Stat. 10, 403–421 (2001)
5. Gentleman, R., Vandal, A.C.: Nonparametric estimation of the bivariate cdf for arbitrarily censored data. Can. J. Stat. 30, 557–571 (2002)
6. Gil, M.A., López-Diaz, M., Ralescu, D.A.: Overview on the development of fuzzy random variables. Fuzzy Set Syst. 157, 2546–2557 (2006)
7. Klir, G.J.: Uncertainty and Information: Foundations of Generalized Information Theory. Wiley, New York (2006)
8. Kruse, R., Meyer, K.D.: Statistics with Vague Data. Reidel, Dordrecht (1987)
9. Kwakernaak, H.: Fuzzy random variables—I. definitions and theorems. Inform. Sciences 15, 1–29 (1978)
10. Kwakernaak, H.: Fuzzy random variables—II. Algorithms and examples for the discrete case. Inform. Sciences 17, 253–278 (1979)
11. Leekwijck, W.V., Kerre, E.E.: Defuzzification: criteria and classification. Fuzzy Set Syst. 108, 159–178 (1999)
12. Liu, X.: Nonparametric Estimation with Censored Data: a Discrete Approach. PhD Thesis, McGill University, Montreal, Canada (2005)
13. Liu, X.: CDF nonparametric maximal likelihood estimation with fuzzy data: a preliminary exploration (in preparation)
14. Maathuis, M.H.: Reduction Algorithm for the NPMLE for the Distribution Function of Bivariate Interval-Censored Data. J. Comput. Graph. Stat. 14, 352–363 (2005)

15. Peto, R.: Experimental survival curves for interval-censored data. Appl. Stat-J. Roy. St. C 22, 86–91 (1973)
16. Puri, M.L., Ralescu, D.A.: Fuzzy random variables. J. Math. Anal. Appl. 114, 409–422 (1986)
17. Turnbull, B.W.: The empirical distribution function with arbitrarily grouped, censored and truncated data. J. Roy. Stat. Soc. B Met. 38, 290–295 (1976)
18. van der Vaart, A., Wellner, J.: Weak Convergence and Empirical Processes with applictions to Statistics. Springer US, New York (1996)
19. Viertl, R.: Statistical Methods for Fuzzy Data. Wiley, New York (2011)
20. Zadeh, L.A.: Fuzzy sets. Inform Control 8, 338–353 (1965)
21. Zadeh, L.A.: Probability measures of fuzzy events. J. Math. Anal. Appl. 23, 421–427 (1968)
22. Zadeh, L.A.: Fuzzy sets as a basis for a theory of possibility. Fuzzy Set Syst. 1, 3–28 (1978)

15. Peto, R.: Experimental survival curves for interval-censored data. Appl. Statist. J. Roy. Stat. Soc. C 22, 86–91 (1973)
16. Puri, M.L., Ralescu, D.A.: Fuzzy random variables. J. Math. Anal. Appl. 114, 409–422 (1986)
17. Turnbull, B.W.: The empirical distribution function with arbitrarily grouped, censored and truncated data. J. Roy. Stat. Soc. B Met. 38, 290–295 (1976)
18. van der Vaart, A.W., Wellner, J.A.: Weak convergence and empirical processes with applications to statistics. Springer, New York (1996)
19. Viertl, R.: Statistical methods for fuzzy data. Wiley, New York (2011)
20. Zadeh, L.A.: Fuzzy sets. Inform. Control 8, 338–353 (1965)
21. Zadeh, L.A.: Probability measures of fuzzy events. J. Math. Anal. Appl. 23, 421–427 (1968)
22. Zadeh, L.A.: Fuzzy sets as a basis for a theory of possibility. Fuzzy Set. Syst. 1, 3–28 (1978)

On Equations with a Fuzzy Stochastic Integral with Respect to Semimartingales

Marek T. Malinowski

Abstract. We consider some equations in a metric space of fuzzy sets with basis of square integrable random vectors. These equations generalize the single-valued stochastic differential equations and set-valued stochastic integral equations as well. A main object is a fuzzy stochastic trajectory integral with respect to a semimartingale. We obtain the existence and uniqueness of global solutions to fuzzy stochastic integral equations driven by continuous semimartingales. Also, we present a stability of solutions under changes of equation's data.

Keywords: Fuzzy stochastic integral, fuzzy stochastic integral equation.

1 Introduction

A typical feature of real-world phenomena is uncertainty. This term is mostly understood as stochastic uncertainty and the methods of probability theory and stochastic analysis are utilized in its investigations (see e.g. [1, 21]). However, the term uncertainty has the second aspect: vagueness (sometimes called imprecision, fuzziness, ambiguity, softness). This second type of uncertainty is appropriately treated by fuzzy set theory.

The dynamical systems and differential equations subjected to two kinds of uncertainties (fuzziness and randomness) are extensively studied nowadays [3, 4, 5, 6, 7, 8, 10, 11, 9, 13, 12, 16, 14, 15, 17, 18, 19, 20, 23]. In [18] one can find the studies on stochastic fuzzy trajectory integral with semimartingale integrator. Also, a fuzzy stochastic integral equation driven by a

Marek T. Malinowski
Faculty of Mathematics, Computer Science and Econometrics,
University of Zielona Góra, 65-516 Zielona Góra, Poland
e-mail: m.malinowski@wmie.uz.zgora.pl

R. Kruse et al. (Eds.): Synergies of Soft Computing and Statistics, AISC 190, pp. 93–101.
springerlink.com © Springer-Verlag Berlin Heidelberg 2013

Brownian motion (a particular semimartingale) was studied. Such equation is understood in [18] as a family of set-valued stochastic differential inclusions and its solution is a fuzzy set with basis of a set of probability measures. In this paper we treat the fuzzy stochastic integral equations driven by semimartingales as the equations in a metric space of fuzzy sets with basis of square integrable random vectors. Our approach is similar to that one presented in [12, 14]. Hence, we do not look for selections of the right-hand side of the equation, but we solve the fuzzy stochastic integral equation directly. In this way we consider the solutions being the fuzzy-set-valued mappings. To obtain continuity of solutions we assume that the semimartingale integrator is continuous. We present a stability property for the solutions.

2 Preliminaries

Let $\mathcal{X}$ be a separable Banach space, $\mathcal{K}_c^b(\mathcal{X})$ the family of all nonempty closed, bounded and convex subsets of $\mathcal{X}$. The Hausdorff metric $H_{\mathcal{X}}$ in $\mathcal{K}_c^b(\mathcal{X})$ is defined by

$$H_{\mathcal{X}}(A,B) = \max\Big\{\sup_{a\in A} \operatorname{dist}_{\mathcal{X}}(a,B), \sup_{b\in B} \operatorname{dist}_{\mathcal{X}}(b,A)\Big\},$$

where $\operatorname{dist}_{\mathcal{X}}(a,B) = \inf_{b\in B} \|a-b\|_{\mathcal{X}}$ and $\|\cdot\|_{\mathcal{X}}$ denotes a norm in $\mathcal{X}$.

It is known that $(\mathcal{K}_c^b(\mathcal{X}), H_{\mathcal{X}})$ is a Polish metric space.

2.1 Measurable Multifunctions

Let $(U,\mathcal{U},\mu)$ be a measure space. A set-valued mapping (multifunction) $F\colon U \to \mathcal{K}_c^b(\mathcal{X})$ is said to be $\mathcal{U}$-measurable (or measurable, for short) if it satisfies: $\{u \in U : F(u) \cap O \neq \emptyset\} \in \mathcal{U}$ for every open set $O \subset \mathcal{X}$.

A measurable multifunction $F\colon U \to \mathcal{K}_c^b(\mathcal{X})$ is said to be $L^p_{\mathcal{U}}(\mu)$-integrably bounded ($p \geq 1$), if there exists $h \in L^p(U,\mathcal{U},\mu;\mathbb{R}_+)$ such that the inequality $|||F|||_{\mathcal{X}} \leq h$ holds μ-a.e., where $|||A|||_{\mathcal{X}} = H_{\mathcal{X}}(A,\{0\}) = \sup_{a\in A} \|a\|_{\mathcal{X}}$ for $A \in \mathcal{K}_c^b(\mathcal{X})$, and $\mathbb{R}_+ = [0,\infty)$.

Let $(\Omega, \mathcal{A}, \{\mathcal{A}_t\}_{t\geq 0}, P)$ be a complete filtered probability space satisfying the usual hypotheses.

At this moment we put $\mathcal{X} = \mathbb{R}^d$, $U = \mathbb{R}_+ \times \Omega$, $\mathcal{U} = \mathcal{P}$, where $\mathcal{P}$ denotes the σ-algebra of the predictable elements in $\mathbb{R}_+ \times \Omega$. A stochastic process $f\colon \mathbb{R}_+ \times \Omega \to \mathbb{R}^d$ is called predictable if f is $\mathcal{P}$-measurable.

A mapping $F\colon \mathbb{R}_+ \times \Omega \to \mathcal{K}_c^b(\mathbb{R}^d)$ is said to be a set-valued stochastic process, if for every $t \in \mathbb{R}_+$ the mapping $F(t,\cdot)\colon \Omega \to \mathcal{K}_c^b(\mathbb{R}^d)$ is an $\mathcal{A}$-measurable multifunction. If for every fixed $t \in \mathbb{R}_+$ the mapping $F(t,\cdot)\colon \Omega \to \mathcal{K}_c^b(\mathbb{R}^d)$ is an $\mathcal{A}_t$-measurable multifunction then F is called $\{\mathcal{A}_t\}$-adapted. A set-valued stochastic process F is predictable if it is a $\mathcal{P}$-measurable multifunction.

2.2 Fuzzy Stochastic Processes

By a fuzzy set u with basis of a space $\mathcal{X}$ we mean a function $u\colon \mathcal{X} \to [0,1]$. By $\mathcal{F}(\mathcal{X})$ we denote the set of all fuzzy sets with basis $\mathcal{X}$. For $\alpha \in (0,1]$ denote $[u]^\alpha := \{x \in \mathcal{X} : u(x) \geq \alpha\}$ and let $[u]^0 := \mathrm{cl}_{\mathcal{X}}\{x \in \mathcal{X} : u(x) > 0\}$, where $\mathrm{cl}_{\mathcal{X}}$ denotes the closure in $(\mathcal{X}, \|\cdot\|_{\mathcal{X}})$. The sets $[u]^\alpha$ are called the α-level sets of fuzzy set u, and 0-level set $[u]^0$ is called the support of u.

Denote $\mathcal{F}_c^b(\mathcal{X}) = \{u \in \mathcal{F}(\mathcal{X}) : [u]^\alpha \in \mathcal{K}_c^b(\mathcal{X}) \text{ for every } \alpha \in [0,1]\}$. In this set we consider two metrics: the generalized Hausdorff metric

$$D_{\mathcal{X}}(u,v) := \sup_{\alpha\in[0,1]} H_{\mathcal{X}}([u]^\alpha, [v]^\alpha), \quad \text{and the Skorohod metric}$$

$$D_S^{\mathcal{X}}(u,v) := \inf_{\lambda\in\Lambda} \max\Big\{ \sup_{t\in[0,1]} |\lambda(t) - t|,\ \sup_{t\in[0,1]} H_{\mathcal{X}}(x_u(t), x_v(\lambda(t)))\Big\},$$

where Λ denotes the set of strictly increasing continuous functions $\lambda\colon [0,1] \to [0,1]$ such that $\lambda(0) = 0$, $\lambda(1) = 1$, and $x_u, x_v\colon [0,1] \to \mathcal{K}_c^b(\mathcal{X})$ are the càdlàg representations for the fuzzy sets $u, v \in \mathcal{F}_c^b(\mathcal{X})$, see [2] for details. The space $(\mathcal{F}_c^b(\mathcal{X}), D_{\mathcal{X}})$ is complete and non-separable, and the space $(\mathcal{F}_c^b(\mathcal{X}), D_S^{\mathcal{X}})$ is Polish.

For our aims we will consider two cases of $\mathcal{X}$. Namely we will take $\mathcal{X} = \mathbb{R}^d$ or $\mathcal{X} = L^2$, where $L^2 = L^2(\Omega, \mathcal{A}, P; \mathbb{R}^d)$ and we assume, from now on, that σ-algebra $\mathcal{A}$ is separable with respect to probability measure P.

Definition 1. ([22]). By a fuzzy random variable we mean a function $\mathfrak{u}\colon \Omega \to \mathcal{F}_c^b(\mathcal{X})$ such that $[\mathfrak{u}(\cdot)]^\alpha\colon \Omega \to \mathcal{K}_c^b(\mathcal{X})$ is an $\mathcal{A}$-measurable multifunction for every $\alpha \in (0,1]$.

It is known (see [2]) that for a mapping $\mathfrak{u}\colon \Omega \to \mathcal{F}_c^b(\mathcal{X})$ it holds:
- $\mathfrak{u}$ is the fuzzy random variable if and only if $\mathfrak{u}$ is $\mathcal{A}|\mathcal{B}_{D_S^{\mathcal{X}}}$-measurable,
- if $\mathfrak{u}$ is $\mathcal{A}|\mathcal{B}_{D_{\mathcal{X}}}$-measurable, then it is the fuzzy random variable; the opposite implication is not true, where $\mathcal{B}_{D_S^{\mathcal{X}}}$ ($\mathcal{B}_{D_{\mathcal{X}}}$, respectively) is the σ-algebra generated by the topology induced by $D_S^{\mathcal{X}}$ ($D_{\mathcal{X}}$, respectively).

A fuzzy-set-valued mapping $\mathfrak{f}\colon \mathbb{R}_+ \times \Omega \to \mathcal{F}_c^b(\mathcal{X})$ is called a fuzzy stochastic process if $\mathfrak{f}(t,\cdot)\colon \Omega \to \mathcal{F}_c^b(\mathcal{X})$ is a fuzzy random variable for every $t \in \mathbb{R}_+$.

The fuzzy stochastic process $\mathfrak{f}\colon \mathbb{R}_+ \times \Omega \to \mathcal{F}_c^b(\mathcal{X})$ is said to be predictable if the set-valued mapping $[\mathfrak{f}]^\alpha\colon \mathbb{R}_+ \times \Omega \to \mathcal{K}_c^b(\mathcal{X})$ is $\mathcal{P}$-measurable for every $\alpha \in (0,1]$.

Let $\mathfrak{f}\colon \mathbb{R}_+ \times \Omega \to \mathcal{F}_c^b(\mathcal{X})$ be a predictable fuzzy stochastic process. The process $\mathfrak{f}$ is said to be $L^2_{\mathcal{P}}(\mu)$-integrably bounded, if $\|\|[\mathfrak{f}]^0\|\|_{\mathcal{X}} \in L^2_{\mathcal{P}}(\mu)$.

2.3 Set-Valued Stochastic Trajectory Integral w.r.t. Continuous Semimartingales

Let $(\Omega, \mathcal{A}, \{\mathcal{A}_t\}_{t\geq 0}, P)$ be a complete, filtered probability space with a filtration $\{\mathcal{A}_t\}_{t\geq 0}$ satisfying usual hypotheses, i.e. $\{\mathcal{A}_t\}_{t\geq 0}$ is right continuous and $\mathcal{A}_0$ contains all P-null sets.

Let $Z\colon \mathbb{R}_+ \times \Omega \to \mathbb{R}$ be a semimartingale with the canonical representation

$$Z = A + M, \quad Z_0 = 0, \quad A_0 = 0, \quad M_0 = 0, \tag{1}$$

where $A\colon \mathbb{R}_+ \times \Omega \to \mathbb{R}$ is an $\{\mathcal{A}_t\}$-adapted cádlág stochastic process of finite vatiation on each compact interval of $\mathbb{R}_+$, and $M\colon \mathbb{R}_+ \times \Omega \to \mathbb{R}$ is a local $\{\mathcal{A}_t\}$-martingale.

If Z is continuous, so are the processes A, M. Also A is predictable and representation (1) is unique.

Since A is of finite variation, almost each (w.r.t. to P) trajectory $A(\cdot,\omega)$ generates a measure $\Gamma_{A(\cdot,\omega)}$ with the total variation on the interval $[0,t]$ given by $|A(\omega)|_t = \int_0^t \Gamma_{A(\cdot,\omega)}(ds)$. For a local martingale M one can define the quadratic variation process $[M]\colon \mathbb{R}_+ \times \Omega \to \mathbb{R}$ (cf. [1]). Now we denote by $\mathcal{H}^2$ the set of all semimartingales $Z\colon \mathbb{R}_+ \times \Omega \to \mathbb{R}$ with finite norm $\|\cdot\|_{\mathcal{H}^2}$, where

$$\|Z\|_{\mathcal{H}^2} := \big\|[M]_\infty^{1/2}\big\|_{L^2} + \big\||A|_\infty\big\|_{L^2},$$

where $[M]_\infty = \lim_{t\to\infty}[M]_t$ P-a.e. and $|A(\omega)|_\infty = \int_0^\infty \Gamma_{A(\cdot,\omega)}(ds)$.

It is known that for a continuous semimartingale $Z \in \mathcal{H}^2$ the process M in (1) is a continuous square integrable martingale and $\mathbb{E}|A|_\infty^2 < \infty$.

Let us consider two measures μ_A, μ_M defined on $(\mathbb{R}_+ \times \Omega, \mathcal{P})$ and induced by the processes A, M from the representation (1) of the semimartingale Z.

The measure μ_A is defined as follows (see [18])

$$\mu_A(C) := \int_\Omega \left(\int_{\mathbb{R}_+} \mathbf{1}_C(t,\omega)|A(\omega)|_\infty \Gamma_{A(\cdot,\omega)}(dt) \right) P(d\omega) \quad \text{for} \quad C \in \mathcal{P}.$$

For $f \in L^2(\mathbb{R}_+ \times \Omega, \mathcal{P}, \mu_A; \mathbb{R}^d)$ one can define the stochasic Lebesgue–Stjeltjes integral $\int_0^t f(s)dA_s$ trajectory-by-trajectory (cf. [21]).

The measure μ_M is the well-known Doléan-Dade measure (cf. [1]) such that

$$\mu_M(\{0\} \times A_0) = 0, \quad \mu_M((s,t] \times A) = \mathbb{E}\mathbf{1}_A(M_t - M_s)^2,$$

where $A_0 \in \mathcal{A}_0$, $0 \leq s < t$, $A \in \mathcal{A}_s$.

For $f \in L^2(\mathbb{R}_+ \times \Omega, \mathcal{P}, \mu_M; \mathbb{R}^d)$ and $t \in \mathbb{R}_+$ one can define the stochastic integral $\int_0^t f(s)dM_s$ and we have (cf. [1])

$$\int_{[0,t]\times\Omega} |f|^2 d\mu_M = \mathbb{E}\int_0^t |f(s)|^2 d[M]_s = \mathbb{E}\Big|\int_0^t f(s)dM_s\Big|^2$$

For a semimartingale $Z \in \mathcal{H}^2$ with the representation (1) one can define a finite measure μ_Z on $(\mathbb{R}_+ \times \Omega, \mathcal{P})$ as

$$\mu_Z(C) := \mu_A(C) + \mu_M(C), \quad C \in \mathcal{P}.$$

Denote $L^2_{\mathcal{P}}(\mu_Z) := L^2(\mathbb{R}_+ \times \Omega, \mathcal{P}, \mu_Z; \mathbb{R}^d)$. For $f \in L^2_{\mathcal{P}}(\mu_Z)$ one can define the single-valued stochastic integral $\int_0^t f(s)dZ_s$ with respect to semimartingale Z as follows

$$\int_0^t f(s)dZ_s := \int_0^t f(s)dA_s + \int_0^t f(s)dM_s.$$

Let $F\colon \mathbb{R}_+ \times \Omega \to \mathcal{K}^b_c(\mathbb{R}^d)$ be a predictable set-valued stochastic process which is $L^2_{\mathcal{P}}(\mu_Z)$-integrably bounded. For such a process let us define the set $S^2_{\mathcal{P}}(F, \mu_Z) := \{f \in L^2_{\mathcal{P}}(\mu_Z) : f \in F, \ \mu_Z\text{-a.e.}\}$. Due to the Kuratowski and Ryll-Nardzewski Selection Theorem we have $S^2_{\mathcal{P}}(F, \mu_Z) \neq \emptyset$.

Definition 2. ([18]) For a predictable and $L^2_{\mathcal{P}}(\mu_Z)$-integrably bounded set-valued stochastic process $F\colon \mathbb{R}_+ \times \Omega \to \mathcal{K}^b_c(\mathbb{R}^d)$ and for $\tau, t \in \mathbb{R}_+$, $\tau < t$ the set-valued stochastic trajectory integral (over interval $[\tau, t]$) of F with respect to the semimartingale Z is the following subset of $L^2(\Omega, \mathcal{A}_t, P; \mathbb{R}^d)$

$$(S)\int_{[\tau,t]} F(s)dZ_s := \Big\{\int_\tau^t f(s)dZ_s : f \in S^2_{\mathcal{P}}(F, \mu_Z)\Big\}.$$

In the rest of the paper, for the sake of convenience, we will write L^2 instead of $L^2(\Omega, \mathcal{A}, P; \mathbb{R}^d)$ and L^2_t instead of $L^2(\Omega, \mathcal{A}_t, P; \mathbb{R}^d)$. Moreover, since Z is considered to be continuous, the integrals $(S)\int_{[\tau,t]} F(s)dZ_s$, $(S)\int_{(\tau,t]} F(s)dZ_s$ coincide. For their common value we will write $(S)\int_\tau^t F(s)dZ_s$. It is known that $(S)\int_\tau^t F(s)dZ_s \in \mathcal{K}^b_c(L^2_t)$.

Theorem 1. *([18]) Let $F_n\colon \mathbb{R}_+ \times \Omega \to \mathcal{K}^b_c(\mathbb{R}^d)$ be the predictable set-valued stochastic processes such that F_1 is $L^2_{\mathcal{P}}(\mu_Z)$-integrably bounded and $F_1 \supset F_2 \supset \ldots \supset F$ μ_Z-a.e., where $F := \bigcap_{n=1}^\infty F_n$ μ_Z-a.e. Then for every $\tau, t \in \mathbb{R}_+$, $\tau < t$ it holds $(S)\int_\tau^t F(s)dZ_s = \bigcap_{n=1}^\infty (S)\int_\tau^t F_n(s)dZ_s$.*

3 Fuzzy Stochastic Trajectory Integral w.r.t. Continuous Semimartingales

Using Thm. 1 and the Representation Theorem of Negoita–Ralescu one can define a notion of a fuzzy stochastic trajectory integral with a semimartingale integrator.

Proposition 1. *([18]) Assume that $\mathfrak{f}\colon \mathbb{R}_+ \times \Omega \to \mathcal{F}^b_c(\mathbb{R}^d)$ is a predictable and $L^2_{\mathcal{P}}(\mu_Z)$-integrably bounded fuzzy stochastic process. Then for every $\tau, t \in \mathbb{R}_+$,*

$\tau < t$ there exists a unique fuzzy set in $\mathcal{F}_c^b(L_t^2)$ denoted by $(F)\int_\tau^t \mathfrak{f}(s)dZ_s$ such that for every $\alpha \in (0,1]$ it holds $\left[(F)\int_\tau^t \mathfrak{f}(s)dZ_s\right]^\alpha = (S)\int_\tau^t [\mathfrak{f}(s)]^\alpha dZ_s$, and $\left[(F)\int_\tau^t \mathfrak{f}(s)dZ_s\right]^0 \subset (S)\int_\tau^t [\mathfrak{f}(s)]^0 dZ_s$.

Definition 3. ([18]) The fuzzy set $(F)\int_\tau^t \mathfrak{f}(s)dZ_s \in \mathcal{F}_c^b(L_t^2)$ from Prop. 1 is said to be the fuzzy stochastic trajectory integral (over interval $[\tau, t]$) of $\mathfrak{f}$ with respect to the semimartingale Z.

Since $\mathcal{F}_c^b(L_t^2) \subset \mathcal{F}_c^b(L^2)$, we have $(F)\int_\tau^t \mathfrak{f}(s)dZ_s \in \mathcal{F}_c^b(L^2)$.

The following properties will be useful in the context of the fuzzy equations considered in the next section.

Theorem 2. *Let $\mathfrak{f}_1, \mathfrak{f}_2 \colon \mathbb{R}_+ \times \Omega \to \mathcal{F}_c^b(\mathbb{R}^d)$ be the predictable and $L_{\mathcal{P}}^2(\mu_Z)$-integrably bounded fuzzy stochastic processes. Then*

(i) for every $\tau, a, t \in \mathbb{R}_+$, $\tau \leq a \leq t$ it holds $(F)\int_\tau^t \mathfrak{f}_1(s)dZ_s = (F)\int_\tau^a \mathfrak{f}_1(s)dZ_s + (F)\int_a^t \mathfrak{f}_1(s)dZ_s$,

(ii) for every $\tau, t \in \mathbb{R}_+$, $\tau < t$ it holds $D_{L^2}^2\left((F)\int_\tau^t \mathfrak{f}_1(s)dZ_s, (F)\int_\tau^t \mathfrak{f}_2(s)dZ_s\right) \leq 2\int_{[\tau,t]\times\Omega} D_{\mathbb{R}^d}^2(\mathfrak{f}_1, \mathfrak{f}_2)d\mu_Z$,

(iii) for every $\tau \in \mathbb{R}_+$ the mapping $[\tau, \infty) \ni t \mapsto (F)\int_\tau^t \mathfrak{f}_1(s)dZ_s \in \mathcal{F}_c^b(L^2)$ is continuous with respect to the metric D_{L^2}.

4 Fuzzy Stochastic Integral Equations Driven by Semimartingales

In this section we consider equations with the fuzzy stochastic trajectory integrals with a continuous semimartingale integrator.

We consider the following relation in the metric space $\left(\mathcal{F}_c^b(L^2), D_{L^2}\right)$ and call it the fuzzy stochasitc integral equation driven by continuous semimartingale:

$$X(t) = X_0 + (F)\int_0^t \mathfrak{f}(s, X(s))dZ_s \quad \text{for } t \in \mathbb{R}_+, \tag{2}$$

where $\mathfrak{f}\colon \mathbb{R}_+ \times \Omega \times \mathcal{F}_c^b(L^2) \to \mathcal{F}_c^b(\mathbb{R}^d)$ and $X_0 \in \mathcal{F}_c^b(L_0^2)$.

Such the equations are generalizations of set-valued stochastic integral equations driven by continuous semimartingales. The results presented in this paper apply instantly to set-valued equations.

Definition 4. By a global solution to (2) we mean a D_{L^2}-continuous mapping $X\colon \mathbb{R}_+ \to \mathcal{F}_c^b(L^2)$ that satisfies (2) for $t \in \mathbb{R}_+$.

Definition 5. A global solution $X\colon \mathbb{R}_+ \to \mathcal{F}_c^b(L^2)$ to (2) is unique if $X(t) = Y(t)$ for every $t \in \mathbb{R}_+$, where $Y\colon \mathbb{R}_+ \to \mathcal{F}_c^b(L^2)$ is any global solution to (2) existing on $\mathbb{R}_+$.

Remark 1. If $X: \mathbb{R}_+ \to \mathcal{F}_c^b(L^2)$ is a global solution to (2), then $X(t) \in \mathcal{F}_c^b(L_t^2)$ for every $t \in \mathbb{R}_+$.

To obtain the existence of a global solution we assume that the nonlinearity $\mathfrak{f}\colon \mathbb{R}_+ \times \Omega \times \mathcal{F}_c^b(L^2) \to \mathcal{F}_c^b(\mathbb{R}^d)$ satisfies:

(G1) it is a $\mathcal{P} \otimes \mathcal{B}_{D_S^{L^2}} | \mathcal{B}_{D_S^{\mathbb{R}^d}}$-measurable mapping,

(G2) $\exists\, K > 0 \quad \forall\, (t,\omega) \in \mathbb{R}_+ \times \Omega \quad \forall\, u, v \in \mathcal{F}_c^b(L^2)$

$$D_{\mathbb{R}^d}\big(\mathfrak{f}(t,\omega,u), \mathfrak{f}(t,\omega,v)\big) \le K D_{L^2}(u,v),$$

(G3) $\exists\, C > 0 \quad \forall\, (t,\omega) \in \mathbb{R}_+ \times \Omega \quad \forall\, u \in \mathcal{F}_c^b(L^2)$

$$D_{\mathbb{R}^d}\big(\mathfrak{f}(t,\omega,u), \hat{\theta}\big) \le C\big(1 + D_{L^2}(u, \hat{\Theta})\big).$$

The descriptions of the symbols $\hat{\theta}$, $\hat{\Theta}$ appearing in (G3) are as follows: let θ, Θ denote the zero elements in $\mathbb{R}^d$ and L^2, respectively, the symbols $\hat{\theta}$, $\hat{\Theta}$ are their fuzzy counterparts, i.e. $\hat{\theta} \in \mathcal{F}_c^b(\mathbb{R}^d)$ and $[\hat{\theta}]^\alpha = \{\theta\}$ for every $\alpha \in [0,1]$, also $\hat{\Theta} \in \mathcal{F}_c^b(L^2)$ and $[\hat{\Theta}]^\alpha = \{\Theta\}$ for every $\alpha \in [0,1]$.

Theorem 3. *Let $X_0 \in \mathcal{F}_c^b(L_0^2)$, and $\mathfrak{f}\colon \mathbb{R}_+ \times \Omega \times \mathcal{F}_c^b(L^2) \to \mathcal{F}_c^b(\mathbb{R}^d)$ satisfy conditions (G1)-(G3). Then Eq. (2) has a unique global solution.*

A stability of solution with respect to initial value is a desired property. Now, we give a result of this type. Therefore let us consider Eq. (2) and equation with another initial value $Y_0 \in \mathcal{F}_c^b(L_0^2)$, i.e.

$$Y(t) = Y_0 + (F)\int_0^t \mathfrak{f}(s, Y(s))dZ_s, \quad t \in \mathbb{R}_+, \tag{3}$$

and let X, Y denote the solutions to (2) and (3), respectively.

By μ_Z^Ω we denote a measure on $(\mathbb{R}_+, \beta_{\mathbb{R}_+})$ defined as $\mu_Z^\Omega(B) = \mu_Z(B \times \Omega)$ for $B \in \beta_{\mathbb{R}_+}$. In the rest of the paper we assume that a continuous semimartingale Z is such that μ_Z^Ω is absolutely continuous with respect to the Lebesgue measure.

Theorem 4. *Let $X_0, Y_0 \in \mathcal{F}_c^b(L_0^2)$. Assume that $\mathfrak{f}\colon \mathbb{R}_+ \times \Omega \times \mathcal{F}_c^b(L^2) \to \mathcal{F}_c^b(\mathbb{R}^d)$ satisfies conditions (G1)-(G3). Then*

$$D_{L^2}\big(X(t), Y(t)\big) \le \sqrt{2}e^{2K^2\mu_Z^\Omega([0,t])} D_{L^2}(X_0, Y_0) \quad \textit{for every } t \in \mathbb{R}_+.$$

Corollary 1. *Under assumptions of Thm. 4 we have*

$$\sup_{t \in \mathbb{R}_+} D_{L^2}\big(X(t), Y(t)\big) \le \sqrt{2}e^{2K^2\mu_Z^\Omega(\mathbb{R}_+)} D_{L^2}(X_0, Y_0).$$

For a stability of a solution with respect to the nonlinearity $\mathfrak{f}$, let us consider Eq. (2) and the following equations (for $n = 1, 2, \ldots$)

$$X_n(t) = X_0 + (F)\int_0^t \mathfrak{f}_n(s, X_n(s))dZ_s, \quad t \in \mathbb{R}_+, \tag{4}$$

and denote by X, X_n the corresponding solutions to (2) and (4).

Theorem 5. *Let $X_0 \in \mathcal{F}_c^b(L_0^2)$. Assume that $\mathfrak{f}, \mathfrak{f}_n\colon \mathbb{R}_+ \times \Omega \times \mathcal{F}_c^b(L^2) \to \mathcal{F}_c^b(\mathbb{R}^d)$ satisfy conditions (G1)-(G3) with the same Lipschitz constant K. If for every $t \in \mathbb{R}_+$ and every $u \in \mathcal{F}_c^b(L^2)$*

$$\int_{[0,t]\times\Omega} D^2_{\mathbb{R}^d}\big(\mathfrak{f}_n(s,u), \mathfrak{f}(s,u)\big)d\mu_Z \longrightarrow 0, \quad \text{as } n \to \infty,$$

then for every $t \in \mathbb{R}_+$

$$D_{L^2}\big(X_n(t), X(t)\big) \longrightarrow 0, \quad \text{as } n \to \infty.$$

References

1. Chung, K.L., Williams, R.J.: Introduction to Stochastic Integration. Birkhäuser, Boston (1983)
2. Colubi, A., Domínguez-Menchero, J.S., López-Díaz, M., Ralescu, D.A.: A $D_E[0,1]$ representation of random upper semicontinuous functions. Proc. Amer. Math. Soc. 130, 3237–3242 (2002)
3. Fei, W.: Existence and uniqueness of solution for fuzzy random differential equations with non-Lipschitz coefficients. Inform. Sciences 177, 4329–4337 (2007)
4. Feng, Y.: Fuzzy stochastic differential systems. Fuzzy Set Syst. 115, 351–363 (2000)
5. Kozaryn, M., Malinowski, M.T., Michta, M., Świątek, K.Ł.: On multivalued stochastic integral equations driven by a Wiener process in the plane. Dynam. Syst. Appl. 21, 293–318 (2012)
6. Li, J., Li, S., Ogura, Y.: Strong solution of Itô type set-valued stochastic differential equation. Acta Math. Sinica Engl. Ser. 26, 1739–1748 (2010)
7. Malinowski, M.T.: On random fuzzy differential equations. Fuzzy Set Syst. 160, 3152–3165 (2009)
8. Malinowski, M.T.: Existence theorems for solutions to random fuzzy differential equations. Nonlinear Anal. TMA 73, 1515–1532 (2010)
9. Malinowski, M.T.: Itô type stochastic fuzzy differential equations with delay. Systems Control Lett. 61, 692–701 (2012)
10. Malinowski, M.T.: Random fuzzy differential equations under generalized Lipschitz condition. Nonlinear Anal-Real. 13, 860–881 (2012)
11. Malinowski, M.T.: Strong solutions to stochastic fuzzy differential equations of Itô type. Math. Comput. Model 55, 918–928 (2012)
12. Malinowski, M.T., Michta, M.: Fuzzy stochastic integral equations. Dynam. Syst. Appl. 19, 473–494 (2010)
13. Malinowski, M.T., Michta, M.: Stochastic set differential equations. Nonlinear Anal-Theor. 72, 1247–1256 (2010)
14. Malinowski, M.T., Michta, M.: Fuzzy Stochastic Integral Equations Driven by Martingales. In: Li, S., Wang, X., Okazaki, Y., Kawabe, J., Murofushi, T., Guan, L. (eds.) Nonlinear Mathematics for Uncertainty and its Applications. AISC, vol. 100, pp. 143–150. Springer, Heidelberg (2011)

15. Malinowski, M.T., Michta, M.: Set-valued stochastic integral equations. Dynam. Cont. Dis. Ser. B 18, 473–492 (2011)
16. Malinowski, M.T., Michta, M.: Stochastic fuzzy differential equations with an application. Kybernetika 47, 123–143 (2011)
17. Malinowski, M.T., Michta, M.: Set-valued stochastic integral equations driven by martingales. J. Math. Anal. Appl. 394, 30–47 (2012)
18. Michta, M.: On set-valued stochastic integrals and fuzzy stochastic equations. Fuzzy Set Syst. 177, 1–19 (2011)
19. Mitoma, I., Okazaki, Y., Zhang, J.: Set-valued stochastic differential equation in M-type 2 Banach space. Comm. Stoch. Anal. 4, 215–237 (2010)
20. Ogura, Y.: On stochastic differential equations with fuzzy set coefficients. In: Soft Methods for Handling Variability and Imprecision. AISC, vol. 48, pp. 263–270. Springer, Berlin (2008)
21. Protter, P.E.: Stochastic Integration and Differential Equations. Springer, Berlin (2005)
22. Puri, M.L., Ralescu, D.A.: Differentials of fuzzy functions. J. Math. Anal. Appl. 91, 552–558 (1983)
23. Zhang, J., Li, S., Mitoma, I., Okazaki, Y.: On the solutions of set-valued stochastic differential equations in M-type 2 Banach spaces. Tohoku Math. J. 61, 417–440 (2009)

Part III
Statistical Methods

A Linear Regression Model for Interval-Valued Response Based on Set Arithmetic

Angela Blanco-Fernández, Ana Colubi,
Marta García-Bárzana, and Manuel Montenegro

Abstract. Several linear regression models involving interval-valued variables have been formalized based on the interval arithmetic. In this work, a new linear regression model with interval-valued response and real predictor based on the interval arithmetic is formally described. The least-squares estimation of the model is solved by means of a constrained minimization problem which guarantees the coherency of the estimators with the regression parameters. The practical applicability of the estimation method is checked on a real-life example, and the empirical behaviour of the procedure is shown by means of some simulation studies.

Keywords: Interval data, least-squares estimation, linear regression, set arithmetic.

1 Introduction

The statistical treatment of imprecise-valued data has been deeply investigated over the last years. The consideration of experimental outcomes associated with not real values, but elements in a more flexible scale allows us to better capture the inherent imprecision/diversity of certain data. The scale of nonempty compact real intervals is commonly used when the characteristic under study corresponds to the fluctuation of a magnitude over a certain period of time, a numerical range, or when working with grouped or censored data (see [3, 5, 8], among others). Real intervals are also useful for

Angela Blanco-Fernández · Ana Colubi ·
Marta García-Bárzana · Manuel Montenegro
Department of Statistics and Operational Research,
University of Oviedo, 33007, Oviedo, Spain
e-mail: {blancoangela,colubi,garciabarmarta.uo,mmontenegro}@uniovi.es

R. Kruse et al. (Eds.): Synergies of Soft Computing and Statistics, AISC 190, pp. 105–113.
springerlink.com © Springer-Verlag Berlin Heidelberg 2013

modelling exact values being imprecisely identified due to any reason (hidden for confidentiality, uncertainty in the measurement, etc). In this case, the interval that contains a representative data point instead of its precise value is considered in the statistical processing (see, for instance, [7]).

The study of a functional dependence between interval variables may be done by considering (interval-valued) functions of the (interval-valued) variables which relates suitably the interval data. In the case of considering a linear function, it can be formalized in terms of the natural interval arithmetic. Several linear regression models between interval data based on the interval arithmetic have been proposed (see [2] for a detailed review).

Real-valued data can obviously be considered as real compact intervals with equal endpoints. Thus, the interval linear models can also be applied to situations in which one or several variables involved in the model are real valued. However, it can be shown that these models are not able to express suitably an interval variable in terms of a real predictor. In this paper a new linear regression model for interval response and a real predictor is proposed, which overcomes the drawbacks of the existing interval models. The rest of the paper is organized as follows: In Sec. 2 some basic concepts and results for intervals are recalled, and the basic linear model for interval data is presented. Its limitations when the predictor is real valued are shown. The new proposed linear model is introduced in Sec. 3. The main theoretical properties of the model and a coherent least squares estimation of its parameters are studied. Both the practical applicability and the empirical behaviour of the estimation method are shown in Sec. 4. Finally, in Sec. 5 some concluding remarks and future directions are highlighted.

2 Notations and Preliminaries

Let $\mathcal{K}_c(\mathbb{R})$ denote the space of nonempty compact real intervals, i.e. $\mathcal{K}_c(\mathbb{R}) = \{[a,b] : a,b \in \mathbb{R},\ a \leq b\}$. Each element $A \in \mathcal{K}_c(\mathbb{R})$ can be represented by means of its endpoints as $A = [\inf A, \sup A]$, with $\inf A \leq \sup A$. If we define $\text{mid}A = (\sup A + \inf A)/2$ and $\text{spr}A = (\sup A - \inf A)/2$, then A can be equivalently expressed as $A = [\text{mid}A \pm \text{spr}A]$. Obviously, $(\text{mid}A, \text{spr}A) \in \mathbb{R} \times \mathbb{R}^+$ are the mid-point and the spread (radius) of A, respectively. In statistical processing the (*mid*, *spr*)-representation for intervals is most commonly used since it involves only a non-negativity condition for the *spr*. The order constraint between the components in the (*inf*, *sup*)-characterization is more difficult to deal with in general.

Through the operations $A + B = \{a + b : a \in A, b \in B\}$ and $\lambda A = \{\lambda a : a \in A\}$, for every $A, B \in \mathcal{K}_c(\mathbb{R})$ and $\lambda \in \mathbb{R}$, a natural arithmetic is defined on $\mathcal{K}_c(\mathbb{R})$. Due to the general lack of symmetric element with respect to the addition, the space is semilinear with these operations (see [2]). Thus, $A + (-1)B$ has not the usual sense of a difference operator. Sometimes it is possible to consider the so-called Hukuhara difference between intervals A and B, defined as the element $C \in \mathcal{K}_c(\mathbb{R})$ (if it exists) such that $B + C = A$. Whenever $\text{spr}B \leq \text{spr}A$ the Hukuhara difference between A and B exists. In that case, it is denoted by $A -_H B$.

The quantification of distances between intervals is also necessary for statistical developments. Among several metrics defined on $\mathcal{K}_c(\mathbb{R})$ (see, for instance, [1, 9]), in [10] an L_2-type metric with good operative and intuitive properties is introduced. Moreover, it generalizes other well-known metrics between intervals. Given $\theta > 0$,

$$d_\theta(A, B) = \sqrt{(\text{mid}\, A - \text{mid}\, B)^2 + \theta(\text{spr}\, A - \text{spr}\, B)^2}\ , \tag{1}$$

for every $A, B \in \mathcal{K}_c(\mathbb{R})$.

Formally, interval-valued random variables (or random intervals) can be considered in an analogous manner to real-valued random variables associated with a probability space $(\Omega, \mathcal{A}, P)$ i.e. mappings taking values on Ω and whose outcomes are associated with elements of $\mathcal{K}_c(\mathbb{R})$ instead of those of the real line, being $(\mathcal{B}_{d_\theta}|\mathcal{A})$-measurable ($\mathcal{B}_{d_\theta}$ denoting the σ-field generated by the topology induced by d_θ on $\mathcal{K}_c(\mathbb{R})$). A random interval X can be equivalently defined through the real-valued random vector $(\text{mid}X, \text{spr}X) : \Omega \to \mathbb{R}^2$ such that $\text{spr}X \geq 0$ a.s. - [P].

The expected value of a random interval X is defined by means of the Aumann expectation. It satisfies that

$$E(X) = [E(\text{mid}\, X) \pm E(\text{spr}\, X)]\ ,$$

whenever those expected values exist, i.e. $\text{mid}\, X, \text{spr}\, X \in L^1$. For $\{X_i, Y_i\}_{i=1}^n$ a random sample obtained from (X, Y), the sample mean of X is formalized in terms of the interval arithmetic as $\overline{X} = (X_1 + \ldots + X_n)/n$ and it verifies that $\overline{X} = [\overline{\text{mid}X} \pm \overline{\text{spr}X}]$.

2.1 Basic Linear Model for Random Intervals

A linear model between two interval-valued random variables has been previously formalized by means of different expressions. Mimicking the classical simple linear model in the real-valued case, in [6] the basic linear model for relating two random intervals X and Y based on the interval arithmetic has been introduced as follows:

$$Y = aX + \varepsilon\ , \tag{2}$$

where ε is an interval random error such that $E(\varepsilon|X) = B \in \mathcal{K}_c(\mathbb{R})$.

Remark 1. The (interval-valued) independent term B in (2) is included as the expectation of the error in order to allow the error to be an interval-valued random set (see [6] for details).

The estimation of the linear model (2) has been solved in [6] by means of a least squares criterion based on the well-known *Bertoluzza*-metric introduced in [1] (which is equivalent to d_θ when $\theta < 1$). Moreover, the LS estimators of the regression parameters are obtained by assuring their coherence with the interval arithmetic. Let $(\widehat{a}, \widehat{B})$ be the LS estimators of (a, B) in (2) obtained in [6]. The prediction of Y for any pair (X, Y) is computed as $\widehat{Y} = \widehat{a}X + \widehat{B}$. Equivalently, the real components $(\text{mid}\widehat{Y}, \text{spr}\widehat{Y})$ are computed as $\text{mid}\,\widehat{Y} = \widehat{a}\,\text{mid}\,X + \text{mid}\,\widehat{B}$ and $\text{spr}\,\widehat{Y} = |\widehat{a}|\,\text{spr}\,X + \text{spr}\,\widehat{B}$, respectively.

Since any real-valued random variable X can be considered as a random interval with $\text{spr}X = 0$, the interval model (2) is applicable to predict the random interval Y in terms of a real-valued predictor x. In this particular case, the estimation of the model (2) leads to the predicted values for Y,

$$\text{mid}\,\widehat{Y} = \widehat{a}\,x + \text{mid}\,\widehat{B} \quad \text{and} \quad \text{spr}\,\widehat{Y} = \overline{\text{spr}Y}\,, \tag{3}$$

with $\widehat{a} = \widehat{\sigma}_{x,\text{mid}Y}/\widehat{\sigma}^2_x$ and $\widehat{B} = \overline{\text{mid}Y} - \widehat{a}\,x$. Thus, model (2) allows to predict mid$Y$ in terms of x equivalently to the classical linear model for these real-valued variables, but it does not allow to predict spr$Y$ in terms of x. With the aim of getting the possibility for x to predict both *mid* and *spr* components of the interval Y, a new linear model for relating an interval response in terms of a real-valued predictor is formalized.

3 A New Linear Model for Interval Response

Let $x : \Omega \to \mathbb{R}$ be a real random variable and $Y : \Omega \to \mathcal{K}_c(\mathbb{R})$ be a random interval, both associated with the same probability space $(\Omega, \mathcal{A}, P)$. The values of Y can be modelled in terms of the corresponding values of x in a linear fashion by means of the following relationship:

$$Y = Ax + \varepsilon\,, \tag{4}$$

where $A \in \mathcal{K}_c(\mathbb{R})$ and $\varepsilon : \Omega \to \mathcal{K}_c(\mathbb{R})$ with $E(\varepsilon|X) = B \in \mathcal{K}_c(\mathbb{R})$.

The expression in (4) is well-formalized in terms of the interval arithmetic. Moreover, it is straightforward to show that from this interval model the following relationships for the real components of Y are transferred:

$$\text{mid}\,Y = \text{mid}A\;x + \text{mid}\,B \quad \text{and} \quad \text{spr}\,Y = \text{spr}A\;|x| + \text{spr}\,B\,. \tag{5}$$

Thus, the predictor variable x explains both components *mid* and *spr* of the interval response.

3.1 Least Squares Estimation of the Model

As usual, given $\{x_i, Y_i\}_{i=1}^n$ a random sample obtained from (x, Y), the least squares (LS) estimation of the model (4) consists in searching the optimal values for the regression parameters (interval-valued in this case) minimizing the squared distance between sample and estimated values, i.e. the aim is to find $D, C \in \mathcal{K}_c(\mathbb{R})$ such that $\frac{1}{n}\sum_{i=1}^n d_\theta^2(Y_i, Dx_i + C)$ has minimum value.

In [3] a least squares fitting of the affine function $Y = Ax + B$ is presented, without probabilistic assumptions for the data. However, the solution is only valid for particular conditions on the observed data: x_i must be non-negative, for all $i = 1, \dots, n$, and the data set $\{x_i, Y_i\}_{i=1}^n$ must be *cohesive*, i.e. it satisfies that

$$\overline{\text{spr}Y}\sum_{i=1}^n (x_i - \overline{x})^2 \geq \overline{x}\sum_{i=1}^n (x_i - \overline{x})(\text{spr}Y_i - \overline{\text{spr}Y}) \geq 0 \;.$$

This condition restricts the application of the fitting method proposed in [3] in many practical situations (see Rmk. 3).

Analogously to the estimation process developed in [6], some conditions are included in the estimation process in order to obtain solutions being coherent with the interval arithmetic. Namely, the existence of the residuals of the model, which are computed as Hukuhara differences between sample and estimated intervals, is guaranteed. Thus, the LS estimation of the model (4) is solved through the problem

$$\left.\begin{array}{l} \displaystyle\min_{D,C\in\mathcal{K}_c(\mathbb{R})} \frac{1}{n}\sum_{i=1}^n d_\theta^2(Y_i, Dx_i + C) \\ \text{subject to} \\ Y_i -_H Dx_i \ \text{ exists, for all } i = 1,\dots,n \end{array}\right\} \qquad (6)$$

Before solving the minimization problem, the feasible set $S = \{D \in \mathcal{K}_c(\mathbb{R}) : \exists\, Y_i -_H Dx_i, \forall i = 1, \dots, n\}$ has been studied. Since $Y_i -_H Dx_i$ exists if, and only if, $\text{spr}(Dx_i) \leq \text{spr}Y_i$, it is easy to show that the conditions on S only affect the value of $\text{spr}D$. Namely, S can be expressed as

$$S = \{D \in \mathcal{K}_c(\mathbb{R}) : \text{spr}D \in [0, d_0]\}\;, \qquad (7)$$

where $d_0 = \min\left\{\dfrac{\text{spr}Y_i}{|x_i|} : x_i \neq 0\right\}$.

As in the classical linear regression, the estimator of the independent parameter B can be obtained first, in terms of the other estimator $\widehat{A}$ by solving (6) for the unknown quantity C. If $\widehat{A} \in S$, the minimum value is attained in

$$\widehat{B} = \overline{Y} -_H \overline{\widehat{A}\,x}\;. \qquad (8)$$

By substituting C with $\widehat{B}$ in (6), the minimization of the objective function for the unknown quantity D over the feasible set S has been solved, leading to the optimal values for mid$D$ and spr$D$.

As a conclusion, the following analytic expressions for the regression estimators of the linear model (4) are obtained:

$$\widehat{A} = [\text{mid}\widehat{A} \pm \text{spr}\widehat{A}] = \left[\frac{\widehat{\sigma}_{x,\text{mid}Y}}{\widehat{\sigma}_x^2} \pm \min\left\{d_0, \max\left\{0, \frac{\widehat{\sigma}_{|x|,\text{spr}Y}}{\widehat{\sigma}_{|x|}^2}\right\}\right\}\right],$$

$$\widehat{B} = [\text{mid}\widehat{B} \pm \text{spr}\widehat{B}] = \left[\left(\text{mid}\overline{Y} - \text{mid}\widehat{A}\,\overline{x}\right) \pm \left(\text{spr}\overline{Y} - \text{spr}\widehat{A}\,\overline{|x|}\right)\right].$$

4 Empirical Studies

The suitability of the estimation process presented in Sec. 3 is illustrated by means of its application on a real-life example. Moreover, the empirical behaviour of the regression estimators is analized by Monte Carlo method.

Example 1. The paired data in Tbl. 1 correspond to a part of the Retail Trade Sales (in millions of dollars) and the number of employees by kind of business of the U.S. in 2002. The complete data set is available in [4]. The Retail Trade Sales are intervals in the period January 2002 - December 2002 and the number of employees is a real value on each individual. Thus, the characteristics can be modelled by the random interval $Y =$"retail trade sale of a kind of business in 2002" and the real random variable $x =$"number of employees in that kind of business in 2002", respectively. If the aim is to relate linearly the retail trade sales in terms of the number of employees, the model $Y = Ax + \varepsilon$ is defined and the estimation process proposed in Sec. 3 based on the available sample data is applied, leading to the estimated model

$$\widehat{Y} = [0.0168, 0.0193]x + [-1739.35, 393.89]\,. \qquad (9)$$

Table 1 Retail Trade Sales and Number of Employees of kinds of Business in 2002

Kind of Business	Retail Trade Sales	Number of Employees
Automotive parts, acc., and tire stores	4638-5795	453468
Furniture stores	4054-4685	249807
Home furnishings stores	2983-5032	285222
...	...	...

Remark 2. The model in [4] relates an LR fuzzy number in terms of real variables. By considering the interval Y as a particular case of an LR fuzzy number, the model is applicable to Example 1. In this case, separate linear models for the three components characterizing the fuzzy number (midpoint, left and right spreads) are formalized. Moreover, in order to assure the non-negativity condition for the spreads, the corresponding linear equations are defined not for the spreads, but for a transformation of them. Thus, non-linear estimated models for the spreads of the response in terms of the explanatory variables are obtained, but the inverse of the linear estimated models for the considered transformations.

Remark 3. The sample data in Tbl. 1 is *non-cohesive*, so the linear fitting proposed in [3] is not applicable to Example 1.

The estimation of the model is also illustrated in Fig. 1. The sample intervals Y_i are represented as vertical lines, where the crosses are the midpoints, each of them on the corresponding sample value x_i at the x-axis. The continuous line represents the estimated model for mid$Y$ in terms of x and the dashed lines correspond to the estimated models for $\inf Y$ and $\sup Y$, respectively, transferred from the interval model (9).

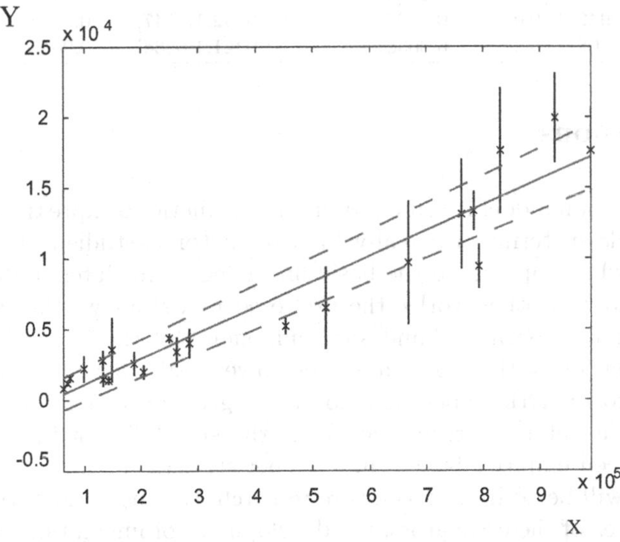

Fig. 1 Sample data and estimated linear relationships in Example 1

The empirical performance of the estimation process of the model (4) is tested by studying the proximity of the estimates to the regression parameters by means of a Monte Carlo simulation. Let $x \sim N(0,1)$ be a real random variable, and let ε be a random interval characterized by mid$\varepsilon \sim N(0,1)$ and

$spr\varepsilon \sim \chi^2_1$ independent from x. Thus, $E(\varepsilon|x) = B = [-1, 1]$. Let us define the random interval Y linearly related with x through the interval model

$$Y = [1, 3]\, x + \varepsilon \,. \tag{10}$$

For different samples sizes n, $k = 10,000$ random samples from (x, Y) verifying (10) have been simulated, computing for each of them the regression estimates of the model, $\widehat{A}$ and $\widehat{B}$. As a description of the performance of the regression estimators, the estimated mean value and the estimated mean squared error of the estimates based on the k iterations have been computed. The results are shown in Tbl. 2. Clearly, the mean values of the estimates are closer to the corresponding regression parameters as the sample size n increases, which shows empirically the asymptotic unbiasedness of the estimators. Moreover, the estimated MSEs go to zero as n increases, which agrees with the empirical consistency of the regression estimators.

Table 2 Empirical validation of the estimation procedure

n	$\widehat{E}(\widehat{A})$	$\widehat{MSE}(\widehat{A})$	$\widehat{E}(\widehat{B})$	$\widehat{MSE}(\widehat{B})$
10	[1.2393,2.7568]	0.2038	[-1.1946,1.1928]	0.2359
50	[1.1337,2.8666]	0.0394	[-1.1042,1.1001]	0.0478
100	[1.0942,2.9048]	0.0194	[-1.0742,1.0747]	0.0234
500	[1.0408,2.9572]	0.0038	[-1.0321,1.0334]	0.0045

5 Conclusions

A linear regression model based on interval arithmetic to express an interval-valued variable in terms of a real-valued predictor is studied in this work. Contrary to what happens to the basic linear model for interval data previously considered in other works, the proposed model allows the real-valued variable to predict both *mid* and *spr* components of the interval response. The LS estimation of the model has been solved, valid for any sample data, and leading to analytic expressions for the regression estimators being coherent with the interval arithmetic. Since the suitability of the estimation process has been empirically shown, a deeper statistical analysis of the proposed model will be addressed as future research; the theoretical study of the main properties of the estimators, the development of inferential studies, the analysis of linear independence, among other studies, could be investigated.

The study in this work could be extended to more complex situations like multiple regression analysis. Moreover, alternative formalizations for a linear model between interval- and real-valued variables could be considered.

Acknowledgements. This research has been partially supported by the Spanish Ministry of Science and Innovation Grant MTM2009-09440-C02-01 and it has been benefited from the COST Action IC0702. All the support is gratefully acknowledged.

References

1. Bertoluzza, C., Corral, N., Salas, A.: On a new class of distances between fuzzy numbers. Mathware Soft. Comp. 2, 71–84 (1995)
2. Blanco-Fernández, A., Colubi, A., González-Rodríguez, G.: Linear Regression Analysis for Interval-Valued Data Based on Set Arithmetic: A Review. In: Borgelt, C., Gil, M.Á., Sousa, J.M.C., Verleysen, M. (eds.) Towards Advanced Data Analysis. STUDFUZZ, vol. 285, pp. 19–31. Springer, Heidelberg (2012)
3. Diamond, P.: Least squares fitting of compact set-valued data. J. Math. Anal. Appl. 147, 531–544 (1990)
4. Ferraro, M.B., Coppi, R., González-Rodríguez, G., Colubi, A.: A linear regression model for imprecise response. Int. J. Approx. Reason. 51, 759–770 (2010)
5. Gil, M.A., Lubiano, A., Montenegro, M., López, M.T.: Least squares fitting of an affine function and strength of association for interval-valued data. Metrika 56, 97–111 (2002)
6. González-Rodríguez, G., Blanco, A., Corral, N., Colubi, A.: Least squares estimation of linear regression models for convex compact random sets. Adv. D Anal. Class. 1, 67–81 (2007)
7. Ham, J., Hsiao, C.: Two-stage estimation of structural labor supply parameters using interval data from the 1971 Canadian Census. J. Econom. 24, 133–158 (1984)
8. Huber, C., Solev, V., Vonta, F.: Interval censored and truncated data: Rate of convergence of NPMLE of the density. J. Stat. Plann. Infer. 139, 1734–1749 (2009)
9. Molchanov, I.: Theory of random sets. Probability and Its Applications. Springer, London (2005)
10. Trutschnig, W., González-Rodríguez, G., Colubi, A., Gil, M.A.: A new family of metrics for compact, convex (fuzzy) sets based on a generalized concept of mid and spread. Inform. Sci. 179, 3964–3972 (2009)

A Proposal of Robust Regression for Random Fuzzy Sets

Maria Brigida Ferraro and Paolo Giordani

Abstract. In standard regression the Least Squares approach may fail to give valid estimates due to the presence of anomalous observations violating the method assumptions. A solution to this problem consists in considering robust variants of the parameter estimates, such as M-, S- and MM-estimators. In this paper, we deal with robustness in the field of regression analysis for imprecise information managed in terms of fuzzy sets. Although several proposals for regression analysis of fuzzy sets can be found in the literature, limited attention has been paid to the management of possible outliers in order to avoid inadequate results. After discussing the concept of outliers for fuzzy sets, a robust regression method is introduced on the basis of one of the proposals available in the literature. The robust regression method is applied to a synthetic data set and a comparison with the non-robust counterpart is given.

Keywords: Imprecise Data, Outliers, Robust Regression.

1 Introduction

Statistical methods usually rely on several assumptions. Unfortunately, in many practical situations data do not fulfill such theoretical assumptions. A common situation is characterized by the presence of atypical observations (outliers) that differ from the main part of the data set. In this case, the performance of standard methods can be very poor. Generally speaking, it is

Maria Brigida Ferraro · Paolo Giordani
Department of Statistical Sciences, Sapienza University of Rome,
00185 Rome, Italy
e-mail: {mariabrigida.ferraro,paolo.giordani}@uniroma1.it

R. Kruse et al. (Eds.): Synergies of Soft Computing and Statistics, AISC 190, pp. 115–123.
springerlink.com © Springer-Verlag Berlin Heidelberg 2013

desirable that statistical methods do not suffer from the presence of outliers. In this connection, the idea of "robust statistics" arises [5, 6, 7].

In this paper, we shall limit our attention to robustness in regression analysis, where the linear relationship between a response variable Y and p explanatory variables, $X_1, X_2, \ldots, X_p$ is studied. The linear regression model is defined as

$$Y = \underline{X}\,\underline{a}' + b + \varepsilon, \tag{1}$$

where $\underline{X} = (X_1, X_2, \ldots, X_p)$ is the row-vector of length p of all the explanatory variables, the error term ε is a random variable with $E(\varepsilon|\underline{X}) = 0$, $\underline{a} = (a^1, a^2, \ldots, a^p)$, is the row-vector of length p of the parameters related to $\underline{X}$ and b is the intercept. Given a sample of size n, $\{Y_i, X_{1i}, X_{2i}, \ldots, X_{pi}\}_{i=1,\ldots,n}$, by means of the Least Squares (LS) approach the aim is to look for $\widehat{\underline{a}}$ and $\widehat{b}$ which minimize

$$\sum_{i=1}^{n} (Y_i - Y_i^*)^2 = \sum_{i=1}^{n} r_i^2, \tag{2}$$

where Y_i, $i = 1, \ldots, n$, is the i-th observed value, $Y_i^* = \underline{X}_i\underline{a}' + b$ is the i-th theoretical value being $\underline{X}_i = (X_{1i}, X_{2i}, \ldots, X_{pi})$. Under the usual assumptions on the errors, the LS estimators of the regression parameters satisfy several desirable statistical properties. Nonetheless, the LS estimators are in general affected by the presence of outliers. Here the concept of outlier refers to observations violating the method assumptions in terms of the response variable (vertical outlier) and/or the explanatory variables (leverage point). The presence of vertical outliers and, especially, leverage points leads to inadequate LS estimates. In fact, LS estimators are strongly affected by anomalous observations because they are characterized by large residuals that noticeably increase the LS loss in (2).

In presence of outliers a robust approach can be considered. The so-called M-estimator is obtained by minimizing the following loss function:

$$\sum_{i=1}^{n} \rho\left(\frac{r_i}{\widehat{\sigma}}\right), \tag{3}$$

where $\widehat{\sigma}$ is a robust scale estimator of the residuals and ρ is a function such that $\rho(x)$ is a non-decreasing function of $|x|$, $\rho(0) = 0$, $\rho(x)$ is increasing for $x > 0$ such that $\rho(x) < \rho(\infty)$ and, if ρ is bounded, it is also assumed that $\rho(\infty) = 1$. An usual choice for ρ is the bisquare family (see, e.g., [5]). In contrast with (2), the loss in (3) is a weighted sum of the residuals where the weights are constructed in such a way to avoid that large residuals dominate the resulting estimators (for more details see [7]). Also note that, differently from (2), in (3) the residuals are normalized by $\widehat{\sigma}$ so that the regression estimators are scale invariant. In order to obtain the minimum of the loss function we have to differentiate it with respect to the regression parameters:

$$\begin{cases} \sum_{i=1}^{n} \psi\left(\frac{r_i}{\widehat{\sigma}}\right) \underline{X}_i = 0, \\ \sum_{i=1}^{n} \psi\left(\frac{r_i}{\widehat{\sigma}}\right) = 0, \end{cases} \tag{4}$$

where $\psi = \rho'$. The above equations could also be written as

$$\begin{cases} \sum_{i=1}^{n} w_i \left(Y_i - \underline{X}_i \underline{a}' - b\right) \underline{X}_i = 0, \\ \sum_{i=1}^{n} w_i \left(Y_i - \underline{X}_i \underline{a}' - b\right) = 0, \end{cases} \tag{5}$$

with $w_i = \psi\left(\frac{r_i}{\widehat{\sigma}}\right) / r_i$. In this way a robustified version of the normal equations is obtained and the solution can be found by Iteratively Reweighted Least Squares (IRWLS). A peculiarity of IRWLS is that the weights of the residuals are updated at every iteration according to the current estimate of the regression parameters.

The robust estimate $\widehat{\sigma}$ in (3) can be found by an M-estimator of scale. In practice, given the constant $\delta \in (0, \rho(\infty))$, this consists in finding $\widehat{\sigma}$ as the solution of

$$\frac{1}{n} \sum_{i=1}^{n} \rho\left(\frac{r_i}{\sigma}\right) = \delta. \tag{6}$$

Such a solution can be found by an iterative procedure starting from an appropriate value σ_0. The so-called S-estimator for the regression coefficients is a particular M-estimator using an M-estimator of scale as described above. Furthermore we can get an MM-estimator according to the following steps. First we find an S-estimator for the regression coefficients and then the M-estimator of scale $\widehat{\sigma}$ as the solution of (6) considering the residuals based on the S-estimator already found. Finally, we perform IRWLS to determine an M-estimator for the regression coefficients using $\widehat{\sigma}$ as robust scale estimator and the previously found S-estimator as the starting point of the algorithm. It can be shown that the performance of the MM-estimator is better than that of M- and S-estimators with respect to both robustness and efficiency [7].

2 Robustness in Case of Imprecise Data

In several occasions, the available information may be affected by imprecision. In these cases it is sensitive to manage such an imprecision in terms of fuzzy sets [10]. In this work, for the sake of simplicity, we limit our attention to the class of symmetric LR fuzzy numbers $\mathcal{F}_S$. A fuzzy number $\widetilde{Z}$ can then be expressed by two parameters, namely the center Z^m and the spread Z^s, and by the following membership function:

$$\mu_{\widetilde{Z}}(z) = \begin{cases} L\left(\frac{Z^m - z}{Z^s}\right) & z \leq Z^m,\, Z^s > 0, \\ 1_{\{Z^m\}}(z) & z \leq Z^m,\, Z^s = 0, \end{cases} \tag{7}$$

where the function L is a particular non-increasing shape function from $\mathbb{R}^+$ to $[0,1]$ such that $L(0) = 1$ and $L(z) = 0, \forall z \in \mathbb{R} \setminus [0,1]$, 1_I is the indicator function of a set I. The membership function value $\mu_{\widetilde{Z}}(z)$ gives the extent to which $z \in U$ (where U denotes a universe of elements) belongs to $\widetilde{Z}$. If $L(z) = 1 - z$, for $0 \leq z \leq 1$, then $\widetilde{Z}$ is a *triangular* fuzzy number. The center of a fuzzy number expresses its location, whereas the spread represents its size. In a fuzzy framework, to assess whether a generic observation is anomalous, one should inspect not only the location, but also the size. It is reasonable to conclude that every observation can be an outlier with respect to its center, its spread or both. In this connection an example is reported in Fig. 1 where the scores of 21 observations on two symmetric fuzzy variables ($\widetilde{Y}$ and $\widetilde{X}$) are displayed. Since a pair of fuzzy variables is involved data are represented as rectangles drawn from the point (X^m, Y^m) and having a width of $2X^s$ and a height of $2Y^s$ (in the plot we did not draw the membership function information). Suppose we are in a fuzzy regression framework aiming at studying the (linear) relationship between a fuzzy-valued response variable ($\widetilde{Y}$) and a set of fuzzy explanatory variables (in this case, one variable, $\widetilde{X}$). In general, this consists in assessing the dependence between response and explanatory

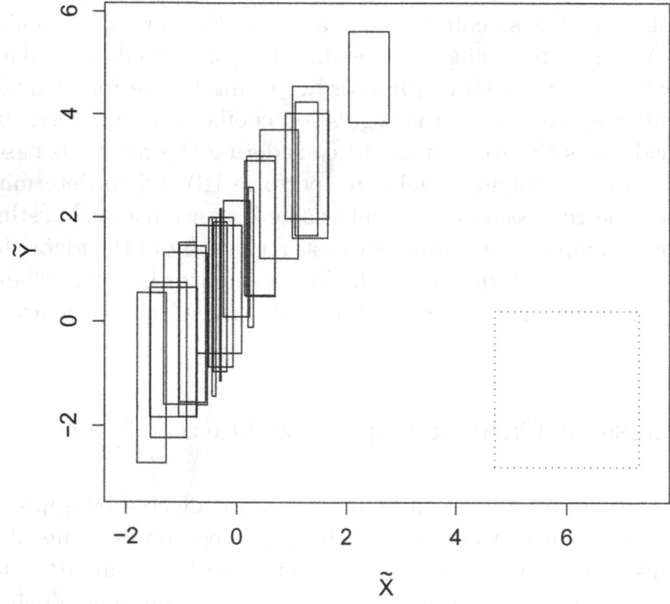

Fig. 1 Fuzzy data set with one leverage point

variables in terms of both the centers and spreads. Although from the figure it is not straightforward to draw conclusions about how, respectively, X^m's and X^s's affect, respectively, Y^s's and Y^m's, some preliminary results can be derived. In particular, twenty regular observations such that a positive relationship between X^m's and Y^m's exists and with X^s values noticeably smaller than the Y^s ones can be found. One observation (dotted rectangle) deviates and can be considered as outlier. It is a leverage point with respect to the location (Y^m consistent with those of the remaining observations and X^m higher than what expected) and the imprecision (high spread of $\widetilde{X}$ in comparison with those of the remaining observations). If we apply standard regression methods for fuzzy data such as the one recalled in the next section, the features of the anomalous observation lead to misleading results (further details shall be given in Sect. 4). In this case a robust approach should be adopted. The circumstance reported in Fig. 1 is not the only one in which the LS approach fails. For instance, we can have leverage points with respect to either the center or the spread of $\widetilde{X}$. With a similar reasoning, we could encounter vertical outliers due to the centers and/or the spreads of $\widetilde{Y}$. For further details on outliers in fuzzy regression analysis see [1].

3 A Robust Regression Method for Imprecise and Random Data

3.1 The Regression Model

Ferraro and Giordani [4] introduce a linear regression model with imprecise and random elements. In particular, suppose that a symmetric LR fuzzy response variable $\widetilde{Y}$ and p symmetric LR fuzzy explanatory variables $\widetilde{X}_1, \widetilde{X}_2, \ldots, \widetilde{X}_p$ are observed on a random sample of n observations, $\{\widetilde{Y}_i, \widetilde{X}_{1i}, \widetilde{X}_{2i}, \ldots, \widetilde{X}_{pi}\}_{i=1,\ldots,n}$. A suitable way to cope simultaneously with imprecision and randomness is given by the concept of Fuzzy Random Variable (FRV). A particular case of FRV is the LR symmetric FRV defined as the random vector $\widetilde{X} : \Omega \rightarrow \mathcal{F}_S$, where $(\Omega, \mathcal{A}, P)$ is a probability space [8]. See also [2, 3] in which a regression model with fuzzy response and non-fuzzy explanatory variables is developed. The shapes of the membership functions are considered fixed, so the fuzzy response and the fuzzy explanatory variables are determined only by means of the centers and the spreads (for instance, the two elements characterize the response, i.e., the center Y^m and the spread Y^s). We introduce an invertible function $g : (0, +\infty) \longrightarrow \mathbb{R}$, in order to avoid the non-negativity constraint of Y^s and make it assuming all the real values. Therefore, every fuzzy datum can also be expressed in terms of its center and its transform of the spread by means of g. The model is then formalized as

$$\begin{cases} Y^m = \underline{X}\,\underline{a}'_m + b_m + \varepsilon_m, \\ g(Y^s) = \underline{X}\,\underline{a}'_s + b_s + \varepsilon_s, \end{cases} \tag{8}$$

where $\underline{X} = (X_1^m, X_1^s, \ldots, X_p^m, X_p^s)$ is the row-vector of length $2p$ of all the components of the explanatory variables, ε_m and ε_s are real-valued random variables with $E(\varepsilon_m|\underline{X}) = E(\varepsilon_s|\underline{X}) = 0$, $\underline{a}_m = (a^1_{mm}, a^1_{ms}, \ldots, a^p_{mm}, a^p_{ms})$ and $\underline{a}_s = (a^1_{sm}, a^1_{ss}, \ldots, a^p_{sm}, a^p_{ss})$ are row-vectors of length $2p$ of the parameters related to $\underline{X}$ for the center model and the spread model, respectively. Finally, b_m and b_s denote the intercepts for the center and spread models, respectively. The idea underlying the model is to explain the center and the spread of $\widetilde{Y}$ by the centers and the spreads of $\widetilde{X}_1, \widetilde{X}_2, \ldots, \widetilde{X}_p$. Therefore, differently from standard regression, we put emphasis on the spread information in order to discover the existing relationship between the response variable and the explanatory ones. For more details refer to [4].

3.2 The Robust Method

In [4] the estimation problem of the regression parameters is faced by means of the LS criterion minimizing the sum of the squared distances between the observed and theoretical values of the fuzzy response variable. A suitable squared distance for fuzzy data has been proposed by Yang and Ko [9]. Since we express fuzzy numbers in terms of the pair (center, g(spread)), a variant of the Yang and Ko distance is considered. Given two symmetric LR fuzzy numbers $\widetilde{A}$ and $\widetilde{B}$, it is

$$D^2_\lambda((A^m, g(A^s)), ((B^m), g(B^s))) = 3(A^m - B^m)^2 + 2\lambda^2(g(A^s) - g(B^s))^2, \tag{9}$$

where $\lambda = \int_0^1 L^{-1}(\omega)d\omega$ is used to take into account the membership function information (in the triangular case $\lambda = \frac{1}{2}$). Accordingly, the parameters of model (8) are estimated by looking for $\widehat{\underline{a}}_m$, $\widehat{\underline{a}}_s$, $\widehat{b}_m$ and $\widehat{b}_s$ which minimize

$$\begin{aligned} \Delta^2_\lambda &= \sum_{i=1}^n D^2_\lambda((Y_i^m, g(Y_i^s)), ((Y_i^m)^*, g(Y_i^s)^*)) \\ &= \sum_{i=1}^n 3(Y_i^m - (Y_i^m)^*)^2 + 2\lambda^2(g(Y_i^s) - g(Y_i^s)^*)^2 \\ &= \sum_{i=1}^n \widetilde{r}_i^2, \end{aligned} \tag{10}$$

where $\underline{Y}^m$ and $g(\underline{Y}^s)$ are the $n \times 1$ vectors of the observed values, $(\underline{Y}^m)^* = \mathbf{X}\underline{a}'_m + \underline{1}b_m$ and $g(\underline{Y}^s)^* = \mathbf{X}\underline{a}'_s + \underline{1}b_s$ are the theoretical ones being $\mathbf{X} = (\underline{X}_1, \underline{X}_2, \ldots, \underline{X}_n)'$ the $n \times 2p$ matrix of the explanatory variables and $\widetilde{r}_i = D_\lambda((Y_i^m, g(Y_i^s)), ((Y_i^m)^*, g(Y_i^s)^*))$ is the residual of the i-th observation. We obtain the following solution:

$$
\begin{aligned}
\widehat{\underline{a}}'_m &= (\mathbf{X}^{c\prime}\mathbf{X}^c)^{-1}\mathbf{X}^{c\prime}\underline{Y}^{mc}, \\
\widehat{\underline{a}}'_s &= (\mathbf{X}^{c\prime}\mathbf{X}^c)^{-1}\mathbf{X}^{c\prime}g(\underline{Y}^s)^c, \\
\widehat{b}_m &= \overline{Y^m} - \overline{\underline{X}}\,\widehat{\underline{a}}'_m, \\
\widehat{b}_s &= \overline{g(Y^s)} - \overline{\underline{X}}\,\widehat{\underline{a}}'_s,
\end{aligned} \tag{11}
$$

where $\underline{Y}^{mc} = \underline{Y}^m - \underline{1}\overline{Y^m}$ and $g(\underline{Y}^s)^c = g(\underline{Y}^s) - \underline{1}\overline{g(Y^s)}$ are the centered values of the response variables and $\mathbf{X}^c = \mathbf{X} - \underline{1}\,\overline{\underline{X}}$ is the centered matrix of the explanatory variables, being $\overline{Y^m}$, $\overline{g(Y^s)}$ and $\overline{\underline{X}}$ the sample means of Y^m, $g(Y^s)$ and $\underline{X}$, respectively. For our purpose, it can be proved that the loss function in (10) can be written in terms of a single squared Euclidean norm $\|\cdot\|^2$. Let $(\mathbf{X}, \underline{1})$ be the matrix obtained juxtaposing $\underline{1}$ next to $\mathbf{X}$ and $\mathbf{0}$ be a matrix of order $(n \times p+1)$ with zero elements. After a little algebra we have

$$
\begin{aligned}
\Delta^2_{\lambda\rho} &= \left\| \begin{pmatrix} 3^{\frac{1}{2}}\underline{Y}^m \\ (2\lambda^2)^{\frac{1}{2}}\underline{Y}^s \end{pmatrix} - \begin{pmatrix} 3^{\frac{1}{2}}(\mathbf{X},\underline{1}) & \mathbf{0} \\ \mathbf{0} & (2\lambda^2)^{\frac{1}{2}}(\mathbf{X},\underline{1}) \end{pmatrix} \begin{pmatrix} \underline{a}_m \\ b_m \\ \underline{a}_s \\ b_s \end{pmatrix} \right\|^2 \\
&= \|\underline{U} - \mathbf{V}\underline{w}\|^2 = \|\underline{U} - \underline{U}^*\|^2 = \sum_{j=1}^{2n} \left(U_j - U_j^*\right)^2 = \sum_{j=1}^{2n} r_j^2,
\end{aligned} \tag{12}
$$

where $\underline{U}$, $\underline{U}^*$, $\mathbf{V}$ and $\underline{w}$ are implicitly defined in (12). We can thus state that, in the symmetric LR case, the minimization problem reduces to a standard regression problem in which the response variable is the non-fuzzy vector $\underline{U}$ and the explanatory variables plus the intercept are stored in the non-fuzzy matrix $\mathbf{V}$. The regression parameters of the models are in the vector $\underline{w}$ of length $2(p+1)$ containing the regression coefficients of model (8). In other words, we suppose to have a sample with $2n$ 'observations', namely n real-valued centers and n real-valued transforms of the spreads. For the generic 'observation' j, we can compute the associate residual r_j according to (12). A high value of r_j denotes an 'observation' (center *or* spread) violating the model assumptions as described in Sect. 2. In this case, the LS solution in (11) can be inadequate and a preferable choice is to obtain M-estimates for the regression parameters. Taking into account (12), robust estimates for the parameters of model (8) can be found by looking for $\widehat{\underline{a}}^R_m$, $\widehat{\underline{a}}^R_s$, $\widehat{b}^R_m$ and $\widehat{b}^R_s$ which minimize

$$
\Psi^2_\lambda = \sum_{j=1}^{2n} \rho\left(\frac{r_j}{\widehat{\sigma}}\right). \tag{13}
$$

The same reasoning adopted for the classical case can be followed and, hence, the M-, S- and MM-estimators of the regression parameters of the fuzzy regression model in (8) can be found.

It is worth noticing that an alternative could be to compute n residuals, one for each fuzzy observation, taking into account the center *and* the spread. However, this choice leads to a loss of information. Suppose for instance that an observation is regular with respect to the center and anomalous with respect to the spread. If so, the regular center plays a less relevant role in comparison with centers belonging to regular observations for both the center and the spread values because the corresponding residual is down-weighted due to the corresponding anomalous spread.

4 An Application

This section is devoted to an application of the fuzzy regression model introduced in the previous section to the data set of Fig. 1. Bearing in mind (8) the regular data were constructed as

$$\begin{cases} Y^m = 1.5X^m + 0.4X^s + 1 + \varepsilon_m, \\ g(Y^s) = log(Y^s) = -0.1X^m - 0.7X^s + 0.5 + \varepsilon_s, \end{cases} \tag{14}$$

where the centers and the spreads of $\widetilde{X}$ were generated randomly from $N(0,1)$ and $U[0,0.5]$, respectively, and both the centers and the spreads of the noise from N(0,0.1). The center of $\widetilde{X}$ for the outlier is equal to 6.0, which is out from the range of values for the regular observations $(-1.5, 2.4)$. Similarly, the spread of $\widetilde{X}$ is 1.3, which is noticeably higher than the spreads of the remaining observations.

If we fit the model to the regular observations, very good results are obtained. According to (11), we get:

$$\begin{cases} \widehat{Y^m} = 1.48X^m + 0.65X^s + 0.94, \\ \widehat{log(Y^s)} = -0.10X^m - 0.63X^s + 0.47, \end{cases} \tag{15}$$

hence the existing relationships are essentially recognized. Problems due to the presence of the outlier arise. In this case we have

$$\begin{cases} \widehat{Y^m} = 0.97X^m - 5.80X^s + 2.47, \\ \widehat{log(Y^s)} = -0.04X^m + 0.11X^s + 0.29. \end{cases} \tag{16}$$

From (16) we can see that the LS estimates are fully inappropriate. To some extent, we could say that the estimates for the spread model are better than the ones for the center model. Thus, the outlier mainly affects the analysis of the relationship between Y^m and $\widetilde{X}$. By considering the robust approach, the MM estimation applied to (13) leads to the following estimated model:

$$\begin{cases} \widehat{Y^{mR}} = 1.48X^m - 0.64X^s + 0.93, \\ \widehat{log(Y^{sR})} = -0.11X^m - 0.65X^s + 0.47, \end{cases} \tag{17}$$

which is a reasonable estimate of the unknown one.

5 Concluding Remarks

In this work, we proposed a robust regression method for fuzzy data exploiting [4] in order to handle situations characterized by the presence of outliers. The obtained results are encouraging and stimulate to further investigate the empirical behavior of the technique by means of real and synthetic data. For instance, it can be interesting to study the sensitivity of the solution with respect to the use of alternative distances.

References

1. D'Urso, P., Massari, R., Santoro, A.: Robust fuzzy regression analysis. Inf. Sci. 181, 4154–4174 (2011)
2. Ferraro, M.B., Coppi, R., González-Rodríguez, G., Colubi, A.: A linear regression model for imprecise response. Int. J. Approx. Reason. 51, 759–770 (2010)
3. Ferraro, M.B., Colubi, A., González-Rodríguez, G., Coppi, R.: A determination coefficient for a linear regression model with imprecise response. Environmetrics 22, 516–529 (2011)
4. Ferraro, M.B., Giordani, P.: A multiple linear regression model for imprecise information. Metrika (in press), doi10.1007/s00184-011-0367-3
5. Filzmoser, P.: Soft Methods in Robust Statistics. In: Borgelt, C., González-Rodríguez, G., Trutschnig, W., Lubiano, M.A., Gil, M.Á., Grzegorzewski, P., Hryniewicz, O. (eds.) Combining Soft Computing & Stats. Methods. AISC, vol. 77, pp. 273–280. Springer, Heidelberg (2010)
6. Huber, P.J., Ronchetti, E.M.: Robust Statistics. Wiley, New York (2009)
7. Maronna, R.A., Martin, D., Yohai, V.J.: Robust Statistics: Theory and Methods. Wiley, New York (2006)
8. Puri, M.L., Ralescu, D.A.: Fuzzy random variables. J. Math. Anal. Appl. 114, 409–422 (1986)
9. Yang, M.S., Ko, C.H.: On a class of fuzzy *c*-numbers clustering procedures for fuzzy data. Fuzzy Set Syst. 84, 49–60 (1996)
10. Zadeh, L.A.: Fuzzy Sets Inf. Control 8, 338–353 (1965)

Bootstrap Comparison of Statistics for Testing the Homoscedasticity of Random Fuzzy Sets

Ana Belén Ramos-Guajardo, María Asunción Lubiano, and Gil González-Rodríguez

Abstract. The problem of testing the equality of variances of k random fuzzy sets has been recently developed on the basis of Levene's classical procedure. Asymptotic and bootstrap approaches have been carried out in this framework, and the proposed test was compared with a Bartlett-type test. In this work, a deeper comparison between some bootstrap statistics based on both Levene's and Bartlett's classical procedures for testing the homoscedasticity of several random fuzzy sets is analyzed. The empirical behaviour of those statistics is investigated by means of simulation studies concerning both type I and type II errors.

Keywords: Bartlett test, bootstrap approach, homoscedasticity, Levene test, random fuzzy sets.

1 Introduction

The concept of *random fuzzy set* (for short RFS) in Puri and Ralescu's sense (see [12]) is an extension of the notion of random set. It was introduced to formalize imprecise experimental data which can be described by means of fuzzy sets.

While the Aumann's mean of an RFS has been introduced as a fuzzy-valued measure to summarize the "central tendency" of the variable (see [1]), the stochastic variability of the fuzzy values of an RFS can be measured by means of the real Fréchet variance inspired on Körner [7] and defined in terms of a generalized metric introduced in [16].

Ana Belén Ramos-Guajardo · María Asunción Lubiano ·
Gil González-Rodríguez
Department of Statistics, Operational Research and Mathematical Didactics, University of Oviedo, Spain
e-mail: {ramosana,lubiano,gil}@uniovi.es

R. Kruse et al. (Eds.): Synergies of Soft Computing and Statistics, AISC 190, pp. 125–133.
springerlink.com © Springer-Verlag Berlin Heidelberg 2013

A procedure for testing the equality of variances (or homoscedasticity) of k populations based on ANOVA techniques has been introduced in Ramos-Guajardo *et al.* (see [14]). The statistic proposed there is a slight modification of the classical Levene's test (see [9]) which entails the application of an ANOVA methodology on the mean-based residuals considering the square of these residuals. Another strategy tackled in the literature (see, for instance, [3, 5, 8, 15]) consists in involving the estimated fourth moment (related with the kurtosis) in a Bartlett-type test.

In this work several statistics based on both Levene's and Bartlett's approaches are empirically compared by applying bootstrap techniques since the theoretical study of most of the proposed statistics is not easy to handle. In addition, the behaviour of the bootstrap approach is analyzed by means of simulation studies.

In Sect. 2 some preliminaries about fuzzy sets and RFS's are gathered. Some Levene's- and Bartlett-based statistics for testing the homoscedasticity of k RFS's are presented in Sect. 3 and the bootstrap approach is introduced in Sect. 4. Some simulation studies are carried out in Sect. 5 in order to compare empirically the behaviour of the proposed statistics. Section 6 closes the paper with some remarks and some current lines of research.

2 Preliminaries

Let $\mathcal{K}_c(\mathbb{R}^p)$ be the family of all non-empty compact convex subsets of $\mathbb{R}^p$ and let $\mathcal{F}_c(\mathbb{R}^p)$ denote the class of fuzzy sets $U : \mathbb{R}^p \to [0,1]$ such that $U_\alpha \in \mathcal{K}_c(\mathbb{R}^p)$ for all $\alpha \in (0,1]$. The α-levels of U are defined as $U_\alpha = \{x \in \mathbb{R}^p | U(x) \geq \alpha\}$ if $\alpha \in (0,1]$.

The studies will be focused here on the space of fuzzy sets

$$\mathcal{F}_c^2(\mathbb{R}^p) = \{U \in \mathcal{F}_c(\mathbb{R}^p) : s_U \in \mathcal{L}^2(\mathbb{S}^{p-1} \times (0,1], \lambda_p \times \lambda)\},$$

where λ_p and λ denote the uniform surface measure on $\mathbb{S}^{p-1}$ and the Lebesgue measure on $(0,1]$, respectively, and $s_U : \mathbb{S}^{p-1} \times (0,1] \to \mathbb{R}$ is the *support function* of $U \in \mathcal{F}_c(\mathbb{R}^p)$ (see, for instance, [11]) which satisfies $s_U(u,\alpha) = \sup_{v \in U_\alpha} \langle u, v\rangle$, where $\mathbb{S}^{p-1}$ is the unit sphere in $\mathbb{R}^p$ (that is, $\mathbb{S}^{p-1} = \{u \in \mathbb{R}^p | |u| = 1\}$).

The usual arithmetic between fuzzy sets in $\mathcal{F}_c^2(\mathbb{R}^p)$ is based on Zadeh's extension principle and it agrees levelwise with the Minkowski addition and the product by a scalar for elements of $\mathcal{K}_c(\mathbb{R}^p)$.

If $\|\cdot\|_2$ denotes the usual functional L_2-norm with respect to the measure λ_p, then the distance between U, $V \in \mathcal{F}_c^2(\mathbb{R}^p)$ (introduced in [16]) is given by

$$(D_\theta^\varphi(U,V))^2 = \int_{(0,1]} (\|\text{mid}_{U_\alpha} - \text{mid}_{V_\alpha}\|_2^2 + \theta\|\text{spr}_{U_\alpha} - \text{spr}_{V_\alpha}\|_2^2)\, d\varphi(\alpha),$$

where $\text{mid}_{U_\alpha}(u) = \big(s_U(u,\alpha) - s_U(-u,\alpha)\big)/2$ is the mid point of the α-level U_α and $\text{spr}_{U_\alpha}(u) = \big(s_U(u,\alpha) + s_U(-u,\alpha)\big)/2$ is its radius (or spread). A more operational expression for the distance is $\big(D_\theta^\varphi(U,V)\big)^2 = \|s_U - s_V\|^2_{\theta,\varphi}$, where $\|\cdot\|_{\theta,\varphi}$ is a norm on the Hilbert space $\mathcal{L}^2(\mathbb{S}^{p-1} \times (0,1])$.

In both expressions of the distance, $\theta > 0$ determines the relative weight of the distance of the spreads against the distance of the mids of the α-levels. In addition, φ is associated with a square integrable bounded density measure and with support $(0,1]$ which allows us to weigh the importance of each α-level.

Let $(\Omega, \mathcal{A}, P)$ be a probability space. A random fuzzy set (also called fuzzy random variable) in Puri and Ralescu's sense (see [12]) can be defined as the element $\mathcal{X}$ satisfying that $s_{\mathcal{X}}$ is a $\mathcal{L}^2(\mathbb{S}^{p-1} \times (0,1])$-valued random element (see [6]). It can also be shown that an RFS is a Borel measurable mapping with respect to D_θ^φ on $\mathcal{F}_c^2(\mathbb{R}^p)$ (see, for instance, [4, 16]).

The *expected value of an RFS* [12] is defined by means of the generalized Aumann integral [1] as the unique fuzzy set $E(\mathcal{X}) \in \mathcal{F}_c^2(\mathbb{R}^p)$ such that $s_{E(\mathcal{X})} = E(s_{\mathcal{X}})$, whenever $E(\|s_{\mathcal{X}}\|_{\theta,\varphi}) < \infty$ is fulfilled and where the last expectation is given in Bochner's or Pettis' sense.

On the other hand, if $E(\|s_{\mathcal{X}}\|^2_{\theta,\varphi}) < \infty$, then the Fréchet variance of an RFS (see, for instance, [7]) is defined as

$$\sigma^2_{\mathcal{X}} = E\big(D_\theta^\varphi(\mathcal{X}, E(\mathcal{X}))\big)^2.$$

3 Statistics for Testing the Homoscedasticity of k RFS's

For the hypothesis testing problem proposed in this work it is useful to define some sample moments of the fuzzy sets. Consider k populations and k independent RFS's, $\mathcal{X}_1, \ldots, \mathcal{X}_k$, associated with them. From each $\mathcal{X}_i$, a simple random sample $\{\mathcal{X}_{i1}, \ldots, \mathcal{X}_{in_i}\}_{j=1}^{n_i}$ is drawn, where the total sample size equals N.

- The sample mean associated with the i-th variable, $\overline{X_{i\cdot}}$, and the total sample mean, $\overline{X_{\cdot\cdot}}$, are defined as usual on the basis of the fuzzy arithmetic.
- The *variance in the i-th sample* is defined as $\widehat{\sigma}^2_{\mathcal{X}_i} = \dfrac{\sum_{j=1}^{n_i} D_\theta^\varphi(\mathcal{X}_{ij}, \overline{\mathcal{X}_{i\cdot}})}{n_i}$.
- The *quasi-variance in the i-th sample* (unbiased and consistent estimate of the population variance as shown in [10]) is $\widehat{S}^2_{\mathcal{X}_i} = n_i \widehat{\sigma}^2_{\mathcal{X}_i}/(n_i - 1)$.
- The *total sample variance* can be expressed as $\widehat{\sigma}^2 = \frac{1}{N}\sum_{i=1}^k n_i \widehat{\sigma}^2_{\mathcal{X}_i}$.

To test the hypotheses

$$\begin{cases} H_0 : \sigma^2_{\mathcal{X}_1} = \ldots = \sigma^2_{\mathcal{X}_k} \quad \text{vs.} \\ H_1 : \exists\, i,j \in \{1,\ldots,k\} \text{ s.t. } \sigma^2_{\mathcal{X}_i} \neq \sigma^2_{\mathcal{X}_j} \end{cases} \tag{3.1}$$

the following statistics are proposed.

Firstly, inspired by the classical Levene's theory [9] the statistic proposed in [14] is considered, namely,

$$T_L^{[1]} = \frac{\displaystyle\sum_{i=1}^{k} n_i \left(\widehat{\sigma}^2_{\mathcal{X}_i} - \frac{1}{N} \sum_{l=1}^{k} n_l \widehat{\sigma}^2_{\mathcal{X}_l} \right)^2}{\displaystyle\sum_{i=1}^{k} \frac{1}{n_i} \sum_{j=1}^{n_i} \left[\left(D_\theta^\varphi \left(\mathcal{X}_{ij}, \overline{\mathcal{X}_{i\cdot}} \right) \right)^2 - \widehat{\sigma}^2_{\mathcal{X}_i} \right]^2}. \tag{3.2}$$

Remark 1. In [14] the asymptotic distribution of the previous statistic is analyzed, which is a combination of normal variables.

In the same way, a statistic inspired on the classical Levene's statistic for RFS's can be established as follows:

$$T_L^{[2]} = \frac{\displaystyle (N-k) \sum_{i=1}^{k} n_i \left(\widehat{\sigma}^2_{\mathcal{X}_i} - \frac{1}{N} \sum_{l=1}^{k} n_l \widehat{\sigma}^2_{\mathcal{X}_l} \right)^2}{\displaystyle (k-1) \sum_{i=1}^{k} \sum_{j=1}^{n_i} \left[\left(D_\theta^\varphi \left(\mathcal{X}_{ij}, \overline{\mathcal{X}_{i\cdot}} \right) \right)^2 - \widehat{\sigma}^2_{\mathcal{X}_i} \right]^2}, \tag{3.3}$$

which coincides with $T_L^{[1]}$ when equal sample sizes are considered.

On the other hand, based on the classical Bartlett's statistic [2] we can defined the following one for RFS's:

$$T_B^{[3]} = (N-k) \ln \left(\frac{1}{N-k} \sum_{i=1}^{k} (n_i - 1) \widehat{S}^2_{\mathcal{X}_i} \right) - \sum_{i=1}^{k} (n_i - 1) \ln \widehat{S}^2_{\mathcal{X}_i}. \tag{3.4}$$

An extension of the Bartlett-type statistic developed by [15] for RFS's is considered. Taking into account Appendix 1 in Shoemaker [15] and Proposition 2 in [13], a Shoemaker-based statistic is proposed to be as follows:

$$T_S^{[4]} = \sum_{i=1}^{k} \left(\frac{n_i \widehat{\sigma}^4_{\mathcal{X}_i} \left(\ln \widehat{S}^2_{\mathcal{X}_i} - \frac{1}{k} \sum_{i=1}^{k} \ln \widehat{S}^2_{\mathcal{X}_i} \right)^2}{\displaystyle \frac{1}{n_i} \sum_{j=1}^{n_i} \left[\left(D_\theta^\varphi \left(\mathcal{X}_{ij}, \overline{\mathcal{X}_{i\cdot}} \right) \right)^2 - \widehat{\sigma}^2_{\mathcal{X}_i} \right]^2 + \frac{2\widehat{\sigma}^4_{\mathcal{X}_i}}{n_i - 1}} \right). \tag{3.5}$$

Finally, a slight modification of $T_S^{[4]}$ is established by considering that $2\sigma^4_{\mathcal{X}_i}/(n_i - 1)$ converges to 0 as $n_i \to \infty$.

$$T_S^{[5]} = \sum_{i=1}^{k} \left(\frac{n_i \widehat{\sigma}^4_{\mathcal{X}_i} \left(\ln \widehat{S}^2_{\mathcal{X}_i} - \frac{1}{k} \sum_{i=1}^{k} \ln \widehat{S}^2_{\mathcal{X}_i} \right)^2}{\frac{1}{n_i} \sum_{j=1}^{n_i} \left[\left(D_\theta^\varphi \left(\mathcal{X}_{ij}, \overline{\mathcal{X}_{i\cdot}} \right) \right)^2 - \widehat{\sigma}^2_{\mathcal{X}_i} \right]^2} \right). \tag{3.6}$$

4 Bootstrap Approach

Due to the difficulties in handling the asymptotic distributions of the previous statistics, bootstrap techniques are applied to approximate those distributions.

Let $\mathcal{X}_1, \ldots, \mathcal{X}_k$ be k independent RFS's and let $\{\mathcal{Y}_i\}_{i=1}^k = \{\widehat{\sigma}^2 \mathcal{X}_i / \widehat{\sigma}^2_{\mathcal{X}_i}\}_{i=1}^k$ be the bootstrap population satisfying $\widehat{\sigma}^2_{\mathcal{Y}_i} = \widehat{\sigma}^2$ for $i \in \{1, \ldots, k\}$, which is the null hypothesis in (3.1). Let $\{\mathcal{X}_{ij}\}_{j=1}^{n_i}$ a simple random sample from $\mathcal{X}_i$ and calculate $\{\mathcal{Y}_{ij}\}_{j=1}^{n_i} = \{\widehat{\sigma}^2 \mathcal{X}_{ij} / \widehat{\sigma}^2_{\mathcal{X}_i}\}_{j=1}^{n_i}$. Let $\{\mathcal{X}^*_{ij}\}_{j=1}^{n_i}$ and $\{\mathcal{Y}^*_{ij}\}_{j=1}^{n_i}$ be bootstrap samples from the previous ones so that $\{\mathcal{Y}^*_{ij}\}_{j=1}^{n_i} = \{\widehat{\sigma}^2 \mathcal{X}^*_{ij} / \widehat{\sigma}^2_{\mathcal{X}_i}\}_{j=1}^{n_i}$ for $i \in \{1, \ldots, k\}$. If $\widehat{\sigma}^2_{\mathcal{X}^*_i} = \sum_{j=1}^{n_i} \left(D_\theta^\varphi (\mathcal{X}^*_{ij}, \overline{\mathcal{X}^*_{i\cdot}}) \right)^2 / n_i$, the corresponding bootstrap statistics are defined as follows:

$$T_L^{[1*]} = \frac{\sum_{i=1}^{k} n_i \left[\frac{\widehat{\sigma}^2_{\mathcal{X}^*_i}}{\widehat{\sigma}^2_{\mathcal{X}_i}} - \frac{1}{N} \sum_{l=1}^{k} n_l \left(\frac{\widehat{\sigma}^2_{\mathcal{X}^*_l}}{\widehat{\sigma}^2_{\mathcal{X}_l}} \right) \right]^2}{\sum_{i=1}^{k} \frac{1}{n_i \widehat{\sigma}^4_{\mathcal{X}_i}} \sum_{j=1}^{n_i} \left((D_\theta^\varphi (\mathcal{X}^*_{ij}, \overline{\mathcal{X}^*_{i\cdot}}))^2 - \widehat{\sigma}^2_{\mathcal{X}^*_i} \right)^2}, \tag{4.7}$$

$$T_L^{[2*]} = \frac{(N-k) \sum_{i=1}^{k} n_i \left[\frac{\widehat{\sigma}^2_{\mathcal{X}^*_i}}{\widehat{\sigma}^2_{\mathcal{X}_i}} - \frac{1}{N} \sum_{l=1}^{k} n_l \left(\frac{\widehat{\sigma}^2_{\mathcal{X}^*_l}}{\widehat{\sigma}^2_{\mathcal{X}_l}} \right) \right]^2}{(k-1) \sum_{i=1}^{k} \frac{1}{\widehat{\sigma}^4_{\mathcal{X}_i}} \sum_{j=1}^{n_i} \left((D_\theta^\varphi (\mathcal{X}^*_{ij}, \overline{\mathcal{X}^*_{i\cdot}}))^2 - \widehat{\sigma}^2_{\mathcal{X}^*_i} \right)^2}, \tag{4.8}$$

$$T_B^{[3*]} = (N-k) \ln \left(\frac{1}{N-k} \sum_{i=1}^{k} (n_i - 1) \frac{\widehat{S}^2_{\mathcal{X}^*_i}}{\widehat{\sigma}^2_{\mathcal{X}_i}} \right) - \sum_{i=1}^{k} (n_i - 1) \ln \left(\frac{\widehat{S}^2_{\mathcal{X}^*_i}}{\widehat{\sigma}^2_{\mathcal{X}_i}} \right), \tag{4.9}$$

$$T_S^{[4*]} = \sum_{i=1}^{k} \left(\frac{n_i \widehat{\sigma}^4_{\mathcal{X}^*_i} \left[\ln \left(\frac{\widehat{S}^2_{\mathcal{X}^*_i}}{\widehat{\sigma}^2_{\mathcal{X}_i}} \right) - \frac{1}{k} \sum_{i=1}^{k} \ln \left(\frac{\widehat{S}^2_{\mathcal{X}^*_i}}{\widehat{\sigma}^2_{\mathcal{X}_i}} \right) \right]^2}{\frac{1}{n_i} \sum_{j=1}^{n_i} \left((D_\theta^\varphi (\mathcal{X}^*_{ij}, \overline{\mathcal{X}^*_{i\cdot}}))^2 - \widehat{\sigma}^2_{\mathcal{X}^*_i} \right)^2 + \frac{2 \widehat{\sigma}^4_{\mathcal{X}^*_i}}{n_i - 1}} \right), \tag{4.10}$$

$$T_S^{[5*]} = \sum_{i=1}^{k} \left(\frac{n_i \widehat{\sigma}^4_{\mathcal{X}^*_i} \left[\ln \left(\frac{\widehat{S}^2_{\mathcal{X}^*_i}}{\widehat{\sigma}^2_{\mathcal{X}_i}} \right) - \frac{1}{k} \sum_{i=1}^{k} \ln \left(\frac{\widehat{S}^2_{\mathcal{X}^*_i}}{\widehat{\sigma}^2_{\mathcal{X}_i}} \right) \right]^2}{\frac{1}{n_i} \sum_{j=1}^{n_i} \left((D_\theta^\varphi (\mathcal{X}^*_{ij}, \overline{\mathcal{X}^*_{i\cdot}}))^2 - \widehat{\sigma}^2_{\mathcal{X}^*_i} \right)^2} \right). \tag{4.11}$$

The Monte Carlo method can be applied to approximate the unknown distributions of the bootstrap statistics as usual. In practice, the bootstrap approach can be applied as follows. Let $\{\mathcal{X}_{ij}\}_{j=1}^{n_i}$ be a realization of a simple random sample from $\mathcal{X}_i$ for each $i \in \{1,\ldots,k\}$.

Step 1. Compute the value of the statistic T.
Step 2. For each $i \in \{1,\ldots,k\}$, obtain a bootstrap sample from $\{\mathcal{X}_{ij}\}_{j=1}^{n_i}$ and compute the value of the bootstrap statistic T^*.
Step 3. Repeat Step 2 a large number B of times to get a set of B values of the bootstrap estimator, denoted by $\{T_1^*,\ldots,T_B^*\}$.
Step 4. Compute the bootstrap p-value as the proportion of values in the set $\{T_1^*,\ldots,T_B^*\}$ which are greater than T.

5 Simulation Studies

Some comparative simulations are developed to illustrate the behaviour of the test by taking into account the previous statistics (although it should be remarked that the methodology proposed in this work can be applied to different real-life situations, as it has been shown, for instance, in [14]). We will consider *trapezoidal fuzzy numbers* which are characterized by four values: the infimum and the supremum of the 1-level (c and d), the left spread (l, the distance between the infima of the 0-level and the 1-level) and the right spread (r, the distance between the suprema of the 0-level and the 1-level). A trapezoidal fuzzy number is denoted by $\Pi(l,c,d,r)$.

Consider three trapezoidal RFS'ss with independent parameters, $\mathcal{X}_i \equiv \Pi(L_i,C_i,D_i,R_i)$ for $i \in \{1,2,3\}$, where C_i, D_i, L_i, R_i are real random variables modeling the corresponding extremes of the 1-level and the spreads, such that:

- $L_1 \equiv \chi_3^2$, $C_1 \equiv U(0,1)$, $D_1 \equiv U(1,2)$ and $R_1 \equiv \chi_8^2$;
- $L_2 \equiv \chi_8^2$, $C_2 \equiv U(-2,-1)$, $D_2 \equiv U(1,2)$ and $R_2 \equiv \chi_3^2$;
- $L_3 \equiv \chi_5^2$, $C_3 \equiv U(1,2)$, $D_3 \equiv U(2,3)$ and $R_3 \equiv \chi_6^2$.

Bootstrap techniques have been applied for testing the equality of variances of $\mathcal{X}_1$, $\mathcal{X}_2$ and $\mathcal{X}_3$ by using all the test statistics proposed in Sec. 3. In addition, the values chosen for θ and φ for computing the sample variances and quasi-variances are $\theta = 1/3$ (which is the weight associated with the Lebesgue measure λ on $[0,1]$ when the equivalence of Bertoluzza's metric and D_θ^φ is considered [16]) and $\varphi = \lambda$ (which assigns all α-levels the same importance).

In this context, 10,000 simulations of the bootstrap tests have been carried out at the significance level $\rho = .05$ and with 1,000 bootstrap replications, which entails a sampling error of 0.427% with a 95% confidence level. The results corresponding to the application of the proposed statistics for equal and different sample sizes are gathered in Table 1.

Table 1 Empirical percentages of rejections under H_0 (Bootstrap approach)

n	$T_L^{[1]} = T_L^{[2]}$	$T_B^{[3]}$	$T_S^{[4]}$	$T_L^{[5]}$	(n_1, n_2, n_3)	$T_L^{[1]}$	$T_L^{[2]}$	$T_B^{[3]}$	$T_S^{[4]}$	$T_L^{[5]}$
10	2.9	10.27	6.79	3.57	(10,15,20)	3.2	3.48	10.64	5.82	4.68
30	3.76	9.74	5.8	5.54	(30,40,30)	3.84	4	9.22	5.44	5.18
50	4.08	7.92	5.19	4.96	(75,55,45)	4.12	4.24	7.46	5.12	5.1
100	4.68	7.28	5.24	5.32	(100,100,150)	5.1	5.04	7.44	5.98	5.96
200	5	6.28	5.2	5.18	(100,180,230)	4.3	4.32	5.90	4.84	4.84

Table 1 shows that in general moderate/large and balanced sample sizes are required in both cases to obtain a percentage of rejections under H_0 close to the nominal significance level. However, the bootstrap version of the Bartlett-based test, $T_B^{[3]}$ shows a bad behaviour since the convergence to the nominal significance level is quite slow in both situations. On the other hand, Levene-based versions, $T_L^{[1]}$ and $T_L^{[2]}$, are more conservative although the second one approximates better the nominal significance level when different sample sizes are considered. It should be remarked that the same occurs in the real framework, since the classical Bartlett test presents a lack of robustness under non-normality while the Levene test behaves better. These differences in robustness would open a new line of research that can be tackled in the future. Finally, Shoemaker-based statistics, $T_S^{[4]}$ and $T_S^{[5]}$, show a good behaviour and the second one approximates the nominal significance level quite good for $n \geq 50$.

Finally, the power function of the homoscedasticity test for RFS's considering the Levene- and Bartlett-based statistics proposed in this work has been empirically analyzed for equal sample sizes ($n = 100$ and $n = 200$) and for $\rho = .05$. In this case, the variable $\mathcal{X}_3$ has been chosen to have a trapezoidal distribution such that $L_3 \equiv \chi_5^2$, $C_3 \equiv U(1,2)$, $D_3 \equiv U(2,3)$ and $R_3 \equiv \chi_{10}^2$. $1,000$ replications of the bootstrap test and $5,000$ simulations have been carried out, involving a sampling error of .604% with a 95% confidence level. The results are gathered in Table 2.

Table 2 shows that the Levene-based test is more powerful in this specific case than the corresponding to the Shoemaker's versions taking into account the considered sampling error. However, the Bartlett-based test is the most

Table 2 Empirical percentages of rejections at level .05 (Power of the bootstrap tests)

n	$T_L^{[1]} = T_L^{[2]}$	$T_B^{[3]}$	$T_S^{[4]}$	$T_S^{[5]}$
100	30.22	34.60	28.46	28.40
200	55.16	56.28	53.58	53.60

powerful although its behaviour under the null hypothesis is not good as it has been shown in Tbl. 1. It could be interesting to develop theoretical results for testing the variance of RFS's based on the classical Bartlett test and its versions.

6 Concluding Remarks

An empirical comparison between several statistics used for testing the equality of variances of k RFS's inspired by some classical tests has been established. Bootstrap techniques have been employed to approximate the distribution of the statistics and some simulations have been carried out which showed their suitability for moderate/large sample sizes. The extension of the Bartlett's and Shoemaker's test for RFS's will be developed theoretically, as well as tests based on other techniques. In this context, a comparison in robustness of the Levene-based and Bartlett-based statistics is required. It could also be interesting to develop a deeper sensitivity analysis taking into account different choices for θ and φ, as well as the shape of the involved distributions and the fuzzy sets chosen for describing the imprecise valuations/perceptions.

Acknowledgements. The research in this paper has been partially supported by the Spanish Ministry of Education and Science Grants MTM2009-09440-C02-01 and MTM2009-09440-C02-02. Their financial support is gratefully acknowledged.

References

1. Aumann, R.J.: Integrals of set-valued functions. J. Math. Anal. Appl. 12, 1–12 (1965)
2. Bartlett, M.S.: Properties of sufficiency and statistical tests. Proc. Roy. Soc. Lond. A 160, 262–282 (1937)
3. Box, G.E.P., Andersen, S.L.: Permutation theory in the derivation of robust criteria and the study of departures from assumptions (with discussion). J. Roy. Statist. Soc. Ser. B 17, 1–34 (1955)
4. Colubi, A., Domínguez-Menchero, J.S., López-Díaz, M., Ralescu, R.: A $D_E[0,1]$ representation of random upper semicontinuous functions. Proc. Amer. Math. Soc. 130, 3237–3242 (2002)
5. Conover, W.J., Johnson, M.E., Johnson, M.M.: A comparative study of tests for homogeneity of variances, with applications to the outer continental shelf bidding data. Technometrics 23, 351–361 (1981)
6. González-Rodríguez, G., Colubi, A., Gil, M.A.: Fuzzy data treated as functional data: a one-way ANOVA test approach. Comput. Statist. Data Anal. 56(4), 943–955 (2012)
7. Körner, R.: On the variance of random fuzzy variables. Fuzzy Set Syst. 92, 83–93 (1997)
8. Layard, M.W.J.: Robust large-sample tests for homogeneity of variances. J. Amer. Stat. Assoc. 68, 195–198 (1973)

9. Levene, H.: Robust Tests for Equality of Variances. In: Contributions to Probability and Statistics, Palo Alto, CA, USA, pp. 278–292. Stanford University Press (1960)
10. Näther, W.: On random fuzzy variables of second order and their application to linear statistical inference with fuzzy data. Metrika 51, 201–221 (2000)
11. Puri, M.L., Ralescu, D.A.: The concept of normality for fuzzy random variables. Ann. Probab. 11, 1373–1379 (1985)
12. Puri, M.L., Ralescu, D.A.: Fuzzy random variables. J. Math. Anal. Appl. 114, 409–422 (1986)
13. Ramos-Guajardo, A.B., Colubi, A., González-Rodríguez, G., Gil, M.A.: One-sample tests for a generalized Fréchet variance of a fuzzy random variable. Metrika 71(2), 185–202 (2010)
14. Ramos-Guajardo, A.B., Lubiano, M.A.: K-sample tests for equality of variances of random fuzzy sets. Comput. Statist. Data Anal. 56(4), 956–966 (2012)
15. Shoemaker, L.H.: Fixing the F test for equal variances. Amer. Statist. 57, 105–114 (2003)
16. Trutschnig, W., González-Rodríguez, G., Colubi, A., Gil, M.A.: A new family of metrics for compact, convex (fuzzy) sets based on a generalized concept of mid and spread. Inf. Sci. 179, 3964–3972 (2009)

Fuzzy Rating *vs.* Fuzzy Conversion Scales: An Empirical Comparison through the MSE

Sara de la Rosa de Sáa, María Ángeles Gil,
María Teresa López García, and María Asunción Lubiano

Abstract. The scale of fuzzy numbers have been used in the literature to measurement of many ratings/perceptions/valuations, expectations, and so on. Among the most common uses one can point out: the so-called 'fuzzy rating', which is based on a free fuzzy numbered response scheme, and the 'fuzzy conversion', which corresponds to the conversion of linguistic (often Likert-type) labels into fuzzy numbers. This paper aims to present an empirical comparison of the two scales. This comparison has been carried out by considering the following steps: fuzzy responses have been first freely simulated; these responses have been 'Likertized' in accordance with a five-point measurement and a plausible criterion; each of the five Likert class has been transformed into a fuzzy number (two fuzzification procedures will be examined); the mean squared error (MSE) has been employed to perform the comparison. On the basis of the simulations we will conclude that for most of the simulated samples the Aumann-type mean is more representative for the fuzzy rating than for the fuzzy conversion scale.

Keywords: Fuzzy conversion scale, fuzzy rating scale, Likert scale, random fuzzy sets.

1 Introduction

In many studies the scale of fuzzy numbers has been applied to express some qualitative elements in human thinking like valuations, ratings, perceptions,

Sara de la Rosa de Sáa · María Ángeles Gil ·
María Teresa López García · María Asunción Lubiano
Departamento de Estadística e I.O. y D.M.,
University of Oviedo, 33007 Oviedo, Spain
e-mail: sara16388@hotmail.com,
{magil,mtlopez,lubiano}@uniovi.es

R. Kruse et al. (Eds.): Synergies of Soft Computing and Statistics, AISC 190, pp. 135–143.
springerlink.com © Springer-Verlag Berlin Heidelberg 2013

judgements, etc. This motivates the interest of such scale in Psychology, Medical Diagnosis, and many other fields.

The most frequent application, the *fuzzy conversion scale*, has consisted of converting each verbal valuation/response within a prefixed list of possible ones (usually a list defined on the basis of a Likert scale) into a fuzzy number by following different conversion/fuzzification procedures.

One of the outstanding reasons supporting such a conversion lies in the fact that fuzzy numbers reflect the inherent imprecision associated with verbal valuations/responses. Furthermore, since fuzzy numbers are real-valued functions one can better handle, explore and exploit mathematically the information contained in them.

Among the many developments involving the fuzzy conversion approach, one can mention the so-called fuzzy SERVQUAL (see, for instance, Aydin and Pakdil [1], Chou *et al.* [6], and Hu *et al.* [14]), or the fuzzification of verbal/linguistic inputs/expressions (see, for instance, Bocklisch *et al.* [4, 5], Herrera [9], Lalla *et al.* [15], and Turksen and Willson [19]). In all these papers one can find several proposals to convert the most usual labels or verbal ratings into fuzzy numbers.

An essentially different approach is the one based on the so-called *fuzzy rating scale* and introduced by Hesketh and Hesketh along with collaborators (cf. [10, 11, 12, 13]). The key difference with the previous approach lies in the whole freedom to describe ratings, valuations, etc. by means of fuzzy numbers. Thus, the rating assessment will not be assumed to be constrained to a prefixed list of potential fuzzy numbered values or responses.

The fuzzy rating scale shares with the fuzzy conversion the ability to reflect the intrinsic imprecision associated with the values/responses. Moreover, new relevant skills for the fuzzy rating scale which are derived from the freedom in valuating should be remarked, namely, the diversity and variability (and hence the subjectivity) of the assessments is definitely much bigger in case we consider the fuzzy rating scale. Consequently, one can better explore and exploit statistically the information contained in them, which could be lost to some extent in case we use the pre-specified list.

In this way, one should emphasize that in spite of the richness of natural languages one could not build a continuous or extremely wide scale by simply using linguistic modifiers and more or less subtle nuances. The potentiality of the whole scale of fuzzy numbers in this respect is really difficult to be improved.

In accordance with practitioners of the fuzzy rating scale, the lack of a statistical methodology to analyze data based on it seems to limit the developments of studies using this approach. In this way, just some few statistical analysis have been carried out, often by considering separate analysis of certain real-valued data characterizing fuzzy numbered data.

The currently available methodology for the statistical analysis of fuzzy data (see, for instance, `http://bellman.ciencias.uniovi.es/SMIRE`), along with methods to be introduced in the future, will allow us to analyze this type of data in a much deeper way. In fact, in such an analysis each fuzzy datum is treated as a whole, instead of as a triple (in the triangular case), as a 4-tuple (in the trapezoidal case), and so on. With the SAFD package (see `http://cran.r-project.org/web/packages/SAFD/index.html`), most of the apparent computational complexities associated with the statistical analysis are overcome, and in case one can consider the fuzzy rating scale the exploited information is definitely richer, diverse and more accurate.

Of course, in many contexts (e.g., when surveys are conducted on the street, by phone, etc.) it would be infeasible to make use of the fuzzy rating scale, and we would be necessarily forced to employ simpler ones like Likert-type or their fuzzy conversion. However, in contexts where the statistical conclusions are quite relevant and it is possible to apply the fuzzy rating scale, we are claiming it is statistically advisable to do it. As an introductory discussion on this matter we will present in this paper an empirical comparison between the two approaches involving the scale of fuzzy numbers, in which the representativeness of the Aumann-type mean of the values/responses is examined by considering a measure of the mean squared error (MSE) based on Bertoluzza *et al.*'s metric between fuzzy numbers [2]. Conclusions are noticeable: the Aumann type mean is much more representative in case we consider the fuzzy rating scale than in case we use the fuzzy conversion.

2 Preliminaries on the Fuzzy Rating and Conversion Scales

The *scale of* (bounded) *fuzzy numbers* is the class $\mathcal{F}_c^*(\mathbb{R})$ of the nonempty compact, convex and normalized fuzzy sets of $\mathbb{R}$, that is, the space of the mappings $\widetilde{U} : \mathbb{R} \to [0,1]$ such that for each $\alpha \in [0,1]$, the α-level $\widetilde{U}_\alpha$ is a nonempty compact interval, where $\widetilde{U}_\alpha = \{x \in \mathbb{R} : \widetilde{U}(x) \geq \alpha\}$ for $\alpha > 0$ and $\widetilde{U}_0 = \text{cl}\{x \in \mathbb{R} : \widetilde{U}(x) > 0\}$.

On one hand, a *fuzzy conversion scale* of verbal values/responses is often either explicitly or implicitly based on the Likert k-point scale, and it consists on converting each of the k points point into a fuzzy number. A conversion scale means a finite subset of $\mathcal{F}_c^*(\mathbb{R})$. Many conversion scales can be found in the literature. Some of them consider special easy-to-draw and easy-to-handle shapes for the fuzzy numbers, like triangular or trapezoidal (see, for instance, [1, 6, 9, 14, 15]). Other ones lie on more sophisticated bases, frequently related to the aggregation of experts' conversions (see, for instance, [4, 5, 19]).

On the other hand, by combining the expertise in computing with that in psychometric evaluation Hesketh and Hesketh have introduced [10, 11, 12, 13] the *fuzzy rating scale* (and its computerized graphic version) as an extension of the semantic differential. This scale provides us with a common method of rating a variety of stimuli and analyzing/comparing the responses meaningfully across the stimuli. This rating consists of choosing in accordance with a free-response scheme the fuzzy number which 'best' represents or describe the response. To ease the posterior analysis, Hesketh and Hesketh have asked the respondents to elicit their responses by using triangular fuzzy numbers for which the "∨" pointer (upper vertex) means the 'preferred point' and the left and right spreads indicate how far to the left or the right a particular rating can be possible/compatible (or, as Hesketh and Hesketh referred to, they determine the tolerable range of preferences).

However, because of the ease and extent of the applicability of the available and ongoing advances in the statistical analysis of fuzzy data (and also of the computational R developments around by Trutschnig *et al.* [18]), there is no need for the values/responses to be triangular. Thus, the preferred point in the fuzzy rating scale can be extended to be the interval of values which are considered to be fully compatible with the respondent rating. From now on, the use of the fuzzy rating scale will be then understood without constraining to triangular fuzzy numbers.

In this paper we aim to show empirically an important statistical advantage of the use of the fuzzy rating scale.

3 Preliminaries on Statistics with Fuzzy Data

Although the fuzzy rating scale has been viewed as a valuable tool, offering clear benefits and odds w.r.t. Likert's or their fuzzy conversion scales, potential users often ignore there is a rather recently introduced statistical methodology which allows us to explore and exploit appropriately the information in fuzzy datasets. This statistical methodology is based on:

- the usual fuzzy arithmetic;
- the choice of a suitable distance between fuzzy numbers (this would be probably enough for a descriptive purpose);
- the well-formalized concept of random fuzzy numbers (or one-dimensional fuzzy random variables in Puri and Ralescu's sense [16]) allowing us to develop inferential procedures.

In conducting statistical analysis of fuzzy data the two basis operations from the *arithmetic of fuzzy numbers* are the sum and the product by scalars, both based on Zadeh's extension principle [20] which can be equivalently stated so that for $\widetilde{U}, \widetilde{V} \in \mathcal{F}_c^*(\mathbb{R})$, $\gamma \in \mathbb{R}$, and whatever the level $\alpha \in [0,1]$ may be,

$$(\widetilde{U} + \widetilde{V})_\alpha = \left[\inf \widetilde{U}_\alpha + \inf \widetilde{V}_\alpha, \sup \widetilde{U}_\alpha + \sup \widetilde{V}_\alpha\right],$$

$$(\gamma \cdot \widetilde{U})_\alpha = \begin{cases} \left[\gamma \cdot \inf \widetilde{U}_\alpha, \gamma \cdot \sup \widetilde{U}_\alpha\right] & \text{if } \gamma \geq 0 \\ \left[\gamma \cdot \sup \widetilde{U}_\alpha, \gamma \cdot \inf \widetilde{U}_\alpha\right] & \text{if } \gamma < 0 \end{cases}$$

It is well-known that when $\mathcal{F}_c^*(\mathbb{R})$ is endowed with this arithmetic we don't have a linear but a semilinear space, so one cannot establish a convenient definition for the difference between fuzzy numbers.

This drawback has been mostly overcome by using a suitable *metric between fuzzy numbers*. In this paper we will consider the L^2 metric introduced by Bertoluzza *et al.* [2]. For $\widetilde{U}, \widetilde{V} \in \mathcal{F}_c^*(\mathbb{R})$ this metric is given by

$$D(\widetilde{U}, \widetilde{V}) = \int_{[0,1]} \int_{[0,1]} \left(\widetilde{U}_\alpha^{[\lambda]} - \widetilde{V}_\alpha^{[\lambda]} \right)^2 d\lambda \, d\alpha,$$

where $\widetilde{U}_\alpha^{[\lambda]} = \lambda \sup \widetilde{U}_\alpha + (1-\lambda) \inf \widetilde{U}_\alpha$.

The probabilistic model we consider to formalize the random mechanism generating fuzzy numbered values/responses is stated as follows: given a probability space $(\Omega, \mathcal{A}, P)$, a mapping $\mathcal{X} : \Omega \to \mathcal{F}_c^*(\mathbb{R})$ is said to be a *random fuzzy number* (for short RFN) if for all $\alpha \in [0,1]$ the mapping $\mathcal{X}_\alpha : \Omega \to \mathcal{P}(\mathbb{R})$ (with $\mathcal{X}_\alpha(\omega) = \left(\mathcal{X}(\omega)\right)_\alpha$) is a compact random interval (i.e., a Borel-measurable mapping w.r.t. the Borel σ-field generated on $\mathcal{F}_c^*(\mathbb{R})$ by the topology associated with D). The Borel-measurability implies that one can properly refer to the distribution induced by an RFN, the statistical independence of RFNs, and so on. The *Aumann-type mean of an RFN* has been defined by Puri and Ralescu [16] as the fuzzy number $\widetilde{E}(\mathcal{X}) \in \mathcal{F}_c^*(\mathbb{R})$ such that for all $\alpha \in [0,1]$

$$\left(\widetilde{E}(\mathcal{X})\right)_\alpha = [E(\inf \mathcal{X}_\alpha), E(\sup \mathcal{X}_\alpha)] \,.$$

$\widetilde{E}$ preserves all the main properties of the mean of a random variable. Several statistical developments, especially inferential ones, about the Aumann-type means of RFNs have been already developed (see, for instance, [3]).

4 Empirical Comparative Study between Fuzzy Rating and Fuzzy Conversion Scales: The Mean Squared Error Associated with the Aumann-Type Mean

To compare fuzzy rating and conversion scales from a statistical perspective, we mimic the situation in which a person is simultaneously allowed to give a free response in the scale of fuzzy numbers (fuzzy rating), and to classify it in accordance with a five-point Likert scale. Later, the Likert labels are converted into fuzzy ones by using one of the two fuzzy conversion scales in Figure 1. The one on the left side (I) corresponds to a strong fuzzy triangular partition of the interval [1,5] which is quite frequently considered in the literature (cf. [7, 9]). The one on the right side (II) is an example of a conversion in a fuzzy SERVQUAL study (see [1]).

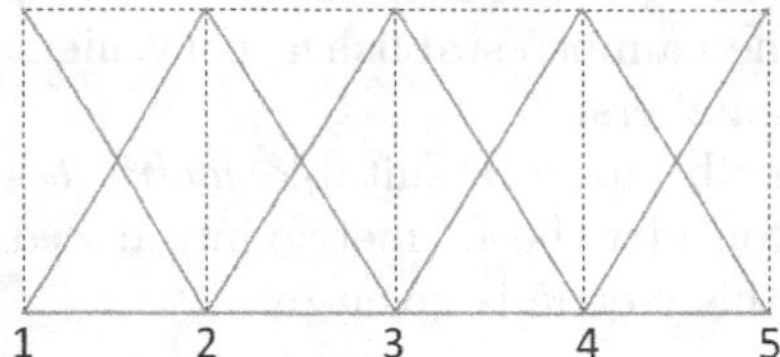

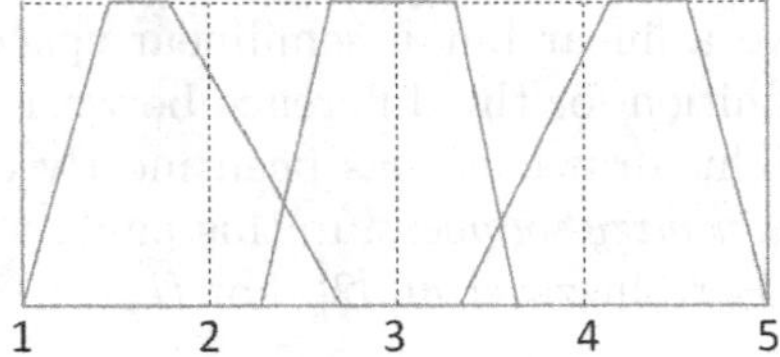

Fig. 1 Examples of fuzzy conversion scales of 5-point Likert labels (Scale I on the left, and Scale II on the right).

The general simulation process is structured as follows: 1000 iterations of samples containing n trapezoidal fuzzy numbers are simulated ($n \in \{10, 30, 50, 100, 300\}$). To generate each trapezoidal fuzzy response, we have followed the steps in De la Rosa and Van Aelst [8], and Sinova *et al.* [17]. In this way,

- One value of the nonstandard (i.e., re-scaled and translated standard) beta distribution $4 \cdot \beta(p, q) + 1$ is generated at random, with (p, q) varying to cover six different situations of distributions with values in $[1, 5]$, namely, uniform, symmetrical weighting extreme values, symmetrical weighting central values, and three asymmetric ones. The generated value is the centre of the 1-level, mid $\widetilde{U}_1$.
- To avoid unrealistic fuzzy values/responses, some constraints on the values have been imposed,by following ideas in [8]; the trapezoidal fuzzy number is finally built from the generated mid-point and deviations.

Once the fuzzy responses are generated, they are 'Likertized' so that if $\widetilde{U}$ is the generated fuzzy number, the criterion associates with it the integer number

$$\imath(\widetilde{U}) = \arg \min_{j \in \{1, \ldots, 5\}} D(\widetilde{U}, \mathbf{1}_{\{j\}}).$$

The comparative analysis has been based on examining the representativeness of the Aumann type mean, the D-mean squared error (MSE) being considered to quantify this representativeness in the fuzzy case. It should be pointed out that the D-MSE minimizes at the Aumann type mean so, as for the real-valued case, the D-MSE w.r.t. the Aumann type mean provides clues as to how representative this mean is of the individual fuzzy data/values. Thus, for each sample we have computed:

- the FRMSE, where if $\widetilde{x}_1, \ldots, \widetilde{x}_n$ are the values of $\mathcal{X}$ in the sample, then

$$\text{FRMSE(sample)} = \frac{1}{n} \sum_{i=1}^{n} \left[D(\widetilde{x}_i, \overline{\widetilde{x}}) \right], \quad \overline{\widetilde{x}} = \frac{1}{n} \cdot (\widetilde{x}_1 + \ldots + \widetilde{x}_n);$$

- the LMSE, which is the MSE of the Likertized sample;

- the FCIMSE, which corresponds to the sample D-MSE for the fuzzy conversion with the scale I of the Likertized sample;
- the FCIIMSE, which corresponds to the sample D-MSE for the fuzzy conversion with the scale II of the Likertized sample.

Along the 1000 simulated samples we have computed the percentage of samples for which the FRMSE was lower than the other ones, which have been gathered in the following table:

Table 1 Percentages of simulated samples for which the use of the fuzzy rating scale produces an MSE lower than that of the Likert and two fuzzy converted scales mid $\widetilde{U}_1 \sim 4\,\beta(p,q)+1$

(p,q)	n	% FRMSE < LMSE	% FRMSE < FCIMSE	% FRMSE < FCIIMSE
$(p,q)=(1,1)$	10	72.9	74.2	83.1
	30	86.9	90	95.5
	50	93.4	95.3	98.7
	100	97.3	99.1	99.9
	300	100	100	100
$(p,q)=(.75,.75)$	10	78.5	72.6	84
	30	93.4	84.7	96.3
	50	96.9	92.6	99.2
	100	100	97.5	100
	300	100	100	100
$(p,q)=(2,2)$	10	74.7	72.5	82.7
	30	89.2	86	95.2
	50	93.6	94.1	98.4
	100	99.6	98.9	99.9
	300	100	100	100
$(p,q)=(4,2)$	10	67.9	78.5	84.4
	30	79.1	90.9	96
	50	87.5	96	99.6
	100	95.4	99.5	99.9
	300	99.7	100	100
$(p,q)=(6,1)$	10	97.1	72.8	83.2
	30	100	86.4	96
	50	100	90.7	99.1
	100	100	98	100
	300	100	100	100
$(p,q)=(6,10)$	10	82.4	85.9	93.5
	30	93	96.9	99.7
	50	98.2	99	100
	100	99.9	99.9	100
	300	100	100	100

Empirical conclusions from the results in Table 1 are very clear: the mean value is mostly better represented by using the fuzzy rating than by using Likert or fuzzy conversion scales.

Acknowledgements. The authors are grateful to Professor Thierry Denoeux for having suggested the comparative study they have introduced in this paper, as well as to Professors Gil González Rodríguez and José Muñiz for their insightful comments and discussions. The research in this paper has been partially supported by/benefited from the Spanish Ministry of Science and Innovation Grant MTM2009-09440-C02-01, and the COST Action IC0702. De la Rosa de Sáa would like also thank the financial coverage of her Short Term Scientific Missions in Gent University through the COST Action, as well as the Contract CP-PA-11-SMIRE from the Principality of Asturias-Universidad de Oviedo. Their financial support is gratefully acknowledged.

References

1. Aydin, O., Pakdil, F.: Fuzzy SERVQUAL Analysis in Airline Services. Organizacija 41, 108–115 (2008)
2. Bertoluzza, C., Corral, N., Salas, A.: On a new class of distances between fuzzy numbers. Math. Soft. Comp. 2, 71–84 (1995)
3. Blanco-Fernández, A., Casals, M.R., Colubi, A., Corral, N., García-Bárzana, M., Gil, M.A., González-Rodríguez, G., López, M.T., Lubiano, M.A., Montenegro, M., Ramos-Guajardo, A.B., de la Rosa de Sáa, S., Sinova, B.: Random fuzzy sets: a mathematical tool to develop statistical fuzzy data analysis. Iran. J. Fuzzy Syst. (in press)
4. Bocklisch, F.A., Bocklisch, S.F., Krems, J.F.: How to Translate Words into Numbers? A Fuzzy Approach for the Numerical Translation of Verbal Probabilities. In: Hüllermeier, E., Kruse, R., Hoffmann, F. (eds.) IPMU 2010. LNCS, vol. 6178, pp. 614–623. Springer, Heidelberg (2010)
5. Bocklisch, F.A.: The vagueness of verbal probability and frequency expressions. Int. J. Adv. Comp. Sci. 1, 52–57 (2011)
6. Chou, C.C., Liu, L.J., Huang, S.F., Yih, J.M., Han, T.C.: An evaluation of airline service quality using the fuzzy weighted SERVQUAL method. Appl. Soft. Comp. 11, 2117–2128 (2011)
7. Colubi, A., González-Rodríguez, G.: Triangular fuzzification of random variables and power of distribution tests: Empirical discussion. Comp. Stat. Data Anal. 51, 4742–4750 (2007)
8. de la Rosa de Sáa, S., van Aelst, S.: Comparing the Representativeness of the 1-Norm Median for Likert and Free-Response Fuzzy Scales. In: Borgelt, C., Gil, M.Á., Sousa, J.M.C., Verleysen, M. (eds.) Towards Advanced Data Analysis. STUDFUZZ, vol. 285, pp. 87–98. Springer, Heidelberg (2012)
9. Herrera, F.: Genetic fuzzy systems: taxonomy, current research trends and prospects. Evol. Intel. 1, 27–46 (2008)
10. Hesketh, B., Griffin, B., Loh, V.: A future-oriented retirement transition adjustment framework. J. Vocat. Behav. 79, 303–314 (2011)
11. Hesketh, B., Hesketh, T., Hansen, J.-I., Goranson, D.: Use of fuzzy variables in developing new scales from strong interest inventory. J. Couns. Psych. 42, 85–99 (1995)

12. Hesketh, T., Hesketh, B.: Computerised fuzzy ratings: the concept of a fuzzy class. Behav. Res. Meth. Inst. Comp. 26, 272–277 (1994)
13. Hesketh, T., Pryor, R.G.L., Hesketh, B.: An application of a computerised fuzzy graphic rating scale to the psychological measurement of individual differences. Int. J. Man Mach. Stud. 29, 21–35 (1988)
14. Hu, H.-Y., Lee, Y.-C., Yen, T.-M.: Service quality gaps analysis based on fuzzy linguistic SERVQUAL with a case study in hospital out-patient services. The TQM J. 22, 499–515 (2010)
15. Lalla, M., Facchinetti, G., Mastroleo, G.: Vagueness evaluation of the crisp output in a fuzzy inference system. Fuzzy Set. Syst. 159, 3297–3312 (2008)
16. Puri, M.L., Ralescu, D.A.: Fuzzy random variables. J. Math. Anal. Appl. 114, 409–422 (1986)
17. Sinova, B., de la Rosa de Sáa, S., Gil, M.A.: A generalized L^1-type metric between fuzzy numbers for an approach to central tendency of fuzzy data (submitted)
18. Trutschnig, W., Lubiano, M.A., Lastra, J.: SAFD—An R Package for Statistical Analysis of Fuzzy Data. In: Borgelt, C., Gil, M.Á., Sousa, J.M.C., Verleysen, M. (eds.) Towards Advanced Data Analysis. STUDFUZZ, vol. 285, pp. 107–118. Springer, Heidelberg (2012)
19. Turksen, I.B., Willson, I.A.: A fuzzy set preference model for consumer choice. Fuzzy Set. Syst. 68, 253–266 (1994)
20. Zadeh, L.A.: The concept of a linguistic variable and its application to approximate reasoning. Part 1. Inform. Sci. 8, 199–249; Part 2. Inform. Sci. 8, 301–353; Part 3. Inform. Sci. 9, 43–80 (1975)

A Law of Large Numbers for Exchangeable Random Variables on Nonadditive Measures

Li Guan and Shoumei Li

Abstract. In this paper, we use the relationship between set-valued random variables and capacity to prove a strong law of large numbers for exchangeable random variables with respect to nonadditive measures.

Keywords: Law of large numbers, nonadditive measure, set-valued random variables.

1 Introduction

The classical laws of large numbers (LLN) play an important role in probability theory and its applications. All the classical results are under the assumption that the measures are additive. But the additive property for measures is not always satisfied because of uncertainty. In recent years, more and more people are interested in nonadditive measures. The pioneer work is Choquet's work [5] in 1953, where nonadditive measures are called capacities.

In 1989 [20], using nonadditive measures and Choquet's expectation utility, Schmeidler successfully explained the Allais paradox and Ellsberg paradox. Maccheroni and Marinacci [16] proved a strong law of large numbers(SLLN) for capacities by using the result on set-valued random variables. In [16] the random variables are assumed to be bounded, pairwise independent and identically distributed. In [4], Chen and Wu proved some strong laws of large numbers for Bernoulli experiments on an upper probability space. In [19], Rebille proved some laws of large numbers for nonadditive measures under weak negatively dependent conditions.

Li Guan · Shoumei Li
College of Applied Sciences, Beijing University of Technology,
100 Pingleyuan, Chaoyang District, Beijing, 100124, China
e-mail: {guanli,lisma}@bjut.edu.cn

R. Kruse et al. (Eds.): Synergies of Soft Computing and Statistics, AISC 190, pp. 145–152.
springerlink.com © Springer-Verlag Berlin Heidelberg 2013

It is not always reasonable to assume the random variables are independent. However it can usually be assumed to be permutation invariant with respect to distributions. This concept of permutation invariance with respect to distributions is formally called *exchangeable*. Exchangeable random variables are widely used in statistics as pointed out by Inoue in [13] and [12]. Thus it is important to study strong law of large numbers for exchangeable random variables. For convergence theorems of set-valued random variables, Taylor *et al.* [21] obtained some strong laws of large numbers by using a measure of non-orthogonality provided by de Finetti's theorem for arrays where each row consists of an infinite sequence of exchangeable random variables. Also, Patterson and Taylor [18] used reverse martingale techniques to prove the same laws of large numbers. In [13] and [12], some strong laws of large numbers were obtained for weighted sums of exchangeable set-valued and fuzzy set-valued random variables. In this paper, we are concerned with the laws of large numbers for rowwise exchangeable random variables under nonadditive measures. We will prove a SLLN for rowwise exchangeable random variables by using the method of [16].

This paper is organized as follows. In Section 2, we briefly recall some concepts and notations on set-valued random variables. In Section 3, we summarize the relationship of set-valued random variable and capacity. In Section 4, we prove the SLLN of rowwise exchangeable random variables under nonadditive measures.

2 Preliminaries on Set-Valued Random Variables

Throughout this paper, we assume that $(\Omega, \mathcal{A}, \mu)$ is a nonatomic complete probability space, $(\mathfrak{X}, \|\cdot\|)$ is a real separable Banach space, $\mathbb{R}$ is the of all real numbers, $\mathbb{N}$ is the set of natural numbers, $\mathbf{K}_k(\mathfrak{X})$ is the family of all nonempty compact subsets of $\mathfrak{X}$, and $\mathbf{K}_{kc}(\mathfrak{X})$ is the family of all nonempty compact convex subsets of $\mathfrak{X}$.

For $A, B \subset \mathfrak{X}$ and $\lambda \in \mathbb{R}$, we define

$$A + B = \{a + b : a \in A, b \in B\},$$

$$\lambda A = \{\lambda a : a \in A\}.$$

The Hausdorff metric on $\mathbf{K}_k(\mathfrak{X})$ is defined by

$$d_H(A, B) = \max\{\sup_{a\in A} \inf_{b\in B} \|a - b\|, \sup_{b\in B} \inf_{a\in A} \|a - b\|\}, \qquad A,\ B \in \mathbf{K}_k(\mathfrak{X}).$$

For $A \in \mathbf{K}_k(\mathfrak{X})$, we set $\|A\|_{\mathbf{K}} = d_H(\{0\}, A)$. The metric space $(\mathbf{K}_k(\mathfrak{X}), d_H)$ is complete and separable, and $\mathbf{K}_{kc}(\mathfrak{X})$ is a closed subset of $(\mathbf{K}_k(\mathfrak{X}), d_H)$ (cf. [15], Theorems 1.1.2 and 1.1.3).

For any $A \in \mathbf{K}_{kc}(\mathfrak{X})$, the support function of A is defined as

$$s(x^*, A) = \sup_{a \in A} < x^*, a >, \quad x^* \in \mathfrak{X}^*,$$

where $\mathfrak{X}^*$ is the dual space of $\mathfrak{X}$.

A set-valued mapping $F : \Omega \to \mathbf{K}_k(\mathfrak{X})$ is called *a set-valued random variable (or a random set, or a multifunction)* if, for each open subset O of $\mathfrak{X}$, $F^{-1}(O) = \{\omega \in \Omega : F(\omega) \cap O \neq \emptyset\} \in \mathcal{A}$.

In fact, set-valued random variables can be defined as a mapping from Ω to the family of all closed subsets of $\mathfrak{X}$. Since our main results only deal with compact set-valued random variables, we limit the above definition to the compact case. For the definition of general set-valued random variables and equivalent characterizations, please refer to [2], [10] and [15].

A set-valued random variable F is called *integrably bounded* (cf. [10] or [15]) if $\int_\Omega \|F(\omega)\|_{\mathbf{K}} d\mu < \infty$.

Let $L^1[\Omega, \mathcal{A}, \mu; \mathbf{K}_k(\mathfrak{X})]$ denote the space of all integrably bounded random variables, and $L^1[\Omega, \mathcal{A}, \mu; \mathbf{K}_{kc}(\mathfrak{X})]$ denote the space of all integraly bounded random variables taking values in $\mathbf{K}_{kc}(\mathfrak{X})$. For $F, G \in L^1[\Omega, \mathcal{A}, \mu; \mathbf{K}_k(\mathfrak{X})]$, $F = G$ if and only if $F(\omega) = G(\omega)$ *a.e.*(μ).

For each set-valued random variable F, *the expectation of* F, denoted by $E[F]$, is defined as

$$E[F] = \Big\{ \int_\Omega f d\mu : f \in S_F \Big\},$$

where $\int_\Omega f d\mu$ is the usual Bochner integral in $L^1[\Omega, \mathfrak{X}]$, the family of integrable $\mathfrak{X}$-valued random variables, and $S_F = \{f \in L^1[\Omega; \mathfrak{X}] : f(\omega) \in F(\omega), a.e.(\mu)\}$. This integral was first introduced by Aumann [1], called Aumann integral in literature.

3 Relationship between Set-Valued Random Variables and Capacity

First we recall the definition of the capacity. Let $\mathcal{B}(\Omega)$ be the family of all the Borel sets of Ω.

Definition 1. We call $\nu : \mathcal{B}(\Omega) \to [0, 1]$ a capacity, if it satisfies the following conditions:

(1)$\nu(\emptyset) = 0, \ \nu(\Omega) = 1$;

(2) For any $A \subseteq B$ and $A, B \in \mathcal{B}(\Omega)$, we have $\nu(A) \le \nu(B)$.

Definition 2. We call a capacity ν totally monotone , if

$$\nu(\bigcup_{i=1}^n A_i) \ge \sum_{\emptyset \neq I \subseteq \{1,\cdots,n\}} (-1)^{|I|+1} \nu(\bigcap_{i \in I} A_i), \quad \forall n \ge 2, \ \forall \{A_1, \cdots, A_n\} \subseteq \mathcal{B}(\Omega).$$

Definition 3. We call a capacity ν infinitely alternating, if

$$\nu(\bigcap_{i=1}^{n} A_i) \leq \sum_{\emptyset \neq I \subseteq \{1,\cdots,n\}} (-1)^{|I|+1} \nu(\bigcup_{i\in I} A_i), \quad \forall n \geq 2,\ \forall \{A_1, \cdots, A_n\} \subseteq \mathcal{B}(\Omega).$$

For any set-valued random variable F and $A \in \mathcal{B}(\Omega)$, we define

$$F^{-1}(A) = \{\omega \in \Omega : F(\omega) \cap A \neq \emptyset\},$$

$$F_{-1}(A) = \{\omega \in \Omega : F(\omega) \subset A\}.$$

Obviously, we have $F^{-1}(A) = \Omega - F_{-1}(A)$.

Furthermore, for $A \in \mathcal{B}(\Omega)$, we define the following two capacities,

$$\underline{\nu}_F(A) = \mu(F_{-1}(A)),$$

$$\overline{\nu}_F(A) = \mu(F^{-1}(A)).$$

Here we call $\underline{\nu}_F(A)$ and $\overline{\nu}_F(A)$ the upper distribution and lower distribution of F respectively. And we obviously have the following dual relation

$$\overline{\nu}_F(A) = 1 - \underline{\nu}_F(A^c), \quad A \in \mathcal{B}(\Omega).$$

$\underline{\nu}_F(A)$ and $\overline{\nu}_F(A)$ are also belief function and plausible function respectively.

For any $\mathfrak{X}$-valued random variable x, the Choquet integral of a bounded random variable x with respect to a totally monotone capacity ν is defined as

$$\int x d\nu = \int_0^{+\infty} \nu\{x > t\} dt + \int_{-\infty}^{0} [\nu\{x > t\} - 1] dt.$$

The Choquet integral is positively homogeneous, monotone and translation invariant(i.e.,$\int x + c d\nu = \int x d\nu + c$ if c is constant). It reduces to the standard integral when ν is an additive probability measure.

Theorem 1. *(cf.[3]) Let $(\Omega, \mathcal{A}, \mu)$ be a probability space, $\mathfrak{X}$ a Polish space, and $F : [0,1] \longrightarrow \mathbf{K}_k(\Omega)$ a set-valued random variable. Then for each $\mathfrak{X}$-valued random variable $x : \Omega \to \mathfrak{X}$,*

$$\int x \circ F d\mu = \Big\{ \int x \circ f d\mu : f \in S_F \Big\}.$$

If μ is nonatomic, the integral $\int x \circ F d\mu$ is convex (cf.[15]), and then we have

$$\int x \circ F d\mu = \Big\{ \int x d\mu_f : f \in S_F \Big\}.$$

For any measurable selection $f \in S_F$, there is an induced measure μ_f such that $\mu_f(A) = \mu\{f \in A\}$ for any Borel subsets A of $\mathfrak{X}$. Since $\underline{\nu}_F \leq \mu_f \leq \overline{\nu}_F$ for any $f \in S_F$. Then when x is both lower and upper Weierstrass, we have

$$\int x \circ F dP = \Big\{ \int x dP_f : f \in S_F \Big\} = \Big[(C) \int x d_{\underline{\nu}_F}, (C) \int x d_{\overline{\nu}_F} \Big].$$

The following lemma is from [3], [16].

Theorem 2. *A set function $\nu : \mathcal{B}(\Omega) \to [0,1]$ is a totally monotone capacity if and only if there exists a set-valued random variable $F : [0,1] \to \mathbf{K}_k(\Omega)$ such that $\nu = \underline{\mu}_F$.*

4 Main Results

In this section, we shall give a strong law of large numbers with respect to nonadditive measures. First we give the definition of exchangeable random variables with respect to nonadditive measures.

Definition 4. The family of $\mathfrak{X}$-valued random variables $\{x_i : 1 \leq i \leq n\}$ is called exchangeable with respect to a nonadditive measure ν, if for any permutation $\pi = (\pi_1, \pi_2, \cdots, \pi_n)$ of $(1, 2, ..., n)$, and any Borel sets $B_1, \cdots, B_n$ of $\mathfrak{X}$, we have

$$\nu\{x_1 \in B_1, \cdots, x_n \in B_n\} = \nu\{x_{\pi_1} \in B_1, x_{\pi_2} \in B_2, \cdots, x_{\pi_n} \in B_n\}.$$

The array $\{x_{nk} : n \geq 1, 1 \leq k \leq n\}$ is called rowwise exchangeable random variables with respect to a nonadditive measure ν, if for any fixed n, $\{x_{nk} : 1 \leq k \leq n\}$ is exchangeable.

The next theorem is the strong law of large numbers for rowwise exchangeable set-valued random variables in the sense of d_H (cf. [11],[8]), which will be used later for the proof of our main result .

Theorem 3. *Let $\{F_{ni} : 1 \leq i \leq n\}$ be a rowwise exchangeable array of set-valued random variables taking values in $\mathbf{K}_{kc}(\mathfrak{X})$, and $\|F_{ni}\|_{\mathbf{K}}$ be bounded by $h(\omega) \in L^1[\Omega, \mathfrak{X}]$. Let $\{A_n\}$ and $\{a_n\}$ be random variables where a_n is a symmetric function of $(F_{n1}, \cdots, F_{nn})$. Then if*

(1) $E\|F_{n1}\|_{\mathbf{K}} \to 0$ as $n \to \infty$,
(2) $\|F_{n1}\|_{\mathbf{K}} \geq \|F_{n+1,1}\|_{\mathbf{K}} \geq \cdots$,
(3) $\|a_n\|_{\mathbf{K}}/A_n < \frac{1}{n}$,
we have

$$d_H\Big(\frac{1}{A_n}\sum_{i=1}^{n} a_n F_{ni}, \{0\}\Big) \to 0, \ a.e.(\mu).$$

Theorem 4. *Let $\{K_n : n \geq 1\}$ be a sequence in $\mathbf{K}_k(\mathfrak{X})$ s.t. $d_H(K_n, [\alpha, \beta]) \to 0$. Then*

$$\alpha \leq \liminf_n k_n \leq \limsup_n k_n \leq \beta$$

for each sequence $\{k_n\}$ *in* $\mathfrak{X}$ *such that* $k_n \in K_n$ *for all* $n \geq 1$.

Proof. The proof is similar to the proof in [16].

The next theorem is a strong law of large numbers for rowwise exchangeable random variables with respect to nonadditive measures.

Theorem 5. *Let* ν *be a totally monotone capacity ,* $\{x_{ni} : n \geq 1\}$ *be* $\mathfrak{X}$*-valued rowwise exchangeable random variables with respect to* ν *such that* $\|x_{ni}\| \leq h(x) \in L^1[\Omega, \mathfrak{X}]$. *Assume that* $\{A_n\}, \{a_n\}$ *are random variables and* a_n *is a symmetric function of* $\{x_{n1}, \cdots, x_{nn}\}$. *If*

(1) x_{n1} *is essential bounded, i.e.* $\|x_{ni}\|_\infty = \inf\{c \geq 0 : \mu\{|x_{ni}| > c\} = 0\}$;

(2) $\|x_{n1}\| \geq \|x_{n+1,1}\| \geq \cdots$;

(3) $\frac{\|a_n\|}{A_n} < \frac{1}{n}$;

then we have

$$\nu\Big\{\omega : \liminf_n \frac{1}{A_n} \sum_{j=1}^{n} a_n x_{nj}(\omega) = \limsup_n \frac{1}{A_n} \sum_{j=1}^{n} a_n x_{nj}(\omega) = 0\Big\} = 1.$$

Proof. By Theorem 2, we know that there exists a set-valued random variable $F : [0,1] \to \mathbf{K}_k(\Omega)$ such that $\nu = \overline{\lambda}_F$, where λ is the Lebesgue measure of [0,1]. Since

$$(x_{ni} \circ F)_{-1}(A) = F_{-1}(x_{ni}^{-1}(A)), \quad \forall A \subset \mathfrak{X}.$$

Thus $x_{ni} \circ F$ is set-valued random variable.

Now we prove that the sequence $\{x_{ni} \circ F : 1 \leq i \leq n\}$ is exchangeable with respect to λ. Indeed, for any Borel sets $B_1, \cdots, B_n \in \mathcal{B}(\mathbf{K_k}(\mathfrak{X}))$, we have

$$\begin{aligned}
\lambda\Big\{x_{n1} \circ F \in B_1, \cdots, x_{nn} \circ F \in B_n\Big\} &= \lambda\Big\{F \subset x_{n1}^{-1}(B_1), \cdots, F \subset x_{nn}^{-1}(B_n)\Big\} \\
&= \underline{\lambda}_F\Big\{x_{n1}^{-1}(B_1), \cdots, x_{nn}^{-1}(B_n)\Big\} \\
&= 1 - \nu\Big\{x_{n1}^{-1}(B_1), \cdots, x_{nn}^{-1}(B_n)\Big\} \\
&= 1 - \nu\Big\{x_{\pi_{n1}}^{-1}(B_1), \cdots, x_{\pi_{nn}}^{-1}(B_n)\Big\} \\
&= \underline{\lambda}_F\Big\{x_{\pi_{n1}}^{-1}(B_1), \cdots, x_{\pi_{nn}}^{-1}(B_n)\Big\} \\
&= \lambda\Big\{F \subset x_{\pi_{n1}}^{-1}(B_1), \cdots, F \subset x_{\pi_{nn}}^{-1}(B_n)\Big\} \\
&= \lambda\Big\{x_{\pi_{n1}} \circ F \in B_1, \cdots, x_{\pi_{nn}} \circ F \in B_n\Big\}.
\end{aligned}$$

Furthermore $\{co(x_{ni} \circ F) : 1 \leq i \leq n\}$ is also exchangeable with respect to λ (cf.[13]). We also have $\|co(x_{ni} \circ F)\|_{\mathbf{K}} = \sup_{f \in F} \|x_{ni} \circ f\| \leq h(\omega) \in L^1[\Omega, \mathfrak{X}]$. Since $\|x_{n1}\| \geq \|x_{n+1,1}\| \geq \cdots$, so for any $f \in F$, we have $\|x_{n1} \circ f\| \geq$

$\|x_{n+1,1}\circ f\| \geq \cdots$, thus we have $\sup_{f\in F}\|x_{n1}\circ f\| \geq \sup_{f\in F}\|x_{n+1,1}\circ f\| \geq \cdots$, that is $\|co(x_{n1}\circ F)\|_{\mathbf{K}} \geq \|co(x_{n+1,1}\circ F)\|_{\mathbf{K}} \geq \cdots$. Hence by theorem 3, we can have

$$d_H\Big(\frac{1}{A_n}\sum_{j=1}^{n} a_n co(x_{nj}\circ F), \{0\}\Big) \longrightarrow 0, \quad a.e.\lambda$$

Then by lemma 3.1.4 of [15], we can have

$$d_H\Big(\frac{1}{A_n}\sum_{j=1}^{n} a_n (x_{nj}\circ F), \{0\}\Big) \longrightarrow 0, \quad a.e.\lambda$$

Let $b_n(\omega) = \frac{1}{A_n}\sum\limits_{j=1}^{n} a_n x_{nj}(\omega)$. Set

$$S_1 = \Big\{s\in[0,1] : \frac{1}{A_n}\sum_{j=1}^{n} a_n x_{nj}(F(s)) \to \{0\}\Big\},$$

$$S_2 = \Big\{s\in[0,1] : \liminf_n b_n(\omega) = \limsup_n b_n(\omega) = 0, \quad \forall\omega\in F(s)\Big\},$$

$$\Omega_2 = \Big\{\omega\in\Omega : \liminf_n b_n(\omega) = \limsup_n b_n(\omega) = 0\Big\}.$$

By theorem 4, we know that $S_1 \subset S_2$. Notice that

$$\nu(\Omega_2) = \lambda\{s\in[0,1] : F(s)\subset\Omega_2\} = \lambda\{S_2\}$$

Therefore $\nu(\Omega_2) = \lambda(S_2) \geq \lambda(S_1) = 1$. This completes the proof of the result.

Acknowledgements. We thank the referees for their helpful comments on the first version of this paper. The research of the authors is supported by Beijing Natural Science Foundation(Stochastic Analysis with uncertainty and applications in finance), PHR (No. 201006102), NSFC(No. 11171010).

References

1. Aumann, R.: Integrals of set valued functions. J. Math. Anal. Appl. 12, 1–12 (1965)
2. Castaing, C., Valadier, M.: Convex Analysis and Measurable Multifunctions. Lecture Notes in Mathemathics, vol. 580. Springer, Berlin (1977)
3. Castaldo, A., Maccheroni, F., Marinacci, M.: Random correspondences as bundles of random variables. Ind. J. Stat. 66, 409–427 (2004)
4. Chen, Z., Wu, P.: Strong Laws of Large Numbers for Bernoulli Experiments under Ambiguity. In: Li, S., Wang, X., Okazaki, Y., Kawabe, J., Murofushi, T., Guan, L. (eds.) Nonlinear Maths for Uncertainty and its Appli. AISC, vol. 100, pp. 19–30. Springer, Heidelberg (2011)

5. Choquet, G.: Theory of capacities. Ann. Inst. Fourier 5, 131–295 (1953)
6. Colubi, A., López-Díaz, M., Domínguez-Menchero, J.S., Gil, M.A.: A generalized strong law of large numbers. Probab. Theory Rel. 114, 401–417 (1999)
7. Feng, Y.: Strong law of large numbers for stationary sequences of random upper semicontinuous functions. Stoch. Anal. Appl. 22, 1067–1083 (2004)
8. Guan, L., Li, S., Inoue, H.: Strong laws of large numbers for rowwise exchangeable fuzzy set-valued random variables. In: Proc. of the 11th World Congress of International Fuzzy Systems Association, pp. 411–415 (2005)
9. Hess, C.: Measurability and integrability of the weak upper limit of a sequence of multifunctions. J. Math. Anal. Appl. 153, 226–249 (1983)
10. Hiai, F., Umegaki, H.: Integrals, conditional expectations and martingales of multivalued functions. J. Multivar. Anal. 7, 149–182 (1977)
11. Inoue, H.: Randomly weighted sums for exchangeable fuzzy random variables. Fuzzy Set. Syst. 69, 347–354 (1995)
12. Inoue, H.: Exchangeability and convergence for random sets. Inf. Sci. 133, 23–37 (2001)
13. Inoue, H., Taylor, R.L.: A SLLN for arrays of rowwise exchangeable fuzzy random sets. Stoch. Anal. Appl. 13, 461–470 (1995)
14. Klein, E., Thompson, A.C.: Theory of correspondences including applications to mathematical economics. John Wiley & Sons (1984)
15. Li, S., Ogura, Y., Kreinovich, V.: Limit Theorems and Applications of Set-Valued and Fuzzy Set-Valued Random Variables. Kluwer Academic Publishers, Dordrecht (2002)
16. Maccheroni, F., Marinacci, M.: A strong law of large numbers for capacities. Ann. Probab. 33, 1171–1178 (2005)
17. Molchanov, I.: On strong laws of large numbers for random upper semicontinuous functions. J. Math. Anal. Appl. 235, 249–355 (1999)
18. Patterson, R.F., Taylor, R.L.: Strong laws of large numbers for striangular arrays of exchangeable random variables. Stoch. Anal. Appl. 3, 171–187 (1985)
19. Rebille, Y.: Law of large numbers for non-additive measures. J. Math. Anal. Appl. 352, 872–879 (2009)
20. Schmeidler, D.: Subjective probability and expectated utility without additivity. Econometrica 57, 571–587 (1989)
21. Taylor, R.L., Daffer, P.Z., Patterson, R.F.: Limit theorems for sums of exchangeable variables. Rowman & Allanheld, Totowa (1985)
22. Taylor, R.L., Inoue, H.: Convergence of weighted sums of random sets. Stoch. Anal. Appl. 3, 379–396 (1985)

Credibility Theory Oriented Sign Test for Imprecise Observations and Imprecise Hypotheses

Gholamreza Hesamian and Seyed Mahmood Taheri

Abstract. This paper extends the sign test to the case where the available observations and underlying hypotheses about the population median are imprecise quantities, rather than crisp. To do this, the associated test statistic is extended, using some elements of credibility theory. Finally, to reject or accept the null hypothesis of interest, we extend the concept of classical p-value.

Keywords: Credibility measure, fuzzy variable, p-value, sign test statistic.

1 Introduction

Nonparametric tests are statistical tests used to analyze data for which an underlying distribution is not assumed. They are advantages over their parametric counterparts because they have fewer underlying assumptions (e.g., data normality, equal variance, etc). A particular class of non-parametric tests is used for the location parameter. The well-known sign test seems to be a good nonparametric alternative to parametric tests for single population location problem [5, 6]. Such tests are commonly based on crisp (precise) observations. But, in the real world, there are many situations in which the available data are imprecise rather than precise. In such cases, we are often faced with two sources of uncertainty; randomness and fuzziness. Randomness is related to the uncertainties in the outcomes of an experiment; fuzziness, on the other hand, involves uncertainties in the meaning of the data. For

Gholamreza Hesamian · Seyed Mahmood Taheri
Dept. of Mathematical Sciences, Isfahan University of Technology,
Isfahan 84156, Iran
e-mail: g.hesamian@math.iut.ac.ir,taheri@cc.iut.ac.ir

R. Kruse et al. (Eds.): Synergies of Soft Computing and Statistics, AISC 190, pp. 153–163.
springerlink.com © Springer-Verlag Berlin Heidelberg 2013

instance, the life of a tire a company recently developed, under some unexpected situations, cannot be measured precisely. We can just obtain the tire life around a number such as "about 3200 miles", "approximately 33000 miles", etc. Therefore, to deal with both types of uncertainties, it is necessary to incorporate uncertain concept into statistical technique.

After introducing fuzzy set theory, there are a lot of attempts for developing statistical methods in imprecise/fuzzy environments. Concerning the purposes of this article, we refer the reader to [2, 7, 8, 9, 11, 10, 12, 13, 20, 21]. It should be mentioned that the method introduced in [2], [10] and [20] are based on a concept of p-value, while the methods investigated in [7, 8, 9, 11, 13] are based on critical regions. For more on fuzzy statistic in fuzzy environment, see, for example, [1, 14, 17, 22].

The present paper, using the elements of credibility measure, aims to develop sign test with imprecise observations and imprecise hypotheses, based on the concept of p-value.

This paper is organized as follows: In Sec. 2, we review the classical sign test. In Sec. 3, we recall some definitions and results from credibility theory. In Sec. 4, based on an index for ranking fuzzy variables, we introduce a method to extend the sign test statistic to imprecise observations. Inside, we also extend the concept of classical p-value. Finally, at a given (crisp) significance level, we introduce a method for testing imprecise hypotheses. To explain the proposed method, we provide a numerical example. Section 5 concludes the paper.

2 Sign Test: A Brief Overview

Suppose that a random sample $X_1, X_2, \ldots, X_N$ is drawn from a population F_X with an unknown median M, where F_X is assumed to be continuous and strictly increasing, at least in the vicinity of M. In other words, the N random variables are independent and identically distributed, with unique median. The hypothesis to be tested concerns the value of the population median $H_0 : M = M_0$ where M_0 is a specified value, against a corresponding one or two-sided alternative. The number of observations larger than M_0,

Table 1 p-value for sign test

Alternative Hypothesis: H_1	$p-value$
$M > M_0$	$p-value = \sum_{i=k_N}^{N} \binom{N}{i}(0.5)^N$
$M < M_0$	$p-value = \sum_{i=0}^{k_N} \binom{N}{i}(0.5)^N$
$M \neq M_0$	$2\min\{\sum_{i=0}^{k_N} \binom{N}{i}(0.5)^N, \sum_{i=k_N}^{N} \binom{N}{i}(0.5)^N\}$

denoted by k_N, is used to test the validity of the null hypothesis. At a given significance level δ, we reject H_0 if $p - value < \delta$ and otherwise, we accept it, in which the corresponding p-value for given alternative hypotheses are shown in Tbl. 1 (for more details, see [5]).

3 Credibility Measure

To perform the sign test based on imprecise observations, we need a suitable method of ranking fuzzy data. Here, we introduce a method for ranking such data using credibility theory, which will be used in this article.

First, let us remark some elementary concepts of credibility theory.

Definition 1. (Liu [15]) Let Ω be a nonempty set, and $\mathcal{A}$ the power set of Ω. Each element in $\mathcal{A}$ is called an event. A set function $Cr : \mathcal{A} \to [0,1]$ is called a credibility measure if it satisfies the following four axioms,

1. Axiom 1. (Normality) $Cr\{\Omega\} = 1$.
2. Axiom 2. (Monotonicity) $Cr\{A\} \leq Cr\{B\}$ whenever $A \subseteq B$.
3. Axiom 3. (Self-Duality) $Cr\{A\} + Cr\{A^c\} = 1$ for any event A, where A^c denotes $\Omega - A$.
4. Axiom 4. (Maximality) $Cr\{\cup_i A_i\} = \sup_i Cr\{A_i\}$ for any events $\{A_i\}$ with $\sup_i Cr\{A_i\} < 0.5$.

The $(\Omega, \mathcal{A}, Cr)$ is called a credibility space.

Definition 2. (Liu [15]) A fuzzy variable is a measurable function from a credibility space $(\Omega, \mathcal{A}, Cr)$ to the set of real numbers.

We briefly say that a fuzzy variable $\widetilde{A}$ is normal if and only if $\sup_{x\in\mathbb{R}} \mu_{\widetilde{A}}(x) = 1$. Since a fuzzy variable $\widetilde{A}$ is a function on a credibility space, note that for any set C of real numbers, the set

$$\{\widetilde{A} \in C\} = \{w \in \Omega : \widetilde{A}(w) \in C\}, \tag{1}$$

is always an element in $\mathcal{A}$.

Definition 3. (Liu [15], p. 179) Let $\widetilde{A}$ be a fuzzy variable defined on the credibility space $(\Omega, \mathcal{A}, Cr)$. Then its membership function is defined based on the credibility measure by

$$\mu_{\widetilde{A}}(x) = \min\{2Cr\{\widetilde{A} \in \{x\}\}, 1\}, x \in \mathbb{R}. \tag{2}$$

By a trapezoidal fuzzy variable we mean the fuzzy variable fully determined by the quadrulpe (a^l, a^c, a^s, a^r) (briefly, $\widetilde{A} = (a^l, a^c, a^s, a^r)_T$) of crisp numbers with $a^l < a^c < a^s < a^r$, whose membership function is given by

$$\mu_{\widetilde{A}}(x) = \begin{cases} 0 & x < a^l, \\ \frac{x-a^l}{a^c-a^l} & a^l \leq x < a^c, \\ 1 & a^c \leq x < a^s, \\ \frac{a^r-x}{a^r-a^s} & a^s \leq x \leq a^r, \\ 0 & x > a^r. \end{cases} \quad \forall x \in \mathbb{R}. \tag{3}$$

If $a^c = a^s$, it is called a triangular fuzzy number and is denoted by $\widetilde{A} = (a^l, a^c, a^r)_T$. For more on fuzzy numbers, we refer the reader to [15].

Definition 4. (Liu [15], p. 178) Let $f : \mathbb{R}^2 \to \mathbb{R}$ be a function, and $\widetilde{A}$ and $\widetilde{B}$ be fuzzy variables on the credibility space $(\Omega, \mathcal{A}, Cr)$. Then $\widetilde{C} = f(\widetilde{A}, \widetilde{B})$ is a fuzzy variable defined as $\widetilde{C}(w) = f(\widetilde{A}(w), \widetilde{B}(w))$, for any $w \in \Omega$.

Lemma 1. *([15], p. 189) Let $\widetilde{A}$ and $\widetilde{B}$ be two normal fuzzy variables with membership functions $\mu_{\widetilde{A}}$ and $\mu_{\widetilde{B}}$, and $f : \mathbb{R}^2 \to \mathbb{R}$ a function. Then for any set C of real numbers, the credibility $Cr\{f(\widetilde{A}, \widetilde{B}) \in C\}$ is*

$$\frac{1}{2}\Big(\sup_{f(x,y)\in C} \min\{\mu_{\widetilde{A}}(x), \mu_{\widetilde{B}}(y)\} + 1 - \sup_{f(x,y)\notin C} \min\{\mu_{\widetilde{A}}(x), \mu_{\widetilde{B}}(y)\}\Big). \tag{4}$$

As an special case of the above equation, let $f(x, y) = x - y$ and $C = (0, \infty)$. Therefore, the credibility measure of $\widetilde{A} - \widetilde{B} \in (0, \infty)$ is obtained as follows

$$Cr\{\widetilde{A}-\widetilde{B} \in (0,\infty)\} = \frac{1}{2}(\sup_{x>y} \min\{\mu_{\widetilde{A}}(x), \mu_{\widetilde{B}}(y)\}+1-\sup_{x\leq y} \min\{\mu_{\widetilde{A}}(x), \mu_{\widetilde{B}}(y)\}). \tag{5}$$

Example 1. ([16], p. 168) Let $\widetilde{A}$ be a normal fuzzy variable and $\mu_{\widetilde{B}}(x) = I(x = k)$, $k \in \mathbb{R}$. Then, the given credibility measure in Eq. (5) reduces as follows

$$Cr\{\widetilde{A} \in (k,\infty)\}\} = \frac{1}{2}(\sup_{x>k} \mu_{\widetilde{A}}(x) + 1 - \sup_{x\leq k} \mu_{\widetilde{A}}(x)). \tag{6}$$

Example 2. Let $\widetilde{A} = (a^l, a^c, a^s, a^r)_T$ be a triangular fuzzy number and k be a real number. Then, the credibility of $\{\widetilde{A} \in (k, \infty)\}$ is

$$Cr\{\widetilde{A} \in (k,\infty)\} = \begin{cases} 1 & if \;\; k \leq a^l, \\ \frac{1}{2}(2 - \frac{k-a^l}{a^c-a^l}) & if \;\; a^l < k < a^c, \\ \frac{1}{2} & if \;\; a^c \leq k \leq a^s, \\ \frac{1}{2}(\frac{a^r-k}{a^r-a^s}) & if \;\; a^s < k < a^r, \\ 0 & if \;\; k \geq a^r. \end{cases} \tag{7}$$

Therefore, $Cr\{\widetilde{A} \in (k, \infty)\} > 0.5$ if and only if $k < a^c$.

Remark 1. For two normal fuzzy variables $\widetilde{A}$ and $\widetilde{B}$, it is well known that the degree of necessity to which $\widetilde{A}$ is larger than $\widetilde{B}$ is fulfilled by $Nec\ (\widetilde{A} \succ \widetilde{B})$

$= 1 - \sup_{x,y;x\leq y} \min\{\mu_{\widetilde{A}}(x), \mu_{\widetilde{B}}(y)\}$ [3]. In addition, the degree of possibility to which $\widetilde{B}$ is "larger than or equal to" the $\widetilde{A}$ is defined to be $Pos\ (\widetilde{B} \succeq \widetilde{A}) = 1 - Nec\ (\widetilde{A} \succ \widetilde{B})$. Since the necessity index has some appropriate properties, natural interpretation, and effectiveness in applied statistical problems, it is employed in many applications based on imprecise observations (see, for example [2, 11, 12]).

To compare Nec and Cr indices, it is worth noting that, we have $Nec\ (\widetilde{A} \succ \widetilde{B}) = 1$ and $Cr\{\widetilde{A} - \widetilde{B} \in (0,\infty)\} = 1$ if and only if $b^r < a^l$. In addition, we can observe that the Nec doesn't have the self-duality property while Cr does. From Example 1, we can also observe that $Cr\{\widetilde{A} - \widetilde{B} \in (0,\infty)\} = \frac{1}{2}(Pos\ (\widetilde{A} \succ \widetilde{B}) + Nec\ (\widetilde{A} \succ \widetilde{B}))$, and $Cr\{\widetilde{A} - \widetilde{B} \in (0,\infty)\} \geq Nec\ (\widetilde{A} \succ \widetilde{B})$.

Remark 2. For two trapezoidal fuzzy variables $\widetilde{A} = (a^l, a^c, a^s, a^r)_T$ and $\widetilde{B} = (b^l, b^c, b^s, b^r)_T$ with $b^s < a^c$, it is natural to say that "$\widetilde{A}$ is larger $\widetilde{B}$" with a reasonable degree. In this case, we have $Cr\{\widetilde{A} - \widetilde{B} \in (0,\infty)\} > 0.5$, however for a such case the necessity index may not has power enough to interpret the sentence "larger than" between $\widetilde{A}$ and $\widetilde{B}$. For instance, consider two trapezoidal fuzzy variables $\widetilde{A} = (-1, 3.3, 4.3, 5.3)_T$ and $\widetilde{B} = (1, 2, 3, 4)_T$. Then, we observe that $Nec\ (\widetilde{A} \succ \widetilde{B}) \simeq 0$, while $Cr\{\widetilde{A} - \widetilde{B} \in (0,\infty)\} \simeq 0.5$ (> 0.5).

Definition 5. Let $\widetilde{A}$ and $\widetilde{B}$ be two fuzzy variables with membership functions $\mu_{\widetilde{A}}$ and $\mu_{\widetilde{B}}$. Then, $\widetilde{A}$ is said to be larger than $\widetilde{B}$, denoted by $\widetilde{A} \succ_{Cr} \widetilde{B}$, if and only if $Cr\{\widetilde{A} - \widetilde{B} \in (0,\infty)\} > 0.5$.

Example 3. Suppose that $\widetilde{A} = (a^l, a^c, a^r)_T$ and $\widetilde{B} = (b^l, b^c, b^r)_T$, where $a^c > b^c$, are two triangular fuzzy numbers. If $b^r \leq a^l$, then it is readily seen that $Cr\{\widetilde{A} - \widetilde{B} \in [0,\infty)\} = 1$. If $b^r > a^l$, then it is easy to verify that $\sup_{x>y} \min\{ \mu_{\widetilde{A}}(x), \mu_{\widetilde{B}}(y)\} = 1$ and $\sup_{x\leq y} \min\{ \mu_{\widetilde{A}}(x), \mu_{\widetilde{B}}(y)\} = \frac{b^r - x^*}{b^r - b^c}$, where $x^* = \frac{b^r(a^c - a^l) + a^l(b^r - b^c)}{(a^c - a^l) + (b^r - b^c)}$. Therefore, from Example 1

$$Cr\{\widetilde{A} - \widetilde{B} \in (0,\infty)\} = \frac{b^r - 2b^c + x^*}{2(b^r - b^c)}. \tag{8}$$

For instance, if $\widetilde{A} = (1.2, 2, 2.3)_T$ and $\widetilde{B} = (0.5, 1, 1.5)_T$, then $Cr\{\widetilde{A} - \widetilde{B} \in (0,\infty)\} = 0.88$. Therefore, based on Def. 5, $\widetilde{A}$ is larger than $\widetilde{B}$.

The following theorem shows that the proposed ranking method has the transitivity property.

Theorem. Let $\widetilde{A}$, $\widetilde{B}$, and $\widetilde{C}$ be some fuzzy variables. Then, $\widetilde{A} \succ_{Cr} \widetilde{B}$ and $\widetilde{B} \succ_{Cr} \widetilde{C}$ imply $\widetilde{A} \succ_{Cr} \widetilde{C}$.

Proof. Write $E = \{\widetilde{A} - \widetilde{B} \in (0,\infty)\}$ and $F = \{\widetilde{B} - \widetilde{C} \in (0,\infty)\}$, then $E \cap F \subseteq \{\widetilde{A} - \widetilde{C} \in (0,\infty)\}$. Since $\widetilde{A} \succ_{Cr} \widetilde{B}$ and $\widetilde{B} \succ_{Cr} \widetilde{C}$, we have

$$Cr\{E\} > 0.5, \quad Cr\{F\} > 0.5.$$

Thus, by self-duality of credibility measure, we obtain

$$Cr\{E^c\} = 1 - Cr\{E\} < 0.5, \quad Cr\{F^c\} = 1 - Cr\{F\} < 0.5.$$

It follows from monotonicity and maximality of credibility measure that

$$\begin{aligned} Cr\{\widetilde{A} - \widetilde{C} \in (0,\infty)\} &\geq Cr\{E \cap F\} = 1 - Cr\{E^c \cup F^c\} \\ &= 1 - \max\{Cr\{E^c\}, Cr\{F^c\}\} > 1 - \max\{0.5, 0.5\} = 0.5. \end{aligned}$$

Therefore, $\widetilde{A} \succ_{Cr}^{\gamma} \widetilde{C}$.

4 Sign Test with Imprecise Observations and Imprecise Hypotheses

In this section, we extend the classical sign test to the case when the available observations are imprecise rather than crisp and the hypotheses of interest are precisely presented.

As we mentioned in Sec. 2, in the classical approach to test the null hypothesis $H_0 : M = M_0$, the observed sign statistic is given by

$$t_N = \sum_{i=1}^{N} I(x_i > M_0), \tag{9}$$

where, I is the indicator function,

$$I(\rho) = \begin{cases} 1 & \text{if } \rho \text{ is true,} \\ 0 & \text{if } \rho \text{ is false.} \end{cases}$$

Now, based on imprecise observations $\widetilde{x}_1, \widetilde{x}_2, \ldots, \widetilde{x}_n$, suppose we want to test the imprecise null hypothesis $H_0 : M \ is \ \widetilde{M}_0$, where $\widetilde{M}_0$ is a fuzzy variable. This situation corresponds to imprecisely formulated hypotheses of the type "the population median is about $\widetilde{M}_0$".

To verify the null hypothesis $H_0 : M \ is \ \widetilde{M}_0$, we have to count those imprecise observations $\widetilde{x}_i$, $i = 1, 2, \ldots, n$, for which "$\widetilde{x}_i$ is larger than $\widetilde{M}_0$". Since expression "larger than" is univocal in fuzzy environment, so we apply the proposed ranking method in this paper for realizing whether $\widetilde{x}_i$ could be regarded as "larger than" $\widetilde{M}_0$. Since, for every $\alpha > 0.5$, the value

$$\sum_{i=1}^{n} I(Cr\{\widetilde{x}_i - \widetilde{M}_0 \in (0,\infty)\} \geq \alpha), . \tag{10}$$

is a candidate of a test statistic, therefore we have a set of values for the classical sign test statistics as follows

$$\widetilde{t}_N = \{\widetilde{t}_N^L, \widetilde{t}_N^L + 1, \ldots, \widetilde{t}_N^U\}, \tag{11}$$

in which

$$\widetilde{t}_N^L = \inf_{\alpha > 0.5} \sum_{i=1}^{n} I(Cr\{\widetilde{x}_i - \widetilde{M}_0 \in (0, \infty)\} \geq \alpha), \tag{12}$$

$$\widetilde{t}_N^U = \sup_{\alpha > 0.5} \sum_{i=1}^{n} I(Cr\{\widetilde{x}_i - \widetilde{M}_0 \in (0, \infty)\} \geq \alpha)\}. \tag{13}$$

Now, to make a decision rule to accept or reject the null hypothesis H_0 : $M\ is\ \widetilde{M}_0$, for imprecise observations, we apply a decision rule based on the concept of p-value (similar to that of Grzegorzewski [10]). However, since our extended sign test statistic is a set, hence p-value corresponding to the output of the extended sign test is an interval. Therefore, for various kinds of alternative hypotheses, and as a counterpart of the traditional p-value, we consider an interval $\widetilde{p} - value = [\widetilde{p}^L - value, \widetilde{p}^U - value]$ as follows

- for testing H_0 against the alternative hypothesis $H_1 : M \succ_{Cr} \widetilde{M}_0$ (i.e., $Cr\{\widetilde{M}_0 - M \in (-\infty, 0) > 0.5$),

$$\widetilde{p}^L - value = \sum_{i=\widetilde{k}_N^U}^{N} \binom{N}{i} (0.5)^N,\ \widetilde{p}^U - value = \sum_{i=\widetilde{k}_N^L}^{N} \binom{N}{i} (0.5)^N, \tag{14}$$

- for testing alternative hypothesis $H_1 : \widetilde{M}_0 \succ_{Cr} M$ (i.e., $Cr\{\widetilde{M}_0 - M \in (0, \infty) > 0.5$),

$$\widetilde{p}^L - value = \sum_{i=0}^{\widetilde{k}_N^L} \binom{N}{i} (0.5)^N,\ \widetilde{p}^U - value = \sum_{i=0}^{\widetilde{k}_N^U} \binom{N}{i} (0.5)^N, \tag{15}$$

- for testing alternative hypothesis $H_1 : \widetilde{M}_0 \neq_{Cr} M$ (i.e., $M \succ_{Cr} \widetilde{M}_0$ or $\widetilde{M}_0 \succ_{Cr} M$),

$$\widetilde{p}^L - value = \min_{w \in \widetilde{t}_N} 2 \min\{\sum_{i=0}^{w} \binom{N}{i} (0.5)^N, \sum_{i=w}^{N} \binom{N}{i} (0.5)^N\}, \tag{16}$$

$$\widetilde{p}^U - value = \max_{w \in \widetilde{t}_N} 2 \min\{\sum_{i=0}^{w} \binom{N}{i} (0.5)^N, \sum_{i=w}^{N} \binom{N}{i} (0.5)^N\}. \tag{17}$$

Remark 3. If the imprecise observations $\widetilde{x}_1, \widetilde{x}_2, \ldots, \widetilde{x}_n$ reduce to the crisp values $x_1, x_2, \ldots, x_n$, then, for any $\alpha \in (0.5, 1]$, we observe that $Cr\{\widetilde{x}_i - \widetilde{M}_0 \in (0, \infty)\} = \frac{1}{2}(I(x_i > M_0) + 1 - I(x_i \leq M_0)) \geq \alpha$ if and only if $x_i \geq M_0$. So,

$\widetilde{k}_N^L = \widetilde{k}_N^U = \sum_{i=1}^N I(x_i > M_0) = k_N$ and therefore, $\widetilde{p} - value = p - value$. In such a case, therefore, the exteded sign test statistic $\widetilde{k}_N$ and $\widetilde{p} - value$ reduce to the classical sign test statistic k_N and $p - value$, respectively (Sec. 2).

Now, consider the problem of the hypothesis test $H_0 : M \; is \; \widetilde{M}_0$ with imprecise observations at a given crisp significance level. One can expect that, for the case of testing H_0 versus $H_1 : M \succ_{Cr} \widetilde{M}_0$ (i.e., $Cr\{\widetilde{M}_0 - M \in (-\infty, 0)\} > 0.5$), if the observed $\widetilde{p} - value$ is bigger than a given significance level δ, then H_0 should be accepted; otherwise, H_0 should be rejected, (the similar argument can be stated for other two cases). Since the proposed $\widetilde{p} - value$ is an interval, by modifying the proposed method given by Grzegorzewski [10], a degree of reject the null hypothesis H_0 would be given as follows at any significance level δ,

$$\phi_\delta(\widetilde{x}_1, \widetilde{x}_2, \ldots, \widetilde{x}_n) = \begin{cases} 0 & if \;\; \delta < \widetilde{p}^L, \\ \frac{\delta - \widetilde{p}^L}{\widetilde{p}^U - \widetilde{p}^L} & if \;\; \widetilde{p}^L \leq \delta \leq \widetilde{p}^U, \\ 1 & if \;\; \delta > \widetilde{p}^U. \end{cases} \tag{18}$$

Therefore, we accept H_0 with degree of $1 - \phi_\delta(\widetilde{x}_1, \widetilde{x}_2, \ldots, \widetilde{x}_n)$ and reject it with degree of rejection $\phi_\delta(\widetilde{x}_1, \widetilde{x}_2, \ldots, \widetilde{x}_n)$.

Remark 4. Using a concept of fuzzy test statistic, Grzegorzewski [9, 11] investigated median tests for vague data with crisp hypotheses tests. He utilized the necessity-index suggested by Dubois and Prade [3], for ranking fuzzy data. At a crisp significance level, he constructed a fuzzy test based on the classical critical region, in which the result of a test is presented by two possibility-necessity-based indices, while our method leads to the classical binary decision. However, as we investigated in Remarks 1 and 2, it seems our proposed ranking method is complied better concerning our intuition to interpret "larger than" to rank imprecise observations.

He also proposed a modification of the classical one-sided upper sign test to cope with vague data modeled by intuitionists fuzzy set for testing crisp or imprecise hypotheses [10]. Let $\mu_G(\widetilde{x}_i)$ shows the degree to which observation $\widetilde{x}_i$ is absolutely greater than $\widetilde{M}_0$ and $\nu_G(\widetilde{x}_i)$ represents the degree to which the above mentioned relationship is not satisfied. Based on his approach, the output of the sign test statistic is an interval $\widetilde{T}(\widetilde{x}_1, \widetilde{x}_2, \ldots, \widetilde{x}_n) = [T^L, T^U]$, where $T^L = \sum_{i=1}^n \mu_G(\widetilde{x}_i)$ and $T^U = \sum_{i=1}^n \nu_G(\widetilde{x}_i)$. As a counterpart of the traditional p-value, he considered an interval $\widetilde{p} = [\widetilde{p}^L, \widetilde{p}^U]$, where

$$\widetilde{p}^L = \sum_{i=\llcorner T^L \lrcorner}^{N} \binom{N}{i} (0.5)^N, \; \widetilde{p}^U = \sum_{i=\ulcorner T^U \urcorner}^{N} \binom{N}{i} (0.5)^N,$$

in which, $\llcorner x \lrcorner$ is the biggest integer smaller or equal to x, while $\ulcorner x \urcorner$ stands for the smallest integer greater than or equal to x (see also, [4, 18, 19] for other approaches to define a concept of p-value in fuzzy environment). For any assumed significance level δ, he made the following decision rules:

if $\widetilde{p}^U \leq \delta$, then we reject H_0,
if $\widetilde{p}^U > \delta$, then we accept H_0,
if $\widetilde{p}^L \leq \delta < \widetilde{p}^U$, then our test is not decisive.

Note that, using the proposed credibility measure in this paper, we have $\nu_G(\widetilde{x}_i) = Cr\{\widetilde{x}_i - \widetilde{M}_0 \in (0,\infty)\}$, and therefore $\widetilde{T}(\widetilde{x}_1, \widetilde{x}_2, \ldots, \widetilde{x}_N) = T^L = T^U = \sum_{i=1}^n Cr\{\widetilde{x}_i - \widetilde{M}_0 \in (0,\infty)\}$. Based on the proposed ranking method in this paper, we considered those observations $\widetilde{x}_i$ in which $\mu_G(\widetilde{x}_i) > 0.5$, and we suggested a different method to that of Grzegorzewski's method by defining a set $\widetilde{T}(\widetilde{x}_1, \widetilde{x}_2, \ldots, \widetilde{x}_N) = \{T^L, \ldots, T^U\}$ of the sign test statistics, where

$$T^L = \inf_{\alpha > 0.5} \sum_{i=1}^n I(Cr\{\widetilde{x}_i - \widetilde{M}_0 \in (0,\infty)\} \geq \alpha),$$

$$T^U = \sup_{\alpha > 0.5} \sum_{i=1}^n I(Cr\{\widetilde{x}_i - \widetilde{M}_0 \in (0,\infty)\} \geq \alpha).$$

Finally, for making decision to reject or accept a given imprecise null hypothesis, we proposed a modified version of Grzegorzewski's method by extending the concept of p-value.

Table 2 Data set in Example 4

$\widetilde{x}_1 = (60, 65, 65, 70)_T$	$\widetilde{x}_2 = (60, 65, 65, 75)_T$	$\widetilde{x}_3 = (80, 85, 85, 90)_T$	$\widetilde{x}_4 = (55, 60, 70, 75)_T$
$\widetilde{x}_5 = (65, 70, 75, 80)_T$	$\widetilde{x}_6 = (60, 75, 75, 80)_T$	$\widetilde{x}_7 = (70, 80, 80, 90)_T$	$\widetilde{x}_8 = (75, 85, 85, 90)_T$
$\widetilde{x}_9 = (60, 65, 65, 75)_T$	$\widetilde{x}_{10} = (75, 80, 85, 90)_T$	$\widetilde{x}_{11} = (80, 90, 90, 100)_T$	$\widetilde{x}_{12} = (40, 50, 55, 75)_T$
$\widetilde{x}_{13} = (50, 55, 55, 70)_T$	$\widetilde{x}_{14} = (60, 65, 65, 70)_T$	$\widetilde{x}_{15} = (70, 75, 75, 80)_T$	$\widetilde{x}_{16} = (60, 65, 75, 80)_T$

Example 4. The performance evaluation in an aerospace firm is made by using a system based on a fuzzy scale. According to this system, the evaluator assigns a fuzzy value of performance to each worker by taking into account some certain criteria such as experience, responsibility, mental, and physical efforts, and so forth. A random sample of 16 workers was selected. The imprecise evaluation values are given in Tbl. 2 [13].

Suppose that we wish to test the hypothesis $H_0 : M \; is \; \widetilde{M}_0 = (70, 80, 90)_T$ versus the alternative hypothesis $H_1 : M \prec_{Cr} \widetilde{M}_0$, at the significance level of $\delta = 0.01$. Using Example 1, from Eqs. (11) and (13), we get $\widetilde{t}_N = \{1, 2, 3\}$, and so, from Eq. (15), we obtain

$$\widetilde{p}-value=[\sum_{i=0}^{1}\binom{16}{i}(0.5)^{16},\sum_{i=0}^{3}\binom{16}{i}(0.5)^{16}]=[0.0002,0.0106].$$

Let $\delta = 0.01$ be the given significance level. From Eq. (18), therefore, we reject H_0 with degree of $\phi_\delta(\widetilde{x}_1, \widetilde{x}_2, \ldots, \widetilde{x}_N) = \frac{0.01-0.0002}{0.0106-0.0002} = 0.94$, and we accept it with degree of 0.06.

5 Conclusions

As a natural generalization of the sign test, we proposed a method based on imprecise observations, when the underlying hypotheses are imprecise, too. To do this, the usual concepts of the sign test statistic and p-value are extended, using some concepts of credibility theory. For rejecting or accepting the null hypothesis of interest, we proposed a method to compare the observed p-value and a given significance level. The proposed method is general and it can be used for other nonparametric rank-based tests.

Acknowledgements. Part of this work were completed during the first author's stay in the Department of Mathematical Sciences of Tsinghua University, China. He is grateful for the hospitality of Professor Baoding Liu and this department.

References

1. Bertoluzza, C., Gil, M.A., Ralescu, D.A. (eds.): Statistical Modeling, Analysis and Management of Fuzzy Data. STUDFUZZ, vol. 87. Physica, Heidelberg (2002)
2. Denœux, T., Masson, M., Hébert, P.: Nonparametric rank-based statistics and significance tests for fuzzy data. Fuzzy Sets Syst. 153, 1–28 (2005)
3. Dubois, D., Prade, H.: Ranking fuzzy numbers in the setting of possibility theory. Inform. Sciences 30, 183–224 (1983)
4. Filzmoser, P., Viertl, R.: Testing hypotheses with fuzzy data: The fuzzy p-value. Metrika 59, 21–29 (2004)
5. Gibbons, J.D., Chakraborti, S.: Non-parametric Statistical Inference. Marcel Dekker, New York (2003)
6. Govindarajulu, Z.: Nonparametric Inference. World Scientific, Hackensack (2007)
7. Grzegorzewski, P.: Statistical inference about the median from vague data. Control Cybern. 27(3), 447–464 (1998)
8. Grzegorzewski, P.: Distribution-free tests for vague data. In: Soft Methodology and Random Information Systems, pp. 495–502. Springer, Heidelberg (2004)
9. Grzegorzewski, P.: Two-sample median test for vague data. In: Proc. EUSFLAT-LFA, Barcelona, Spain, September 7-9, pp. 621–626 (2005)
10. Grzegorzewski, P.: A bi-robust test for vague data. In: Proc. 12th Int. Conf. Information Processing and Management of Uncertainty in Knowledge-Based Systems (IPMU 2008), pp. 138–144 (2008)

11. Grzegorzewski, P.: k-sample median test for vague data. Int. J. Intell. Syst. 24, 529–539 (2009)
12. Hryniewicz, O.: Goodman-Kruskal γ measure of dependence for fuzzy ordered categorical data. Comput. Stat. Data Anal. 51, 323–334 (2006)
13. Kahraman, C., Bozdag, C.E., Ruan, D., Özok, A.F.: Fuzzy sets approaches to statistical parametric and nonparametric tests: Research articles. Int. J. Intell. Syst. 19, 1069–1087 (2004)
14. Kruse, R., Meyer, K.D.: Statistics With Vague Data. D. Reidel, Dordrecht (1987)
15. Liu, B.: Uncertainty Theory, 2nd edn. STUDFUZZ, vol. 145. Springer, Berlin (2007)
16. Liu, B., Liu, Y.: Expected value of fuzzy variable and fuzzy expected value models. IEEE Trans. Fuzzy Syst. 10, 445–450 (2002)
17. Nguyen, H.T., Wu, B.: Fundamentals of Statistics with Fuzzy Data. STUDFUZZ, vol. 198. Springer, Berlin (2006)
18. Parchami, A., Taheri, S., Mashinchi, M.: Fuzzy p-value in testing fuzzy hypotheses with crisp data. Stat. Pap. 51, 209–226 (2010)
19. Parchami, A., Taheri, S., Mashinchi, M.: Testing fuzzy hypotheses based on vague observations: a p-value approach. Stat. Pap. 53, 469–484 (2012)
20. Taheri, S.M., Hesamian, G.: Goodman-Kruskal measure of association for Fuzzy-Categorized variables. Kybernetika 47, 110–122 (2011)
21. Taheri, S.M., Hesamian, G., Viertl, R.: Contingency table with fuzzy categories. In: Proc. 11th Int. Conf. Intelligent Technologies (InTech 2010), pp. 100–104 (2011)
22. Viertl, R.: Statistical Methods for Fuzzy Data. Wiley, Chichester (2011)

A Note on Large Deviations of Random Sets and Random Upper Semicontinuous Functions

Xia Wang

Abstract. In this paper, we show a sufficient condition under which the law of sums of i.i.d. compact random sets in a separable type p Banach space (resp. compact random upper semicontimuous functions) satisfies large deviations if the law of sums of its corresponding convex hull of compact random sets(resp. quasiconcave envelope of compact random upper semicontimuous functions) satisfies large deviations.

Keywords: Large deviations, random sets, random upper semicontinuous functions.

1 Introduction

The theory of large deviation principle (LDP) deals with the asymptotic estimation of probabilities of rare events and provides exponential bound on probability of such events. Some authors have discussed LDP on random sets and random upper semicontinuous functions. In 1999, Cerf [3] proved Cramér type LDP for sums of i.i.d. compact random sets in a separable type p Banach space with respect to the Hausdorff distance d_H. In 2006, Terán obtained Cramér type LDP of compact random upper semicontinuous functions [9], and Bolthausen type LDP of compact convex random upper semicontinuous functions [10] on a separable Banach space in the sense of the uniform Hausdorff distance d_H^∞. In 2009, Ogura and Setokuchi [7] proved a Cramér type LDP for compact random upper semicontiunous functions on the underling separable Banach space with respect to the metric d_Q (see [7] for the notation) in a different method, which is weaker than the uniform Hausdorff distance

Xia Wang
College of Applied Sciences, University of Technology, Beijing, China
e-mail: wangxia@bjut.edu.cn

R. Kruse et al. (Eds.): Synergies of Soft Computing and Statistics, AISC 190, pp. 165–172.
springerlink.com © Springer-Verlag Berlin Heidelberg 2013

d_H^∞. In 2010, Ogura, Li and Wang [6] also discussed LDP for random upper semicontinuous functions whose underlying space is d-dimensional Euclidean space $\mathbb{R}^d$ under various topologies for compact covex random sets and random upper semicontinuous functions, Wang [12] considered functional LDP of compact random sets, Wang and Li [11] obtained LDP for bounded closed convex random sets and related random upper semicontiunous functions. In fact, about these work above, some work of papers extended compact convex random sets(resp. compact convex random upper semicontinuous functions) to the non-convex case in a separable type p Banach space (see [3, 7, 9, 12]). So we hope the LDP of the law of sums of i.i.d. compact random sets(resp. compact random upper semicontimuous functions) still holds if the law of sums of its corresponding convex hull of compact random sets(resp. quasiconcave envelope of compact random upper semicontimuous functions) satisfies large deviations. However, until now, all ideal of "deconvexification" comes from Cerf's basic work(Lemma 2 in [3]). In [3], Cerf gives a sufficient condition for the case of compact random sets : $E[\exp\{\lambda\|X\|_{\mathcal{K}}\}] < \infty$ for any $\lambda > 0$. In [9], Terán gives a sufficient condition(see Lemma 4.4 in [9]) for the case of compact random upper semicontinuous functions : $E[\exp\{\lambda\|X_0\|_{\mathcal{K}}\}] < \infty$ for some $\lambda > 0$. In [9], the author doesn't give the proof of Lemma 4.4, and he said the basic idea is the same as Cerf's paper. I think, if the author use Cerf's idea, the Lemma 4.4 can't be obtained under the condition: $E[\exp\{\lambda\|X_0\|_{\mathcal{K}}\}] < \infty$ for some $\lambda > 0$. So in our paper, we don't use Cerf's idea and use another method to give another condition for compact random sets: $E[\exp\{\lambda\|X\|_{\mathcal{K}}^{\frac{p}{2}}\}] < \infty$ for some $\lambda > 0$, and another condition for compact random upper semicontinuous functions: $E[\exp\{\lambda\|X\|_{\mathcal{F}}^{\frac{p}{2}}\}] < \infty$ for some $\lambda > 0$. Under these conditions, we prove the laws of sums of i.i.d. compact random sets and compact random upper semicontimuous functions satisfy large deviations if the laws of sums of its corresponding convex hull of compact random sets and quasiconcave envelope of compact random upper semicontimuous functions satisfy large deviations.

The paper is structured as follows. Section 2 will give some preliminaries about compact random sets and compact random upper semicontinuous functions. In section 3, we will give and prove our main results.

2 Preliminaries

Throughout this paper, we assume that $(\Omega, \mathcal{A}, P)$ is a complete probability space, $(\mathfrak{X}, \|\cdot\|_{\mathfrak{X}})$ is a real separable Banach space with its dual space $\mathfrak{X}^*$. We suppose that $\mathfrak{X}$ is of type $p > 1$, i.e., there exists a constant c such that

$$E\Big[\|\sum_{i=1}^{n} f_i\|_{\mathfrak{X}}^{\frac{p}{2}}\Big] \leq c\sum_{i=1}^{n} E\Big[\|f_i\|_{\mathfrak{X}}^{\frac{p}{2}}\Big],$$

for any independent random variables $f_1, f_2, \cdots, f_n$ with values in $\mathfrak{X}$ and mean zero. Every Hilbert space is type 2; the space L^p with $1 < p < \infty$ are of type $\min(p, 2)$. However, the space of continuous functions on $[0,1]$ equipped with supremum norm is of type 1 and not of type p for any $p > 1$.

$\mathcal{K}_k(\mathfrak{X})$(resp. $\mathcal{K}_c(\mathfrak{X}), \mathcal{K}_{kc}(\mathfrak{X})$) is the family of all non-empty compact (resp. convex, compact convex) subsets of $\mathfrak{X}$.

Let A and B be two non-empty subsets of $\mathfrak{X}$ and let $\lambda \in \mathbb{R}$, we can define addition and scalar multiplication by $A+B = cl\{a+b : a \in A,\ b \in B\}, \lambda A = \{\lambda a :\ a \in A\}$, where clA is the closure of set A taken in $\mathfrak{X}$. The Hausdorff distance on $\mathcal{K}_k(\mathfrak{X})$ is defined by

$$d_H(A, B) = \max\Big\{ \sup_{a\in A} \inf_{b\in B} \|a-b\|_{\mathfrak{X}}, \sup_{b\in B} \inf_{a\in A} \|a-b\|_{\mathfrak{X}} \Big\}.$$

In particular, we denote $\|A\|_{\mathcal{K}} = d_H(\{0\}, A) = \sup_{a\in A}\{\|a\|_{\mathfrak{X}}\}$.

X is called compact random set (resp. compact convex random set), if it is a measurable mapping from the space $(\Omega, \mathcal{A}, P)$ to $(\mathcal{K}_k(\mathfrak{X}), \mathfrak{B}(\mathcal{K}_k(\mathfrak{X})))$, (resp. $(\mathcal{K}_k(\mathfrak{X}), \mathfrak{B}(\mathcal{K}_{kc}(\mathfrak{X}))))$ where $\mathfrak{B}(\mathcal{K}_k(\mathfrak{X}))$ (resp. $\mathfrak{B}(\mathcal{K}_{kc}(\mathfrak{X}))$) is the Borel σ-field of $\mathcal{K}_k(\mathfrak{X})$ (resp. $\mathcal{K}_{kc}(\mathfrak{X})$) generated by the Hausdorff distance d_H.

In the following, we introduce the definition of a random upper semicontinuous function. Let $\mathcal{F}_k(\mathfrak{X})$ denote the family of all functions $u : \mathfrak{X} \to [0,1]$ satisfying the conditions: (1) the 1-level set $[u]_1 = \{x \in \mathfrak{X} :\ u(x) = 1\} \neq \emptyset$, (2) each u is upper semicontinuous, i.e., for each $\alpha \in (0,1]$, the α level set $[u]_\alpha = \{x \in \mathfrak{X} :\ u(x) \geq \alpha\}$ is a compact subset of $\mathfrak{X}$, (3) the support set $[u]_0 = \mathrm{cl}\{x \in \mathfrak{X} :\ u(x) > 0\}$ is compact.

The subfamily of all u such that $[u]_\alpha$ is in $\mathcal{K}_c(\mathfrak{X})$ for all $\alpha \in [0,1]$ will be denoted of $\mathcal{F}_c(\mathfrak{X})$. Let $\mathcal{F}_{kc}(\mathfrak{X})$ denote the subfamily of all u such that u is in both $\mathcal{F}_k(\mathfrak{X})$ and $\mathcal{F}_c(\mathfrak{X})$. For every $u \in \mathcal{F}_k(\mathfrak{X})$, denote by $cou \in \mathcal{F}_{kc}(\mathfrak{X})$ the quasiconcave envelope of u, we have $[cou]_\alpha = co[u]_\alpha$ for all $\alpha \in (0,1]$.

For any two upper semicontinuous functions u_1, u_2, define

$$(u_1 + u_2)(x) = \sup_{x_1+x_2=x} \min\{u_1(x_1), u_2(x_2)\} \quad \text{for any} \quad x \in \mathfrak{X}.$$

Similarly, for any upper semicontinuous function u and for any $\lambda \geq 0$ and $x \in \mathfrak{X}$, define

$$(\lambda u)(x) = \begin{cases} u(\dfrac{x}{\lambda}), & \text{if } \lambda \neq 0, \\ I_0(x), & \text{if } \lambda = 0, \end{cases}$$

where I_0 is the indicator function of 0.

The following distance is the uniform Hausdorff distance which is extension of the Hausdorff distance d_H : for $u, v \in \mathcal{F}_b(\mathfrak{X})$, $d_H^\infty(u,v) = \sup_{\alpha\in[0,1]} d_H([u]_\alpha, [v]_\alpha)$, this distance is the strongest one considered in the literatures.

X is called a compact random upper semicontinuous function (or random fuzzy set or fuzzy set-valued random variable), if it is a measurable mapping $X: (\Omega, \mathcal{A}, P) \to (\mathcal{F}_k(\mathfrak{X}), \mathfrak{B}(\mathcal{F}_k(\mathfrak{X})))$ (where $\mathfrak{B}(\mathcal{F}_k(\mathfrak{X}))$ is the Borel σ-field of $\mathcal{F}_k(\mathfrak{X})$ generated by the uniform Hausdorff distance d_H^∞).

3 Main Results

Before giving our main results for random sets and random upper semicontinuous functions, we define rate functions and LDP. We refer to the books of Dembo and Zeitouni [4] and Deuschel and Stroock [5] for the general theory on large deviations (also see Yan, Peng, Fang and Wu [13]).

Let E be a regular Hausdorff topological and $\{\mu_n : n \geq 1\}$ be a family of probability measures on $(E, \mathcal{E})$, where $\mathcal{E}$ is the Borel σ-algebra. A *rate function* is a lower semicontinuous mapping $I : E \to [0, \infty]$. A *good rate function* is a rate function such that the level sets $\{x : I(x) \leq \alpha\}$ are compact subset of E. A family of probability measures $\{\mu_n : n \geq 1\}$ on the measurable space $(E, \mathcal{E})$ is said to satisfy the *LDP* with speed $\frac{1}{n}$ and with the rate function I if, for all open set $V \subset \mathcal{E}$, $\liminf_{n\to\infty} \frac{1}{n} \ln \mu_n(V) \geq -\inf_{x\in V} I(x)$, for all closed set $U \subset \mathcal{E}$, $\limsup_{n\to\infty} \frac{1}{n} \ln \mu_n(U) \leq -\inf_{x\in U} I(x)$.

In the following, we give our main two results. We first present LDP for $(\mathcal{K}_k(\mathfrak{X}), d_H)$-valued *i.i.d.* random variables.

Theorem 1. Let $\mathfrak{X}$ be a Banach space of type $p > 1$. And $X_1, X_2, \ldots, X_n$ be $(\mathcal{K}_k(\mathfrak{X}), d_H)$-valued *i.i.d.* random variables satisfying $Ee^{\lambda \|X_1\|_{\mathcal{K}}^p} < \infty$ for some $\lambda > 0$. Let $S_n = \frac{X_1+X_2+\cdots+X_n}{n}$, $coS_n = \frac{coX_1+coX_2+\cdots+coX_n}{n}$. If the law of the random set coS_n satisfies a LDP with the good rate function I_1', then the law of the random set S_n also satisfies a LDP with the good rate function I_1(for $x \in \mathcal{K}_{kc}(\mathfrak{X}), I_1(x) = I_1'(x)$, for $x \in \mathcal{K}_k(\mathfrak{X}) \backslash \mathcal{K}_{kc}(\mathfrak{X}), I_1(x) = +\infty$,) i.e., Then for any open set $U \subset (\mathcal{K}_k(\mathfrak{X}), d_H)$,

$$\liminf_{n\to\infty} \frac{1}{n} \log P\left\{\frac{X_1 + X_2 + \cdots + X_n}{n} \in U\right\} \geq -\inf_{x\in U} I_1(x),$$

any for any closed set $V \subset (\mathcal{K}_k(\mathfrak{X}), d_H)$,

$$\limsup_{n\to\infty} \frac{1}{n} \log P\left\{\frac{X_1 + X_2 + \cdots + X_n}{n} \in V\right\} \leq -\inf_{x\in V} I_1(x).$$

In the following, we give LDP for $(\mathcal{F}_k(\mathfrak{X}), d_H^\infty)$-valued *i.i.d.* random variables.

Theorem 2. Let $\mathfrak{X}$ be a Banach space of type $p > 1$. And $X_1, X_2, \ldots, X_n$ be $(\mathcal{F}_k(\mathfrak{X}), d_H^\infty)$-valued *i.i.d.* random variables satisfying $Ee^{\lambda \|X_1\|_{\mathcal{F}}^p} < \infty$ for some $\lambda > 0$. $S_n = \frac{X_1+X_2+\cdots+X_n}{n}$, $coS_n = \frac{coX_1+coX_2+\cdots+coX_n}{n}$. If the law of the random set coS_n satisfies a LDP with the good rate function I', then the law of the random set S_n also satisfies a LDP with the good rate function I(for $x \in \mathcal{F}_{kc}(\mathfrak{X}), I(x) = I'(x)$, for $x \in \mathcal{F}_k(\mathfrak{X}) \backslash \mathcal{F}_{kc}(\mathfrak{X}), I(x) = +\infty$,) i.e., Then for any open set $U \subset (\mathcal{F}_k(\mathfrak{X}), d_H^\infty)$,

$$\liminf_{n\to\infty} \frac{1}{n} \log P\left\{\frac{X_1 + X_2 + \cdots + X_n}{n} \in U\right\} \geq -\inf_{x\in U} I(x), \tag{1}$$

any for any closed set $V \subset (\mathcal{F}_k(\mathfrak{X}), d_H^\infty)$,

$$\limsup_{n\to\infty} \frac{1}{n} \log P\left\{\frac{X_1 + X_2 + \cdots + X_n}{n} \in V\right\} \leq -\inf_{x\in V} I(x). \tag{2}$$

In order to prove our two main theorems above, we need the following two lemmas.

Lemma 3: Let $\mathfrak{X}$ is of type $p > 1$ and $X_1, X_2, \cdots, X_n$ be $(\mathcal{K}_k(\mathfrak{X}), d_H)$-valued i.i.d. random variables such that $Ee^{\lambda\|X_1\|_{\mathcal{K}}^p} < \infty$ for some $\lambda > 0$, then for any $\delta > 0$,

$$\limsup_{n\to\infty} \frac{1}{n} \ln P(d_H(\frac{X_1 + X_2 + \cdots + X_n}{n}, \frac{coX_1 + coX_2 + \cdots + coX_n}{n}) \geq \delta) = -\infty.$$

This proof is same as those of the following Lemma 4, so we omit it. But here we state the inequality of Puri and Ralescu we use in our proofs of Lemma 3 and Lemma 4.

Let A belong to $\mathcal{K}_k(\mathfrak{X})$, and its inner radius is $r(A)$, and we know $r(A) \leq 2\|A\|_{\mathcal{K}}$. In [8], Puri and Ralescu extended a result of Cassels [2] and proved the following inequality(we call it inequality of Puri and Ralescu): for any $A_1, A_2, \cdots, A_n$ in $\mathcal{K}_k(\mathfrak{X})$,

$$d_H(A_1 + A_2 + \cdots + A_n, coA_1 + coA_2 + \cdots + coA_n) \leq c^{\frac{1}{p}}(r(A_1)^p + r(A_2)^p + \cdots + r(A_n)^p)^{\frac{1}{p}}.$$

Lemma 4: Let $\mathfrak{X}$ is of type $p > 1$ and $X_1, X_2, \cdots, X_n$ be $(\mathcal{F}_k(\mathfrak{X}), d_H^\infty)$-valued i.i.d. random variables such that $Ee^{\lambda\|X_1\|_{\mathcal{F}}^p} < \infty$ for some $\lambda > 0$, then for some $\delta > 0$,

$$\limsup_{n\to\infty} \frac{1}{n} \ln P(d_H^\infty(\frac{X_1 + X_2 + \cdots + X_n}{n}, \frac{coX_1 + coX_2 + \cdots + coX_n}{n}) \geq \delta) = -\infty.$$

Proof: We apply the definition of d_H^∞ and the inequality of Puri and Ralescu and for any $A \in \mathcal{K}_k(\mathfrak{X})$, the inner radius has the property: $r(A) \leq 2\|A\|_{\mathcal{K}}$, then

$$
\begin{aligned}
& d_H^{\infty}(\frac{X_1+X_2+\cdots+X_n}{n}, \frac{coX_1+coX_2+\cdots+coX_n}{n}) \\
&= \frac{1}{n} \sup_{\alpha\in[0,1]} d_H(\sum_{i=1}^{n}[X_i]_\alpha, \sum_{i=1}^{n}[coX_i]_\alpha) \\
&\leq \frac{1}{n}\cdot c^{\frac{1}{p}} \sup_{\alpha\in[0,1]} (r([X_1]_\alpha)^p + r([X_2]_\alpha)^p + \cdots + r([X_n]_\alpha)^p)^{\frac{1}{p}} \\
&\leq \frac{1}{n}\cdot 2c^{\frac{1}{p}}(\|X_1\|_{\mathcal{F}}^{p} + \|X_2\|_{\mathcal{F}}^{p} + \cdots + \|X_n\|_{\mathcal{F}}^{p})^{\frac{1}{p}}.
\end{aligned}
$$

In view of the condition of Lemma 4: $Ee^{\lambda\|X_1\|_{\mathcal{K}}^{p}} < \infty$ for some $\lambda > 0$, then for this positive $\lambda > 0$, we have $Ee^{\lambda\|X_1\|_{\mathcal{K}}^{p}} < \infty$, so we can apply the chebyshev exponential inequality , then we obtain

$$
\begin{aligned}
& \limsup_{n\to\infty} \frac{1}{n} \ln P(d_H^{\infty}(\frac{X_1+X_2+\cdots+X_n}{n}, \frac{coX_1+coX_2+\cdots+coX_n}{n}) \geq \delta) \\
&\leq \limsup_{n\to\infty} \frac{1}{n} \ln P(\|X_1\|_{\mathcal{F}}^{p} + |X_2\|_{\mathcal{F}}^{p} + \cdots + \|X_n\|_{\mathcal{F}}^{p} \geq \frac{n^p\delta^p}{2^p c}) \\
&\leq \limsup_{n\to\infty} \frac{1}{n} \ln[e^{-\frac{\lambda n^p \delta^p}{2^p c}}(Ee^{\lambda\|X_1\|_{\mathcal{F}}^{p}})^n] \\
&= \limsup_{n\to\infty}(-\frac{\lambda n^{p-1}\delta^p}{2^p c} + Ee^{\lambda\|X_1\|_{\mathcal{F}}^{p}}) \\
&= -\infty.
\end{aligned}
$$

So we complete the proof of this lemma.

Since random sets are particular cases of those for fuzzy random variables, then we omit the proof of Theorem 1, and only give the proof of Theorem 2.

Proof of theorem 2: Step 1: First we prove the upper bound of (1). Let $\mathcal{U}$ be a closed subset of $(\mathcal{F}_k(\mathfrak{X}), d_H^{\infty})$. For any $\forall\ \delta > 0$, let

$$
\mathcal{U}^{\delta} = \{x \in \mathcal{F}_k(\mathfrak{X}) :\ d_H^{\infty}(x, \mathcal{U}) = \inf_{y\in\mathcal{U}} d_H^{\infty}(x, y) < \delta\}.
$$

Then $P(S_n \in \mathcal{U}) \leq P(coS_n \in \overline{\mathcal{U}^{\delta}}) + P(d_H^{\infty}(S_n, coS_n) \geq \delta)$. So

$$
\begin{aligned}
& \limsup_{n\to\infty} P(S_n \in \mathcal{U}) \\
&\leq \max\{\limsup_{n\to\infty} P(coS_n \in \overline{\mathcal{U}^{\delta}}), \limsup_{n\to\infty} P(d_H^{\infty}(S_n, coS_n) \geq \delta)\} \\
&= -\inf_{x\in\overline{\mathcal{U}^{\delta}}} I(x).
\end{aligned}
$$

Since $I(x)$ is a good rate function, by [1], we have

$$\lim_{\delta\downarrow 0}\inf_{x\in\overline{\mathcal{U}^\delta}} I(x) = \inf_{x\in\mathcal{U}} I(x).$$

So (1) holds.

Step 2: we prove the lower bound of (2). Let $\mathcal{U}$ be an open subset of $(\mathcal{F}_k(\mathfrak{X}), d_H^\infty)$. $\forall\, x \in \mathcal{U}$, then there exists a $\delta > 0$ and an open subset V of $(\mathcal{F}_k(\mathfrak{X}), d_H^\infty)$ such that $x \in V \subset V^\delta \subset \mathcal{U}$. So

$$P(S_n \in \mathcal{U}) \geq P(S_n \in V^\delta) \geq P(coS_n \in V) - P(d_H^\infty(S_n, coS_n) \geq \delta).$$

Hence $P(S_n \in V) \leq P(S_n \in \mathcal{U}) + P(d_H^\infty(S_n, coS_n) \geq \delta)$. By Lemma 4, we have

$$\liminf_{n\to\infty} P(S_n \in \mathcal{U}) \geq -\inf_{x'\in V} I(x') \geq -I(x).$$

Taking the supermum over all elements x in $\mathcal{U}$, we have

$$\liminf_{n\to\infty} P(S_n \in \mathcal{U}) \geq -\inf_{x\in\mathcal{U}} I(x).$$

This completes the proof of Theorem 2.

Remark: In 2010, Ogura, Li and Wang [6] have proved a Cramér type LDP for compact convex random upper semicontinuous functions whose underlying space is d-dimensional Euclidean space $\mathbb{R}^d$ under the condition $E[\exp\{\lambda\|X\|_{\mathcal{F}}\}] < \infty$, for some $\lambda > 0$ with respect to the metric d_Q(see the detailed notation in [6]). Since the d-dimensional Euclidean space $\mathbb{R}^d$ is type 2, then if $X_1, X_2, \cdots, X_n$ are $(\mathcal{F}_k(\mathbb{R}^d), d_H^\infty)$-valued i.i.d. random variables such that $Ee^{\lambda\|X_1\|_{\mathcal{F}}^2} < \infty$ for some $\lambda > 0$, then Lemma 4 holds. And the condition $E[\exp\{\lambda\|X\|_{\mathcal{F}}\}] < \infty$ also holds for this positive λ. By Theorem 3.4 in [6], we know the law of sums of quasiconcave envelope of compact random upper semicontimuous functions satisfies large deviations, then in view of Theorem 2 in our paper, the law of sums of compact random upper semicontimuous functions satisfies large deviations with the same rate function.

Acknowledgements. This research is partially supported by PHR (No.201006102) and Beijing Natural Science Foundation(Stochastic analysis with uncertainty and applications in finance)and supported by NSFC(11171010).

References

1. Baxter, J.R., Jain, N.C.: A comparison principle for large deviations. Proc. Amer. Math. Soc. 103(4), 1235–1240 (1988)
2. Cassels, J.W.S.: Measures of the Non-Convexity of sets and the Shapley–Folkman–Starr theorem. Math. Proc. Cambridge Philos. Soc. 78(03), 433–436 (1975)

3. Cerf, R.: Large deviations for sums of i.i.d. random compact sets. Proc. Amer. Math. Soc. 127(08), 2431–2436 (1999)
4. Dembo, A., Zeitouni, O.: Large Deviations Techniques and Applications, 2nd edn. Stochastic Modelling and Applied Probability, vol. 38. Springer, Heidelberg (1998)
5. Deuschel, J., Stroock, D.W.: Large deviations, 3rd edn. Academic Press, Inc., San Diego (1989)
6. Ogura, Y., Li, S., Wang, X.: Large and moderate deviations of random upper semicontinuous functions. Stochastic Anal. Appl. 28(2), 350–376 (2010)
7. Ogura, Y., Takayoshi, S.: Large deviations for random upper semicontinuous functions. Tohoku Math. J. 61(2), 213–223 (2009)
8. Puri, M.L., Ralescu, D.A.: Limit theorems for random compact sets in banach space. Math. Proc. Cambridge Philos. Soc. 97(1), 151–158 (1985)
9. Terán, P.: A large deviation principle for random upper semicontinuous functions. Proc. Amer. Math. Soc. 134(02), 571–580 (2005)
10. Terán Agraz, P.: On borel measurability and large deviations for fuzzy random variables. Fuzzy Set. Syst. 157(19), 2558–2568 (2006)
11. Wang, X., Li, S.: Large Deviations of Random Sets and Random Upper Semicontinuous Functions. In: Borgelt, C., González-Rodríguez, G., Trutschnig, W., Lubiano, M.A., Gil, M.Á., Grzegorzewski, P., Hryniewicz, O. (eds.) Combining Soft Computing & Statis. Methods. AISC, vol. 77, pp. 627–634. Springer, Heidelberg (2010)
12. Wang, X.: Sample path large deviations for random compact sets. Int. J. Intell. Technol. Appl. Stat. 3(3), 317–333 (2010)
13. Yan, J., Peng, S., Fang, S., Wu, L.: Several Topics in Stochastic Analysis. Academic Press of China, Beijing (1997)

Conditional Density Estimation Using Fuzzy GARCH Models

Rui Jorge Almeida, Nalan Baştürk,
Uzay Kaymak, and João Miguel da Costa Sousa

Abstract. Time series data exhibits complex behavior including non-linearity and path-dependency. This paper proposes a flexible fuzzy GARCH model that can capture different properties of data, such as skewness, fat tails and multimodality in one single model. Furthermore, additional information and simple understanding of the underlying process can be provided by the linguistic interpretation of the proposed model. The model performance is illustrated using two simulated data examples.

Keywords: Conditional volatility, density estimation, fuzzy GARCH, fuzzy model, time series analysis.

1 Introduction

A conditional density of a random variable is an estimate of the probability distribution of the current value of that variable, given its past values or other variables. Conditional density estimation has an important role in

---

Rui Jorge Almeida · Nalan Baştürk
Erasmus School of Economics, Econometric Institute,
Erasmus University Rotterdam, 3000 DR Rotterdam, The Netherlands
e-mail: {rjalmeida,basturk}@ese.eur.nl

Uzay Kaymak
School of Industrial Engineering, Eindhoven University of Technology,
5600 MB Eindhoven, The Netherlands
e-mail: u.kaymak@ieee.org

João Miguel da Costa Sousa
Instituto Superior Técnico, Dept. of Mechanical Engineering,
CIS/IDMEC – LAETA, Technical University of Lisbon,
1049-001 Lisbon, Portugal
e-mail: jmsousa@ist.utl.pt

R. Kruse et al. (Eds.): Synergies of Soft Computing and Statistics, AISC 190, pp. 173–181.
springerlink.com © Springer-Verlag Berlin Heidelberg 2013

quantitative finance and risk management. Estimating an accurate model for the distribution of financial returns is not a simple task, as financial time-series typically possess non-trivial statistical properties, such as asymmetric distributions and non-constant variability of returns. Many statistical quantiles such as Value-at-Risk or Expected Shortfall, which are directly linked to the tail of the return distribution of a portfolio of financial assets are widely accepted financial risk management tools [9].

Different types of approaches have been proposed for estimating conditional density of returns. A popular approach where volatility changes dynamically is the Generalized Autoregressive Heteroskedasticity (GARCH) model [2]. Extended GARCH models are proposed in the literature to capture different aspects of data behavior: Student-t GARCH models to capture fat tails [3], GJR-GARCH [4] models to capture skewness, regime-switching GARCH models to capture multimodality [1]. These models can be extended to capture all complex data behavior in one model, but the estimation and identification of such models are not trivial [1].

Fuzzy systems have been combined with GARCH models to analyze dynamic processes with time-varying variance. In [10, 7, 6, 8] fuzzy GARCH models are proposed, where linguistic descriptors are combined with GARCH models in a rule-base system and each rule corresponds to an individual GARCH model. In [5] a GARCH model with error terms obtained from a set of fuzzy rules has been proposed.

This paper proposes a flexible fuzzy GARCH model that can capture different properties of data, such as skewness, fat tails and multimodality in one single model. Furthermore, the linguistic interpretation of the model can provide a simple understanding of the underlying returns process. This proposed model is more flexible than the standard GARCH model and previously proposed fuzzy GARCH models. This model can capture data behavior even if the underlying data distribution is misspecified. This is illustrated using simulated data. Specifically, we show that the fuzzy GARCH model can explain data generated from general GARCH-type models, such as the student-t and regime switching GARCH models.

2 Fuzzy GARCH Models

The standard GARCH (p, q) model for $t = 1, \ldots, T$ observations is:

$$y_t \mid h_t \sim N(0, h_t); \;\; h_t = \alpha_0 + \sum_{i=1}^{q} \alpha_i y_{t-i}^2 + \sum_{j=1}^{p} \beta_j h_{t-j} \,, \tag{1}$$

where y_t is the data, h_t is the unobserved conditional variance, and $N(\mu, \sigma^2)$ denotes independent normal distribution with mean μ and variance σ^2. The following restrictions provide positive variance terms h_t at every period

$$\alpha_0 > 0,\ \alpha_i \geq 0,\ \beta_j \geq 0,\ \sum_{i=1}^{q} \alpha_i + \sum_{j=1}^{p} \beta_j < 1, i = 1, \ldots, q. j = 1, \ldots, p. \quad (2)$$

A fuzzy GARCH model, similar to a probabilistic fuzzy system [11], combines two different types of uncertainty, namely fuzziness or linguistic vagueness, and probabilistic uncertainty. This model consists of a set of IF-THEN rules, where the antecedent of each rule are fuzzy sets and the consequents are GARCH models, consisting of l-th rules [10, 7, 6, 8]

$$R_l : \text{If } \mathbf{x} \text{ is } F_l \text{ then } h_t^l = \alpha_0^l + \sum_{i=1}^{q} \alpha_i^l y_{t-i}^2 + \sum_{j=1}^{p} \beta_j^l h_{t-j}\,, \quad (3)$$

where $\mathbf{x} \in \mathbb{R}^n$ is an input vector, $F_l : X \longrightarrow [0,1]$ is a multidimensional fuzzy set defined on a continuous sample space X. Parameter restrictions (2) should hold for every rule to ensure positive conditional variance h_t^l. The combination of h_t^l in (3) provides the *unobserved* conditional variance h_t. The density of output y_t is based on h_t

$$h_t = \sum_{l=1}^{L} g_{l,t} h_t^l\,, \quad (4)$$

$$y_t \mid h_t, x_t \sim N(\mu, h_t)\,, \quad (5)$$

where $g_{l,t} = u_{l,t} / \sum_{l=1}^{L} u_{l,t}$ are membership functions with $u_{l,t} \geq 0$ for $l = 1, \ldots, L$, $\sum_{l=1}^{L} u_{l,t} > 0$, and by definition $g_{l,t} \geq 0$ and $\sum_{l=1}^{L} g_{l,t} = 1$. This model can capture changing conditional variance, but not skewness or multimodality.

In [10] the parameters of the model were estimated in a two step approach. First the antecedents were obtained using a fuzzy clustering heuristic, followed by the estimation of the GARCH parameters using maximum likelihood estimation. In this work, the GARCH models were constrained, such that $\beta_j = 0$ in the variance term h_t^l (GARCH(p,0)) or the conditional variance is not given by a GARCH model but constant over time $h_t^l = h^l, \forall t$. In [7, 6] the parameters of the fuzzy GARCH model are obtained using a genetic algorithm, while in [8] particle swarm optimization is used. The objective function, as defined in all these work, is the mean squared error between the output density $\sqrt{h_t}\epsilon_t$ and observation y_t. To the best of our knowledge, the calculation of this objective function is not possible.

In this paper we propose a different type of fuzzy GARCH model where the output y_t and conditional variance h_t are defined by each of l-th fuzzy rule

$$R_l : \text{If } \mathbf{x} \text{ is } F_l \text{ then } y_t^l \mid x_t, h_t^l \sim N(\mu^l, h_t^l)\,, \tag{6}$$

$$\text{with } h_t^l = \alpha_0^l + \sum_{i=1}^{q} \alpha_i^l y_{t-i}^2 + \sum_{j=1}^{p} \beta_j^l h_{t-j}\,, \tag{7}$$

where h_t is given by (4). Using (7) and (6), the output of the system is

$$y_t \sim \sum_{l=1}^{L} g_{l,t} N(\mu^l, h_t^l)\,. \tag{8}$$

Parameter restrictions (2) provide positive variance terms h_t at every period, and (a positive) mean of the variance term h_t to exist.

The proposed model is more general than the existing fuzzy GARCH models: In (6) and (7), output y shows a smooth transition between normal densities, with possible different mean and variances. Hence the density of each observation might be multimodal or skewed, while in the previous fuzzy GARCH models the output density in (5) is a unimodal and symmetric normal density. In the proposed model the combination of normal densities in the rule output can lead to unimodal or skewed distributions depending on model parameters:

1. If $\mu^l = \mu^{l^\star}$ for all $l, l^\star \in \{1, \ldots, L\}$, output y comes from a normal distribution and conditional variance h changes over time. This case leads to the the previous fuzzy GARCH models as defined in (5).
2. If mean parameters μ^l are relatively different and h_t^l are relatively small and similar across $l = \{1, \ldots, L\}$, output y is likely to have a multimodal distribution,
3. If mean parameters μ^l are relatively close to each other and h_t^l are relatively different across $l = \{1, \ldots, L\}$, output y is likely to have a skewed distribution.

We illustrate the difference between both fuzzy GARCH model definition using simulated data. Figure 1 shows the conditional density of output y for simulated data from the proposed model (6) and model (3) from [7]. The latter model leads to unimodal and symmetric conditional densities while simulated data from the proposed model has a more complex behavior with skewed, asymmetric and bimodal conditional densities.

It is possible to estimate the model in (6) by maximum likelihood method, given that input x_t is included in the information set at time $t-1$, i.e. x is *predetermined* with respect to y. More specifically, x_t can for instance take past y values or can be an exogenous variable. Given that the type and number of membership functions $g_{l,t}$ are known, the log-likelihood of data $y = \{y_{t^\star}, \ldots, y_T\}$ is:

$$\ln \ell(y \mid I_{t-1}) = \ln \prod_{t=t^\star}^{T} \ell(y_t \mid x_t, h_t) = \sum_{t=t^\star}^{T} \ln \left(\sum_{l=1}^{L} g_{l,t} \phi(y_t, \mu^l, h_t^l) \right), \tag{9}$$

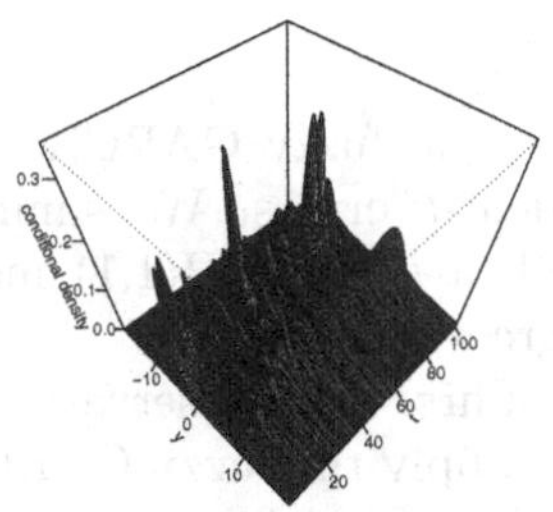
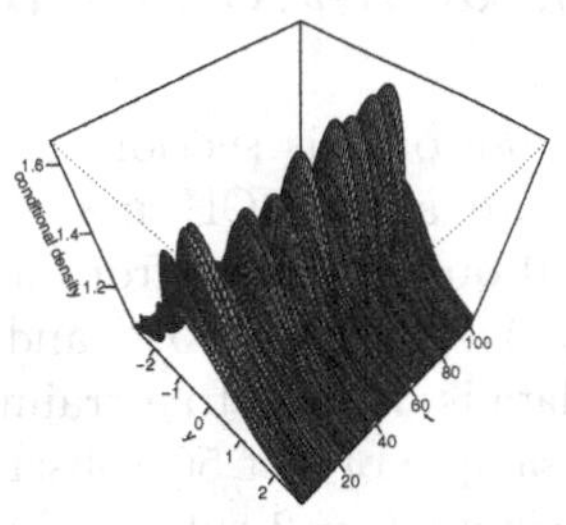

proposed model with $(\mu^1, \mu^2) = (-6, 6)$ model proposed by [7] with $\mu = 0$

Fig. 1 Conditional distributions of simulated data from Fuzzy GARCH models

where h_t is calculated from (7), and $t^* = \max(p+q) + 1$. In the remaining of this paper we consider that $g_{l,t}$ are trapezoidal membership functions but other types of parametric membership functions can be used.

Given trapezoidal membership functions, the likelihood in (9) is optimized with respect to all model parameters (GARCH and fuzzy membership), using a gradient search method. We constrain the search space to solutions satisfying the positive variance condition and membership functions that cover the universe of the input variables in the antecedent space.

3 Applications

In this section we illustrate the capabilities to estimate the conditional data density using the proposed fuzzy GARCH model with two applications. In both applications the underlying process is not a fuzzy GARCH model or an individual rule in the model. We show that the combination of the rule-base system in the fuzzy GARCH model however, overcomes this misspecification problem. In both applications we apply the fuzzy GARCH model where the antecedent $\mathbf{x}_t = y_{t-1}$, the previous value in the time series data. This antecedent is a natural choice, since the GARCH model builds on the relation between the data density and its past value. We compare the performance of the proposed model with that of the standard GARCH model as defined in (1). Since both these models estimate an output density rather than a point estimate, an intuitive model validation method is to consider the quantiles of the estimated output distribution, and compare it with the actual number of observations within the estimated quantiles. Due to space limitations, we focus on the model performance instead of parameter estimates and the linguistic description.

3.1 Student-t GARCH Data

The purpose of this section is to show that the fuzzy GARCH model can approximate a GARCH model with Student-t errors. We simulate data with 3000 observations, from a single Student-t GARCH(1,1) model with $(\alpha_0, \alpha_1, \beta_1) = (0.5, 0.07, 0.8)$ and with 4 degrees of freedom.

The data is divided to a training sample of first 2500 observations and the forecast sample of last 500 observations. We apply the fuzzy GARCH model in (6) defining $L = 2$ rules and a trapezoidal membership function. The left panel in Fig. 2 shows the 10%, 5% and 1% tails of the estimated output y density in the forecast sample using the fuzzy GARCH model. Fig. 2 shows that estimated tails of the density are quite different across time, indicating the complex data behavior and follow the extreme data values quite closely in the forecast sample.

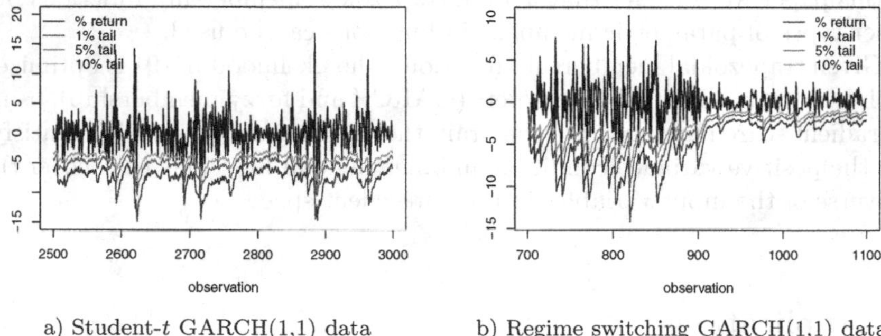

a) Student-t GARCH(1,1) data b) Regime switching GARCH(1,1) data

Fig. 2 Forecast data and fuzzy GARCH model quantile estimates

Table 1 summarizes the estimated output density using the fuzzy GARCH model and the standard GARCH model. For the standard GARCH model, none of the observations fall below the estimated 1% tail in the forecast sample. Hence the misspecified GARCH model overestimates the data variance. The fuzzy GARCH model, however, leads to substantially close percentage of observations in the respective estimated distribution tails. We conclude that the standard GARCH model cannot capture the tails of the distribution. The fuzzy GARCH model on the other hand, captures the data distribution accurately despite the misspecification in each rule which assumes error terms with normal distribution. We note that the combination of the rules in the proposed fuzzy GARCH model leads to error terms which are not necessarily normal distributed.

Table 1 Simulated data from a Student-t GARCH model: percentage of observations in respective distribution tails

	Fuzzy GARCH model			standard GARCH model		
	1 % tail	5 % tail	10 % tail	1 % tail	5 % tail	10 % tail
training sample	0.02	0.04	0.07	0.01	0.03	0.07
forecast sample	0.01	0.04	0.09	0.00	0.03	0.09

3.2 Regime-Switching GARCH Data

In this section we analyze simulated data from a different misspecified GARCH model, namely a regime-switching GARCH(1,1) model, and show that the Fuzzy GARCH model can capture the data behavior despite the original misspecification. The data is simulated from the following model:

$$\begin{aligned} h_t &= 0.8 + 0.13y_{t-i}^2 + 0.8h_{t-j} \ (\textit{regime 1}), \\ h_t &= 0.4 + 0.13y_{t-i}^2 + 0.5h_{t-j} \ (\textit{regime 2}), \\ \epsilon_t &\sim N(0,1),\, y_t = \sqrt{h_t}\epsilon_t , \end{aligned} \tag{10}$$

where the data comes from *regime 1* for $T = 1, \ldots, 200, 501, \ldots, 900$ and from *regime 2* for $T = 201, \ldots, 500, 901, \ldots, 1100$.

We apply the Fuzzy GARCH model with $L = 2$ rules on this data. The training and forecast samples consist of the first 700 and last 400 observations, respectively. The forecast sample and estimated 1%, 5% and 10% tails of the output distribution obtained by the Fuzzy GARCH model are given in the right panel of Fig. 2. Although the underlying GARCH model is misspecified, estimated distribution's tail values adjust to the high and low variance regimes at the beginning and the end of the forecast sample, respectively.

Table 2 summarizes the estimated output density using the fuzzy GARCH model and the standard GARCH model. For the standard GARCH model, none of the observations fall below the estimated 1% tail in the training and forecast samples and the percentage of observations in the 5% tails of the estimated distribution are 2%, substantially smaller than the target value. Hence the misspecified GARCH model overestimates the data variance. The fuzzy GARCH model on the other hand leads to similar percentage of observations in the respective estimated distribution tails. We conclude that although the underlying GARCH model is misspecified, fuzzy GARCH model can capture the data properties accurately. The misspecified GARCH model on the other hand, is not appropriate for this data. Note that in both applications considered, the proposed fuzzy GARCH model uses only two rules, and it captures the varying data properties accurately.

Table 2 Simulated data from a regime switching GARCH model: percentage of observations in respective distribution tails

	Fuzzy GARCH model			Standard GARCH model		
	1 % tail	5 % tail	10 % tail	1 % tail	5 % tail	10 % tail
training sample	0.02	0.05	0.11	0.00	0.02	0.08
forecast sample	0.01	0.05	0.08	0.00	0.02	0.07

4 Conclusion

We propose a new fuzzy GARCH model that can capture complex data behavior, such as skewness, multimodality and fat tail distributions where all model parameters are estimated using the maximum likelihood approach. We illustrate the model capabilities using two simulated datasets, exhibiting different data properties. We show that the proposed model captures the underlying data distribution in both cases. In future work, we plan to generalize the model to multiple inputs and discuss the linguistic interpretation of the model.

Acknowledgements. This work was partially supported by the European Science Foundation through COST Action IC0702 and by the Netherlands Organisation for Scientific Research (NWO) Secondment Grant number 400-07-703.

References

1. Bauwens, L., Preminger, A., Rombouts, J.: Theory and inference for a Markov switching GARCH model. Economet. J. 13(2), 218–244 (2010)
2. Bollerslev, T.: Generalized autoregressive conditional heteroskedasticity. J. Econometrics 31(3), 307–327 (1986)
3. Bollerslev, T.: A conditionally heteroskedastic time series model for speculative prices and rates of return. Rev. Econ. Stat. 69(3), 542–547 (1987)
4. Glosten, L., Jagannathan, R., Runkle, D.: On the relation between the expected value and the volatility of the nominal excess return on stocks. J. Financ. 48(5), 1779–1801 (1993)
5. Helin, T., Koivisto, H.: The garch-fuzzydensity method for density forecasting. Appl. Soft. Comput. 11(6), 4212–4225 (2011)
6. Hung, J.C.: A fuzzy asymmetric GARCH model applied to stock markets. Inf. Sci. 179(22), 3930–3943 (2009a)
7. Hung, J.C.: A fuzzy GARCH model applied to stock market scenario using a genetic algorithm. Expert Syst. with Appl. 36(9), 11,710–11,717 (2009b)
8. Hung, J.C.: Adaptive fuzzy-GARCH model applied to forecasting the volatility of stock markets using particle swarm optimization. Inf. Sci. 181(20), 4673–4683 (2011)

9. Jorion, P.: Value at Risk: the new benchmark for managing financial risk, 3rd edn. McGraw-Hill, New York (2006)
10. Popov, A.A., Bykhanov, K.V.: Modeling volatility of time series using fuzzy GARCH models. In: 9th Russian–Korean Int. Symp. on Science and Technology, KORUS 2005, pp. 687–692 (2005)
11. van den Berg, J., Kaymak, U., Almeida, R.J.: Function approximation using probabilistic fuzzy systems. ERIM Report Series ERS-2011-026-LIS, Erasmus Research Institute of Management ERIM (2012)

LP Methods for Fuzzy Regression and a New Approach

Bekir Çetintav and Firat Özdemir

Abstract. Linear Programming (LP) methods are commonly used to construct fuzzy linear regression (FLR) models. Probabilistic Fuzzy Linear Regression (PFLR) [9] and Unrestricted Fuzzy Linear Regression (UFLR) [3] are two of the mostly applied models that employ LP methods. In this study, a modified fuzzy linear regression model which use LP methods is proposed. PFLR, UFLR and proposed model compared in terms of mean squared error (MSE) and total fuzziness by using two simulated and one real data set.

Keywords: Fuzzy linear regression, linear programming methods.

1 Introduction

Regression analysis is a commonly used methodology for analyzing relationships between a response variable, also called dependent variable, and one or more explanatory variables, independent variables. In classical linear regression model; the deviation between the observed value and estimated value of dependent variable Y_i is generally regarded as error and that error is normally distributed with zero mean. Among several methods, the least square method is frequently used for estimation of parameters.

After improvements of fuzzy set theory, it has been successfully demonstrated in many applications, such as: reliability, quality control, econometrics, engineering applications, etc. The common point of these different areas is that there are data with vagueness (or fuzzy data). So special tools for applications of these data are needed. Because the original vagueness is not taken into account in the analysis when the fuzzy data is analyzed through

Bekir Çetintav · Firat Özdemir
Dokuz Eylul University, Izmir, Turkey
e-mail: {bekir.cetintav,firat.ozdemir}@deu.edu.tr

R. Kruse et al. (Eds.): Synergies of Soft Computing and Statistics, AISC 190, pp. 183–191.
springerlink.com © Springer-Verlag Berlin Heidelberg 2013

nonfuzzy techniques and it makes the model inaccurate. Therefore Fuzzy Regression (FR) models have been constructed to restore regression analysis for fuzzy space. Although it makes the model precise, FR models could be used for analyzing the crisp data. Because the crisp data is also a kind of fuzzy data (Even though it is degenerated). For example some FR models could be used when some properties of CLR are not maintained.

Recent years, many kinds of fuzzy regression models have been constructed to restore regression analysis. These models can be roughly categorized into three groups, linear programming (LP) methods, multi-objective (MO) techniques and least square (LS) methods. The LP methods are commonly used for fuzzy linear regression (FLR) because they are simple and easy to apply. Also it needs nearly no assumption. But it doesn't mean these methods are appropriate for all kinds of data sets. They also have some weaknesses; (i) they are extremely sensitive to outliers [2]; (ii) when there is an outlier they don't allow all observations for estimation and (iii) estimated fuzziness per unit increases as number of observations increase [7]. The multi-objective (MO) techniques are proposed to solve some of these weaknesses [4], but these techniques are not as simple as LP methods. Also they are not as good as the other methods (especially LS methods) for predictability. In this study; a new LP model which is more predictable is proposed.

In Sec. 2, we try to introduce some of frequently-used LP models and their characteristics. A new FLR model is proposed in Sec. 3. Its application and comparison with other models by numerical examples are in Sec 4. Section 5 gives our conclusions.

2 LP Methods for FLR

The LP methods are the first approaches for FLR. Therefore they are the most famous ones. As can be understood from the name, the LP models are used to estimate the parameters in these methods and the main aim is to minimize the fuzziness of the estimated regression model. Therefore they are also called The Minimum Uncertainty methods. (Since this study focuses on the predictability of LP methods, the details of the models which are about outlier problem[6, 1, 2] are ignored.)

2.1 Tanaka's Models

Tanaka *et al.* [10] proposed the first FLR model in 1982. According to that article, the deviation between the observed value and estimated value of dependent variable Y_i can be defined as *fuzziness* and it depends on the fuzziness of the system structure [10]. That is also the main idea of the LP methods. The fuzzy model is;(for $i = 1, \ldots, n$ (is # of ind. variables) and $j = 1, \ldots, N$ (is # of obs.))

$$\mathbf{Y_i^*} = \mathbf{A_i} \cdot \mathbf{X_{ij}} \tag{1}$$

The model consists of fuzzy parameters such as $A_i = (a_i, c_i)$ and dependent variable $Y_i = (y_i, e_i)$. They both have triangular membership functions. a_i is the center and c_i is the fuzziness of the fuzzy parameter A_i and observed Y_i has center y_i and fuzziness e_i .Also estimated Y_i^* is similar. Tanaka [10] proposed a linear programming model to obtain the estimations of parameters. Basic ideas of this model; (i)It should minimize the total fuzziness of the parameters. (Sum of c_i), (ii)The (membership function of) estimated Y_i^* should include the (membership function of) observed Y_i.(iii)There should be a threshold value H, which presents the degree of fitting value of estimated Y_i^* to observed Y_i.(iv)The fuzziness of a parameter should be nonnegative. Tanaka modified his first model in 1987 and 1989 [8, 9]. The total *fuzziness* of the parameters (sum of c_i) was minimized in the first model. On the contrary, the second model try to minimize the total fuzziness of the model .That model is called Possibilistic Fuzzy Linear Regression (PFLR).They modified only the objective function by multiplying *fuzziness* of the parameters (c_i) to absolute value of independent variable(s) (x_i). All other parts are the same with his first model.That modification reduced the fuzziness of the model significantly and brought it to the level required to be. But Tanaka's basic ideas (approach) did not change. The linear programing model:

$$\min z = \sum_{i=1}^{N} (\mathbf{c_0} \cdot |\mathbf{x_0}|) + (\mathbf{c_1} \cdot |\mathbf{x_1}|) + \cdots + (\mathbf{c_n} \cdot |\mathbf{x_n}|) \tag{2}$$

subject to

$$\mathbf{a^t x_i} + (1 - H) \cdot \mathbf{c^t} |\mathbf{x_i}| \leq \mathbf{y_i} + (1 - H) \cdot \mathbf{e_i} \tag{3}$$

$$\mathbf{a^t x_i} - (1 - H) \cdot \mathbf{c^t} |\mathbf{x_i}| \geq \mathbf{y_i} - (1 - H) \cdot \mathbf{e_i} \tag{4}$$

$$\mathbf{c_i} \geq 0 \ [for \ \ i = 1, 2, \ldots, N.] \tag{5}$$

The first two constraints are *density constraints* which make the estimated Y_i^* to include observed Y_i in the model. So they should be generate for all data (total number of data is N).The last one is *constraint of sign* that makes the fuzziness parameters c_i nonnegative.

2.2 Lee and Chang's Model (UFLR)

Another problem in PFLR is conflicting trends. In the cases where shrinking or expanding trends in the observations exist, PFLR frequently misinterprets the model. In order to avoid that problem Lee and Chang suggested canceling the constraint of sign ($c_i > 0$) in the PFLR model and called new model Unrestricted in Sign Fuzzy Linear Regression (UFLR) [3]. The UFLR Model is very similar with PFLR. Only difference is there is no constraint for fuzziness of parameters (c_i), those could be negative in this model. That means some independent variables could affect the fuzziness of the model negatively. In other word some independent variables decrease the total fuzziness of the model. With that change model could capture the different trends.

UFLR model works well in the data sets which have trend; however there is confusion about *negative fuzziness* and also outliers create problems like in the PFLR model.

3 Proposed New Model

Since it is simple and easy to apply, the most widely used approach while constructing FLR models is linear programming. However, there are some points that should be discussed in detail: Redden and Woodall [7] have stated that (i) they are extremely sensitive to outliers; (ii) when there is an outlier, they don't allow all observations for estimation (iii) as the number of observations increase, estimated fuzziness per unit also increases. Peter's and Chen's models which were given in Chap. 2 have tried to solve this problem.

Although prediction and estimation are the two main goals in regression analysis, these two models are not satisfactory enough in this respect [5]. And that makes them a little bit inadequate.

In FLR based on LP methods, the deviation between the observed value and estimated value of dependent variable Y_i can be defined as *vagueness* and it depends on the fuzziness of the system structure. In other words, vagueness results from the system parameters included in the model. The main goal of LP methods (for FLR) is to minimize that vagueness. However, there might be several problems in a linear regression model like model specification, variable selection or lack of fit. Vagueness caused by problems given above and some other similar problems may be defined as *unexplained vagueness.* In literature, FLR models based on LP methods ignore this unexplained part and focus on vagueness resulted from the parameters in the model. But it is not *fair.*

3.1 Fair Fuzzy Linear Regression (FFLR)

Proposed model FFLR divides total vagueness into two parts as explained and unexplained. Explained vagueness (or fuzziness) is caused by independent variables which are included in the model. And unexplained vagueness (or fuzziness) is caused by problems mentioned above.

A new parameter F is added to the model to represent the unexplained vagueness part. It has a triangular membership function with center 0 and fuzziness f, $F = (0, f)$. Then the regression function becomes as follows.

$$\mathbf{Y_i^*} = \mathbf{A_i} \cdot \mathbf{X_{ij}} + F \tag{6}$$

FFLR and other LP based models in the literature are very similar in estimating the model parameters. The only difference is that boundary constraints are modified and there is an additional constraint for the new parameter F. The LP model is as follows;

$$\min z = \sum_{i=1}^{N} (\mathbf{c_0} \cdot |\mathbf{x_0}|) + \ (\mathbf{c_1} \cdot |\mathbf{x_1}|) + \cdots + \ (\mathbf{c_n} \cdot |\mathbf{x_n}|) \tag{7}$$

subject to

$$\mathbf{a^t x_i} + \ (1-H) \cdot \mathbf{c^t} \, |\mathbf{x_i}| + (1-H) f \leq \mathbf{y_i} + (1-H) \cdot \mathbf{e_i} \tag{8}$$

$$\mathbf{a^t x_i} - \ (1-H) \cdot \mathbf{c^t} \, |\mathbf{x_i}| - (1-H) f \geq \mathbf{y_i} - (1-H) \cdot \mathbf{e_i} \tag{9}$$

$$N \cdot f \leq \sum_{i=1}^{N} (\mathbf{c_i} \cdot |\mathbf{x_i}|) \tag{10}$$

$$\mathbf{c_i} \geq \ 0 \ \ [for \ \ i = 1, 2, \ldots, N.] \tag{11}$$

3.2 Some Remarks on FFLR Model

FFLR can be introduced as a modified version of PFLR and UFLR. Although the objective function of the FFLR is the same with these two model's objective functions, it only tries to minimize the explained vagueness not the unexplained one represented by F.

In general, a model can minimize the vagueness caused by its independent parameters (explained vagueness). Therefore the proposed FFLR model aims to minimize explained vagueness but optimize the remaining unexplained vagueness part. That is why the objective function does not include f, the vagueness of the unexplained part.

Except for the new parameter f, the boundary constraints of FFLR are similar to PFLR and UFLR. The main idea doesn't change. All models aim to get the estimated Y_i^* to include observed Y_i.

There is a new constraint (10) which makes the model meaningful by limiting F in that the unexplained vagueness part could not be greater than the explained vagueness part.

The constraint of sign is optional. It could be used if the vagueness of the parameters are considered nonnegative as in the PFLR model, or it may be cancelled to catch the trend (if it exists), as in Lee and Chang's UFLR model.

4 Numerical Examples

In this section, two simulated data sets and one real-world data set are used to illustrate how the proposed model (FFLR) performs. In first data set, all parameters are positive. There is a negative parameter in the second one. The third data set is from Tanaka's article to see how it works for real-world data. There are different kinds of independent variables and we have no idea which kind of distribution they have.

The results of FFLR model are compared with Tanaka's PFLR and Lee and Chang's UFLR models. For simplicity, the observations are assumed to be symmetric triangular fuzzy numbers and are denoted by $Y_i = (y_i, e_i)$. Also estimated fuzzy parameters are same as, $A_i = (a_i, c_i)$. The threshold value is $H = 0,5$ for all models.

NOTE: All details about these applications and data sets can be provided by first author.(bekir.cetintav@deu.edu.tr)

Example 1

The data set is obtained by a simulation study with R program. The distributions of independent variables are $X_1, X_2 \in N(2,1)$ and $X_3 \in N(4,1)$. The dependent variable is calculated from following equation, $Y = 2 * X_1 + 3 * X_2 + 2 * X_3 + e$ where $e \in N(0,1)$. The fuzziness of the independent variable is $e_i \in N(4,1)$.

There are two cases for comparisons of predictability and fuzziness, FFLR-PFLR and FFLR-UFLR. Because the fuzzy part of the parameters (c_i) are must be nonnegative in PFLR model, but they are unrestricted in UFLR. Proposed FFLR model modifies PFLR and UFLR models on their own conditions. Final results are given.

Table 1 Comparison PFLR-FFLR, data set-1

n	SST(PFLR)	MSE(PFLR)	Vag(PFLR)	SST(FFLR)	MSE(FFLR)	Vag(FFLR)
10	7.341	0.744	62.921	6.399	0.640	59.467
13	6.476	0.498	88.552	5.905	0.454	84.770
16	8.835	0.552	109.181	8.303	0.519	104.506
18	9.347	0.519	125.071	8.919	0.496	119.163
20	9.740	0.487	138.759	0.107	0.005	132.258

[a]SST:Total Sum of Squares, MSE:Mean Squared Error, Vag:Total Vagueness of Model

The results are compared for different number of data size. $n = 10$ means that the first ten observations in the data set are used and there is no descending or ascending order in the data set.

Table 2 Comparison UFLR-FFLR, data set-1

n	SST(UFLR)	MSE(UFLR)	Vag(UFLR)	SST(FFLR)	MSE(FFLR)	Vag(FFLR)
10	6.470	0.647	62.158	7.584	0.758	58.781
13	11.555	0.889	86.437	7.938	0.611	83.569
16	10.425	0.652	108.026	10.189	0.637	103.776
18	10.935	0.607	123.678	10.548	0.586	117.910
20	11.697	0.585	137.725	11.165	0.558	131.323

[a]SST:Total Sum of Squares, MSE:Mean Squared Error, Vag:Total Vagueness of Model

As can be shown below; proposed FFLR model gives better results from PFLR for both predictability and fuzziness. Also FFLR model gives better results from UFLR for sample sizes 13, 16, 18, 20. Only for sample size 10, UFLR model is better than FFLR for predictability.

Example 2

The data set, as in Tbl. 2, is obtained by a simulation study with R program, too. The distributions of independent variables are $X_1, X_2, X_3 \in N(2,1)$ .The dependent variable is calculated from following equation, $Y = 2 * X_1 - 3 * X_2 + 4 * X_3 + e$ where $e \in N(0,1)$. The fuzziness of the independent variable is $e_i \in N(4,1)$. Different from data set-1, there is an independent variable which has negative effect on the dependent variable.

Table 3 Comparison PFLR-FFLR, data set-2

n	SST(PFLR)	MSE(PFLR)	Vag(PFLR)	SST(FFLR)	MSE(FFLR)	Vag(FFLR)
10	29.627	2.963	71.675	12.267	1.227	63.705
13	11.341	0.872	97.442	9.560	0.735	89.748
16	15.415	0.963	121.750	11.428	0.714	110.332
18	15.525	0.862	141.638	12.468	0.693	126.424
20	16.644	0.832	157.089	13.196	0.660	140.760

[a]SST:Total Sum of Squares, MSE:Mean Squared Error, Vag:Total Vagueness of Model

As can be shown below; proposed FFLR model gives better results from PFLR for both predictability and fuzziness. Also FFLR model gives better results from UFLR for sample sizes 13, 16, 18, 20. Only for sample size 10, UFLR model is better than FFLR for predictability.

Table 4 Comparison UFLR-FFLR, data set-2

n	SST(UFLR)	MSE(UFLR)	Vag(UFLR)	SST(FFLR)	MSE(FFLR)	Vag(FFLR)
10	13.088	1.309	66.818	16.644	1.664	61.066
13	11.052	0.850	96.603	9.542	0.734	87.583
16	15.415	0.963	121.750	11.981	0.749	109.320
18	16.052	0.892	141.614	12.799	0.711	126.021
20	17.294	0.865	156.926	15.043	0.752	140.535

[a]SST:Total Sum of Squares, MSE:Mean Squared Error, Vag:Total Vagueness of Model

Table 5 ComparisonS, data set-3

Model	SST	MSE	Vag	Model	SST	MSE	Vag
PFLR	2,099,050	139,936.7	19,107.68	UFLR	1,692,787	112,852.5	16,617.56
FFLR	1,754,053	116,936.9	19,107.68	FFLR	1,692,787	112,852.5	16,617.57

[a]SST:Total Sum of Squares, MSE:Mean Squared Error, Vag:Total Vagueness of Model

Example 3

Data set-3 is from [10] which is the first article of FLR. There are 5 independent variables (X_1 represents the constant), which are rank of material, first floor space (m^2), second floor space (m^2), number of rooms, number of Japanese-style rooms. Independent variable Y is fuzzy prices of the houses.

Proposed FFLR model gives better results from PFLR for both predictability and fuzziness. However it gives same results with UFLR for data set-3.

5 Conclusions

In this study, a new LP model is proposed for FLR. This new model modifies previous LP models by dividing total vagueness into two parts as explained and unexplained and aims to minimize only explained vagueness. So the estimations of parameters (centers of parameters) and unexplained vagueness are optimized. The results from three examples indicate that the proposed method generally has better performance than PFLR and UFLR in terms of MSE and total fuzziness and it also improves predictability of LP methods. It gives same results with other models only for a few cases and it may be a subject of a future study to find which conditions cause it.

References

1. Chen, Y.: Outliers detection and confidence interval modification in fuzzy regression. Fuzzy Sets Syst. 119(2), 259–272 (2001)
2. Hung, W., Yang, M.: An omission approach for detecting outliers in fuzzy regression models. Fuzzy Sets Syst. 157(23), 3109–3122 (2006)
3. Lee, E.S., Chang, P.: Fuzzy linear regression analysis with spread unconstrained in sign. Comp. Math. Appl. 28(4), 61–70 (1994)
4. Lu, J., Wang, R.: An enhanced fuzzy linear regression model with more flexible spreads. Fuzzy Sets Syst. 160(17), 2505–2523 (2009)
5. Modarres, M., Nasrabadi, E., Nasrabadi, M.: Fuzzy linear regression models with least square errors. Appl. Math. Comput. 163(2), 977–989 (2005)
6. Peters, G.: Fuzzy linear regression with fuzzy intervals. Fuzzy Sets Syst. 63(1), 45–55 (1994)
7. Redden, D.T., Woodall, W.H.: Properties of certain fuzzy linear regression methods. Fuzzy Sets Syst. 64(3), 361–375 (1994)
8. Tanaka, H.: Fuzzy data analysis by possibilistic linear models. Fuzzy Sets Syst. 24(3), 363–375 (1987)
9. Tanaka, H., Hayashi, I., Watada, J.: Possibilistic linear regression analysis for fuzzy data. Eur. J. Oper. Res. 40(3), 389–396 (1989)
10. Tanaka, H., Uejima, S., Asai, K.: Linear regression analysis with fuzzy model. IEEE Trans. Syst. Man Cyb. 12(6), 903–907 (1982)

References

1. Chen, Y.S.: Outliers detection and confidence interval modification in fuzzy regression. Fuzzy Sets Syst. 119(2), 259–272 (2001)
2. Hung, W., Yang, M.: An omission approach for detecting outliers in fuzzy regression models. Fuzzy Sets Syst. 157, 3109–3122 (2006)
3. Lee, E.S., Chang, P.: Fuzzy linear regression analysis with spread unconstrained in sign. Comput. Math. Appl. 28(4), 61–70 (1994)
4. Lu, J., Wang, R.: An enhanced fuzzy linear regression model with more flexible spreads. Fuzzy Sets Syst. 160(17), 2505–2523 (2009)
5. Modarres, M., Nasrabadi, E., Nasrabadi, M.M.: Fuzzy linear regression models with least square errors. Appl. Math. Comput. 163(2), 977–989 (2005)
6. Peters, G.: Fuzzy linear regression with fuzzy intervals. Fuzzy Sets Syst. 63(1), 45–55 (1994)
7. Redden, D.T., Woodall, W.H.: Properties of certain fuzzy linear regression methods. Fuzzy Sets Syst. 64(3), 361–375 (1994)
8. Tanaka, H.: Fuzzy data analysis by possibilistic linear models. Fuzzy Sets Syst. 24(3), 363–375 (1987)
9. Tanaka, H., Hayashi, I., Watada, J.: Possibilistic linear regression analysis for fuzzy data. Eur. J. Oper. Res. 40(3), 389–396 (1989)
10. Tanaka, H., Uejima, S., Asai, K.: Linear regression analysis with fuzzy model. IEEE Trans. Syst. Man Cybern. 12(6), 903–907 (1982)

Fuzzy Least Squares Estimation with New Fuzzy Operations

Jin Hee Yoon and Seung Hoe Choi

Abstract. This paper deals with fuzzy least squares estimation of the fuzzy linear regression model with fuzzy input-output data that has an error structure. The paper proposes fuzzy least squares estimators (FLSEs) for regression parameters based on a suitable metric, and shows that the estimators are fuzzy-type linear estimators. To find these estimators, we first defined a notion of triangular fuzzy matrices whose elements are given as triangular fuzzy numbers, and also provided some operations among all triangular fuzzy matrices. Simple computational examples of this applications are given.

Keywords: Fuzzy least squares estimator, fuzzy random variable, triangular fuzzy matrix.

1 Introduction

The least squares method is the most widely used statistical technique to find the unknown parameters of regression model. But there are many situations where observations cannot be describe accurately. To record these data, we need some approach to handle the uncertainty. Zadeh [23] first introduced the concept of fuzzy sets to explain such uncertainty or vagueness. Tanaka *et al.* [18] introduced fuzzy concept to regression analysis.

Diamond [5] introduced fuzzy least squares estimations for triangular fuzzy numbers. He considered two types of fuzzy linear regression models:the fuzzy

Jin Hee Yoon
School of Economics, Yonsei University,
Seoul 120-749, South Korea
e-mail: `jin9135@yonsei.ac.kr`

Seung Hoe Choi
School of Liberal Arts and Sciences, Korea Aerospace University,
Koyang, 412-791, South Korea
e-mail: `shchoi@kau.ac.kr`

R. Kruse et al. (Eds.): Synergies of Soft Computing and Statistics, AISC 190, pp. 193–202.
springerlink.com © Springer-Verlag Berlin Heidelberg 2013

input-output regression model and the crisp input, fuzzy output model. After that many authors have addressed and attempted to resolve the fuzzy least squares problems. But many studies have emphasized the fuzziness of the response alone, so they deal with crisp input, fuzzy output model. Some authors have discussed the situation in which both the response and the explanatory variables are fuzzy [2, 17, 21]. A common characteristic of these studies is that they treated the regression coefficients as fuzzy numbers. But this approach has a weakness because the spread of the estimated responses widens as the magnitude of the explanatory variables increases, even though the spreads of the observed responses remain roughly constant, or even decrease. So some authors have been studying the fuzzy input-output model with crisp parameters of the model [1, 5, 6, 9, 10, 11, 12, 13, 14, 19, 20].

It is not easy to express the least squares estimators in one formula. Some authors use $\alpha-$level sets to express the estimators [11, 16, 17, 19, 20], and others separate the estimators into three parts, the mode and two spreads [3, 4, 22]. Moreover, some authors do not express the estimators. They find the estimates directly from normal equations in each cases, not from the estimators [5, 14] because it is hard to express the estimators in one compact form. It is important to express the estimators in one formula because that makes it easy to prove the optimal properties or asymptotic theories [13] of the estimators. To overcome this problems, we introduce some operations which are defined later. With the operations, we express the estimators in one compact formula. To find the least squares estimators of the model, we define a new matrix whose elements are triangular fuzzy numbers. It will be called the *triangular fuzzy matrix.* We also have provided suitable operations among all triangular fuzzy matrices. Using the operations, we express the estimators of the fuzzy regression model in one formula.

Note that model fuzzy regression simultaneously considers two different kinds of uncertainty, vagueness and randomness. As such, experimental data can be regarded as sampled from a fuzzy random variable which has both fuzziness and randomness. Taking the new model into account, we introduce some definitions [5, 7, 8, 23] regarding the fuzzy sets and the fuzzy numbers, as well some basic results of the fuzzy theory.

2 Preliminaries

A fuzzy subset of $\mathbb{R}$ is a map, so called the membership function, from $\mathbb{R}$ into $[0,1]$. Thus fuzzy subset A is identified with its membership function $\mu_A(x)$. For any $\alpha \in (0,1]$ the crisp set $A_\alpha = \{x \in \mathbb{R} : \mu_A(x) \geq \alpha\}$ is called the α-cut or α-level set of A. The set of all fuzzy numbers will be denoted by $\mathcal{F}_c(\mathbb{R})$. In fact, there are no general rules to obtain the membership function of a fuzzy observation. As a special case, we often use the following parametric class of fuzzy numbers, the so called LR-fuzzy numbers:

$$\mu_A(x) = \begin{cases} L((m-x)/l) & \text{if } x \le m, \\ R((x-m)/r) & \text{if } x > m \end{cases} \qquad \text{for } x \in \mathbb{R},$$

where $L, R : \mathbb{R}^+ \to [0,1]$ are fixed left-continuous and non-increasing functions with $R(0) = L(0) = 1$ and $R(1) = L(1) = 0$. L and R are called left and right shape functions of X, m the mode of A and l, $r > 0$ are left, right spread of X. We abbreviate an LR-fuzzy number by $A = \langle m, l, r \rangle_{LR}$. The spreads l and r represent the fuzziness of the number and could be symmetric or non-symmetric. If $l = r = 0$, there is no fuzziness of the number, and it is a crisp number. The α-cuts of the fuzzy numbers are given by the intervals

$$A_\alpha = [m - L^{-1}(\alpha)l, \;\; m + R^{-1}(\alpha)r], \quad \alpha \in (0,1].$$

We denote the set of all LR-fuzzy numbers as $\mathcal{F}_{LR}(R)$. In particular, if $L(x) = R(x) = [1-x]^+$ in $A = \langle m, l, r \rangle_{LR}$ then A is called a triangular fuzzy number and denoted by $A = \langle m, l, r \rangle_\Delta$.

Basic operations of fuzzy numbers are defined via the well known *Zadeh's extension principle* [23]. The advantage of LR-fuzzy numbers is that $\oplus$ and $\cdot$ can be expressed by simple operations w.r.t. the parameters m, l, r as following.

$$\langle m_1,\; l_1,\; r_1 \rangle_{LR} \oplus \langle m_2,\; l_2,\; r_2 \rangle_{LR} = \langle m_1 + m_2,\; l_1 + l_2,\; r_1 + r_2 \rangle_{LR}$$

and

$$\lambda \langle m,\; l,\; r \rangle_{LR} = \begin{cases} \langle \lambda m, \;\; \lambda l, \;\; \lambda r \rangle_{LR} & \text{if } \lambda > 0, \\ \langle \lambda m, -\lambda r, -\lambda l \rangle_{LR} & \text{if } \lambda < 0, \\ \langle 0, 0, 0 \rangle_{LR} & \text{if } \lambda = 0. \end{cases}$$

Diamond [5] introduced a metric in the set of all triangular fuzzy numbers. Let $\mathcal{F}_T(\mathbb{R})$ denote the set of all triangular fuzzy numbers in $\mathbb{R}$. For $X, Y \in \mathcal{F}_T(\mathbb{R})$, define

$$d^2(X,Y) = D_2^2(\text{supp}X, \text{supp}Y) + [m(X) - m(Y)]^2, \tag{1}$$

where supp$X$ denotes the compact interval of support of X, and $m(X)$ its mode. If $X = \langle x, \xi^l, \xi^r \rangle_\Delta$, $Y = \langle y, \eta^l, \eta^r \rangle_\Delta$, then

$$d^2(X,Y) = [y - \eta^l - (x - \xi^l)]^2 + [y + \eta^r - (x + \xi^r)]^2 + (y - x)^2. \tag{2}$$

3 Triangular Fuzzy Matrices and Their Operations

Let A be a triangular fuzzy number, then $A = \langle m, l, r \rangle_\Delta$, where m is the mode, l and r are the left and the right spreads of A. We have another representation, $A = (l_a, a, r_a)$, of A. In this case, $m = a$ and l_a and r_a are the left and the right end points of A. The latter expression will be called *vector representation* of A. To define fuzzy element matrices and their operations, we use vector representations of fuzzy numbers.

We denote $M_{\mathbb{R}}$ as the set of all $n \times n$ real crisp matrices. Let $\mathbb{R}^+$ be a set of all non-negative real numbers. Then, we define $\mathcal{F}_{\mathcal{T}}(\mathbb{R}^+)$ is the set of all triangular fuzzy numbers on the non-negative real numbers and $M_{\mathcal{F}_{\mathcal{T}}(\mathbb{R}^+)}$ is the set of all fuzzy element matrices on $\mathcal{F}_{\mathcal{T}}(\mathbb{R}^+)$. From now on, we denote simply $\mathcal{F}_{\mathcal{T}}(\mathbb{R}^+)$ by $\mathcal{F}_{\mathcal{T}}$ and $M_{\mathcal{F}_{\mathcal{T}}(\mathbb{R}^+)}$ by $M_{\mathcal{F}_{\mathcal{T}}}$. We define two types of products in $\mathcal{F}_{\mathcal{T}}$ as following.

Definition 1. For $X = (l_x, x, r_x) \in \mathcal{F}_{\mathcal{T}}$ and $Y = (l_y, y, r_y) \in \mathcal{F}_{\mathcal{T}}$,

$$X \diamond Y = l_x l_y + xy + r_x r_y,$$
$$X \otimes Y = (l_x l_y,\ xy,\ r_x r_y).$$

Clearly, $X \diamond Y \in \mathbb{R}^*$ and $X \otimes Y \in \mathcal{F}_{\mathcal{T}}$. Note that a triangular fuzzy number can be regarded as a vector in $\mathbb{R}^3$, so we define the multiplications of a crisp vector and a triangular fuzzy number. That is, even though X is a crisp vector and Y is a triangular fuzzy number, we also define $\diamond$ and $\otimes$ as above.

We also define operations of among the matrices. For the convenience, we express only the case of $n \times n$.

Definition 2. A *triangular fuzzy matrix (t.f.m.)* is the matrix whose elements are triangular fuzzy numbers. For given two $n \times n$ *t.f.m's*, $\tilde{\Gamma} = [X_{ij}]$, and $\tilde{\Lambda} = [Y_{ij}]$, their addition $\tilde{\Gamma} \oplus \tilde{\Lambda}$ is defined by the $n \times n$ *t.f.m.* $\tilde{\Sigma} = [Z_{ij}]$, where $Z_{ij} = X_{ij} \oplus Y_{ij}$. And two products $\tilde{\Gamma} \diamond \tilde{\Lambda}$ and $\tilde{\Gamma} \otimes \tilde{\Lambda}$ are defined, respectively, as follows:

$$\tilde{\Gamma} \diamond \tilde{\Lambda} = [\sum_{k=1}^{n} X_{ik} \diamond Y_{kj}], \qquad \tilde{\Gamma} \otimes \tilde{\Lambda} = [\bigoplus_{k=1}^{n} X_{ik} \otimes Y_{kj}].$$

Moreover, the product of crisp matrix $A = [a_{ij}]$ and *t.f.m.* $\tilde{\Gamma}$, $A\tilde{\Gamma}$, and scalar multiplication, $k\tilde{\Gamma}(k \in \mathbb{R})$, are defined respectively as follows:

$$\tilde{A}\tilde{\Gamma} = [\bigoplus_{k=1}^{n} a_{ik} X_{kj}], \qquad k\tilde{\Gamma} = [kX_{ij}].$$

Note that $\tilde{\Gamma} \oplus \tilde{\Lambda}$, $\tilde{\Gamma} \otimes \tilde{\Lambda}$, $\tilde{A}\tilde{\Gamma} \in M_{\mathcal{F}_{\mathcal{T}}}$ and $\tilde{\Gamma} \diamond \tilde{\Lambda} \in M_{\mathbb{R}^*}$.

Here we define three types of fuzzy scalar multiplications of crisp matrix.

Definition 3. For given $X \in \mathcal{F}_{\mathcal{T}}$, $\tilde{A} = [A_{ij}] \in M_{\mathbb{R}^*}$ and $\tilde{\Gamma} = [X_{ij}] \in M_{\mathcal{F}_{\mathcal{T}}}$, we define three fuzzy scalar multiplications, XA, $X \diamond \tilde{\Gamma}$ and $X \otimes \tilde{\Gamma}$.

$$X\tilde{A} = [a_{ij}X], \qquad X \diamond \tilde{\Gamma} = [X \diamond X_{ij}], \qquad X \otimes \tilde{\Gamma} = [X \otimes X_{ij}].$$

Note that $X\tilde{A}$, $X \otimes \tilde{\Gamma} \in M_{\mathcal{F}_{\mathcal{T}}}$ and $X \diamond \tilde{\Gamma} \in M_{\mathbb{R}^*}$.

4 Fuzzy Least Squares Estimation

Now we consider the multiple regression model

$$Y_i = \beta_0 \oplus \beta_1 X_{i1} \oplus \cdots \oplus \beta_p X_{ip} \oplus \Phi_i, \qquad i = 1, \cdots, n, \tag{3}$$

where X_{ij}, Y_i $(j = 1, \cdots, p)$ are fuzzy random variables respectively, β_j are unknown regression crisp parameters to be estimated on the basis of fuzzy observations on Y_i and X_{ij}. And Φ_i are assumed to be fuzzy error terms, which express randomness and fuzziness. They are represented by $X_{ij} = (l_{x_{ij}}, x_{ij}, r_{x_{ij}})$ and $Y_i = (l_{y_i}, y_i, r_{y_i})$ for $i = 1, \cdots, n$, $j = 1, \cdots, p$. We assume that Φ_i are the fuzzy random errors for expressing fuzziness and randomness, which are represented by $\Phi_i = (\theta_i^l,\ \epsilon_i,\ \theta_i^r)$ with crisp random variables ϵ_i, θ_i^l, θ_i^r, and we regard ϵ_i, θ_i^l, θ_i^r as crisp random variables [13]. And $\Phi_i = (\theta_i^l,\ \epsilon_i,\ \theta_i^r)$ is constrained that the spreads of Y_i to be positive a.s.

Note that in (1), there are 2^p distinct representations of $\sum_{j=0}^{p} \beta_j X_{ij}$ which depend on the signs of β_j for $j = 1, \cdots, p$. We can encompass all cases by

$$l_{x_{ij}} = \begin{cases} x_{ij} - \xi_{l_{ij}} & \text{if} \quad \beta_j \geq 0, \\ x_{ij} + \xi_{r_{ij}} & \text{if} \quad \beta_j < 0 \end{cases}$$

$$r_{x_{ij}} = \begin{cases} x_{ij} + \xi_{r_{ij}} & \text{if} \quad \beta_j \geq 0, \\ x_{ij} - \xi_{l_{ij}} & \text{if} \quad \beta_j < 0, \end{cases}$$

where $\xi_{l_{ij}}$ and $\xi_{r_{ij}}$ are the left and right spreads of X_{ij}, respectively.

On the other hand, by the metric d in (2) and (3), we obtain

$$d^2\left(Y_i,\ \sum_{j=0}^{p} \beta_j X_{ij}\right) = \left(l_{y_i} - \sum_{j=0}^{p} \beta_j l_{x_{ij}}\right)^2 + \left(y_i - \sum_{j=0}^{p} \beta_j x_{ij}\right)^2 + \left(r_{y_i} - \sum_{j=0}^{p} \beta_j r_{x_{ij}}\right)^2$$

for $i = 1, \cdots, n$. In case of $j = 0$, $l_{x_{0i}} \equiv x_{01} \equiv r_{x_{0i}} \equiv 1$. Now, we want to minimize following objective function $Q = Q(\beta_0, \beta_1, \cdots, \beta_p)$.

$$Q(\beta_0, \beta_1, \cdots, \beta_p) = \sum_{i=1}^{n} d^2\left(Y_i,\ \sum_{j=0}^{p} \beta_j X_{ij}\right)$$

For $k = 0, 1, \cdots, p$, we obtain

$$\frac{\partial Q}{\partial \beta_k} = 2\sum_{i=1}^{n}\left(l_{y_i} - \sum_{j=0}^{p}\beta_j l_{x_{ij}}\right) l_{x_{ik}} + 2\sum_{i=1}^{n}\left(y_i - \sum_{j=0}^{p}\beta_j x_{ij}\right) x_{ik} + 2\sum_{i=1}^{n}\left(r_{y_i} - \sum_{j=0}^{p}\beta_j r_{x_{ij}}\right) r_{x_{ik}}.$$

For each $k = 0, 1, \cdots, p$, $\frac{\partial Q}{\partial \beta_k} = 0$ results the normal equation, which has $\hat{\beta}_j$ as solutions,

$$\sum_{j=0}^{p}\hat{\beta}_j \sum_{i=1}^{n}\left(l_{x_{ik}} l_{x_{ij}} + x_{ik}x_{ij} + r_{x_{ik}} r_{x_{ij}}\right) = \sum_{i=1}^{n}\left(l_{x_{ik}} l_{y_i} + x_{ik}y_i + r_{x_{ik}} r_{y_i}\right). \quad (4)$$

Here, we define the design matrix $\tilde{X}$ as $[(l_{x_{ik}}, x_{ik}, r_{x_{ik}})]_{n\times(p+1)}$, i.e.,

$$\tilde{X} = \begin{bmatrix} (1,1,1) & (l_{x_{11}}, x_{11}, r_{x_{11}}) & \cdots & (l_{x_{1p}}, x_{1p}, r_{x_{1p}}) \\ \vdots & \vdots & \ddots & \vdots \\ (1,1,1) & (l_{x_{n1}}, x_{n1}, r_{x_{n1}}) & \cdots & (l_{x_{np}}, x_{np}, r_{x_{np}}) \end{bmatrix},$$

and define $\tilde{\mathbf{y}}$ as $[(l_{y_i}, y_i, r_{y_i})]_{n\times 1} = [(l_{y_1}, y_1, r_{y_1}), \cdots, (l_{y_n}, y_n, r_{y_n})]^t$. Then, the coefficient matrix of the system of normal equations which consists of (5) can be represented by

$$\tilde{X}^t \diamond \tilde{X} = \left[\sum_{i=1}^{n}(l_{x_{ik}} l_{x_{ij}} + x_{ik}x_{ij} + r_{x_{ik}} r_{x_{ij}})\right]_{(p+1)\times(p+1)}.$$

And the right-hand side of the system is represented by

$$\tilde{X}^t \diamond \tilde{\mathbf{y}} = \left[\sum_{i=1}^{n}(l_{x_{ik}} l_{y_i} + x_{ik}y_i + r_{x_{ik}} r_{y_i})\right]_{(p+1)\times 1}$$

for $k = 0, 1, \cdots, p$. Consequently, if $det(\tilde{X}^t \diamond \tilde{X}) \neq 0$, we have

$$\hat{\boldsymbol{\beta}} = (\tilde{X}^t \diamond \tilde{X})^{-1}\tilde{X}^t \diamond \tilde{\mathbf{y}}. \quad (5)$$

Next, we apply above estimation to some examples and compare our method with several methods through R-square and RMSE (Root Mean Squared Error) that will be proposed.

Example 1. A simple example is taken from [17]. We compare the performance of the estimators of some various methods: Kao and Chyu [11, 12],

Nasrabadi and Nasrabadi [15], Bargiela *et al.* [1] and our proposed least squares estimation. The estimated models are followings, respectively.

$$Y_{KC} = 3.565 + 0.522X + (-0.962, -0.011, 0.938),$$

$$Y_{NN} = 3.5767\tilde{\oplus}(0.5467, 1) \cdot X,$$

$$Y_{BPN} = 3.4467 + 0.5360X$$

and we get

$$Y = -4.2047 \oplus 1.569X$$

using our proposed least squares estimation.

We introduce the *coefficient of determination* [6] for fuzzy regression model as a measure of the quality of the best fit which is given by

$$R^2 = 1 - \frac{\sum_{i=1}^{n} d^2(\sum_{j=0}^{p} \hat{\beta}_j X_{ij}, Y_i)}{\sum_{i=1}^{n} d^2(Y_i, \overline{Y})}, \tag{6}$$

with $\overline{Y} = \frac{1}{n} \bigoplus_{i=1}^{n} Y_i$.

Using (6), the calculated R-squares of Kao and Chyu [11, 12], Nasrabadi and Nasrabadi [15], Bargiela *et al.* [1] and our proposed estimations are 0.4314, 0.4684, 0.4596 and 0.8289, respectively. In addition to, for a measure of the prediction accuracy, we modify RMSE in [1] for fuzzy regression models as follows.

$$RMSE = \sqrt{\frac{1}{n} \sum_{i=1}^{n} d^2(Y_i, \hat{Y}_i)}, \tag{7}$$

where $d(Y_i, \hat{Y}_i)$ is the residual error.

Table 1 Numerical data with RMSE and R-square for Example 1

Input	Output	Residual Errors			
(l_{x_i}, x_i, r_{x_i})	(l_{y_i}, y_i, r_{y_i})	Kao('03)	Nasrabadi('04)	Bargiela('07)	Proposed
(3.5, 4.0, 4.5)	(1.5, 2.0, 2.5)	6.387	6.519	6.228	0.421
(5.0, 5.5, 6.0)	(3.0, 3.5, 4.0)	5.164	5.341	5.024	1.652
(6.5, 7.5, 8.5)	(4.5, 5.5, 6.5)	3.474	3.836	3.469	3.662
(6.0, 6.5, 7.0)	(6.5, 7.0, 7.5)	1.010	0.226	0.349	1.789
(8.0, 8.5, 9.0)	(8.0, 8.5, 9.0)	1.338	0.479	0.922	1.166
(7.0, 8.0, 9.0)	(9.5, 10.5, 11.5)	4.845	4.472	4.834	3.814
(10.0, 10.5, 11.0)	(10.5, 11, 11.5)	3.550	2.915	3.351	2.236
(9.0, 9.5, 10.0)	(12.0, 12.5, 13.0)	6.980	6.460	6.869	3.142
RMSE		20.891	19.531	19.853	6.287
R-square		0.4314	0.4684	0.4596	0.8289

The RMSEs of Kao and Chyu [11, 12], Nasrabadi and Nasrabadi [15], Bargiela et al. [1] and our proposed estimations are 20.891, 19.531, 19.853 and 6.287, respectively, which show the excellence of our proposed estimation.

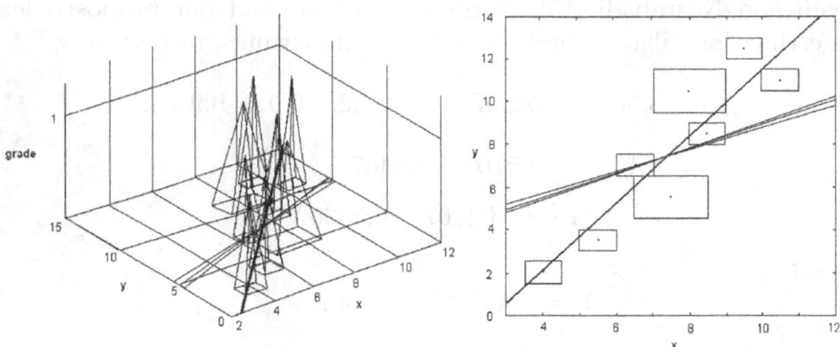

Fig. 1 Fuzzy linear regression in Example 1

Table 2 Numerical data with RMSE and R-square for in Example 2

Example 2.

Input $(l_{x_{i1}}, x_{i1}, r_{x_{i1}})$	$(l_{x_{i2}}, x_{i2}, r_{x_{i2}})$	Output (l_{y_i}, y_i, r_{y_i})	Residual Errors Wu('03)	Proposed
(151, 274, 322)	(1432, 2450, 3461)	(111, 162, 194)	19.504	9.720
(101, 180, 291)	(2448, 3254, 4463)	(88, 120, 161)	30.411	23.812
(221, 375, 539)	(2592, 3802, 5116)	(161, 223, 288)	38.392	27.741
(128, 205, 313)	(1414, 2838, 3252)	(83, 131, 194)	6.192	10.478
(62, 86, 112)	(1024, 2347, 3766)	(51, 67, 83)	12.972	14.280
(132, 265, 362)	(2163, 3782, 5091)	(124, 169, 213)	39.006	28.600
(66, 98, 152)	(1687, 3008, 4325)	(62, 81, 102)	19.600	13.025
(151, 330, 463)	(1524, 2450, 3864)	(138, 192, 241)	53.405	42.753
(115, 195, 291)	(1216, 2137, 3161)	(82, 116, 159)	20.988	22.883
(35, 53, 71)	(1432, 2560, 3782)	(41, 55, 71)	7.627	6.873
(307, 430, 584)	(2592, 4020, 5562)	(168, 252, 367)	25.550	38.570
(284, 372, 498)	(2792, 4427, 6163)	(178, 232, 346)	39.796	54.705
(121, 236, 370)	(1734, 2660, 4094)	(111, 144, 198)	41.324	32.086
(103, 157, 211)	(1426, 2088, , 3312)	(78, 103, 148)	14.206	9.938
(216, 370, 516)	(1785, 2605, 4042)	(167, 212, 267)	49.749	40.106
RMSE			990.866	823.202
R-square			0.9223	0.9355

Wu [20] designed an example for multiple fuzzy linear regression as in Table 2. The estimated model of Wu is

$$Y_{Wu} = 3.4526 + 0.4960X_1 + 0.0092X_2$$

and our estimated model is

$$Y = 23.2368 \oplus 0.4917X_1 \oplus 0.0039X_2.$$

Using the proposed R-square in (6), we get 0.9223 for estimated model of Wu and 0.9355 for our proposed estimation. Additionally, the RMSE of Wu's is 990.866 and 823.202 for our proposed estimation which performs better.

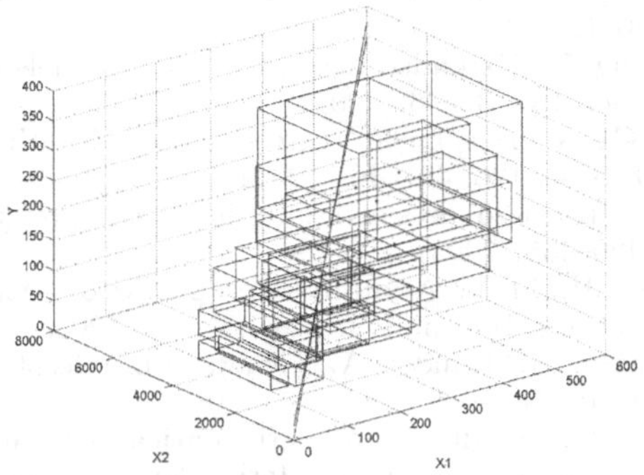

Fig. 2 Fuzzy linear regression in Example 2

5 Conclusions

Many authors have proposed fuzzy least squares estimators in different ways, but it is hard to represent those estimators concisely. For this purpose, this paper defined a suitable matrix which called the triangular fuzzy matrix (*t.f.m*), and defined some operations for *t.f.m*. If we use *t.f.m* and the proposed operations, we can represent the fuzzy least squares estimators briefly and we can also represent the estimators as fuzzy-type linear estimators. And we presented some examples in other papers and compared them with our results through r-squares. We discussed the case of fuzzy input-output model, but if we expand our study to general case, then we can apply these estimators to various models. So, further research needs to generalize the model and we need to investigate more properties of these estimators.

References

1. Bargiela, A., Pedrycz, W., Nakashima, T.: Multiple regression with fuzzy data. Fuzzy Set. Syst. 158, 2169–2188 (2007)
2. Celminš, A.: Least squares model fitting to fuzzy vector data. Fuzzy Set. Syst. 22, 245–269 (1987)
3. Chang, P.-T., Lee, E.S.: Fuzzy least absolute deviations regression and the conflicting trends in fuzzy parameters. Comput. Math. Appl. 28, 89–101 (1994)
4. Choi, S.H., Yoon, J.H.: General fuzzy regression using least squares method. Int. J. Syst. Sci. (accepted)
5. Diamond, P.: Fuzzy least squares. Inform. Sci. 46, 141–157 (1988)
6. Diamond, P., Körner, R.: Extended fuzzy linear models and least squares estimates. Comput. Math. Appl. 33, 15–32 (1997)
7. Diamond, P., Kloeden, P.: Metric Spaces of Fuzzy Sets: Theory and Applications. World Scientific (1989)
8. Dubois, D., Prade, H.: Fuzzy sets and systems: theory and applications. Academic Press, New York (1980)

9. Hong, D.H., Hwang, C.: Support vector fuzzy regression machines. Fuzzy Set. Syst. 138, 271–281 (2003)
10. Hong, D.H., Hwang, C.: Extended fuzzy regression models using regularization method. Inform. Sci. 164, 31–46 (2004)
11. Kao, C., Chyu, C.: A fuzzy linear regression model with better explanatory power. Fuzzy Set. Syst. 126, 401–409 (2002)
12. Kao, C., Chyu, C.: Least-squares estimates in fuzzy regression analysis. Eur. J. Oper. Res. 148, 426–435 (2003)
13. Kim, H.K., Yoon, J.H., Li, Y.: Asymptotic properties of least squares estimation with fuzzy observations. Inform. Sci. 178, 439–451 (2008)
14. Ming, M., Friedman, M., Kandel, A.: General fuzzy least squares. Fuzzy Set. Syst. 88, 107–118 (1997)
15. Nasrabadi, M.M., Nasrabadi, E.: A mathematical-programming approach to fuzzy linear regression analysis. Appl. Math. Comput. 155, 873–881 (2004)
16. Parchami, A., Mashinchi, M.: Fuzzy estimation for process capability indices. Inform. Sci. 177, 1452–1462 (2006)
17. Sakawa, M., Yano, H.: Multiobjective fuzzy linear regression analysis for fuzzy input-output data. Fuzzy Set. Syst. 47, 173–181 (1992)
18. Tanaka, H., Uejima, S., Asai, K.: Linear regression analysis with fuzzy model. IEEE Trans. Syst. Man Cyb. 12(6), 903–907 (1982)
19. Wu, H.-C.: Linear regression analysis for fuzzy input and output data using the extension principle. Comput. Math. Appl. 45, 1849–1859 (2003)
20. Wu, H.-C.: Fuzzy estimates of regression parameters in linear regression models for imprecise input and output data. Comput. Stat. Data Anal. 42, 203–217 (2003)
21. Yang, M., Lin, T.: Fuzzy least-squares linear regression analysis for fuzzy input-output data. Fuzzy Set. Syst. 126, 389–399 (2002)
22. Yoon, J.H., Choi, S.H.: Separate fuzzy regression with crisp input and fuzzy output. J. Korean Data Inform. Sci. 18, 301–314 (2007)
23. Zadeh, L.A.: Fuzzy sets. Inform. Control 8, 338–353 (1965)

Efficient Calculation of Kendall's τ for Interval Data

Olgierd Hryniewicz and Karol Opara

Abstract. Calculation of the strength of dependence in the case of interval data is computation-wise a very demanding task. We consider the case of Kendall's τ statistic, and calculate approximations of its minimal and maximal values using very easy to compute heuristic approximations. Using Monte Carlo simulations and more accurate calculations based on an evolutionary algorithm we have evaluated the effectiveness of proposed heuristics.

Keywords: Dependence measure, heuristics, interval data.

1 Introduction

Statistical analysis of dependencies between random variables requires calculation of appropriate statistics. Except for the case of the bivariate normal distribution, nonparametric statistics, like Spearman's ρ or Kendall's τ, are recommended for the analysis of monotone (not only linear!) dependence. These statistics are based on the idea of ranks, and thus belong to a general class of rank statistics.

In case of imprecise statistical data, described by fuzzy sets, the problem of the calculation of fuzzy-valued rank statistics becomes very hard. The exact algorithm proposed in Hébert *et al.* [5] is effective only for very small samples. Denœux *et al.* [2] presented a general approach for the calculation of fuzzy rank statistics using imprecise data, and proposed to use for this purpose the algorithm for the random generation of linear extensions of a partial order originally introduced by Bubley and Dyer. Unfortunately, this improved algorithm performs well only for relatively small samples (not larger than 30,

Olgierd Hryniewicz · Karol Opara
Systems Research Institute, Warsaw, Poland
e-mail: {hryniewi,opara}@ibspan.waw.pl

R. Kruse et al. (Eds.): Synergies of Soft Computing and Statistics, AISC 190, pp. 203–210.
springerlink.com © Springer-Verlag Berlin Heidelberg 2013

according to the authors of [2]). The usage of faster computers and more efficient numerical algorithms (e.g. parallelization) may extend this limit, but not too much due to the complexity of the problem.

Hryniewicz and Szediw [7] calculated of Kendall's τ statistic for imprecise autocorrelated data. It was used in an application of this statistic requiring relatively large samples (not less than 50 observations). For this application they proposed a heuristic algorithm for the calculation of approximate values of both, lower and upper, limits of Kendall's τ in the case of interval data. Monte Carlo experiments have shown that those approximations are usually sufficient from a practical point of view.

In this paper we use the same approach as in [7] for the calculation of the approximate values of lower and upper limits of Kendall's τ assuming independence of the pairs of observations. We consider 28 different heuristics (14 for the calculation of the lower limit, and 14 for the calculation of the upper limit) that are described in the second section of the paper. The necessity to use such many heuristics stems from the fact that neither of them dominates the others. This have been shown in simulation experiments presented in the third section of the paper.

The heuristics used for the calculation of the interval values of Kendall's τ consist in generation of certain sets of crisp data that belong to intervals constituting original imprecise interval data. These artificial sets of data may be used as the initial population for the calculation of more precise limits using biologically inspired optimization algorithms such as evolutionary algorithms. The paper is concluded in the last section where the directions for future work are also indicated.

2 Calculation of the Interval-Valued Kendall's τ

Analysis of statistical dependence is one of the most important areas of mathematical statistics. For data given by a p-dimensional random vector $X_1, X_2, \ldots, X_p$ the full description of statistical dependencies between the components of this vector is implied by the knowledge of its p-dimensional cumulative distribution function $F(x_1, x_2, \ldots, x_p)$. In his seminal paper Sklar [11] has proved that for every two-dimensional cumulative probability distribution function $H(x, y)$ with one-dimensional marginal cumulative probability functions denoted by $F(x)$ and $G(y)$, respectively, there exists a unique function C, called a *copula*, such that $H(x, y) = C(F(x), G(y))$. Later on, the concept of the copula has been generalized for the case of any p-dimensional probability distribution. The formal definition of the copula can be found in many sources, such as e.g. the monograph by Nelsen [8].

The coefficient of association (dependence) τ was introduced by Kendall. Let $(X_i, Y_i), i = 1, \ldots, n$ be a random sample representing n independent pairs of observations of dependent random variables X and Y. Then,

Kendall's τ_n sample statistic which measures the association between random variables X and Y is given by the following alternative formula proposed by Genest and Rivest [4]

$$\tau_n = \frac{4}{n-1}\sum_{i=1}^{n-1} V_i - 1, \tag{1}$$

where

$$V_i = \frac{card\{j : X_j < X_i, Y_j < Y_i\}}{n-2}, i = 1, \dots, n. \tag{2}$$

When the vectors $(X_1, X_2, \dots, X_n)$ and $(Y_1, Y_2, \dots, Y_n)$ are mutually independent, the pairs of observations (X_i, Y_i), $i = 1, \dots, n$ are also independent, and the probability distribution of (1) is known. Its expected value is equal to $E(\tau_n) = 0$, and its variance is equal to $Var(\tau_n) = \frac{2(2n+5)}{9n(n-1)}$. For sufficiently large sample size n Kendall's τ_n has the normal distribution with these parameters. Its easy to show that (1) is also a rank statistic.

Genest and MacKay [3] considered the population version of the Kendall's τ. Let, for a given copula $C(x, y)$, $K(t)$ be the cumulative probability function of the random variable $T = C(U_1, U_2)$, where U_1 and U_2 are random variables uniformly distributed on $[0, 1]$. The following relation links a copula with Kendall's τ:

$$\tau = 3 - 4\int_0^1 K(t)dt \tag{3}$$

The special cases of (3) for different specific copulas are given in many papers and textbooks, such as e.g. [8].

Now, let us assume that instead of crisp values of (X_i, Y_i), $i = 1, \dots, n$ we observe imprecise values $(\mathbf{X}_i, \mathbf{Y}_i)$, $i = 1, \dots, n$ where $\mathbf{X}_i : [X_{i,L}, X_{i,U}]$ and $\mathbf{Y}_i : [Y_{i,L}, Y_{i,U}]$. For such observed data the observed value of Kendall's τ will be also imprecise, and given as the interval $\boldsymbol{\tau}_n = [\tau_{n,L}, \tau_{n,U}]$, where the values of $\tau_{n,L}$ and $\tau_{n,U}$ are obtained by inserting in (1) instead of V_i the respective values

$$V_{i,L} = \min_{\substack{X_j \in [X_{j,L}, X_{j,U}] \\ Y_j \in [Y_{j,L}, Y_{j,U}]}} \frac{card\{j : X_j < X_i, Y_j < Y_i\}}{n-2}, \tag{4}$$

$$V_{i,U} = \max_{\substack{X_j \in [X_{j,L}, X_{j,U}] \\ Y_j \in [Y_{j,L}, Y_{j,U}]}} \frac{card\{j : X_j < X_i, Y_j < Y_i\}}{n-2}. \tag{5}$$

An example of imprecise data considered in this paper is shown in Fig. 1. Each of four rectangles represents a pair of interval data vectors $(\mathbf{x}_i, \mathbf{y}_i)$, while diamonds, circles and asterisks denote observations from three possible

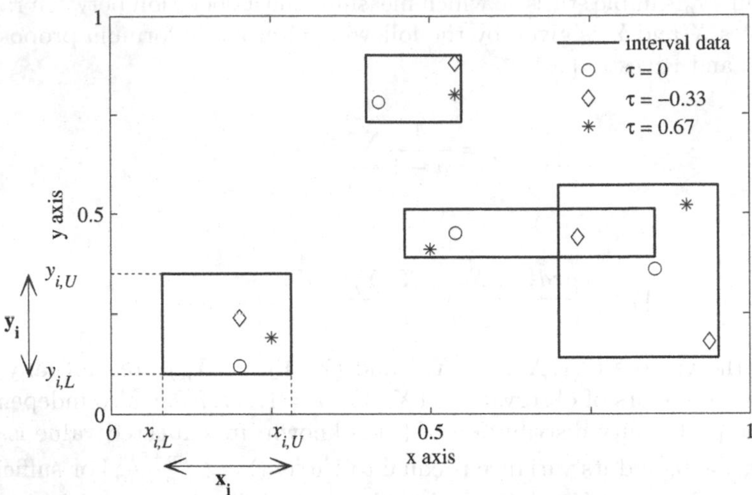

Fig. 1 Interval data for $n = 4$ imprecise observations and three possible crisp observations resulting in different values of Kendall's τ

crisp data vectors. Each of these vectors represents four points (x_i, y_i) such that $x_i \in \mathbf{x}_i$ and $y_i \in \mathbf{y}_i$ for $i = 1, \ldots, 4$ and each gives a different value of Kendall's τ. Optimization tasks considered in this study (4)–(5) consist in finding such possible crisp observations, which yield minimal and maximal value of the correlation coefficient, i.e. $\tau_{n,L}$ and $\tau_{n,U}$. The search space contains all possible crisp observations $([x_{1,L}, x_{1,U}] \times \cdots \times [x_{n,L}, x_{n,U}]) \times ([y_{1,L}, y_{1,U}] \times \cdots \times [y_{n,L}, y_{n,U}]) \subset \mathbb{R}^{2n}$. This is a $2n$-dimensional continuous optimization problem with box constraints, which can be solved with general purpose black-box optimization algorithms.

Denœux *et al.* [2] represented this problem in the form of integer programming, exploiting the fact that Kendall's τ is a rank statistic with a finite number of possible values. This allows for exact calculation of minimal and maximal values of the correlation coefficient for small samples. Finding approximate solutions for larger samples requires, however, implementing a problem-specific combinatorial optimization algorithm.

3 Heuristic Solutions

Hryniewicz and Szediw [7] have found that specific patterns depicting strongly correlated observations can be used for the construction of heuristic algorithms that find minimal and maximal values of measures of dependence in presence of interval data. Close look at (1)–(2) reveals that in the case of strongly positively correlated data there exists a permutation of observed vectors $(x_{(i)}, y_{(i)}), i = 1, \ldots, n$ such that both sets $x_{(1)}, \ldots, x_{(n)}$ and $y_{(1)}, \ldots, y_{(n)}$ form either decreasing (or nearly decreasing) or increasing (or nearly increasing) sequences of points. Similarly, in the case of strongly negatively correlated data one of these permuted sets form an decreasing sequence of points, and the second one forms an increasing sequence. If we look at (4)–(5) we can see that we have to find similar sets of points $x^{\star}_{(1)}, \ldots, x^{\star}_{(n)}$ and $y^{\star}_{(1)}, \ldots, y^{\star}_{(n)}$ such that for all $i = 1, \ldots, n$ $x^{\star}_{(i)} \in \mathbf{x}_i = [x_{(i),L}, x_{(i),U}]$ and $y^{\star}_{(i)} \in \mathbf{y}_i = [y_{(i),L}, y_{(i),U}]$. This can be done using a simple algorithm, whose pseudocode is presented below.

Algorithm 1. Minimization (maximization) heuristic—finding $\hat{\tau}_{n,L}$ $(\hat{\tau}_{n,U})$

```
Step 1: order first interval variable to obtain O_p^t
x*_(i) ← sort first variable x_i = [x_(i),L, x_(i),U] decreasing for i = 1,...,n
Step 2: compute values for the second interval variable to obtain T^t
y*_(1) ← y_(1),U                        (for maximization use y*_(1) ← y_(1),L)
for k = 1,2,...,n− do
    y' ← y*_(k) − ε                     (for maximization use y' ← y*_(k) + ε)
    y'' ← min (y', y_(k+1),U)
    y*_(k+1) ← max (y'', y_(k+1),L)
end for
return pair of series (x*_(i), y*_(i)) for i = 1,...,n as (O_p^t, T^t)
```

In this algorithm we first order pairs of interval data vectors $(\mathbf{x}_i, \mathbf{y}_i)$ in such a way that certain points $x^{\star}_{(1)}, \ldots, x^{\star}_{(n)}$ (or $y^{\star}_{(1)}, \ldots, y^{\star}_{(n)}$) form a non-increasing (non-decreasing) series. Then we find the respective values $y^{\star}_{(1)}, \ldots, y^{\star}_{(n)}$ (or $x^{\star}_{(1)}, \ldots, x^{\star}_{(n)}$). Let's denote the n-dimensional variable ordered according to some defined values by O^t_p. The lower index p can take three values: u (if the upper limit of the data interval is taken), c (if the center of the data interval is taken), and l (if the lower limit of the data interval is taken). The upper index t can take two values: d (if points are ordered in a non-increasing series), and a (if points are ordered in a non-decreasing series). The variable whose values are computed using our algorithm is denoted by T^t, where the upper index indicates the direction of the trend (d or a).

For finding the maximal value of the interval-valued $\boldsymbol{\tau}$ statistic we considered six types of heuristics described as: (O^t_p, T^t) (where $p = u, c, l$ and $t = d, a$). Because of symmetric usage of variables $\mathbf{X}$ and $\mathbf{Y}$ we have used altogether 12 heuristics of these types. Moreover, we used heuristics (T^t, T^t) (where $t = d, a$), described together as (T, T), for which the values of both variables have been calculated using our heuristic algorithms. For finding the minimal endpoint of $\boldsymbol{\tau}$ we considered the following heuristics: (O^a_p, T^d), (O^d_p, T^a), and (T^a, T^d). Because of symmetric usage of variables $\mathbf{X}$ and $\mathbf{Y}$ we have used altogether 14 heuristics of these types.

Calculation of approximate minimal and maximal value of Kendall's $\boldsymbol{\tau}$ using the heuristics described above is very simple. Therefore, in practice one can use all of them, and find the best solution. However, it may be interesting to investigate their effectiveness in different settings. In order to do so we have performed exhaustive simulation experiments. The crisp data were simulated from several copulas (normal, Clayton, Gumbel, Frank, FGM) with standard normal marginals. Parameter setting of the copulas ensured obtaining requested strength of dependence measured by Kendall's τ. Then, the crisp data were replaced with intervals of random length and different location around crisp points. The level of imprecision was defined by setting the maximum width of an interval, measured as the multiplicity z of the standard deviations of the marginals. In the experiment we calculated percentages of cases in which proposed heuristics yielded the optimal (i.e., respectively, minimal and maximal) results. These percentages can be used for the indication of the effectiveness of the proposed heuristics in given settings. The simulation experiment revealed that *the type of copula does not influence the effectiveness of considered heuristics.* However, this effectiveness *strongly depends upon the strength of dependence, sample size, and the level of imprecision.*

Table 1 Percentages of best results for different heuristics; Clayton copula, $n = 100$

	τ	z	(O_u^d, T^d)	(O_c^d, T^d)	(O_l^l, T^d)	(O_u^a, T^a)	(O_c^a, T^a)	(O_l^a, T^a)	(T, T)
Min.	0.8	2.0	2.2	38.8	2.9	3.6	49.7	2.8	0.0
	0.8	0.5	9.9	22.5	9.9	13.3	30.6	13.8	0.0
	−0.1	0.5	15.4	9.2	15.3	12.6	7.7	13.4	26.4
	−0.8	2.0	0.9	0.1	0.7	0.1	0.0	0.3	99.2
Max.	0.8	2.0	0.4	0.0	0.2	0.2	0.0	0.2	99.0
	−0.1	0.5	16.5	10.9	16.2	18.8	13.2	19.8	4.6
	−0.8	2.0	1.6	16.1	1.3	6.2	72.1	2.7	0.0

Consider, for example, the data generated using the Clayton copula $\left(C(u,v) = \max\left(\left[u^{-\alpha} + v^{-\alpha} - 1\right]^{-1/\alpha}, 0\right), \alpha \in [-1, \infty) \setminus 0\right)$. Table 1 shows the percentages of best results for calculation of the maximal endpoint of the interval $\boldsymbol{\tau}$ and a few parameter settings. It seems that neither of the considered types of heuristics is significantly better than others. However, in special cases (e.g. for strong dependencies) some of them are significantly better.

4 Comparison of Efficiency

In the case of large samples of imprecise data, calculating the exact value of correlation coefficient $\boldsymbol{\tau}$ has unacceptable computational complexity. Therefore, in order to evaluate the efficiency of the proposed heuristic solutions it is necessary to compare them with other approximate solutions obtained with larger computational effort, and thus more accurate. For this purpose

Table 2 Estimates of $\boldsymbol{\tau} = [\tau_{n,L}, \tau_{n,U}]$ obtained by heuristic, evolutionary algorithm and Monte Carlo search; the best results are typed in bold; Clayton copula, $n = 100$

		Minimal endpoint $\hat{\tau}_{n,L}$				Maximal endpoint $\hat{\tau}_{n,U}$			
τ	z	Heur.	Heur. DE	Only DE	Only MC	Heur.	Heur. DE	Only DE	Only MC
−0.8	0.5	**−0.86**	**−0.86**	−0.85	−0.84	−0.81	**−0.76**	**−0.76**	−0.77
	2.0	**−0.85**	**−0.85**	−0.70	−0.65	−0.51	**−0.35**	**−0.35**	−0.41
−0.1	0.5	−0.17	**−0.20**	**−0.20**	−0.18	−0.13	**−0.09**	**−0.09**	−0.11
	2.0	−0.13	**−0.26**	**−0.26**	−0.19	0.03	**0.14**	**0.14**	0.07
0.1	0.5	0.12	**0.07**	**0.07**	0.08	0.15	**0.17**	**0.17**	0.15
	2.0	0.00	**−0.13**	**−0.13**	−0.06	0.17	0.23	**0.24**	0.18
0.8	0.5	0.78	**0.73**	**0.73**	0.74	**0.84**	**0.84**	0.83	0.81
	2.0	0.51	**0.39**	**0.40**	0.45	**0.77**	**0.77**	0.73	0.68

we compared results of the heuristic solution with uniform random sampling and an evolutionary algorithm. We used a variant of a state-of-the-art optimization method [1] called Differential Evolution (DE) [10]. Description of this algorithm and parameter setting used in this study is given in paper [9] as DE/rand/∞/bin (except for population size, which was increased to $4 \cdot n$ to account for higher problem dimension).

Performance of each of these methods was tested on eight test problems generated with use of Clayton copula for sample size $n = 100$, crisp origins of interval data $\tau \in \{-0.8, -0.1, 0.1, 0.8\}$ and uncertainty parameter $z \in \{0.5, 2\}$. Table 2 presents estimates of endpoints of interval $\boldsymbol{\tau}$ coefficient for:

- the best of 14 heuristic solutions (Heur.),
- optimization with evolutionary algorithm initialized with use of heuristic solutions (Heur. DE),
- evolutionary algorithm initialized randomly (Only DE),
- uniform random (Monte Carlo) samping (Only MC).

Since both Differential Evolution and Monte Carlo are nondeterministic methods, in Table 2 we report medians over 100 independent runs of each. Variability of final values was however small both absolutely and relatively. In each case the interquartile range of final $\tau_{n,L}$ and $\tau_{n,U}$ estimates was lower than 0.02 and in most of them lower than 0.005. Consequently, to save space we do not report it in Table 2. The stopping criterion of both DE and MC methods was set to exceeding $4 \cdot 10^4$ calculations of Kendall's τ for a pair of crisp vectors. This accounted to 100 generations of an evolutionary algorithm.

Statistical significance of differences between reported median values was tested with Wilcoxon rank sum test adjusted for multiple comparisons with Bonferroni correction, i.e. at confidence level $1 - \frac{0.05}{4}$. The best results are typed in bold. In most cases, they were obtained by either variants of Differential Evolution, followed by Monte Carlo sampling. This suggests, that using heuristics for initialization of evolutionary algorithm do not improve performance significantly, except for the case of strong negative (positive) dependence and heuristic for finding the value of minimal (maximal) endpoint of τ. In these cases the approximate results obtained using our heuristics are exactly the same as the optimal ones found by evolutionary algorithm with much larger computational effort. This result should be credited to heuristic (T, T), as it nearly always outperforms others in case of strong dependencies, see Table 1. For weak dependences, both positive and negative, the proposed heuristics do not perform well. It is understandable, as their construction mimics the behavior of strongly dependent data.

5 Conclusions

We proposed very easy to compute approximate solutions of the problem of the estimation of Kendall's τ statistic for data having imprecise, interval form. Simulation experiments (only few of which are described here due to space limitation) show their applicability in case of strongly dependent data.

Acknowledgements. K. Opara's study was supported by research fellowship within "Information technologies: research and their interdisciplinary applications" agreement number POKL.04.01.01-00-051/10-00.

References

1. Arabas, J.: Evolutionary computation for global optimization – current trends. J. Telecommun. Inf. Technol. 4, 5–10 (2011)
2. Denœux, T., Masson, M.-H., Hébert, P.A.: Nonparametric rank-based statistics and significance tests for fuzzy data. Fuzzy Set. Syst. 153, 1–28 (2005)
3. Genest, C., McKay, R.J.: The joy of copulas: Bivariate distributions with uniform marginals. Am. Stat. 88, 1034–1043 (1986)
4. Genest, C., Rivest, L.-P.: Statistical inference procedures for bivariate Archimedean copulas. J. Am. Stat. Assoc. 88, 1034–1043 (1993)
5. Hébert, P.-A., Masson, M.H., Denœux, T.: Fuzzy rank correlation between fuzzy numbers. In: Proc. of IFSA World Congress, Istanbul, Turkey, pp. 224–227 (2003)
6. Hryniewicz, O., Opara, K.: Computation of the measures of dependence for imprecise data. In: New Developments in Fuzzy Sets, Intuitionistic Fuzzy Sets, Generalized Nets and Related Topics, Volume I: Foundations. SRI PAS, Warsaw (in press)
7. Hryniewicz, O., Szediw, A.: Fuzzy Kendall τ statistic for autocorrelated data. In: Soft Methods for Handling Variability and Imprecision, pp. 155–162. Springer, Berlin (2008)
8. Nelsen, R.B.: Introduction to Copulas. Springer US, New York (1999)
9. Opara, K., Arabas, J.: Differential Mutation Based on Population Covariance Matrix. In: Schaefer, R., Cotta, C., Kołodziej, J., Rudolph, G. (eds.) PPSN XI. LNCS, vol. 6238, pp. 114–123. Springer, Heidelberg (2010)
10. Price, K.V., Storn, R.M., Lampien, J.A.: Differential evolution a practical approach to global optimization. Natural Computing Series. Springer (2005)
11. Sklar, A.: Fonctions de répartition à n dimensions et leurs marges. Publications de l'Institut de Statistique de l'Université de Paris 8, 229–231 (1959)

Continuous Gaussian Estimation of Distribution Algorithm

Shahram Shahraki and Mohammad Reza Akbarzadeh Tutunchy

Abstract. Metaheuristics algorithms such as Estimation of Distribution Algorithms use probabilistic modeling to generate candidate solutions in optimization problems. The probabilistic presentation and modeling allows the algorithms to climb the hills in the search space. Similarly in this paper, Continuous Gaussian Estimation of Distribution Algorithm (CGEDA) which is kind of multivariate EDAs is proposed for real coded problems. The proposed CGEDA needs no initialization of parameters; mean and standard deviation of solution is extracted from population information during optimization processing adaptively. Gaussian Data distribution and dependent Individuals are two assumptions that are considered in CGEDA. The fitting task model in CGEDA is based on maximum likelihood procedure to estimate parameters of assumed Gaussian distribution for data distribution. The proposed algorithm is evaluated and compared experimentally with Univariate Marginal Distribution Algorithm (UMDA), Particle Swarm Optimization (PSO) and Cellular Probabilistic Optimization Algorithm (CPOA). Experimental results show superior performance of CGEDA V.S. the other algorithms.

Keywords: Evolutionary algorithms, particle swarm optimization.

1 Introduction

Many important problems in science, commerce and industry can be expressed into the optimization problems categories. These problems would be solved if we can find a solution that maximizes or minimizes some important and

Shahram Shahraki
Mashhad Branch, Islamic Azad University,
Department of AI and Computer Sciences
e-mail: shahraki@mshdiu.ac.ir

Mohammad Reza Akbarzadeh Tutunchy
Ferdowsi University of Mashhad, Departments of Electrical and
Computer Engineering
e-mail: akbarzadeh@ieee.org

R. Kruse et al. (Eds.): Synergies of Soft Computing and Statistics, AISC 190, pp. 211–218.
springerlink.com © Springer-Verlag Berlin Heidelberg 2013

measurable properties or attributes, such as cost or profit function. Evolutionary search algorithms are important population based optimization techniques in the recent years as a consequence of computation ability increment to solve this kind of problems. Compared to traditional optimization methods, these techniques have demonstrated their potentials through many areas such as machine learning or industry and also show that they are robust and global. These techniques search through many possible solutions which operate on a set of potential individuals to get better estimation of solution by using the principle of survival of the fittest, as in natural evolution. Genetic algorithms (GAs) developed by Fraser [3], Bremermann [1], and Holland [4], evolutionary programming (EP) developed by Fogel [2], and evolution strategies (ES) developed by Rechenberg [6] and Schwefel [7] establish the backbone of evolutionary computation which have been formed for the past 50 years. Estimation of Distribution Algorithms (EDAs), or Probabilistic Model-Building Genetic Algorithms, or Iterated Density Estimation Algorithms have been proposed by Mühlenbein and Paaß [5] are as an extension of genetic algorithms which are one of the main and basic methods in evolutionary techniques. EDAs generate their new offspring based on the probability distribution defined by the selected points Instead of performing recombination of individuals. The main advantages of EDAs over genetic algorithms are the explanatory and transparency of the probabilistic model that guides the search process. In traditional version of EDAs, they are inherently defined for problems with binary representation. So, for the problem in the real domain it must be first mapped to a binary coding before being optimized for real coded problems. This approximation might lead to undesirable limitations and errors on real coded problems [8]. The bottleneck of EDAs lies in estimating the joint probability distribution associated with the population that contains the selected individuals. Accordingly EDAs can be essentially divided to univariate, bivariate or multivariate approaches. As interdependencies between the variables that are captured increase, exponentially complexity and computations of EDAs increase as well. In this paper a new kind of multivariate EDAs is announced called Continuous Gaussian Estimation of Distribution Algorithm (CGEDA). CGEDA has been designed for real coded problems. The proposed algorithm assumed Gaussian distribution of data to model and estimate the joint distribution of promising solutions based on maximum likelihood technique on every dimension of search space. This type of probabilistic representation of CGEDA allows the algorithm to escape from local optimums and move free through fitness function.

2 Continuous Gaussian Estimation of Distribution Algorithm

Continuous Gaussian Estimation of Distribution Algorithm (CGEDA) which is a subset of multivariate EDAs and has been designed for real coded problems is introduced in this paper. The most important and crucial step of EDAs is the construction of probabilistic model for estimation of probability distribution, to do this step of CGEDA, Gaussian distribution of individuals is assumed to model and estimate the joint distribution of promising solutions in every dimension of the problem. The following estimation is used to generate new candidate solutions.

$$f(X) = \frac{1}{2\pi^{k/2}|\Sigma|^{1/2}} e^{[-\frac{(x-\mu)^T \Sigma^{-1}(x-\mu)}{2}]} \tag{1}$$

Where μ is mean and Σ is covariance matrix which can be written for k-dimensional random vector $X = [X1, X2 \ldots Xk]$ in the following notation:

$$\Sigma = E\left[(x-\mu)(x-\mu)^T\right] = \begin{bmatrix} E[(x_1-\mu_1)(x_1-\mu_1)] & \cdots & E[(x_1-\mu_1)(x_k-\mu_k)] \\ \vdots & \ddots & \vdots \\ E[(x_k-\mu_k)(x_1-\mu_1)] & \cdots & E[(x_k-\mu_k)(x_k-\mu_k)] \end{bmatrix} \tag{2}$$

For two-dimensional problems Eq. (1) is reduced to:

$$f(x,y) = \frac{1}{2\pi\delta_x\delta_y\sqrt{1-\rho^2}} \times e^{[-\frac{1}{2(1-\rho^2)}[\frac{(x-\mu_x)^2}{\delta_x^2} + \frac{(y-\mu_y)^2}{\delta_y^2} - \frac{2\rho(x-\mu_x)(y-\mu_y)}{\delta_x\delta_x}]} \tag{3}$$

Where ρ is the correlation between x and y.

So, mean and standard deviation parameters of promising population are required which computed adaptively by maximum likelihood technique to model the data (population).

One of the advantages of EDAs against other EAs is in exploration of search space. But CGEDA can reinforce the exploitation of EDAs by different Gaussian distribution estimations for every dimensions of solution.

The proposed algorithm has two implicit parameters: mean and standard deviation, where extracted from promising population adaptively, this means that no need to set manually the parameters in CGEDA.

The procedure of proposed algorithm is described as below:

Step 1 Initializes first generation randomly with uniform distributed random numbers in all dimensions.

Step 2 Evaluates the fitness function of all the real valued individuals.

Step 3 Is the main loop of algorithm. Continue until termination condition (max generation production) meets.

Step 4 In this step, based on truncation selection model, top evaluated of individuals are selected to estimate parameters of distribution over them for achieving better offspring. On the other hand, weak individuals are eliminated to not participate in the estimation.

Step 5 Distribution parameters are estimated based on maximum likelihood estimation technique.

Suppose a training set with n patterns $X = (x_1, x_2, \ldots, x_n)$, characterized by a distribution function $p(X \mid \theta)$, where θ is a parameter vector of the distribution (e.g. in our case the mean and standard deviation vector of a Gaussian distribution). An interesting approach of obtaining sample estimates of the parameter vector θ is to maximize $p(X \mid \theta)$, which viewed as a function of θ and is called the likelihood of θ for the given training set.

Step 6 Based on the estimated mean and standard deviation for every dimension, a new population is sampled as

$$\mathrm{x_{ij}} = \mathrm{G}(\mu_\mathrm{i}, \Sigma_\mathrm{i}) \tag{4}$$

Where μ, Σ, are estimated parameters of population based on top evaluated individuals and $G(.,.)$ is a Gaussian random number generator. In addition, $i = 1, 2, \ldots, d$ (d dimensions problem) is the dimension indicator and $j = 1, 2, \ldots, k$ (max population size is k) is the population size indicator.

Step 7 This is consistency check step.

$$x_{ij} = \begin{cases} x_{ij} & l_i < x_{ij} < u_i \\ \mathrm{G}(\mu_i, \Sigma_\mathrm{i}) & \text{else} \end{cases} \tag{5}$$

Step 8 In this step all the real valued individuals are evaluated by the fitness function.

Step 9 So far, two generations are created, one is current generation and the other one is the offspring of them based on CGEDA procedure. Next generation will be selected from specified populations based on truncation selection model.

This type of probabilistic representation of CGEDA allows the algorithm to escape from local optimums. CGEDA is guided to global optimum based on adaptive estimated standard deviation. Clearly high values of standard deviation make CGEDA focuses on exploration and low values of standard deviation make it focuses on exploitation of search space and so CGEDA decides to emphasize on exploration or exploitation of search space based on problem conditions and estimated standard deviations. In fact one of the superiority of CGEDA is due to the adaptive Gaussian distribution parameters extraction.

Because of the fact that no coding or decoding procedure is essential in CGEDA, it is a faster algorithm versus classic EDA or GA or other algorithms which code their populations

3 Experimental Result

In this paper the dimension of the problems is set to $m = 50, 100$. The population size for all of the experiments is set to 36, and the maximum generation termination condition which is used, equals 50. All results are averaged over 20 runs. To measure our work with other evolutionary algorithm we implement CPOA (Cellular Probabilistic Optimization Algorithm) and PSO (Particle Swarm Optimization) and UMDA (Univariate Marginal Distribution Algorithm) which is a univariate type of EDAs. Parameters that are used in this algorithms are δ^S, α, β, Rmu, Rdel, S and Mutate for CPOA and w, c1 and c2 are used for PSO, It is obvious that the best parameters for every algorithm are problem dependent, these parameters which are set by an expert and experiment, are summarized in Tab. 1. The parameters of PSO are equivalent for all problems and are equal to:

$$W = 0.9, C1 = 0.1, C2 = 0.2. \tag{6}$$

In this paper, the benchmark problems that have been used to evaluate our algorithm are numerical function optimization problems that contain Schwefel, Ackley, Griewank, Rosenbrock, G1, Kennedy, Rastrigin, and Michalewics.

Table 2, summarizes the experimental results of PSO, CGEDA, UMDA and CPOA for specified benchmark functions. Note that discussed functions are proposed for minimization and have a global minimum with some local minimum, in spite the proposed algorithm is designed for maximization. As a result, we redefine them to maximize $f(x)$. As it seems in most of benchmark

Table 1 Parameter settings for specified algorithms

	CPOA						
	δ^S	α	β	**Mutate**	**Rmu**	**Rdel**	**S**
Schwefel	0.3	0.05	0.03	0.005	0.002	0.002	6
Ackley	0.3	0.05	0.03	0.005	0.002	0.002	6
Griewank	0.3	0.5	0.03	0.005	0.002	0.002	6
Rosenbrock	0.3	0.5	0.2	0.005	0.002	0.002	6
G1	0.3	0.05	0.03	0.005	0.002	0.002	6
Kennedy	0.3	0.05	0.03	0.005	0.002	0.002	6
Rastrigin	0.3	0.03	0.03	0.005	0.002	0.002	6
Michalewics	0.3	0.05	0.03	0.005	0.002	0.002	6

Table 2 Result for 100 Dimension Problems, Averaged Over 20 runs, Best results mark as BOLD

	PSO				CGEDA			
	MEAN	*STD*	*BEST*	*Worst*	*MEAN*	*STD*	*BEST*	*Worst*
Schwefel	-4.173e+04	16.58	-4.17e+04	-4.18e+04	**-4.17 e+04**	4.3	-4.14 e+04	-4.195 e+04
Ackley	-9.12	0.91	-7.84	-10.65	**-0.95**	0.065	-0.0263	-1.076
Griewank	-1.12	0.03	-1.08	-1.16	**-1.0053**	0.12	-1.0012	-1.1087
Rosenbrock	-8.168e+05	3.15e+05	-3.66e+05	-1.26e+06	**-1.895e+04**	1.37e+03	-5.32e+03	-1.054e+04
G1	18.53	0.02	18.55	18.5	**18.542**	0.0009	18.55	18.535
Kennedy	-3.96e+04	2.68e+04	-7.37e+03	-1.011e+05	**-598.469**	15.67	-292.46	-894.4765
Rastrigin	-1.51e+03	100	-1.36e+03	-1.66e+03	**-964.02**	59.64	-875.69	-996.367
Michalewicsz	9.095	0.9596	10.558	7.47	**9.34**	1.152	11.22	7.254

	UMDA				CPOA			
	MEAN	*STD*	*BEST*	*Worst*	*MEAN*	*STD*	*BEST*	*Worst*
Schwefel	-4.172e+04	3.1	-4.17e+04	-4.181e+04	-4.1933 e+04	1.00025	-4.19 e+04	-4.2082 e+04
Ackley	-2.3	1.1	-1.01	-5.91	-1.5987	0.58	-0.96	-2.689
Griewank	-1.14	0.05	-1.1	-1.16	-1.616	0.05	-1.55	-1.69
Rosenbrock	-2.36e+04	4.36e+4	-1.84e+04	-3.6e+04	-2.28e+06	9.434e+04	-2.154e+06	-2.474e+06
G1	18.26	0.09	18.534	18.06	18.035	0.7724	18.549	16.86
Kennedy	-2.78e+03	1.06e+03	-2.53e+03	-2.96e+03	-1.69e+05	3.12e+04	-1.738e+05	-3.41e+05
Rastrigin	-1.53e+3	186	-1156	-2.658e+3	-1.69e+03	43.32	-1.60e+03	-1.76e+03
Michalewicsz	8.67	1.01	9.56	6.642	8.94	1.27	12.3207	8.017

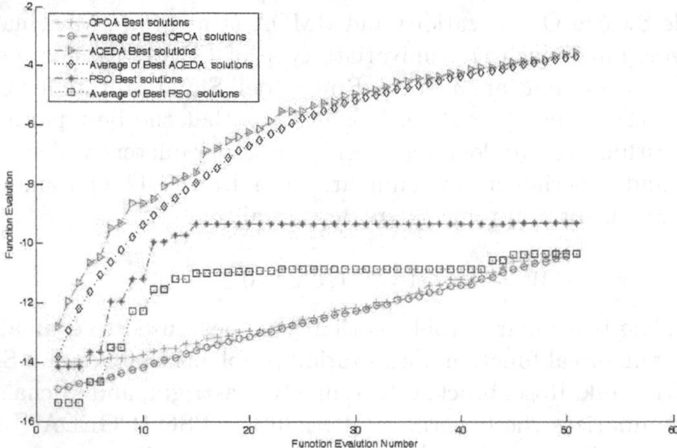

Fig. 1 Function evaluation for CGEDA and PSO and CPOA for 100 dimensions over 50 function evaluations

functions, the performance of CGEDA is better than PSO and is very close in Michalewics. In addition the performance of CGEDA in all the problems is better than CPOA. In Tab. 2, best results are marked as Bold and as it shows, CGEDA could find best results in all problems.

Experimental results show that initially CGEDA has a good speed versus the other algorithms (see Figs. 1 and 2).

While closing to the global optimum, CGEDA tries to find the best solution with its power in local search or exploitation. This superiority is due to the adaptive Gaussian distribution parameters extraction, and this type of probabilistic representation of the algorithm. Standard deviation often has large value at first, but as the algorithm close to the global optimal, its value decreases adaptively. This means that CGEDA has more exploration at first and more exploitation at later. Also CGEDA has better performance than algorithms which are designed for binary coded problems [9] such as UMDA. Also Fig. 1 illustrates that PSO has faster convergence than CGEDA which may causes premature convergence. Figure 2 illustrates the best and average of the best solutions in every function evaluation for CGEDA, PSO and CPOA. PSO with twice function evaluation cannot reach as good as result of CGEDA which is shown in Fig. 2. The proposed algorithm will trap

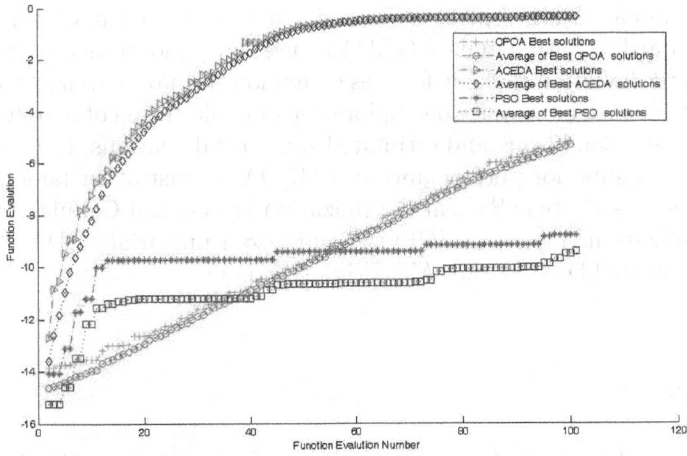

Fig. 2 Function evaluation for CGEDA and PSO and CPOA for 100 dimensions over 100 function evaluations

in local optimum on more conditions with less probability than the others because of probabilistic nature and Gaussian parameters of CGEDA. Therefore, standard deviation is an important parameter which set adaptively.

Finding best parameter for EAa is hard work for beginners and can be a new optimization problem to solve. A way to find optimal parameters is to use prior information of data if it is possible which CGEDA did, in the other hand, in this algorithm standard deviation is extracted from information of population adaptively that can be regarded as strength of CGEDA.

4 Conclusion

This paper proposed a novel EA, inspired by the Estimation of distribution algorithms. The Continuous Gaussian Estimation of Distribution Algorithm (CGEDA) is designed for real codec problems. This presentation of EDAs can improve exploitation of search space that is corporate with great nature exploration of EDAs. To achieve this ability in CGEDA, different Gaussian distribution estimation is used for every dimension of individuals, so to represent the individuals, means and standard deviations of Gaussian distributions are used. These are estimated by maximum likelihood technique to reach as real as results. Then based on this information, new population is sampled. The next generation is generated with these two populations: the current generation and the sampled population.

The probabilistic representation of the CGEDA enables it to climb up or down the hills in the search space. This type of probabilistic representation allows the algorithm to escape from local optimums. CGEDA is guided to global optimum based on adaptive estimated standard deviation. Clearly high values of standard deviation make CGEDA focuses on exploration and low values of standard deviation make it focuses on exploitation of search space and so CGEDA decides to emphasize on exploration or exploitation of search space based on problem conditions and estimated standard deviations. Experimental results shows superior performance of CGEDA against other famous EA algorithms such as Particle Swarm Optimization (PSO) and Cellular Probabilistic Optimization Algorithm (CPOA) and also a univariate EDA named Univariate Marginal Distribution Algorithm (UMDA).

References

1. Bremermann, H.J.: Optimization through evolution and recombination. In: Yovits, M.C., Jacobi, G.T., Goldstine, G.D. (eds.) Self-Organizing Systems, pp. 93–106. Spartan, Washington, DC (1962)
2. Fogel, L.J., Owens, A.J., Walsh, M.J.: Artificial Intelligence through Simulated Evolution. Wiley, New York (1966)
3. Fraser, A.S.: Simulation of genetic systems by automatic digital computers. Aust. J. Biol. Sci. 10, 484–491 (1957)
4. Holland, J.H.: Adaptation in Natural and Artificial Systems. University of Michigan Press, Ann Arbor (1975)
5. Mühlenbein, H., Paaß, G.: From Recombination of Genes to the Estimation of Distributions. In: Ebeling, W., Rechenberg, I., Voigt, H.-M., Schwefel, H.-P. (eds.) PPSN 1996. LNCS, vol. 1141, pp. 178–187. Springer, Heidelberg (1996)
6. Rechenberg, I.: Evolutionsstrategie — Optimierung technischer Systeme nach Prinzipien der biologischen Evolution. Frommann-Holzboog, Stuttgart-Bad Cannstatt, Germany (1973)
7. Schwefel, H.P.: Evolution and Optimum Seeking. John Wiley & Sons, Ltd., New York (1995)
8. Tayarani-N., M.-H., Akbarzadeh-T., M.-R.: Probabilistic optimization algorithms for numerical function optimization problems. In: IEEE Conf. on Cybernetics and Intelligent Systems, pp. 1204–1209 (2008)
9. Zinchenko, L., Radecker, M., Bisogno, F.: Multi-Objective Univariate Marginal Distribution Optimization of Mixed Analogue-Digital Signal Circuits. In: GECCO 2007. ACM, London (2007)

Regional Spatial Analysis Combining Fuzzy Clustering and Non-parametric Correlation

Bülent Tütmez and Uzay Kaymak

Abstract. In this study, regional analysis based on a limited number of data, which is an important real problem in some disciplines such as geosciences and environmental science, was considered for evaluating spatial data. A combination of fuzzy clustering and non-parametrical statistical analysis is made. In this direction, the partitioning performance of a fuzzy clustering on different types of spatial systems was examined. In this way, a regional projection approach has been constructed. The results show that the combination produces reliable results and also presents possibilities for future works.

Keywords: Fuzzy clustering, rank correlation, spatial data.

1 Introduction

In spatial analysis, each observation is associated with a location and there is at least an implied connection between the location and the observation. Geostatistical (probabilistic) and soft computing methods can be applied for assessing spatial distributions in a site [1]. When observations are made in space, the data can exhibit complex correlation structures. The correlation can be two-dimensional if the data are taken only over a spatial surface [11].

Bülent Tütmez
School of Engineering, İnönü University,
44280 Malatya, Turkey
e-mail: bulent.tutmez@inonu.edu.tr

Uzay Kaymak
School of Industrial Engineering, Eindhoven University of Technology,
5600 MB, Eindhoven, The Netherlands
e-mail: u.kaymak@ieee.org

R. Kruse et al. (Eds.): Synergies of Soft Computing and Statistics, AISC 190, pp. 219–227.
springerlink.com © Springer-Verlag Berlin Heidelberg 2013

It is obvious that the spatial patterns of individual sampling locations in any study area have different patterns and observations depend on the relative positions of observed locations within the site. The classical geostatistical tools such as variogram, although suitable for irregularly-spaced data, have practical difficulties. One of the main drawbacks is that it is insufficient to analyze the regional heterogeneous behavior of a spatial parameter [5]. In general, spatial systems have heterogeneous properties rather than homogeneous structures. Heterogeneity means that the properties observed at different locations do not have the same value, and that different zones are observed in the site.

One of the practical problems encountered in spatial systems such as in geosciences, ecology and geography is the limited number of data. Often, correlations are estimated from a small number of observations. The correlation coefficient is particularly important in cases with sparse data such as pollution and offshore petroleum data [10]. In these cases, because the measure is expensive and time consuming, it may be necessary to work with limited number of data. Hence, a regional analysis with limited data becomes an important task in spatial systems.

The main objective of a cluster analysis is to partition a given data set of data or objects into clusters [9]. Because most of the clustering algorithms employ the distances between the observations, for a spatial system, the clusters provided by clustering can be considered as distinguished regions [12]. Analyzing a spatial system based on structural properties is a difficult task and applicability of clustering for this purpose should be examined. In this study, the performance of the Fuzzy c-means Algorithm (FCM), which is the well-known clustering algorithm, in conditioned spatial systems is investigated. The partitioning capacity of the algorithm with limited number of data is appraised using Rank Correlation Method (RCM) that is also a well-known non-parametric method.

The rest of the paper is structured as follows. Sect. 2 describes the basics of weighted fuzzy arithmetic and the hybrid fuzzy least-squares regression. Confidence interval-based approach for coefficients and predictions is presented in Sect. 3. Finally, Sect. 4 gives the conclusions.

2 Methodology

Fuzzy clustering and non-parametric correlation analysis are well-known methods. The algorithm proposed in this study aims a combination to appraise a spatial system based on an areal analysis. In this section, a brief review and the basis of the combination is presented.

2.1 Fuzzy Clustering

The main purpose of clustering is to recognize natural groupings of data from a large data set to produce a concise representation of a system's behavior. The FCM is a well-known data clustering method in which a data set is grouped into clusters (regions) with every data point in the data set belonging to every cluster to a certain degree. As a suitable algorithm, the FCM was also proposed to make spatial evaluations [2].

Let $\{\mathbf{x}_1, \mathbf{x}_2, \ldots, \mathbf{x}_N\}$ be a set of N data objects represented by p-dimensional feature vectors $\mathbf{x}_k = [x_{1k}, \ldots, x_{pk}]^T \in \mathbb{R}^p$. A set of N feature vectors is then represented as $p \times N$ data matrix $\mathbf{X}$. A fuzzy clustering algorithm partitions the data $\mathbf{X}$ into M fuzzy clusters, forming a fuzzy partition in $\mathbf{X}$. A fuzzy partition can be conveniently represented as a matrix $\mathbf{U}$, whose elements $u_{ik} \in [0, 1]$ represent the membership degree of $\mathbf{x}_k$ in cluster i. Hence, the i-th row of $\mathbf{U}$ contains values of the i-th membership function in the fuzzy partition.

Objective function based fuzzy clustering algorithms minimize an objective function of the type:

$$J(\mathbf{X}; \mathbf{U}, \mathbf{V}) = \sum_{i=1}^{M} \sum_{k=1}^{N} (u_{ik})^m \; d^2(\mathbf{x}_k, \mathbf{v}_i), \tag{1}$$

where $\mathbf{V} = [\mathbf{v}_1, \mathbf{v}_2, \ldots, \mathbf{v}_M]$, $\mathbf{v}_i \in \mathbb{R}^p$ is M-tuple centers which have to be computed, and $m \in (1, \infty)$ is a weighting exponent which defines the fuzziness of the clusters. The conventional FCM uses Euclidean distance. The optimization is constrained, amongst others, by the constraint

$$\sum_{i=1}^{M} u_{ik} = 1, \qquad \forall k. \tag{2}$$

2.2 Non-parametric Rank Correlation

Nonparametric statistics can be an effective tool when data is observed on a discrete scale of values or when the assumptions required by parametric statistics can not be satisfied. This time we cannot rely on the central limit theorem which is a concept to justify use of parametric tests and we must turn to a category of alternative procedures named nonparametric techniques. The nonparametric tests use information of a lower rank, such as nominal or ordinal observations. No assumptions about the form of the parent population are required [6].

Spearman's rank correlation is one of the statistical tools to calculate non-parametric correlations between pairs of samples. If we make two sets of ordinal observations on a number of objects, we can designate one of the sets as x and the other as y. We then rank each observation and call the two sets of ranks $R(x_i)$ and $R(y_i)$. Spearman's coefficient measures the similarity between these two ranks [4],

$$r_s = 1 - \frac{6\sum_{i=1}^{n}[R(x_i) - R(y_i)]^2}{n(n^2-1)}. \tag{3}$$

The term inside the brackets of the numerator is simply the difference between the rank of property x and the rank of property y as observed on the i-th object. The following assumptions can be given for conducting the implementation.

- The correlation between the variables should be linear.
- The two variables have been reduced to an ordinal scale of observation.
- If a test of significance is applied, the sample has been selected randomly from the population.

The rank correlation r_s, is analogous to simple correlation r in that it varies from +1.0 (perfect correspondence between the ranks) to −1.0 (perfect inverse relationship between the ranks). A rank correlation of $r_s = 0$ shows that the two sets of ranks are independent. Note that the rank correlation analysis is insufficient, if the number of observations is bigger than 60 [8].

2.3 Regional Appraisal with Memberships

Generally, in natural world spatial systems have heterogeneous property and different zones are observed in a site. Due to these available separate regions, from a clustering algorithm a better partition is expected for heterogeneous sites rather than homogeneous sites. From this point, it could be anticipated that the correlations provided between the clusters should be bigger for a heterogeneous system than a homogeneous system.

In some circumstances, a relatively small sample, whose size cannot be increased and whose underlying population may be distinctly non-normal, has to be studied. When the sample size is small, the uncertainty about the value of the true correlation can be very large, particularly when the estimated correlation is low [10]. Considering this condition, to measure the correlations between the clusters, membership values and their ranks could be used on the ground of a non-parametric correlation analysis. The algorithm of the analysis can be presented by a flowchart as in Fig. 1.

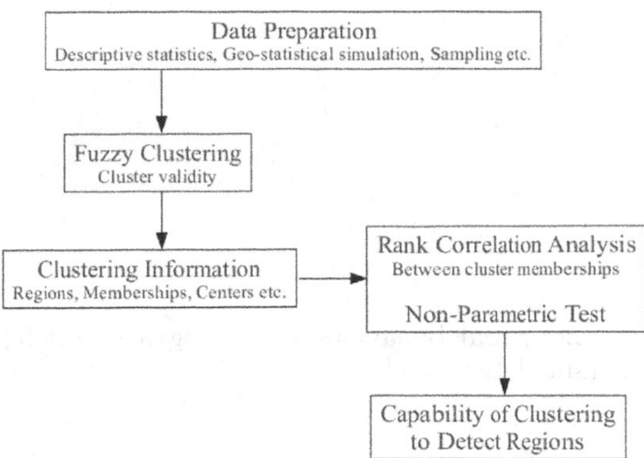

Fig. 1 Flow chart of the analysis

3 Simulation Studies

3.1 Data Set

Experimental studies have been carried out using two simulated data sets. In the applications, the effectiveness and partitioning capacity of the FCM algorithm on different types of spatial systems has been investigated. The spatial real data set (108 observations) used in [13] was handled. This data set comprised of Elasticity Modulus (EM) values of rock samples collected from an Andesite quarry in Ankara.

To perform the simulation studies, the real set was conditioned by a geostatistical simulation technique which is lower-upper (LU) decomposition technique [7]. For the first case study two simulated sets, one of which has homogeneous and other has heterogeneous properties, were provided based on conditional simulation. In the heterogeneous site, the EM values generate different zones and the spatial variability of the site can be modeled by a function such as Spherical or Gaussian type functions.

Each simulation is conducted on a 21×21 regular grid, yielding a total of 441 values. After that, a similar procedure was followed for the second case study. This time, a simulation was carried out on a 20×20 regular grid and a total of 400 values.

3.2 Simulation Study 1

Two sample sets (49 records) were randomly drawn from the simulated data sets, each of them including 441 observations. To illustrate the different spatial characteristics, a semi-variance analysis, which is a well-known

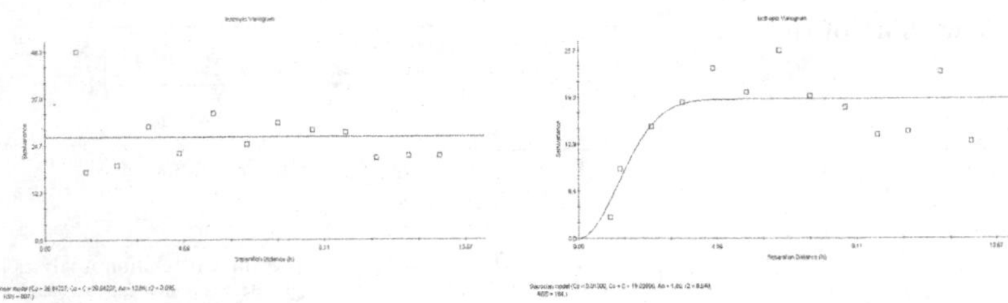

Fig. 2 Spatial behaviors for homogeneous (left) and heterogeneous (right) data sets (simulation study 1)

geostatistical analysis, has been performed. Figure 2 shows the variogram models provided for the sampled distinguished sets. As can be seen in the Fig. 3, although the homogeneous site can not show a spatial relationship (pure nugget model), the heterogeneous site has a Gaussian character. To specify the regions in the sites, fuzzy clustering applications have been performed both for homogeneous and heterogeneous data sets. These different data sets used by the clustering algorithm have the same coordinates and different EM values. Therefore, data matrix **X** contains three dimensions (spatial positions and EM). As a result of the clustering validity studies [3], the optimal number of clusters was defined as four for both sites.

Statistically, if the coefficient of skewness S_f is zero, then the distribution is symmetrical and must be zero for the normal distribution. Similarly, if the Kurtosis is zero, then the distribution of data is approximately normal [14]. Based on these criteria, the memberships have been appraised and use of a nonparametric rank correlation analysis method is decided. Table 1 summarizes the non parametrical (cross) correlation coefficients with the average values for both homogeneous and heterogeneous sets. The values under $N(0,1)$ describe the approximated values of the coefficients required from the large number of data.

Table 1 Rank correlation coefficients among the clusters

Cross Correlation	Homogeneous	Homogeneous $N(0,1)$	Heterogeneous	Heterogeneous $N(0,1)$
r_{12}	−0.380	−2.630	0.101	0.696
r_{13}	−0.023	−0.158	−0.487	−3.370
r_{14}	0.028	0.192	0.168	1.162
r_{23}	0.042	0.288	−0.089	−0.619
r_{24}	0.132	0.913	−0.482	−3.338
r_{34}	−0.377	−2.610	0.010	0.072
Average Correlation	−0.096	−0.668	−0.130	−0.900

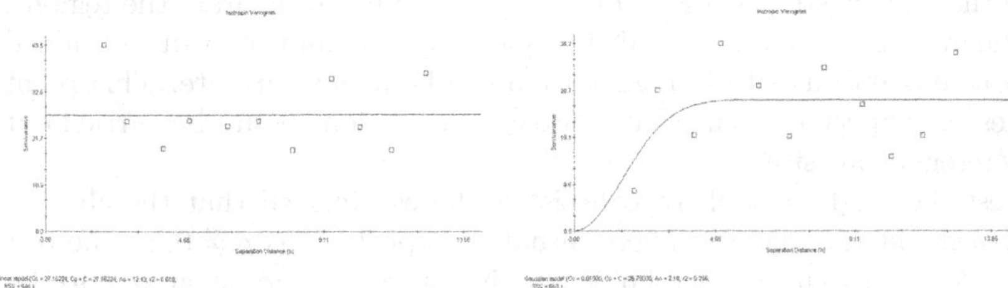

Fig. 3 Spatial behaviors for homogeneous (top) and heterogeneous (bottom) data sets (simulation study 2)

3.3 Simulation Study 2

For the second application, a similar procedure to the one followed in the first application is performed. Firstly, two data sets (each of 25 records) were randomly sampled from the simulated data sets, including 400 observations each one. In order to measure the spatial variability of the observations variogram models have been obtained. Figure 3 illustrates the models. In the homogeneous site, no meaningful spatial dependence is recorded. On the other hand, the heterogeneous site shows a spatial model that is Gaussian.

Based on clustering validity, the optimal number of clusters has been determined as four for both data structures. By using the memberships provided from the clustering application, the nonparametric rank correlation analysis method is applied. Table 2 indicates the cross correlation coefficients with the average values for both data sets.

Table 2 Absolute rank correlation coefficients among the clusters

Cross Correlation	Homogeneous	Heterogeneous
r_{12}	0.439	0.132
r_{13}	0.070	0.125
r_{14}	0.013	0.476
r_{23}	0.237	0.402
r_{24}	0.350	0.066
r_{34}	0.385	0.335
Average Absolute Correlation	0.249	0.256

3.4 Results and Discussion

Because limited number of data may not be increased and the underlying population may be distinctly non-normal in spatial environmental systems, the applications were conducted in the proposed manner. First application showed that the clustering algorithm has a capability to separate the regions. Both the average correlation coefficients are negative and the value obtained for the heterogeneous site is bigger than the homogeneous site. This point indicates the expected result that more clear partition should be carried out for a heterogeneous site.

To test the study, a null hypothesis can be established that the clusters are independent (i.e. $\rho = 0$). The alternative hypothesis is $\rho \neq 0$, so the test is two-tailed, with either very large positive or very large negative correlations leading to rejection. Our analysis shows that the null hypothesis is not rejected both for the homogeneous and the heterogeneous case, indicating independence of clusters.

Second case study was performed by relatively small data sets. Both the average correlation coefficients address the inverse correlations and the clustering algorithm has a capability to determine the regions. In this application, to overcome a possible compensation that may be resulted from pairs close to $+1$ and -1, the study has been carried out using the absolute values. The null hypothesis is that cluster memberships are independent, or that $\rho = 0$. The alternative hypothesis is $\rho \neq 0$, the test is one-tailed. Again, it is found that the null hypothesis is not rejected. Depending on the limited number of data, a crisp difference between two data sets has not been recorded.

4 Conclusions

The partitioning performance of a fuzzy clustering algorithm on different type spatial systems is examined. To appraise the conditioned spatial systems via limited number of data, fuzzy clustering and non-parametric rank correlation method is integrated. By this way, a regional projection method has been constructed. In conclusion, the combination of fuzzy clustering and non-parametric correlation analysis has produced some reliable results and provide possibilities for future studies in depth.

References

1. Bardossy, G., Fodor, J.: Evaluation of Uncertainties and Risks in Geology. Springer, Berlin (2004)
2. Bezdek, J.C., Ehrlich, R., Full, W.: FCM: the fuzzy c-means clustering algorithm. Computers & Geosciences 10(2-3), 191–203 (1984)
3. Bezdek, J.C., Pal, N.R.: Some new indexes of cluster validity. IEEE Transactions on Systems, Man and Cybernetics, Part B 28(3), 301–315 (1998)
4. Conover, W.J.: Practical Nonparametric Statistics. Wiley, New York (1999)
5. Şen, Z.: Spatial Modelling Principles in Earth Sciences. Springer, New York (2009)
6. Davis, J.: Statistics and Data Analysis in Geology. Wiley, New York (2002)

7. Deutsch, C.V., Journel, A.G.: GSLIB: Geostatistical Software Library and User's Guide. Oxford University Press, New York (1998)
8. Dudzic, S.: Companion to Advanced Mathematics and Statistics. Hodder Education, London (2007)
9. Höppner, F., Klawonn, F., Kruse, R., Runkler, T.: Fuzzy Cluster Analysis: methods for classification, data analysis and image recognition. Wiley, New York (1999)
10. Niven, E.B., Deutsch, C.V.: Calculating a robust correlation coefficient and quantifying its uncertainty. Computers & Geosciences 40, 1–9 (2012)
11. Piegorsch, W.W., Bailer, A.J.: Analyzing Environmental Data. Wiley, Chichester (2005)
12. Tutmez, B.: Spatial dependence-based fuzzy regression clustering. Applied Soft Computing 12(1), 1–13 (2012)
13. Tutmez, B., Tercan, A.E.: Spatial estimation of some mechanical properties of rocks by fuzzy modelling. Computers and Geotechnics 34, 10–18 (2006)
14. Wellmer, F.W.: Statistical Evaluations in Exploration for Mineral Deposits. Springer, Heidelberg (1998)

The p-value Line: A Way to Choose from Different Test Results

Alfonso García-Pérez

Abstract. It is common practice to perform an Exploratory Data Analysis to decide whether to use a classical or a robust test; i.e., to choose between a classical or a robust result. What is not so clear is how to choose among the results provided by different competing robust tests. In this paper we propose to use the function *p-value line*, that we shall define later, to compare the results obtained by different tests in order to choose one: the result with largest p-value line. This function takes into account the usual trade-off between robustness and power that is present in most, if not all, robust tests, trade-off that is expressed thought a parameter fixed in a subjective way. With our proposal we can fix it in an objective manner. We shall apply this proposal to choose the trimming fraction in the location test based on the trimmed mean.

Keywords: Robust tests, von Mises expansion.

1 The p-value Line: Definition and Properties

In the paper we consider tests for the null hypothesis $H_0 : \theta \in \Theta_0$ against the alternative $H_1 : \theta \in \Theta_1$, being $\Theta = \Theta_0 \cup \Theta_1$ the parameter space, tests that reject H_0 for large values of the test statistic $T_n = T_n(X_1, ..., X_n)$ (although the results can be extended to other situations) and where the observable random variables X_i, for $i = 1, ..., n$, follow the model F_θ, $\theta \in \Theta$.

Alfonso García-Pérez
Departamento de Estadística, I. O. y C. N., Universidad Nacional de Educación a Distancia (UNED), 28040 Madrid, Spain
e-mail: agar-per@ccia.uned.es

R. Kruse et al. (Eds.): Synergies of Soft Computing and Statistics, AISC 190, pp. 229–236.
springerlink.com © Springer-Verlag Berlin Heidelberg 2013

Although in this section we write T_n, it must be understood as its standardized version that we use in the computation of the test, i.e., $(T_n - E[T_n])/\sigma(T_n)$, as we shall do in the applications section.

The ordinary p-value p_n, i.e., the p-value that we compute in the daily use of statistical methods, is defined as the maximum probability, under the null hypothesis, of observing a value of the test statistic at least as extreme as the one obtained $t_n = T_n(x_1, ..., x_n)$, where $(x_1, ..., x_n)$ is the observed sample; i.e., $p_n = \sup_{\theta \in \Theta_0} P_\theta\{T_n > t_n\} = \sup_{\theta \in \Theta_0}(1 - F_{n;\theta}(t_n))$, where $F_{n;\theta}$ is the cdf of T_n under F_θ. We compute p_n in a test because it is a measure of evidence against the null hypothesis.

On the other hand, the random variable p-value P_n is defined as $P_n = \sup_{\theta \in \Theta_0}(1 - F_{n;\theta}(T_n))$ and, although this is a random variable with uniform distribution on $[0, 1]$ if the null hypothesis is true, we use only one of its values (the observed one p_n) to accept or reject the null hypothesis.

For a fixed alternative $\theta_1 \in \Theta_1$, we propose to use one value of $1 - F_{n;\theta_1}(T_n)$ (the observed one $P_{\theta_1}\{T_n > t_n\}$) to choose from different results because this quantity is a measure of evidence against the null hypothesis assuming that $\theta_1 \in \Theta_1$ is true.

Hence if we have two results s_n and t_n on the same test, we should choose result t_n instead of s_n if $P_{\theta_1}\{T_n > t_n\} > P_{\theta_1}\{S_n > s_n\}$.

In general, because H_1 could be composite, we propose to use the function of the alternative $\theta_1 \in \Theta_1$ $l(\theta_1) = P_{\theta_1}\{T_n > t_n\}$ that we call the *p-value line*, to choose from different results. Namely, we propose to choose the result with largest p-value line because this is a measure of evidence against the null hypothesis assuming that the alternative is true.

Moreover, the p-value line is very important in robustness studies because, as we shall see in the next paragraphs, it takes into account the usual trade-off between robustness and power that is present in most, if not all, robust tests, balance that is expressed through a parameter that is usually fixed in a very complicated and/or subjective manner and that can fixed objectively with the p-value line.

1.1 Power Function and the p-value Line

The p-value line is very closely related to the power function of a test. Let us call θ_0 to the boundary case in the null hypothesis Θ_0, i.e., the parameter value where the size α of the test is reached, i.e., $P_{\theta_0}\{T_n > k_n^\alpha\} = \sup_{\theta \in \Theta_0} P_\theta\{T_n > k_n^\alpha\} = \alpha$, so being k_n^α the critical value.

Proposition 1. *The p-value line of a test is its power function when we consider the particular significance level* $P_{\theta_0}\{T_n > t_n\}$.

Hence, when we obtain results t_n and s_n from two competing tests and compare the associated p-value lines, we are really comparing the power functions of the tests where the results come, in particular (and usually different) significance levels. And, on the power function comparison of tests with different significance levels, theory says that we should prefer the test with steeper ascent to power 1, from the boundary of the null hypothesis, because the steeper the power function the better it will be to detect the alternative. Hence, we should match them at θ_0 in order to choose easily the p-value line with steepest ascent, i.e., with the largest slope.

In this paragraph we have supposed, in fact, that the ordinary p-values p_n of the competing tests are somewhat similar. If this were not the case it would be very likelihood to think that some outliers should be present in the data and we would be in the next paragraph situation.

1.2 Test Robustness and the p-value Line

If there are no outliers in the sample and, for instance, we are testing $H_0 : \theta = \theta_0 = 0$ against $H_1 : \theta > 0$, the usual situation of p-value lines is shown in the left side of Figure 1, in which the solid line would represent the most powerful test (the classical) and the dashed line other least powerful one (the robust): at θ_0 both p-value lines are very similar and for $\theta > \theta_0$ the most powerful has steepest ascent to power 1.

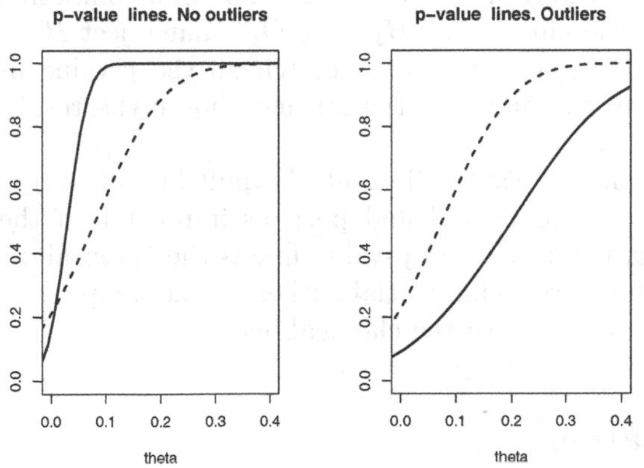

Fig. 1 Usual classical (solid) and robust (dashed) p-value lines

If there are outliers in the sample, these will affect more to t_n (the classical statistics) than s_n (the robust one). If we consider first the situation in which outliers affect the test statistic value by increasing it, these will affect the p-value line of a classical test $P_\theta\{T_n > t_n\}$ (decreasing it) more than the robust one $P_\theta\{S_n > s_n\}$, changing even the sign of the expected inequality from $P_\theta\{T_n > t_n\} > P_\theta\{S_n > s_n\}$ to $P_\theta\{T_n > t_n\} < P_\theta\{S_n > s_n\}$ if outliers are very influential.

Hence, if there are outliers in the sample the typical situation is shown in the right side of Figure 1 and, again, if we match them at θ_0, we should choose the p-value line with steepest ascent, i.e., with largest slope.

In some practical cases (e.g., tests on the regression coefficients) the test statistics are used in absolute value and outliers affect them only by increasing them.

In the application of hypothesis testing problems, we establish the two hypotheses to test because of the observed values of the test statistics. For instance, we test that the location parameter is 0 against the alternative that it is greater than 0, if the observed values of the two (standardized) statistics t_n and s_n that we consider, take values around 0 or greater than 0. Hence, only outliers with positive influence should be considered; otherwise we should change the hypotheses to test and we should have the symmetric situation.

Next we establish our proposal to choose from different competing test results, robust and/or classical, on a common null hypothesis.

Proposal:

To choose among the results obtained from several competing tests of $H_0 : \theta \in \Theta_0$ against the alternative $H_1 : \theta \in \Theta_1$, that reject H_0 for large values of the test statistic, we propose to match all the p-value lines at θ_0 (the boundary case in the null hypothesis) and choose the result with steepest ascent from θ_0.

As with standard p-values (i.e., at the null hypothesis), sometimes the decision could be complicated and perhaps impossible if the p-value lines cross. Moreover, if the selected p-value line is the classical one, we can also conclude that there are no influential outliers in the sample because otherwise we would not conclude with the classical one.

2 Computation

One advantage of using the p-value line to decide among several results is that it is easier to compute than the power function, mainly because it does not depend on the significance level. Nevertheless, we saw before that the p-value line is very closely related with the power function and hence, in some cases its computation is not easy, especially if we consider small sample sizes. In this situation, a great effort has been made in recent years to find accurate approximations for the ordinary p-value; see for instance, [3, 6, 4]. Nevertheless, nearly nothing has been made in the computation of the power. Next, in the computation of p-value lines we propose to use the ideas given in [2] to transfer computations under the alternative hypothesis to computations under the null.

If we use expression (1) in Corollary 2 of [2], considering as distribution G the underlying model under the null hypothesis, $G \equiv F_{\theta_0}$, and as distribution F the underlying model under an alternative $\theta \in \Theta_1$, $F \equiv F_\theta$, we have (in the general case of non iid models)

$$\begin{aligned}
\text{p-value}_{H_1} &= P_{(F_1,\dots,F_n)_\theta}\{T_n(\mathbf{X}_1,\mathbf{X}_2,\dots,\mathbf{X}_n) > t_n\} \\
&\simeq (1-n)P_{(F_1,\dots,F_n)_{\theta_0}}\{T_n(\mathbf{X}_1,\mathbf{X}_2,\dots,\mathbf{X}_n) > t_n\} \\
&\quad + \int_{\mathcal{X}} P_{(F_2,\dots,F_n)_{\theta_0}}\{T_n(\mathbf{x},\mathbf{X}_2,\dots,\mathbf{X}_n) > t_n\}\, dF_{1;\theta}(\mathbf{x}) \\
&\quad + \cdots + \int_{\mathcal{X}} P_{(F_1,\dots,F_{n-1})_{\theta_0}}\{T_n(\mathbf{X}_1,\dots,\mathbf{X}_{n-1},\mathbf{x}) > t_n\}\, dF_{n;\theta}(\mathbf{x})
\end{aligned}$$

that allows an approximation of the tail probability under the alternative hypothesis, knowing the value of this tail probability under the null.

And, with the same arguments as in Section 3 of [2], in the case that $F_{1;\theta}$,.., $F_{n;\theta}$ are location families

$$\begin{aligned}
\text{p-value}_{H_1} &\simeq (1-n)\text{p-value}_{H_0} + P_{F_{1;\theta_0},\dots,F_{n;\theta_0}}\{T_n(\mathbf{X}_1+(\theta-\theta_0),\mathbf{X}_2,\dots,\mathbf{X}_n) > t_n\} \\
&\quad + \cdots + P_{F_{1;\theta_0},\dots,F_{n;\theta_0}}\{T_n(\mathbf{X}_1,\dots,\mathbf{X}_{n-1},\mathbf{X}_n+(\theta-\theta_0)) > t_n\} \\
&= (1-n)P_{F_{1;\theta_0},\dots,F_{n;\theta_0}}\{T_n(\mathbf{X}_1,\dots,\mathbf{X}_n) > T_n(\mathbf{x}_1,\dots,\mathbf{x}_n)\} \\
&\quad + P_{F_{1;\theta_0},\dots,F_{n;\theta_0}}\{T_n(\mathbf{X}_1,\dots,\mathbf{X}_n) > T_n(\mathbf{x}_1-(\theta-\theta_0),\dots,\mathbf{x}_n)\} \\
&\quad + \cdots + P_{F_{1;\theta_0},\dots,F_{n;\theta_0}}\{T_n(\mathbf{X}_1,\dots,\mathbf{X}_n) > T_n(\mathbf{x}_1,\dots,\mathbf{x}_n-(\theta-\theta_0))\}.
\end{aligned}$$

Hence, the p-value under the alternative can be approximated by computing $(1-n)$ times the usual p-value under the null, i.e., the p-value with the observed data set, plus n p-values (always under the null) computed for n shifted data, all of them usually given by common statistical software, mainly the R statistical software [5].

As we argued in [2], the previous linear approximation is accurate when θ is close to θ_0. If we want to extend the approximation to θ's away from θ_0, we can use an iterative procedure as in Section 4 of [2] obtaining, with k iterations,

$$
\begin{aligned}
\text{p-value}_{H_1} \simeq\ & \text{p-value}_{H_0} + \sum_{j=1}^{k+1} \left[P_{H_0} \left\{ T_n(\mathbf{X}_1 + c_{2j}, \mathbf{X}_2 + c_{1j}, ..., \mathbf{X}_n + c_{1j}) > t_n \right\} \right. \\
& + \cdots + P_{H_0} \left\{ T_n(\mathbf{X}_1 + c_{1j}, ..., \mathbf{X}_{n-1} + c_{1j}, \mathbf{X}_n + c_{2j}) > t_n \right\} \\
& \left. - nP_{H_0} \left\{ T_n(\mathbf{X}_1 + c_{1j}, ..., \mathbf{X}_n + c_{1j}) > t_n \right\} \right] \\
=\ & P_{F_{1;\theta_0}, ..., F_{n;\theta_0}} \left\{ T_n(\mathbf{X}_1, ..., \mathbf{X}_n) > T_n(\mathbf{x}_1, ..., \mathbf{x}_n) \right\} \\
& + \sum_{j=1}^{k+1} \left[P_{F_{1;\theta_0}, ..., F_{n;\theta_0}} \left\{ T_n(\mathbf{X}_1, ..., \mathbf{X}_n) > T_n(\mathbf{x}_1 - c_{2j}, \mathbf{x}_2 - c_{1j}, ..., \mathbf{x}_n - c_{1j}) \right\} \right. \\
& + \cdots + P_{F_{1;\theta_0}, ..., F_{n;\theta_0}} \left\{ T_n(\mathbf{X}_1, ..., \mathbf{X}_n) > T_n(\mathbf{x}_1 - c_{1j}, ..., \mathbf{x}_{n-1} - c_{1j}, \mathbf{x}_n - c_{2j}) \right\} \\
& \left. - nP_{F_{1;\theta_0}, ..., F_{n;\theta_0}} \left\{ T_n(\mathbf{X}_1, ..., \mathbf{X}_n) > T_n(\mathbf{x}_1 - c_{1j}, ..., \mathbf{x}_n - c_{1j}) \right\} \right] \qquad (1)
\end{aligned}
$$

where $c_{1j} = (j-1)(\theta - \theta_0)/(k+1)$ and $c_{2j} = j(\theta - \theta_0)/(k+1)$.

3 Applications

In this section we shall use the p-value line to fix the parameter that represents the usual trade-off between robustness and power and is present in robust tests, because the p-value line *orders* them considering these two factors. In this paper we shall consider the location test based on the trimming mean, choosing the result (the trimming fraction) with highest p-value line, but other applications, as the location test based on the Huber statistic, or tests in robust Generalized Linear Models or robust Generalized Additive Models are possible, because the proposal is based on transferring computations under the alternative to computations under the null, for which we have available libraries of R, such as `robustbase` or `rgam`.

We shall also consider here the null hypothesis $H_0 : \theta = \theta_0 = 0$, against $H_1 : \theta > 0$ and a contaminated normal distribution $(1-\epsilon)N(\theta_0, 1) + \epsilon N(\theta, 1)$ as underlying model, although other hypotheses and models are possible.

3.1 The Trimmed Mean

To test the two hypotheses mentioned before, let us consider the test statistic $T_n = (\overline{X}_\alpha - E[\overline{X}_\alpha])/\sqrt{\widehat{V}(\overline{X}_\alpha)}$, where $\overline{X}_\alpha$ is the trimmed mean. It is known [7, p. 156] that if the underlying model follows a normal distribution, and $H_0 : E[\overline{X}_\alpha] = 0$, then T_n follows asymptotically a Student's t distribution.

And, although with the previous results, we transfer computations under the alternative to computations under the null, the null hypothesis here is $H_0 : \theta = 0$ and the underlying model, $(1-\epsilon)N(0,1) + \epsilon N(\theta,1)$, not a normal distribution. Nevertheless, in [1] a linear approximation to the trimmed mean was obtained, especially useful for small sample sizes.

Under the null hypothesis the test statistic is $T_n = (\overline{X}_\alpha - \epsilon\,\theta)/\sqrt{\widehat{V}(\overline{X}_\alpha)}$ and we can use the iterative linear approximation obtained by [1] to compute the tail probabilities under the null, probabilities that are included in (1).

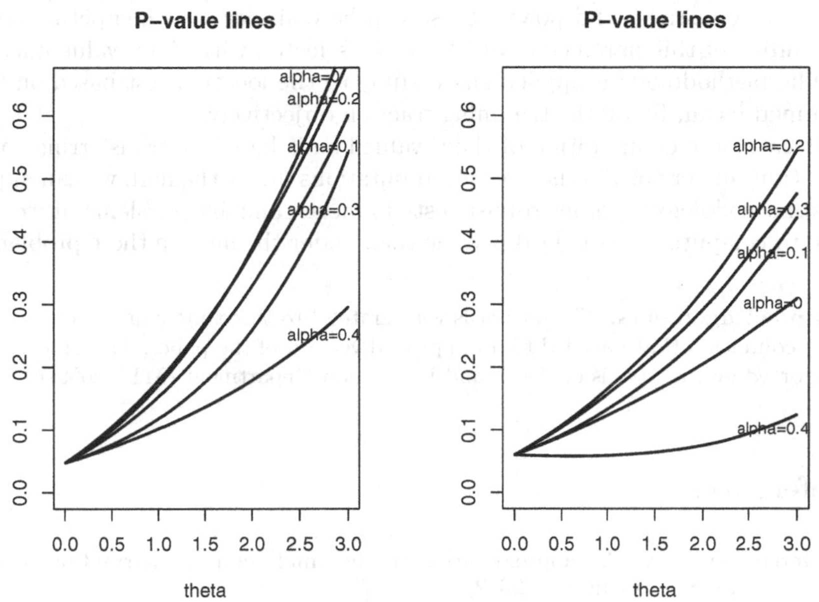

Fig. 2 Comparison of p-value lines for several trimming fractions with no contaminated data (left) and contaminated data (right)

To show some numerical results, let us assume as underlying model $0.95N(\theta_0,1) + 0.05N(\theta,1)$, and that there are no outliers in the observed sample, that we suppose is

```
-0.04243192  0.35124862  1.79327771 -1.01262733  1.37481913
 0.76204004  1.53280552 -0.50457646  0.41478589  0.61417601
```

We see in the left side of Figure 2 (with only 15 iterations) that the p-value line of the usual sample mean is higher than other p-value lines (matched at the null hypothesis), suggesting no trimming at all in the sample.

If we now artificially replace the last value of the previous sample by `6.14176014`, we observe in the right side of Figure 2 (with 15 iterations again) the p-value lines of different sample trimmed means, concluding with the use of the trimming fraction of 0.2.

4 Conclusions

With the function *p-value line* defined in the paper we can compare the results obtained by different competing tests in order to choose one of them. Because most of the robust tests depend on a parameter that represents the trade-off between robustness and power, these can be considered as competing tests depending of this parameter, which can be selected with the p-value line.

The methodology is applied successfully to the location test based on the trimmed mean, fixing the trimming fraction objectively.

Because the computation of the p-value line is based on transferring computations under the alternative to computations under the null, we can apply this methodology to other robust tests in more complex problems, if we are able to compute the standard p-value (i.e., under the null) in these problems.

Acknowledgements. The author is very grateful to three anonymous referees for their comments that have led to an improved version of the paper. This research is supported by the Spanish Science and Innovation Department (MTM2009-10072).

References

1. García-Pérez, A.: An approximation to the small sample distribution of the trimmed mean (submitted, 2012)
2. García-Pérez, A.: A linear approximation to the power function of a test. Metrika 76 (2012)
3. Jing, B., Robinson, J.: Saddlepoint approximations for marginal and conditional probabilities of transformed variables. Ann. Statist. 22(3), 1115–1132 (1994)
4. Lô, S.N., Ronchetti, E.: Robust and accurate inference for generalized linear models. J. Multivariate Anal. 100(9), 2126–2136 (2009)
5. R Development Core Team: R: A Language and Environment for Statistical Computing, Vienna, Austria (2011) ISBN 3-900051-07-0, `http://www.R-project.org`
6. Robinson, J., Ronchetti, E., Young, G.A.: Saddlepoint approximations and tests based on multivariate m-estimates. Ann. Statist. 31(4), 1154–1169 (2003)
7. Staudte, R.G., Sheather, S.J.: Robust Estimation and Testing. Wiley Series in Probability and Statistics, vol. 220. John Wiley & Sons, Ltd., New York (1990)

Outlier Detection in High Dimension Using Regularization

Moritz Gschwandtner and Peter Filzmoser

Abstract. An outlier detection method for high dimensional data is presented in this paper. It makes use of a robust and regularized estimation of the covariance matrix which is achieved by maximization of a penalized version of the likelihood function for joint location and inverse scatter. A penalty parameter controls the amount of regularization.

The algorithm is computation intensive but provides higher efficiency than other methods. This fact will be demonstrated in an example with simulated data, in which the presented method is compared to another algorithm for high dimensional data.

Keywords: Outliers, regularization, robust statistics.

1 Introduction

Outlier detection is a well known field in today's statistics. In order to increase the efficiency and outcome of statistical analysis methods, it is often necessary to prepend a data preparation step, in which outliers are either excluded from the data, or downweighted, in order to avoid the bias of subsequent algorithms.

The higher the dimension of the data is, the more complex the challenge of outlier detection gets. This can easily be understood by means of the following example: Consider the unit square $[0, 1] \times [0, 1]$ in the two dimensional Euclidean vector space. 100 uniformly distributed observations will be

Moritz Gschwandtner · Peter Filzmoser
Department of Statistics and Probability Theory, Vienna University of Technology, 1040 Vienna, Austria
e-mail: moritz.gschwandtner@chello.at,
P.Filzmoser@tuwien.ac.at

R. Kruse et al. (Eds.): Synergies of Soft Computing and Statistics, AISC 190, pp. 237–244.
springerlink.com © Springer-Verlag Berlin Heidelberg 2013

enough in order to sample the square. If we repeat the experiment in a ten dimensional space $[0,1]^{10}$, we would need many more observations in order to achieve the same level of covering. This phenomenon, which is referred to as *Curse of Dimensionality*, also affects statistical methods. Especially the case where the dimension exceeds the number of observations ($n < p$) is a topic of main interest. Thus, it is crucial for the field of robust statistics to develop outlier detection methods which can handle this case.

In this article we want to investigate in more detail a method which was proposed in [1] and [2], respectively. It is based on the regularization of the estimated data covariance matrix. The regularization is achieved by the application of an L1 penalty term in the log-likelihood function of joint location and inverse scatter. The penalty term leads to a *sparse* inverse covariance matrix, which can be an advantage if many noise variables are included in the data. It also results in a lower number of parameters to be estimated, and this may yield more stability and efficiency for outlier detection in high dimension. In addition, a robustness parameter controls the number of observations which are included in the maximization of the likelihood function.

The rest of the article is organized as follows: The following section provides a description of the investigated method, which is referred to as *RegMCD*. In Sect. 3, a comparison of *RegMCD* and another outlier detection method *PCOut* is given by means of a simulated data example. A brief description of *PCOut* is provided as well. Finally, Sect. 4 concludes the article.

2 The RegMCD Method

The RegMCD estimator was introduced in [2] and is further characterized in [1] as a regularized extension of the well-known MCD estimator [6]. As described in [2], it can be used as an outlier detection tool. This chapter summarizes the most important facts.

2.1 Basics

The maximum likelihood method is a well known parameter estimation approach. Given a data sample $\{\mathbf{x}_1, \ldots, \mathbf{x}_n\}$, with $\mathbf{x}_i \in \mathbb{R}^p$, the likelihood function is maximized in order to derive an estimation for the unknown parameter that is most likely in case of the given data. The L1 penalized log-likelihood function of joint location and inverse scatter can be written as

$$\mathcal{L}(\boldsymbol{\mu}, \boldsymbol{\Theta}) = \log\det(\boldsymbol{\Theta}) - \frac{1}{n}\sum_{i=1}^{n}(\mathbf{x}_i - \boldsymbol{\mu})^\top \boldsymbol{\Theta}(\mathbf{x}_i - \boldsymbol{\mu}) - \lambda\|\boldsymbol{\Theta}\|_1, \qquad (1)$$

where $\boldsymbol{\mu}$ denotes the p dimensional mean vector of the data and $\boldsymbol{\Theta} = \boldsymbol{\Sigma}^{-1}$ is the inverse covariance matrix. $\lambda > 0$ is a penalty parameter which controls the amount of regularization and $\|\cdot\|$ denotes the L1 Norm:

$$\|\boldsymbol{\Theta}\|_1 = \sum_{i,j} |\theta_{ij}|, \tag{2}$$

where θ_{ij} are the elements of the matrix $\boldsymbol{\Theta}$. For $\lambda = 0$, this coincides with the classical log-likelihood function. Larger values of λ lead to *sparse* estimations of $\boldsymbol{\Theta}$, which means that $\boldsymbol{\Theta}$ contains many zeroes. The maximization of (1) cannot be done analytically, but requires an iterative regression based algorithm called *glasso* (see [5]).

As a matter of fact, solutions of the maximization problem are not robust. This is due to the fact that the sum in Eq. (1) contains all observations $\mathbf{x}_i$ equally-weighted. The *RegMCD* algorithm counteracts this fact by searching for a data subset H_0 of size $h < n$, for which the maximized likelihood value is larger than for any other subset of size h:

$$\mathcal{L}(\boldsymbol{\mu}, \boldsymbol{\Theta}, H) = \log\det(\boldsymbol{\Theta}) - \frac{1}{h}\sum_{\mathbf{x}\in H} (\mathbf{x} - \boldsymbol{\mu})^\top \boldsymbol{\Theta}(\mathbf{x} - \boldsymbol{\mu}) - \lambda\|\boldsymbol{\Theta}\|_1 \to \max \tag{3}$$

The additional robustness parameter h is typically chosen as $h = 0.75 \cdot n$, which has turned out to be a good compromise between robustness and efficiency, see [6]. In order to solve the maximization problem (3) and to obtain the estimations $\hat{\boldsymbol{\Theta}}$ and $\hat{\boldsymbol{\mu}}$, the *RegMCD* algorithm uses an iterative concentration step procedure as it was suggested in [6].

2.2 The Penalty Parameter

While the robustness parameter h can be set to $h = 0.75 \cdot n$ for most settings, the penalty parameter λ has to be chosen more carefully. The proposed method uses an adapted BIC criterion:

$$BIC(\lambda) = -2 \cdot \tilde{\mathcal{L}}(H_0, \hat{\boldsymbol{\Theta}}, \hat{\boldsymbol{\mu}}) + \kappa(\lambda)\log h, \tag{4}$$

where $\tilde{\mathcal{L}}(H_0, \hat{\boldsymbol{\Theta}}, \hat{\boldsymbol{\mu}})$ denotes the likelihood value without the penalty term, and $\kappa(\lambda)$ measures the total number of estimated parameters:

$$\kappa(\lambda) = k \cdot p + \sum_{i<j} \mathbb{1}_{\{\hat{\theta}_{ij} \neq 0\}}, \tag{5}$$

where $\hat{\theta}_{ij}$ is the element (i, j) of the matrix $\hat{\boldsymbol{\Theta}}$, and $\mathbb{1}$ denotes the indicator function. Finally, λ_0 can be chosen as

$$\lambda_0 = \arg\min_{\lambda} BIC(\lambda). \tag{6}$$

Note that the term (4) consists of two parts. First, the likelihood term $-2 \cdot \tilde{\mathcal{L}}(H_0, \hat{\boldsymbol{\Theta}}, \hat{\boldsymbol{\mu}})$ will be small for small values of λ and increase with an increasing λ. The second part, though, measures the sparseness and will be smaller for large values of λ, as the entries in the inverse covariance matrix are more strongly penalized. The final value of λ_0 will therefore be somewhere in between small and large values and the criterion (6) can be regarded as a compromise between likelihood and sparseness.

2.3 Outlier Detection

A well known distance measure for multivariate data is the squared Mahalanobis distance:

$$d^2(\mathbf{x}, \boldsymbol{\mu}, \boldsymbol{\Sigma}^{-1}) = (\mathbf{x} - \boldsymbol{\mu})^\top \boldsymbol{\Sigma}^{-1}(\mathbf{x} - \boldsymbol{\mu}), \tag{7}$$

where $\boldsymbol{\Sigma}^{-1}$ is the inverse covariance matrix. In contrast to the Euclidean distance, the Mahalanobis distance takes the covariance structure of the data into account. If the data is distributed according to a multivariate normal distribution, the Mahalanobis distance will follow a chi-square distribution with p degrees of freedom.

Using the *RegMCD* estimates $\hat{\boldsymbol{\Theta}}$ and $\hat{\boldsymbol{\mu}}$, robust Mahalanobis distances d_i can be computed:

$$d_i^2(\mathbf{x}_i, \hat{\boldsymbol{\mu}}, \hat{\boldsymbol{\Theta}}) = (\mathbf{x}_i - \hat{\boldsymbol{\mu}})^\top \hat{\boldsymbol{\Theta}}(\mathbf{x}_i - \hat{\boldsymbol{\mu}}) \tag{8}$$

An observation $\mathbf{x}_i$ is classified as an outlier, if

$$d_i^2(\mathbf{x}_i, \hat{\boldsymbol{\mu}}, \hat{\boldsymbol{\Theta}}) > \chi^2_{p,0.975}, \tag{9}$$

with $\chi^2_{p,q}$ being the q-quantile of the χ^2_p distribution.

3 Simulation Studies

In this section, the proposed method is compared to another outlier detection method called *PCOut*, which shall briefly be described.

3.1 PCOut

PCOut was introduced in [4] and is available in the `R` package `mvoutlier` [3]. The main idea of *PCOut* is to give weights to the observations $\mathbf{x}_i$, indicating if one is likely to be an outlier or not. The weights consist of two parts:

- a location weight $w_{1i} \leq 1$, indicating if the observation is a location outlier
- a scatter weight $w_{2i} \leq 1$, indicating if the observation is a scatter outlier

Both weights are combined to receive an overall weight w_i for an observation $\mathbf{x}_i$:

$$w_i = \frac{(w_{1i} + s)(w_{2i} + s)}{(1 + s)^2} \tag{10}$$

An observation is classified as an outlier, if $w_i < 0.25$. The scaling constant s is typically chosen $s = 0.25$. This assures that a small weight w_i is only received, if both w_{1i} and w_{2i} are small themselves.

PCOut was chosen as the competing method as it is known to work well in the high-dimensional setting. *PCOut* is more efficient than other methods for high-dimensional outlier detection, and at the same time it is fast to compute, see [4].

3.2 Simulation Setup

The data setup is closely related to the one in [4] and was produced as follows: In the p-dimensional vector space, $n = 100$ clean observations $\mathbf{x}_i$ were created according to a multivariate normal distribution $\mathbf{N}(\mathbf{0}, \mathbf{\Sigma}_p^*)$, where $\mathbf{\Sigma}_p^*$ denotes the $p \times p$ identity matrix with additional covariance $\sigma_{12}^* = \sigma_{21}^* = 0.7$ between the first and the second variable. The remaining variables do not contain any covariance information and can thus be regarded as noise variables, which favors the application of the *RegMCD* method. The data was then contaminated with $n_{out} = 10$ outliers, distributed according to $\mathbf{N}(\boldsymbol{\mu}_{out}, \sigma_{out} \cdot \mathbf{I}_p)$, where $\mathbf{I}_p$ denotes the $p \times p$ identity matrix.

In order to receive a large variety of settings, the parameters p, $\boldsymbol{\mu}_{out}$, and σ_{out} were varied in the following way:

- $p \in \{30, 50, 100, 200\}$
- $\boldsymbol{\mu}_{out} \in \{(0, \ldots, 0)^\top, (2, \ldots, 2)^\top, (5, \ldots, 5)^\top\}$
- $\sigma_{out} \in \{0.1, 0.5, 1, 2, 5\}$

The values of $\boldsymbol{\mu}_{out}$ and σ_{out} assure that both location and scatter outliers are included in the simulation.

For each combination of the three parameters, the simulation was executed ten times and the proposed outlier detection methods were applied. As a measure of performance, the average false negative (FN) and false positive (FP) rates were computed. False negative means that an observation is classified as a regular observation, although it is in fact an outlier. A false positive denotes a clean observation which is classified as an outlier instead. The simulation was fully executed in R.

The results are shown in Table 1. There we can see that for both methods the percentages of false negatives are mostly 100% for $\boldsymbol{\mu}_{out} = (0, \ldots, 0)^\top$ and

Table 1 RegMCD vs. PCOut: Average percentage of false negatives (FN) and false positives (FP). $n = 100$, $n_{out} = 10$, 10 iterations.

Method	μ_{out}	$\sigma_{out} = 0.1$ %FN	%FP	$\sigma_{out} = 0.5$ %FN	%FP	$\sigma_{out} = 1$ %FN	%FP	$\sigma_{out} = 2$ %FN	%FP	$\sigma_{out} = 5$ %FN	%FP
					$p = 30$						
RegMCD	$(0,\ldots,0)^\top$	100.0	4.2	100.0	0.0	-	3.0	50.0	2.0	0.0	0.0
PCOut	$(0,\ldots,0)^\top$	100.0	8.0	100.0	13.0	-	14.0	30.0	8.0	0.0	11.0
RegMCD	$(2,\ldots,2)^\top$	0.0	0.0	0.0	2.0	0.0	1.0	0.0	0.8	0.0	2.0
PCOut	$(2,\ldots,2)^\top$	0.0	2.0	0.0	7.0	0.0	5.0	0.0	4.0	0.0	7.0
RegMCD	$(5,\ldots,5)^\top$	0.0	2.0	0.0	0.0	0.0	0.0	0.0	1.0	0.0	0.0
PCOut	$(5,\ldots,5)^\top$	0.0	11.0	0.0	8.0	0.0	5.0	0.0	8.0	0.0	6.0
					$p = 50$						
RegMCD	$(0,\ldots,0)^\top$	100.0	2.0	100.0	0.0	-	0.0	10.0	0.0	0.0	0.0
PCOut	$(0,\ldots,0)^\top$	100.0	10.0	100.0	9.0	-	25.0	10.0	7.0	0.0	4.0
RegMCD	$(2,\ldots,2)^\top$	0.0	0.0	0.0	1.0	0.0	0.0	0.0	0.0	0.0	0.0
PCOut	$(2,\ldots,2)^\top$	0.0	6.0	0.0	7.0	0.0	6.0	0.0	8.0	0.0	8.0
RegMCD	$(5,\ldots,5)^\top$	0.0	0.0	0.0	0.0	0.0	0.0	0.0	0.0	0.0	0.2
PCOut	$(5,\ldots,5)^\top$	0.0	5.0	0.0	6.0	0.0	10.0	0.0	8.0	0.0	4.0
					$p = 100$						
RegMCD	$(0,\ldots,0)^\top$	100.0	0.9	100.0	0.0	-	0.1	10.0	0.0	0.0	0.0
PCOut	$(0,\ldots,0)^\top$	100.0	6.0	100.0	7.0	-	7.0	60.0	8.0	0.0	8.0
RegMCD	$(2,\ldots,2)^\top$	0.0	0.0	0.0	0.0	0.0	0.0	0.0	0.0	0.0	0.0
PCOut	$(2,\ldots,2)^\top$	0.0	6.0	0.0	3.0	0.0	5.0	0.0	3.0	0.0	2.0
RegMCD	$(5,\ldots,5)^\top$	0.0	0.0	0.0	0.0	0.0	0.0	0.0	0.0	0.0	0.0
PCOut	$(5,\ldots,5)^\top$	0.0	5.0	0.0	6.0	0.0	0.0	0.0	3.0	0.0	1.0
					$p = 200$						
RegMCD	$(0,\ldots,0)^\top$	100.0	0.0	100.0	0.0	-	0.0	0.0	0.0	0.0	0.0
PCOut	$(0,\ldots,0)^\top$	100.0	6.0	90.0	2.0	-	2.0	20.0	4.0	0.0	4.0
RegMCD	$(2,\ldots,2)^\top$	0.0	0.0	0.0	0.0	0.0	0.0	0.0	0.0	0.0	0.0
PCOut	$(2,\ldots,2)^\top$	0.0	3.0	0.0	0.0	0.0	5.0	0.0	3.0	0.0	4.0
RegMCD	$(5,\ldots,5)^\top$	0.0	0.0	0.0	0.0	0.0	0.0	0.0	0.0	0.0	0.0
PCOut	$(5,\ldots,5)^\top$	0.0	3.0	0.0	3.0	0.0	1.0	0.0	3.0	0.0	7.0

$\sigma_{out} \in \{0.1, 0.5\}$. This is due to the fact that in these cases the outliers are perfectly masked by the rest of the data. For $\boldsymbol{\mu}_{out} = (0, \ldots, 0)^\top$ and $\sigma_{out} = 1$, the outliers are distributed like the non-outliers, which makes a comparison meaningless. For this reason we do not report the corresponding results of FN. For the remaining settings (containing either shift and/or location outliers), we can see that most of the time, the *RegMCD* method performs better than *PCOut*, especially as far as the false positives are concerned. We also want to point out the fact that both methods perform very well for high dimensional data.

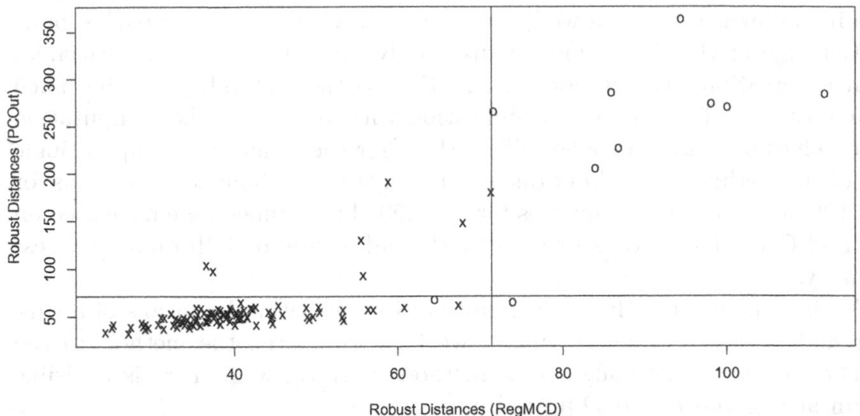

Fig. 1 Distance Distance Plot showing robust squared Mahalanobis distances computed by means of *RegMCD* and *PCOut*. Outliers are flagged with an 'o' symbol. The horizontal and vertical lines denote the 0.975-quantile of the χ^2_{50} distribution. $p = 50$, $\boldsymbol{\mu}_{out} = (0, \dots, 0)^\top$, $\sigma_{out} = 2$.

For the configuration $p = 50$, $\boldsymbol{\mu}_{out} = (0, \dots, 0)^\top$, and $\sigma_{out} = 2$, we created an additional distance-distance plot, which is shown in Fig. 1. The robust distances of the *RegMCD* method were computed according to equation (8), whereas the *PCOut* distances were computed according to

$$d_i^2(\mathbf{x}_i, \hat{\boldsymbol{\mu}}, \hat{\boldsymbol{\Sigma}}^{-1}) = (\mathbf{x}_i - \hat{\boldsymbol{\mu}})^\top \hat{\boldsymbol{\Sigma}}^{-1}(\mathbf{x}_i - \hat{\boldsymbol{\mu}}) \ . \tag{11}$$

In this case, $\hat{\boldsymbol{\mu}}$ and $\hat{\boldsymbol{\Sigma}}$ denote the classical mean and covariance matrix based on those observations $\mathbf{x}_i$ where $w_i > 0.25$ (i.e. those that were flagged by the *PCOut* procedure as regular observations). The horizontal and vertical lines denote the 0.975-quantile of the χ^2_{50} distribution. Outliers are plotted with an 'o' symbol.

The image confirms the results from Table 1: Some of the non-outliers have large distances in terms of *PCOut*, which corresponds to the higher rate of false positives. The *RegMCD* method, though, identifies all non-outliers correctly. As far as the false negatives are concerned, both methods perform equally well.

4 Conclusion

We have investigated in detail the *RegMCD* method for outlier detection; the method is based on a robust and regularized estimation of the inverse covariance matrix. The regularization parameter λ can be chosen according to an adapted BIC criterion in order to achieve optimal results. The method handles high dimensional data well, even if $n < p$, where other approaches fail.

Although in the simulations we used only up to twice as many variables than observations, this is not a restriction to the method; p can be much higher than n. There is only a limitation with respect to the computation time, which increases exponentially with p. For the simulation setup, a single run of the method takes about one second for $p = 50$, about seven seconds for $p = 100$, and up to three minutes for $p = 200$. These times were measured on an Intel Core2 Duo processor with 3GHz and a total of 4Gb random access memory.

We have illustrated the performance of the algorithm by means of a simulated data example. Furthermore, we have compared the method to the *PCOut* procedure. Although both methods perform well, there is a visible advantage for the *RegMCD* procedure.

References

1. Croux, C., Gelper, S., Haesbroeck, G.: The regularized minimum covariance determinant estimator (preprint), `http://www.econ.kuleuven.be/public/NDBAE06/public.htm`
2. Croux, C., Haesbroeck, G.: Robust scatter regularization. Compstat, Book of Abstracts, Paris, Conservatoire National des Arts et Métiers (CNAM) and the French National Institute for Research in Computer Science and Control, INRIA (2010)
3. Filzmoser, P., Gschwandtner, M.: mvoutlier: Multivariate outlier detection based on robust methods. R package version 1.9.4 (2011)
4. Filzmoser, P., Maronna, R., Werner, M.: Outlier identification in high dimensions. Comput. Stat. Data An. 52, 1694–1711 (2008)
5. Friedman, J.H., Hastie, T., Tibshirani, R.: Sparse inverse covariance estimation with the graphical lasso. Biostat. 9, 432–441 (2007)
6. Rousseeuw, P.J., Van Driessen, K.: A fast algorithm for the minimum covariance determinant estimator. Technometrics 41, 212–223 (1999)

Robust Diagnostics of Fuzzy Clustering Results Using the Compositional Approach

Karel Hron and Peter Filzmoser

Abstract. Fuzzy clustering, like the known fuzzy k-means method, allows to incorporate imprecision when classifying multivariate observations into clusters. In contrast to hard clustering, when the data are divided into distinct clusters and each data point belongs to exactly one cluster, in fuzzy clustering the observations can belong to more than one cluster. The strength of the association to each cluster is measured by a vector of membership coefficients. Usually, an observation is assigned to a cluster with the highest membership coefficient. On the other hand, the refinement of the hard membership coefficients enables to consider also the possibility of assigning to another cluster according to prior knowledge or specific data structure of the membership coefficients. The aim of the paper is to introduce a methodology to reveal the real data structure of multivariate membership coefficient vectors, based on the logratio approach to compositional data, and show how to display them in presence of outlying observations using loadings and scores of robust principal component analysis.

Keywords: Compositional biplot, compositional data, fuzzy clustering, robust principal component analysis.

1 Overview of Fuzzy Clustering

In fuzzy clustering, each assignment of an object is distributed proportionally to all clusters through membership coefficients according to the similarity to

Karel Hron
Palacký University, 77146 Olomouc, Czech Republic
e-mail: hronk@seznam.cz

Peter Filzmoser
Vienna University of Technology, 1040 Vienna, Austria
e-mail: p.filzmoser@tuwien.ac.at

R. Kruse et al. (Eds.): Synergies of Soft Computing and Statistics, AISC 190, pp. 245–253.
springerlink.com © Springer-Verlag Berlin Heidelberg 2013

each of the clusters. The number of clusters k for the n objects needs to be provided in advance. Then an objective function

$$\sum_{v=1}^{k} \frac{\sum_{i=1}^{n} \sum_{j=1}^{n} u_{iv}^2 u_{jv}^2 d(i,j)}{2 \sum_{j=1}^{n} u_{jv}^2}, \tag{1}$$

that contains only the similarity measure $d(i,j)$ and the desired membership coefficients u_{iv} of the i-th object to the v-th cluster, needs to be minimized. The measure $d(i,j)$ can be chosen e.g. as squared Euclidean distance when the fuzzy k-means method is applied [3, 4]; an alternative choice is described in [10]. Each object is usually assigned to a cluster with the highest membership coefficient. On the other hand, the refinement of the hard clustering result enables to consider also the possibility of assigning to another cluster according to prior knowledge or specific data structure of the membership coefficients. It means that although an observation belongs to a certain cluster according to the classification rule, the data structure of the membership coefficients implies its pertinence rather to another cluster.

Obviously, the sum of the membership coefficients equals 1 or 100 (in case of proportions or percentages, respectively), so their sample space can be considered to be a k-part simplex,

$$\mathcal{S}^k = \{\mathbf{u} = (u_1, \ldots, u_k)',\, u_i > 0,\, \sum_{i=1}^{k} u_i = 1\}, \tag{2}$$

the prime stands for a transpose. Here we have excluded the case of zero membership values since then the predefined number of clusters obviously needs to be revisited. The important difference of fuzzy clustering to hard clustering methods is contained in the fact that with the latter we obtain a detailed information about the data structure. On the other hand, with an increasing number of the involved groups the results become quite complex so that the obtained information cannot be easily processed further.

For this reason, in this paper we focus on the case of more clusters involved into the analysis and provide a tool to display the multivariate data structure of the membership coefficients using a biplot of loadings and scores from principal component analysis [9]. Hereat we consider in particular a specific data structure of the coefficients, that contain naturally only relative information, and can thus be identified with the concept of compositional data [1]. In addition, we apply a robust counterpart of principal component analysis to ensure that the obtained diagnostics tool will not be influenced by outlying observations. The next section provides a brief review on compositional data and the log-ratio approach for their statistical analysis. Then we introduce classical and robust principal component analysis to construct a biplot and demonstrate how it can be applied in case of compositional data. Finally, the theoretical results will be applied to a real-world example.

2 Relative Information and Compositional Data

Each vector of membership coefficients contains exclusively relative information, thus only ratios between its parts are informative. In the context of fuzzy clustering, the coefficients are normalized to a prescribed constant sum constraint (proportions, percentages). However, this is not a necessary condition but rather a proper representation of the observations, also a positive constant multiple of the vector would provide exactly the same information. In addition, also the concept of relative scale plays an important role here: if a membership coefficient of a certain group increases from 0.1 to 0.2 (two times), it is not the same as an increase from 0.5 to 0.6 (1.2 times), although the Euclidean distances are the same in both cases. All these above properties can be found in the concept of compositional data as introduced in the early 1980s by John Aitchison [1]. The properties of this kind of observations induce a special geometry of compositional data, the Aitchison geometry on the simplex [6] that forms for k-part compositional data, a Euclidean space of dimension $k-1$. Then the main goal is to represent compositional data in orthonormal coordinates with respect to the Aitchison geometry and to perform usual multivariate methods for their statistical analysis. This concept is closely connected with the family of isometric log-ratio (ilr) transformations from the $\mathcal{S}^k$ to the $(k-1)$-dimensional real space $\mathbf{R}^{k-1}$ [5]. One popular choice results for a composition $\mathbf{u} = (u_1, \ldots, u_k)'$ in ilr coordinates $\mathbf{z} = (z_1, \ldots, z_{k-1})'$, where

$$z_i = \sqrt{\frac{k-i}{k-i+1}} \ln \frac{u_i}{\sqrt[k-i]{\prod_{j=i+1}^{k} u_j}}, \quad i = 1, \ldots, k-1. \tag{3}$$

Obviously, the ilr transformations move the Aitchison geometry on the simplex isometrically to the usual Euclidean geometry in real space, i.e. to the geometry that we are used to work in. This has also consequences for visualization of the compositional data structure. Three-part compositions are traditionally displayed in a ternary diagram. The ternary diagram is an equilateral triangle $U_1U_2U_3$ such that a composition $\mathbf{u} = (u_1, u_2, u_3)'$ is plotted at a distance u_1 from the opposite side of vertex U_1, at a distance u_2 from the opposite side of vertex U_2, and at a distance u_3 from the opposite side of the vertex U_3 (see, e.g., [1, 12]).

An example can be seen in Fig. 1 with the well-known Iris data set [8] that contains measurements for 50 flowers from each of 3 species of iris. Fuzzy k-means clustering was applied with $k = 3$. The ternary diagram (left) shows the resulting membership coefficients, where the lines correspond to equal coefficients in two groups. The lines can thus be considered as separation lines for a hard cluster assignment. The plot symbols correspond to the true group memberships. One of the clusters (circles) is clearly distinguishable, but the other two clusters show some overlap that leads to a misclassification. The

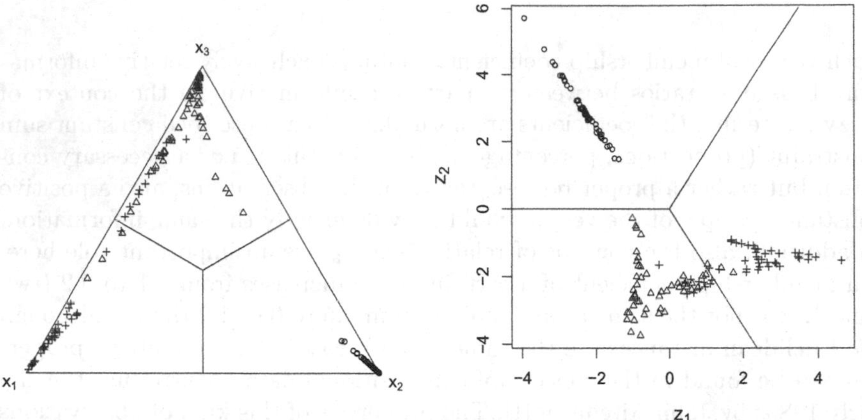

Fig. 1 Membership coefficients of the Iris data in the ternary diagram (left) and after ilr transformation (right) are displayed together with borders (lines) for the classification rule. The symbols correspond to the true memberships.

right plot panel shows the ilr-transformed results, again with the separating lines. The misclassified observations are of course still the same, but the data structure is much better visible in the overlapping region. In this plot the distances are in terms of the usual Euclidean geometry, while in the ternary diagram one has to think in the Aitchison geometry.

Although the ilr transformation has nice geometrical properties, an interpretation of the orthonormal coordinates is sometimes quite complex. Thus, for the purpose of a compositional biplot introduced in the next section, a representation of compositions in a special generating system is more appropriate. The resulting coordinates correspond to the centred logratio (clr) transformation [1], given for a k-part composition $\mathbf{u}$ as

$$(y_1, \ldots, y_k)' = \left(\frac{u_1}{\sqrt[k]{\prod_{i=1}^k u_i}}, \ldots, \frac{u_k}{\sqrt[k]{\prod_{i=1}^k u_i}} \right)'. \tag{4}$$

The clr transformation seems easier to handle than the ilr transformation, however, it leads to a singular covariance matrix, because the sum of y_i, $i = 1, \ldots, k$, equals zero. This makes the use of robust statistical methods not possible. In the next section we show how the ilr transformation can be utilized in this case.

3 Diagnostics Using a Robust Compositional Biplot

Unfortunately, for more than three-part compositional data it is not possible to visualize them in a planar graph without dimension reduction. A proper tool for this purpose seems to be the compositional biplot [2]. It displays both samples and variables of a data matrix graphically in the form of scores and loadings of the first two principal components [9]. Note that the well-known principal component analysis is appropriate for this purpose, because it explains most of the variability of the original multivariate data by only few new variables (the mentioned principal components). Usually, samples in the biplot are displayed as points while variables are displayed either as vectors or rays. For compositional data, one would intuitively construct the biplot for ilr-transformed data, however, due to the complex interpretation of the new variables it is common to construct the compositional biplot for clr-transformed compositions as proposed in [2]. The scores represent the structure of the compositional data set in the Euclidean real space, so they can be used to see patterns and clusters in the data. The loadings (rays) represent the corresponding clr-variables. In the compositional biplot, the main interest is concentrated to links (distances between vertices of the rays); concretely, for the rays i and j, $i, j = 1, \dots, k$, the link approximates the (usual) variance $\mathrm{var}(\ln \frac{u_i}{u_j})$ of the logratio between the compositional parts (clusters) u_i and u_j. Hence, when the vertices coincide, or nearly so, then the ratio between u_i and u_j is constant, or nearly so, and the corresponding clusters are redundant. In addition, directions of the rays signalize where observations with dominance of the clusters are located. Although the dimension reduction, caused by taking only the first two principal components, naturally leads to some inconsistencies (observations from different clusters may overlap, also the display of classification boundaries is not meaningful), the biplot can be used to reconstruct the multivariate data structure and reveal reasons for misclassification within fuzzy clustering.

However, through all the advantages of the compositional biplot, outliers can substantially affect results of the underlying principal component analysis and depreciate the predicative value of the biplot. For this reason, a robust version of the biplot is needed. Because the principal component analysis is based on the estimation of location and covariance, we need to find proper alternatives to the standard choice, represented by the arithmetic mean and the sample covariance matrix that can be strongly influenced by outlying observations. Among the various proposed robust estimators of multivariate location and covariance, the MCD (Minimum Covariance Determinant) estimator (see, e.g., [11]) became very popular because of its good robustness properties and a fast algorithm for its computation [13]. The MCD estimator looks for a subset h out of n observations with the smallest determinant of their sample covariance matrix. A robust estimator of location is the arithmetic mean of these observations, and a robust estimator of covariance is the sample covariance matrix of the h observations, multiplied by a factor for consistency at normal distribution. The subset size h can vary between half the sample size and n, and it will determine the robustness of the estimates, but also their efficiency.

Besides robustness properties the property of affine equivariance of the estimators of location and covariance plays an important role. The location estimator T and the covariance estimator C are called affine equivariant, if for a sample $\mathbf{z}_1, \ldots, \mathbf{z}_n$ of n observations (e.g. ilr-transformed membership vectors) in $\mathbf{R}^{D-1}$, any nonsingular $(D-1) \times (D-1)$ matrix $\mathbf{A}$ and for any vector $\mathbf{b} \in \mathbf{R}^{D-1}$ the conditions

$$T(\mathbf{A}\mathbf{z}_1 + \mathbf{b}, \ldots, \mathbf{A}\mathbf{z}_n + \mathbf{b}) = \mathbf{A}T(\mathbf{z}_1, \ldots, \mathbf{z}_n) + \mathbf{b},$$
$$C(\mathbf{A}\mathbf{z}_1 + \mathbf{b}, \ldots, \mathbf{A}\mathbf{z}_n + \mathbf{b}) = \mathbf{A}C(\mathbf{z}_1, \ldots, \mathbf{z}_n)\mathbf{A}'$$

are fulfilled. The MCD estimator shares the property of affine equivariance for both the resulting location and covariance estimator.

Because the robust statistical methods cannot work with singular data, the robust scores and loadings must be computed from ilr-transformed compositions before their representation in the clr space. Below we provide some technical details according to paper [7].

Given an $n \times k$ data matrix $\mathbf{U}_{n,k}$ with n membership coefficient vectors $\mathbf{u}_i'$, $i = 1, \ldots, n$, in its rows. Applying the clr transformation to each row results in the clr-transformed matrix $\mathbf{Y}$. The relation

$$\mathbf{Z} = \mathbf{Y}\mathbf{V} \tag{5}$$

for the ilr-transformed data matrix $\mathbf{Z}$ of dimension $n \times (k-1)$ follows from the relation between clr and ilr transformations where the columns of the $k \times (k-1)$ matrix $\mathbf{V}$ contain orthonormal basis vectors of the hyperplane $y_1 + \cdots + y_k = 0$, $\mathbf{V}'\mathbf{V} = \mathbf{I}_{k-1}$ (identity matrix of order $k-1$) [5]. Using the location estimator $T(\mathbf{Z})$ and the covariance estimator $C(\mathbf{Z})$ for the ilr-transformed data, the principal component analysis transformation is defined as

$$\mathbf{Z}^* = [\mathbf{Z} - \mathbf{1}T(\mathbf{Z})']\mathbf{G}_{\mathbf{z}}. \tag{6}$$

The $(k-1) \times (k-1)$ matrix $\mathbf{G}_{\mathbf{z}}$ results from the spectral decomposition of

$$C(\mathbf{Z}) = \mathbf{G}_{\mathbf{z}}\mathbf{L}_{\mathbf{z}}\mathbf{G}_{\mathbf{z}}', \tag{7}$$

where the matrix $\mathbf{L}_{\mathbf{z}}$ is made up of the sorted eigenvalues of matrix $C(\mathbf{Z})$.

If the original data matrix has rank $k-1$, the matrix $\mathbf{Z}$ will also have full rank $k-1$, and an affine equivariant estimator like MCD can be used for $T(\mathbf{Z})$ and $C(\mathbf{Z})$, resulting in robust principal component scores $\mathbf{Z}^*$ and loadings $\mathbf{G}_{\mathbf{z}}$. However, since these are no longer easily interpretable, we have to back-transform the results to the clr space. The scores in the clr space, $\mathbf{Y}^*$, are identical to the scores $\mathbf{Z}^*$ of the ilr space, except that the additional last column of the clr score matrix has entries of zero. For obtaining the back-transformed loading matrix we can use relation (5). For an affine equivariant scatter estimator we have

$$C(\mathbf{Y}) = C(\mathbf{ZV}') = \mathbf{V}\,C(\mathbf{Z})\,\mathbf{V}' = \mathbf{V}\,\mathbf{G_z}\mathbf{L_z}\mathbf{G}'_\mathbf{z}\,\mathbf{V}', \tag{8}$$

and thus the matrix

$$\mathbf{G_y} = \mathbf{V}\mathbf{G_z} \tag{9}$$

represents the matrix of eigenvectors to the *nonzero* eigenvalues of $C(\mathbf{Y})$ (with the property $\mathbf{G}'_\mathbf{y}\mathbf{G_y} = \mathbf{I}_{k-1}$). The nonzero eigenvalues of $C(\mathbf{Y})$ are the same as for $C(\mathbf{Z})$ and consequently the explained variance with the chosen number of principal components remains unchanged. Finally, the robust loadings and scores can be used to obtain a robust biplot for compositional data.

The above introduced theoretical framework is applied to geochemical data originated from a 120 km transect running through Oslo. In total, 360 samples from nine different plant species (40 samples for each species) were analyzed for the concentration of 25 chemical elements. The data set is available in the R package `rrcov` as object `OsloTransect`. Here we only used the variables with reasonable data quality, namely Ba, Ca, Cr, Cu, La, LOI, Mg, Mn, P, Pb, Sr and Zn. Since the data set is of compositional nature itself, we first used the ilr-transformation and afterwards applied fuzzy k-means clustering with $k = 9$ (number of different plant species in the data set). This results in nine-part membership coefficients, and thus their visualization in a ternary diagram is no longer possible.

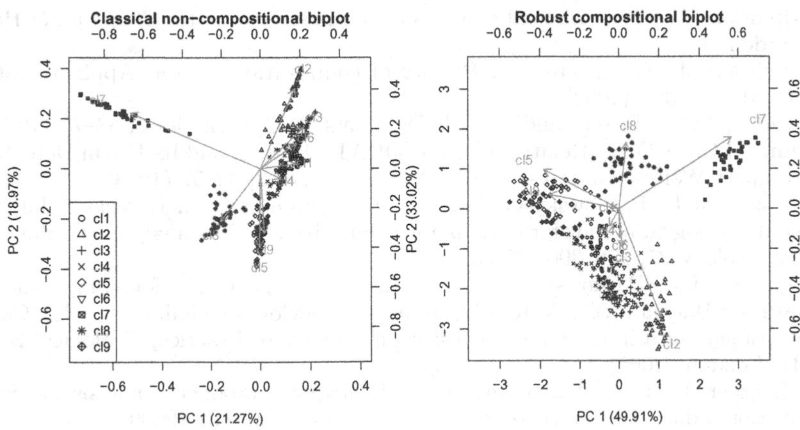

Fig. 2 Biplot resulting from an application to the untransformed membership coefficients (left), and robust biplot resulting from transformed membership coefficients (right)

Without being aware of the above approach based on compositional data analysis, one would probably try to summarize the information contained in the matrix of membership coefficients by principal component analysis (PCA). This procedure is applied here for comparison, and the resulting biplot is presented in Fig. 2 left. The symbols refer to the clusters that have been found with k-means clustering. One can see that there is a certain grouping structure, but there is a lot of overlap of the groups. This is due to an application of PCA in a inappropriate space, the simplex sample space. Note that a robust PCA applied in this space would not lead to an improvement.

Next we apply the procedure as proposed above, by first transforming the membership coefficients, and then applying robust PCA. The resulting robust compositional biplot is displayed in Fig. 2 right. This plot allows for a much better visual inspection. In contrast to the previous biplot, here the first two principal components explain more than 80% of the total variance. It can be seen that fuzzy k-means clustering indeed gave membership coefficients that correspond to relatively clearly separated groups. This also verifies that the algorithm worked well, and that the clustering structure in the data is clearly present. Here we do not further analyse if the correct groups (plant species) were identified, since we are not evaluating the clustering procedure itself.

Acknowledgements. This work was supported by the grant Matematické modely a struktury, PrF_2011_022.

References

1. Aitchison, J.: The statistical analysis of compositional data. Chapman & Hall, London (1986)
2. Aitchison, J., Greenacre, M.: Biplots of compositional data. Applied Statistics 51, 375–392 (2002)
3. Bezdek, J.C.: Cluster validity with fuzzy sets. J. Cybernetics 3, 58–73 (1973)
4. Dunn, J.C.: A Fuzzy Relative of the ISODATA Process and Its Use in Detecting Compact Well-Separated Clusters. J. Cybernetics 3, 32–57 (1973)
5. Egozcue, J.J., Pawlowsky-Glahn, V., Mateu-Figueras, G., Barceló-Vidal, C.: Isometric logratio transformations for compositional data analysis. Mathematical Geology 35, 279–300 (2003)
6. Egozcue, J.J., Pawlowsky-Glahn, V.: Simplicial geometry for compositional data. In: Buccianti, A., Mateu-Figueras, G., Pawlowsky-Glahn, V. (eds.) Compositional Data in the Geosciences: From Theory to Practice, Geological Society, London (2006)
7. Filzmoser, P., Hron, K., Reimann, C.: Principal component analysis for compositional data with outliers. Environmetrics 20, 621–632 (2009)
8. Fisher, R.A.: The use of multiple measurements in axonomic problems. Annals of Eugenics 7, 179–188 (1936)
9. Gabriel, K.R.: The biplot graphic display of matrices with application to principal component analysis. Biometrika 58, 453–467 (1971)

10. Kaufman, L., Rousseeuw, P.J.: Finding groups in data. John Wiley & Sons, New York (1990)
11. Maronna, R., Martin, R.D., Yohai, V.J.: Robust statistics: theory and methods. John Wiley & Sons, New York (2006)
12. Mocz, G.: Fuzzy cluster analysis of simple physicochemical properties of amino acids for recognizing secondary structure in proteins. Protein Science 4, 1178–1187 (1995)
13. Rousseeuw, P., Van Driessen, K.: A fast algorithm for the minimum covariance determinant estimator. Technometrics 41, 212–223 (1999)

Rank Tests under Uncertainty: Regression and Local Heteroscedasticity

Jana Jurečková and Radim Navrátil

Abstract. Data are often affected by unknown heteroscedasticity, which can stretch the conclusions. This is even more serious in regression models, when data cannot be visualized. We show that the rank tests for regression significance are resistant to some types of local heteroscedasticity in the symmetric situation, provided the basic density of errors is symmetric and the score-generating function of the rank test is skew-symmetric. The performance of tests is illustrated numerically.

Keywords: Heteroscedasticity, linear regression, rank test.

1 Introduction and Basic Assumptions

Consider the model

$$Y_i = \beta_0 + \mathbf{x}_{ni}^\top \boldsymbol{\beta} + \sigma_{ni} U_i, \quad i = 1, \ldots, n \tag{1}$$

where $\mathbf{Y}_n = (Y_1, \ldots, Y_n)^\top$ is the vector of observations, $\mathbf{x}_{ni} \in \mathbb{R}_p$, $1 \leq i \leq n$ are known or observable regressors, $\beta_0 \in \mathbb{R}_1$, $\boldsymbol{\beta} \in \mathbb{R}_p$ and $\boldsymbol{\sigma}_n = (\sigma_{n1}, \ldots, \sigma_{nn})^\top \in \mathbb{R}_n^+$ are unknown parameters and $\mathbf{U}_n = (U_1, \ldots, U_n)^\top$ are the i.i.d. errors with the joint but unknown distribution function F. The problem is to test the hypothesis $\mathbf{H}: \ \boldsymbol{\beta} = \mathbf{0}$ with β_0, $\boldsymbol{\sigma}$ unspecified. An alternative for $\mathbf{H}$ is not only $\boldsymbol{\beta} \neq \mathbf{0}$; one must consider also some structure for $\boldsymbol{\sigma}$. In practical problems, many authors consider the regression in scale of the form

Jana Jurečková · Radim Navrátil
Charles University, Department of Probability and Statistics (MFF UK), 196 75 Prague 8, Czech Republic
e-mail: {jurecko,navratil}@karlin.mff.cuni.cz

R. Kruse et al. (Eds.): Synergies of Soft Computing and Statistics, AISC 190, pp. 255–261.
springerlink.com © Springer-Verlag Berlin Heidelberg 2013

$$\sigma_{ni} = \exp\{\mathbf{z}_{ni}^\top \gamma\}, \quad i = 1, \ldots, n \tag{2}$$

with known or observable $\mathbf{z}_{ni} \in \mathbb{R}_q$, $1 \leq i \leq n$, and unknown parameter $\gamma \in \mathbb{R}^q$. Such model was considered by Akritas and Albers [1], who constructed an aligned rank test on some components of parameter $\boldsymbol{\beta}$, with γ replaced by a suitable R-estimator. Gutenbrunner [4] constructed a test of heteroscedasticity, i.e. of $\mathbf{H}^* : \ \gamma = \mathbf{0}$, when $\mathbf{z}_{ni}$ in (2) was partitioned as $\mathbf{z}_{ni}^\top = (1, \mathbf{x}_{ni}^\top, \boldsymbol{\xi}_{ni}^\top)^\top$, with $\mathbf{x}_{ni}$ from (1) and $\boldsymbol{\xi}_{ni}$ was an an external vector, $i = 1, \ldots, n$. His test was based on a combination of regression rank scores for the ξ_{ni} and regression quantile estimator of $\boldsymbol{\beta}$. This test was then modified in [3], see also the review paper [7]. Estimation problem in model (1) was studied by Dixon and McKean [2], who modeled the scale as $\sigma_{ni} = \exp\{\theta h(\mathbf{x}_{ni}^\top \boldsymbol{\beta})\}$, $i = 1, \ldots, n$, with a known function h; they estimated $\boldsymbol{\beta}$ and θ iteratively by means of suitable R-estimates.

In model (1) with scale (2), we want to test the hypothesis

$$\mathbf{H}_0 : \ \boldsymbol{\beta} = \mathbf{0}, \ \beta_0 \ \text{and} \ \ \gamma \ \text{unspecified.}$$

We try to use the rank tests for $\mathbf{H}_0$, profiting from their wide invariance properties. Tests of $\mathbf{H}_0 : \ \boldsymbol{\beta} = \mathbf{0}$ under a nuisance heteroscedasticity, based on regression rank scores, are studied by authors in [6], along with the tests of $\mathbf{H}^* : \ \gamma = \mathbf{0}$ of no heteroscedasticity, with $\boldsymbol{\beta}$ unspecified.

In the present paper, we consider standard rank tests and investigate, up to which extent the ignorance of heteroscedasticity affects the result of the test. We consider the *local heteroscedasticity*, meaning that $\gamma_n = n^{-1/2}\boldsymbol{\delta}$, $\mathbf{0} \neq \boldsymbol{\delta} \in \mathbb{R}_q$. It turns out that the local heteroscedasticity does not worsen the asymptotic efficiency of the test, provided that either the $\mathbf{x}$- and $\mathbf{z}$-regressors are orthogonal, or the errors U_i have a symmetric density f, and the score function of our rank test is skew-symmetric.

Hence, we assume that F has an absolutely continuous symmetric density f and finite Fisher's informations with respect to the location and scale,

$$0 < \mathcal{I}(f) = \int \left(\frac{f'(x)}{f(x)}\right)^2 f(x)dx < \infty \tag{3}$$

$$0 < \mathcal{I}_1(f) = \int \left[-1 - x\frac{f'(x)}{f(x)}\right]^2 f(x)dx < \infty.$$

Let $\mathbf{X}_n = \begin{bmatrix} \mathbf{x}_{n1}^\top \\ \cdots \\ \mathbf{x}_{nn}^\top \end{bmatrix}$, $\quad \widetilde{\mathbf{X}}_n = \begin{bmatrix} (\mathbf{x}_{n1} - \bar{\mathbf{x}}_n)^\top \\ \cdots \\ (\mathbf{x}_{nn} - \bar{\mathbf{x}}_n)^\top \end{bmatrix}$, $\quad \mathbf{Z}_n = \begin{bmatrix} \mathbf{z}_1^\top \\ \cdots \\ \mathbf{z}_n^\top \end{bmatrix}$

be $(n \times p)$ and $(n \times q)$ matrices, respectively, $\bar{\mathbf{x}}_n = \frac{1}{n}\sum_{i=1}^n \mathbf{x}_{ni}$. We assume that the regressors satisfy

$$\begin{aligned}
&\max_{1\le i\le n}\|\mathbf{x}_{ni}\| = o(n^{\frac{1}{2}}), \quad \max_{1\le i\le n}\|\mathbf{z}_{ni}\| = o(n^{\frac{1}{2}}) \ \text{ as } \ n\to\infty,\\
&\lim_{n\to\infty}\widetilde{\mathbf{D}}_n = \lim_{n\to\infty}\frac{1}{n}\widetilde{\mathbf{X}}_n^\top\widetilde{\mathbf{X}}_n = \widetilde{\mathbf{D}}, \qquad \lim_{n\to\infty}\mathbf{Q}_n = \lim_{n\to\infty}\frac{1}{n}\mathbf{Z}_n^\top\mathbf{Z}_n = \mathbf{Q},\\
&\lim_{n\to\infty}\frac{1}{n}\widetilde{\mathbf{X}}_n^\top\mathbf{Z}_n = \mathbf{B} \qquad\qquad (4)\\
&\lim_{n\to\infty}\left[\max_{1\le i\le n}\left\{(\mathbf{x}_{ni}-\bar{\mathbf{x}}_n)^\top(\widetilde{\mathbf{X}}_n^\top\widetilde{\mathbf{X}}_n)^{-1}(\mathbf{x}_{ni}-\bar{\mathbf{x}}_n)\right\}\right] = 0,\\
&\lim_{n\to\infty}\left[\max_{1\le i\le n}\left\{\mathbf{z}_{ni}^\top(\mathbf{Z}_n^\top\mathbf{Z}_n)^{-1}\mathbf{z}_{ni}\right\}\right] = 0
\end{aligned}$$

where $\widetilde{\mathbf{D}}$ and $\mathbf{Q}$ are positive definite $(p\times p)$ and $(q\times q)$ matrices, respectively, and $\mathbf{B}$ is a $(p\times q)$ matrix.

2 Rank Tests and Local Heteroscedasticity

The homoscedasticity in model (1) and (2) means that $\boldsymbol{\gamma} = \mathbf{0}$. We speak on the *local heteroscedasticity*, when

$$\boldsymbol{\gamma} = \boldsymbol{\gamma}_n = n^{-\frac{1}{2}}\boldsymbol{\delta}, \quad \boldsymbol{\delta}\in\mathbb{R}_q,\ \boldsymbol{\delta}\neq\mathbf{0},\ \|\boldsymbol{\delta}\|\le C<\infty. \tag{5}$$

We intend to use a rank test for $\mathbf{H}_0$ based on the ranks $R_{n1},\ldots,R_{nn}$ of $Y_1,\ldots,Y_n$. If we are not aware of the heteroscedasticity, we use the standard rank test based on the vector of linear rank statistics

$$\mathbf{S}_n = n^{-\frac{1}{2}}\sum_{i=1}^n(\mathbf{x}_{ni}-\bar{\mathbf{x}}_n)a_n(R_{ni}) = n^{-\frac{1}{2}}\sum_{i=1}^n(\mathbf{x}_{ni}-\bar{\mathbf{x}}_n)\varphi\left(\frac{R_{ni}}{n+1}\right) \tag{6}$$

where $\varphi : (0,1)\mapsto\mathbb{R}_1$ is a nondecreasing and square-integrable score-generating function. The test criterion for $\mathbf{H}_0$ is the quadratic form in $\mathbf{S}_n$,

$$\mathcal{T}_n^2 = \frac{\mathbf{S}_n^\top\widetilde{\mathbf{D}}_n^{-1}\mathbf{S}_n}{A^2(\varphi)}, \quad A^2(\varphi) = \int_0^1\varphi^2(t)dt - \bar{\varphi}^2,\ \bar{\varphi} = \int_0^1\varphi(t)dt. \tag{7}$$

The test rejects $\mathbf{H}_0$ in favor of $\boldsymbol{\beta}\neq\mathbf{0}$ if $\mathcal{T}_n^2 > C_\alpha$ where C_α is the critical value such that

$$I\!P_{H_0}\left(\mathcal{T}_n^2 > C_\alpha\right) + \tau I\!P_{H_0}\left(\mathcal{T}_n^2 = C_\alpha\right) = \alpha, \quad 0\le\tau<1.$$

In the absence of heteroscedasticity, the null distribution of the test criterion, and hence C_α, does not depend on the specific f. The critical value C_α can be obtained by calculating $\mathcal{T}_n^2$ for all or for an appropriate fraction of the $n!$ permutations of $\{1,\ldots,n\}$ in the role of ranks. For large number n of observations, we use the asymptotic distribution: Under (3) and (4), the asymptotic null distribution of $\mathcal{T}_n^2$, as $n\to\infty$, is χ^2-distribution with

p degrees of freedom. The asymptotic Pitman efficiency of the test follows from [8], Chapter 5. Our problem of interest is to find how the eventual heteroscedasticity affects the efficiency of this test.

Under hypothesis $\mathbf{H}_0$ and under heteroscedasticity (2), the random vector $\mathbf{Y}$ has density

$$q_{n,\gamma}(y_1,\ldots,y_n) = \prod_{i=1}^{n} \exp\{\mathbf{z}_{ni}^\top\gamma\} f\left(y_i \exp\{\mathbf{z}_{ni}^\top\gamma\}\right). \tag{8}$$

Under the local heteroscedasticity (5), the sequence of densities $\{q_{n\gamma}\}$ is contiguous to the sequence $\{q_{n0}\}$, corresponding to $\gamma = \mathbf{0}$ (for contiguity see [5], Chapter VI). It further follows from [5] and from [8], that the asymptotic distribution of $\mathbf{S}_n$ under $\mathbf{H}_0$ and under (5) is normal $\mathcal{N}_p\left(\boldsymbol{\mu}_\delta, A^2(\varphi)\,\widetilde{\mathbf{D}}\right)$, where

$$\boldsymbol{\mu}_\delta = \mathbf{B}\,\boldsymbol{\delta}\int_0^1 \varphi(u)\varphi_1(u,f)du,$$
$$\varphi_1(u,f) = -1 - F^{-1}(u)\frac{f'(F^{-1}(u))}{f(F^{-1}(u))}, \quad 0 < u < 1. \tag{9}$$

Hence, the criterion $\mathcal{T}_n^2$ has, under $\mathbf{H}_0$ and under the local heteroscedasticity, asymptotically noncentral χ_p^2 distribution with noncentrality parameter

$$\eta^2 = \boldsymbol{\delta}^\top\mathbf{B}^\top\mathbf{D}^{-1}\mathbf{B}\boldsymbol{\delta}\,\frac{[\int_0^1 \varphi(u)\varphi_1(u,f)du]^2}{A^2(\varphi)}. \tag{10}$$

Particularly, the noncentrality parameter vanishes if either $\mathbf{Z}_n$ is asymptotically orthogonal to $\widetilde{\mathbf{X}}_n$, i.e. $\widetilde{\mathbf{X}}_n^\top\mathbf{Z}_n \to \mathbf{0}$ as $n \to \infty$, or if f is symmetric and φ is skew-symmetric, i.e.

$$f(x) = f(-x), \quad x \in \mathbb{R}_1 \quad \text{and} \quad \varphi(u) = -\varphi(1-u),\ 0 < u < 1.$$

If this happens, the asymptotic distribution of $\mathcal{T}_n^2$ is still central χ_p^2 distribution, regardless $\boldsymbol{\gamma}$. Hence, if the heteroscedasticity in model (1) is only local, the asymptotic distribution of $\mathcal{T}_n^2$ under $\mathbf{H}_0$ is not changed if either the $\mathbf{x}$-regressors are orthogonal to the $\mathbf{z}$-regressors, or if f is symmetric and φ skew-symmetric. In such case, we reject $\mathbf{H}_0$ if $\mathcal{T}_n^2 \geq \chi^2(\alpha)$ where $\chi^2(\alpha)$ is the $(1-\alpha)$-quantile of the central χ_p^2 distribution. It further follows from the behavior of tests under contiguous alternatives (see [5]), that then even the asymptotic relative efficiency of the test, corresponding to small values of $\|]boldgreek\beta]\| = O(n^{-1/2})$, does not change, either.

If we can expect a symmetry of f or orthogonality of $\widetilde{\mathbf{X}}$ and $\mathbf{Z}$, then we take the rank test with a skew-symmetric score generating function, e.g. the Wilcoxon $\varphi(u) = 2u - 1$, or the median $\varphi(u) = \mathrm{sign}(\mathrm{u} - \frac{1}{2}),\ 0 \leq \mathrm{u} \leq 1$. For a more general case, we can recommend tests based on regression rank scores; this is a subject of the forthcoming study [6].

3 Numerical Illustration

The following simulation study illustrates how the procedures work in finite sample situation for various choices of score function φ and for various model errors U_i.

Consider the model of regression line with a possible heteroscedasticity

$$Y_i = \beta_0 + \beta x_i + \exp\{z_i\gamma\}U_i, \quad i = 1, \ldots, n,$$

and the problem of testing $\mathbf{H}_0 : \beta = 0$ against two-sided alternative $\beta \neq 0$, considering β_0 and γ as nuisance parameters. Nuisance β_0 does not affect power of the tests, because rank tests are invariant to the location, so that further on it will be considered fixed: $\beta_0 = 2$.

We take the following three choices of score function φ:

$\varphi^{(1)}(t) = 2t - 1$ Wilcoxon scores,
$\varphi^{(2)}(t) = \Phi^{-1}(t)$ van der Waerden scores,
$\varphi^{(3)}(t) = \text{sign}(t - 1/2)$ median scores,

where $\Phi^{-1}(t)$ is quantile function of standard normal distribution $N(0,1)$. First we compared the powers of the test (7) for these score functions. The regressors x_i and z_i were generated from independent samples of sizes $n = 100$ from uniform $(-2, 10)$ distribution. Model errors U_i were generated from normal, logistic, Laplace and t-distribution with 6 degrees of freedom, respectively, always with 0 mean a variance 3/2. The empirical powers of tests were

Table 1 Percentage of rejections of hypothesis $\mathbf{H}_0 : \beta = 0$ for various model errors U_i by Wilcoxon, van der Waerden and median tests - (in this order), $\gamma = 0.05$

$\beta \setminus U_i$	normal			logistic			Laplace			t-distribution		
0	4.94	4.32	5.26	5.05	5.21	5.13	5.05	4.21	5.14	5.07	4.51	5.52
−0.03	10.03	8.92	8.75	11.13	9.87	9.92	13.28	11.10	15.00	11.10	9.76	9.89
0.03	10.29	9.20	9.31	10.70	9.20	9.60	13.02	10.59	14.70	11.01	9.72	9.74
0.05	19.33	17.84	15.46	21.76	19.13	18.03	27.73	22.80	30.49	22.92	20.03	19.87
0.07	33.26	31.12	25.38	37.49	33.62	31.04	47.62	40.44	50.46	40.02	35.89	32.89
0.1	59.59	57.55	45.42	65.17	60.46	55.05	76.17	68.45	76.70	67.22	62.39	56.66

Table 2 Percentage of rejections of hypothesis $\mathbf{H}_0$: $\beta = 0$ for various model errors U_i by Wilcoxon, van der Waerden and median tests - (in this order), $\gamma = -0.05$

$\beta \setminus U_i$	normal	logistic	Laplace	t-distribution
0	5.01 4.22 5.20	4.84 4.38 4.98	4.82 4.33 5.02	5.03 4.41 5.54
−0.03	16.62 15.20 13.55	18.79 16.44 16.02	23.98 19.71 26.46	19.52 17.18 16.56
0.03	16.70 15.07 13.91	18.25 15.97 15.41	23.28 18.88 25.99	19.17 16.67 16.35
0.05	37.88 35.27 28.52	41.40 37.10 34.26	52.08 44.48 55.09	43.65 39.03 36.81
0.07	63.09 60.77 48.94	68.16 63.87 58.23	78.68 71.25 80.00	71.26 66.70 61.02
0.1	90.25 89.54 77.51	93.11 90.83 85.61	96.85 94.25 96.34	94.03 92.03 87.60

Table 3 Percentage of rejections of hypothesis $\mathbf{H}_0$: $\beta = 0$ by Wilcoxon test

$\gamma \setminus \beta$	0	0.01	0.03	0.05	0.07	0.1
0	5.03	6.21	16.92	38.33	63.91	90.86
−0.01	4.97	6.50	17.54	40.83	66.82	92.69
0.01	4.76	6.01	15.31	35.92	60.80	88.76
−0.02	4.81	6.30	19.36	45.54	70.28	94.39
0.02	4.88	5.68	15.35	32.97	55.97	86.09
−0.03	4.38	5.86	20.09	45.79	73.97	95.64
0.03	5.02	5.79	14.40	31.03	53.63	83.38

computed as percentages of rejections of $\mathbf{H}_0$ among 10 000 replications, at significance level $\alpha = 0.05$. The results are summarized in Tbl. 1 ($\gamma = 0.05$) and Tbl. 2 ($\gamma = -0.05$).

Notice that the power of Wilcoxon test is comparable with that of the van der Waerden test, even for normally distributed errors. The median test shows better performace for Laplace model errors, for which it is optimal, though the power of Wilcoxon test is only slightly smaller even in this case. Generally, Wilcoxon test achieves the best results for all choices of model errors, despite its simple form.

Table 3 compares the empirical powers of Wilcoxon test for various β and γ. Design of the simulation is considered the same as in the previous situation, model errors U_i were generated from standard normal distribution $N(0,1)$.

Since all the considered score functions are skew-symmetric and all the model errors are symmetric, the rank tests preserve prescribed probability of the error of the first kind α. However, in the finite-sample situation, the powers of tests can still depend on the nuisance parameter γ, compared with the homoscedastic situation $\gamma = 0$. As we have chosen regressors z_i symmetric around 4, rather than around 0, the variance of the model errors is higher for γ positive than for γ negative, what has an effect on the power.

Acknowledgements. The research of the first author was supported by the Grant GAČR 201/12/0083. The research of the second author was supported by the Charles University Grant 105610 and by the Grant SVV 265 315.

References

1. Akritas, M.G., Albers, W.: Aligned rank tests for the linear model with heteroscedastic errors. J. Statist. Planning and Inference 37, 23–41 (1993)
2. Dixon, S.L., McKean, J.W.: Rank-Based Analysis of the Heteroscedastic Linear Model. J. Amer. Statist. Assoc. 91, 699–712 (1996)
3. Gutenbrunner, C., Jurečková, J., Koenker, R.: Regression rank test for heteroscedasticity (preprint)
4. Gutenbrunner, C.: Tests for heteroscedasticity based on regression quantiles and regression rank scores. In: Mandl, P., Hušková, M. (eds.) Asymptotic Statistics: Proc. 5th Prague Symp., pp. 249–260. Physica-Verlag, Heidelberg (1994)
5. Hájek, J., Šidák, Z.: Theory of Rank Tests. Academic Press, New York (1967)
6. Jurečková, J., Navrátil, R.: Rank tests in regression model under a nuisance heteroscedasticity (in preparation)
7. Koenker, R.: Rank tests for linear models. In: Maddala, G.S., Rao, C.R. (eds.) Handbook of Statistics, vol. 15, pp. 175–199 (1997)
8. Puri, M.L., Sen, P.K.: Nonparametric Methods in General Linear Models. John Wiley & Sons, New York (1985)

Robustness Issues in Text Mining

Marco Turchi, Domenico Perrotta, Marco Riani, and Andrea Cerioli

Abstract. We extend the Forward Search approach for robust data analysis to address problems in text mining. In this domain, datasets are collections of an arbitrary number of documents, which are represented as vectors of thousands of elements according to the vector space model. When the number of variables v is so large and the dataset size n is smaller by order of magnitudes, the traditional Mahalanobis metric cannot be used as a similarity distance between documents. We show that by monitoring the cosine (dis)similarity measure with the Forward Search approach it is possible to perform robust estimation for a document collection and order the documents so that the most dissimilar (possibly outliers, for that collection) are left at the end. We also show that the presence of more groups of documents in the collection is clearly detected with multiple starts of the Forward Search.

Keywords: Cosine similarity, document classification, forward search.

1 Introduction

In text mining, where large collections of textual documents are analyzed by automatic tools such as document classifiers or indexers to help human beings to better understand their contents, the most used document representation schema is the vector space model (VSM), introduced by [11] in information

Marco Turchi · Domenico Perrotta
European Commission, Joint Research Centre
e-mail: marco.turchi@jrc.ec.europa.eu,
domenico.perrotta@ec.europa.eu

Marco Riani · Andrea Cerioli
University of Parma, Department of Economics
e-mail: {mriani,andrea.cerioli}@unipr.it

R. Kruse et al. (Eds.): Synergies of Soft Computing and Statistics, AISC 190, pp. 263–272.
springerlink.com © Springer-Verlag Berlin Heidelberg 2013

retrieval. This model transforms a text in a machine readable vector assigning words to numeric vector components. Datasets are collections with an arbitrary, sometimes large, number of units n (the documents) and each unit is identified by dozens of thousands of VSM variables (the v document word identifiers). In several text mining applications there is the need of estimating a centroid for a given document collection, and to define an ordering of the documents with respect to the centroid, from the most to the least representative one. This ordering can be used to identify documents which have a weak semantic relation with the dominant subject(s) in the collection.

Outlying documents are likely to be present in most text mining applications, either because they correspond to documents which are inconsistent with the rest of the collection, or because of human mistakes in document labeling. Three popular strategies for robust estimation in presence of outliers are the following (see, e.g. [7] for a review):

1. Use a reduced number of units in order to exclude outliers from the estimation process;
2. Down-weight each unit according to its deviation from the centroid;
3. Optimize a robust objective function.

Disadvantages of these approaches are the fact that the percentage of units to be discarded needs to be fixed in advance (strategy 1), that there is no universally accepted way to down-weight observations (strategy 2) and that optimization of complex functions may cause severe computational problems (strategy 3). In addition, these strategies cannot be easily extended to heterogeneous datasets, with the purpose of identifying subgroups of similar documents in the collection. A different approach is followed by a fourth robust strategy, the Forward Search (FS) [1, 3]: instead of choosing just one subsample, a sequence of subsets of increasing size is fit and a problem-specific diagnostic is monitored in order to reveal if a new observation is in agreement with those previously included. Outliers are left at the end of the subset sequence and the effect of each unit, once it is introduced into the subset, can be measured and appraised.

With VSM, where the number of variables v is so large and the dataset size n is perhaps smaller by order of magnitudes, none of the above strategies can use traditional metrics such as the Mahalanobis distance to measure the similarity between documents, as well as the distance from an estimated centroid of the collection and any of the documents. The same drawback also affects other robust distance-based methods for cluster analysis, like TCLUST [5]. In this work, we extend the FS method to VSMs by adopting the cosine similarity [14], a metric widely used in text mining. This metric is the cosine of the angle between two vectors, which is therefore non-negative and bounded between 0 and 1. It is also independent of the vector length. More precisely, we propose to monitor the progression of the complement to one of the minimum value of the cosine similarity between the subset centroid and all units outside the subset. We will refer to this diagnostic as to the *minimum cosine dissimilarity*. Documents will be ordered with the FS in such a way that the most dissimilar (possibly outliers, for that collection) are left at the end of sequence. We will see that the extended FS algorithm preserves the good properties shown by the FS in more traditional statistical domains, such as regression [1], multivariate analysis and clustering [3].

The paper is structured as follows. Section 2 introduces the practical motivations for the work and gives details of the data. For our demonstrations, we have used documents from a very rich source: the EuroVoc corpus. Then, since the work relies on two choices, the VSM to represent documents and the cosine similarity to measure their distance, Sect. 3 describes such choices and some related work. Section 4 provides the results and shows the potential of the FS for text mining applications. In particular, Sect. 4.1 contextualises the FS approach to text mining. All computations and simulations have been performed by extending the robust routines included in the FSDA toolbox of Matlab, downloadable from `http://www.riani.it/matlab.htm` and `http://fsda.jrc.ec.europa.eu` [10].

2 The EuroVoc Corpus

This research is driven by a real need in the development of the JRC EuroVoc Indexer (JEX) [13], a freely available multi-label categorization tool[1]. JEX is a system which automatically assigns a set of category labels from a thesaurus to a textual document. This software is based on the supervised profile ranking algorithm proposed by [9], which uses the EuroVoc thesaurus.

The EuroVoc thesaurus[2] is a multilingual, multidisciplinary thesaurus with currently about 6800 categories, covering all activities of the European Union (EU). EuroVoc's category labels have been translated one-to-one into currently 27 languages. It was developed for the purpose of manual (human) categorisation of all important documents in order to allow multilingual and cross-lingual search and retrieval in potentially very large document collections. As EuroVoc has been used to classify legal documents manually for many years, there are now tens of thousands of manually labelled documents per language that can be used to train automatic categorisation systems [9]. This collection of documents is available for download at `http://eur-lex.europa.eu/`.

The number of documents inside each category is highly unbalanced and follows the Zipf's law distribution: few categories contain more than 3000 documents, and a large number of categories has few documents. Categories belong to different domains and they can be very specific (e.g. *Fishery Management*) or very generic (e.g. *Radioactivity*). In both cases, we cannot exclude the presence of groups in the documents.

---

1 `http://langtech.jrc.ec.europa.eu/JRC_Resources.html`

2 `http://Eurovoc.europa.eu/`

Each English document of the corpus has to be preprocessed with an ordered series of operations. These include lowercasing each word (e.g. "*The White House, the*" → "*the white house, the*"), tokenizing the text ("*the white house, the*" → "*the white house , the*") and removing high frequent words (stopwords) using an external list of more than 2500 words ("*the white house , the*" → "*white house*"). This process reduces the vocabulary size and the sparseness in the data. Then we translate the documents into their VSM representations. For this purpose we count all the words in the full collection after pre-processing and we keep a variable for each corpus term, including those with zero frequency. The pre-processing work for the EuroVoc corpus thus results in a VSM vector of 119,112 variables, which is still sparse.

Models for thousands of categories are trained using only human labelled samples for each category. The training process consists in identifying a list of representative terms and associating to each of them a log-likelihood weight, with the training set used as the reference corpus. A new document is represented as a vector of terms with their frequency in the document. The most appropriate categories for the new document are found by ranking the category vector representations (called profiles), according to their similarity to the vector representation of the new document.

Despite the good performance provided by JEX, human label documents are affected by the presence of outliers: documents which are either wrongly assigned to a category or weakly correlated to the other documents into the category. The main motivation of the proposed extension of the FS is the automatic detection of these outliers, which have to be removed from the training data used by JEX.

3 Similarity in the Vector Space Model

In information retrieval the VSM was proposed to automatically retrieve documents which are similar to an input query [11]. In the VSM, a document d is represented in a high-dimensional space, in which each dimension corresponds to a term in the document. Formally, a document is a vector of v components $d = (t_1, t_2, ..., t_v)'$. A component, called *term weight*, measures how a term is important and representative. In general, v can be the vocabulary containing all terms of a natural language or all specific terms in a collection of documents. This representation produces very sparse vectors, which have only few non-zero terms.

Different options for the term weight are possible, most of which are discussed in [12]. The most used is the frequency count of a term in a document (*term frequency*). The higher the count the more likely it is that the term is a good descriptor of the content of the document. Other, more complex, approaches exist that take into account the distribution of a term in all the available documents. However, despite its limitations, the term frequency measure is easy to compute and is still the most popular choice in text mining applications. Therefore, we restrict ourselves to a VSM where each component of d is defined as a frequency count. Similarly, in this work we do not explore possible extensions of the basic model, such as the Phrase-based VSM [8], or the Context VSM [4].

[12] arguments that the similarity between two documents may be obtained, as a first approximation, by applying the standard dot product formula on the boolean vector representation of the two documents. This representation would measure the number of terms that jointly appear in the two documents. In practice, it is preferable to use weights lying in the range $[0,1]$, in order to provide a more refined discrimination among terms, with weights closer to 1 for the more important (frequent) terms. This naturally yields to take as a similarity measure the cosine of the angle between two VSM vectors:

$$\cos(d_1, d_2) = \frac{\sum_{i=1}^{v} d_1(i) d_2(i)}{\sqrt{\sum_{i=1}^{v} d_1^2(i)} \sqrt{\sum_{i=1}^{v} d_2^2(i)}}. \tag{1}$$

Index (1) is called the *cosine similarity* between d_1 and d_2, while $1 - \cos(d_1, d_2)$ represents the cosine dissimilarity. The value of $\cos(d_1, d_2)$ is 0 if the two vectors are orthogonal, and 1 if they are identical. By definition, the numerator takes into account only the non-zero terms of both vectors, while the denominator is affected by all components of the vectors. Note that the cosine similarity between large documents in general results in small values, because they have poor similarity values (a small scalar product and a large dimensionality).

Since its introduction, the cosine similarity has been the dominant document similarity measure in information retrieval and text mining (see e.g. [6]). A key factor for its success is its capacity of working with high-dimensional vectors, as it projects the vectors into the first quadrant of the circle of radius one. This goes at the expenses of the information lost in the drastic reduction of dimensionality. A potential drawback of working with the pairwise measure (1) is its lack of invariance under different correlation models for the v term frequencies appearing in d_1 and d_2. However, a Mahalanobis-type approach is unfeasible in text mining applications, except in very particular situations. This is the price to pay when we work with $v \gg n$.

4 Data Analysis with the Forward Search

4.1 Steps of Forward Search for Text Mining

The FS builds subsets of increasing size m, starting from a small number of units (the VSM vectors), e.g. $m_0 = 5$, until all units are included. The subsets are built using this ordering criterion: at step m, compute the centroid of the m units in the subset and select for the next subset the $m+1$ units with smaller cosine dissimilarity from the centroid. Then, as m goes from m_0 to n, we monitor the evolution of the minimum cosine dissimilarity. In absence of outliers we expect a rather constant or smoothly increasing statistic progression. On the contrary the entry of outliers, which by construction will happen in the last subsets, will be revealed by appreciable changes of the minimum cosine dissimilarity trajectory. A similar behaviour is observed in presence of different groups when we look at the data from the perspective of a centroid fitted to one group. While for outlier detection a single forward search from a good starting subset is sufficient to reveal possible isolated outliers, for cluster identification many searches are needed. Those starting in a same group, will reveal the group presence in the form of converging group trajectories, such as those highlighted in Fig. 2. The precise identification of outliers and groups, with given statistical significance, is possible using confidence envelopes for the cosine dissimilarity, that can be found along the lines of [2]. Refer to [2] also for details on the key concepts recalled in this section.

4.2 Synthetic Data

The distribution of the terms in a corpus, which typically follows a power law (Zipf's distribution), can be easily estimated once the documents are translated into their VSM representation. Based on the estimated distribution parameters of the EuroVoc corpus, we have built synthetic datasets of 100 units and 119112 variables having cosine similarity for each pair of vectors around 0.8. Such synthetic datasets are used to study the properties of the proposed statistical analysis for a collection of documents with features mimicking those of the EuroVoc corpus.

The left panel of Fig. 1 shows the monitoring of the minimum cosine dissimilarity trajectories of 500 randomly started forward searches, for one of these synthetic datasets. A prototype trajectory is displayed by a black solid line. It is uneventful and well included within the bootstrap bands obtained by random selection of the starting point. Therefore, this plot provides evidence of what we can expect from the FS under the null hypothesis of an homogeneous collection of documents.

On the right panel of Fig. 1 five units of the same dataset have been shuffled. In the VSM this corresponds to considering 5 documents with completely different cosine similarity values from the rest of the documents in the collection. Outlyingness of these observations is clearly reflected in the plot by the large peak at the end of the searches, when the anomalous units enter into the fitting subset regardless of the actual starting point. It may also occasionally happen that a search is randomly initialised with one of such units,

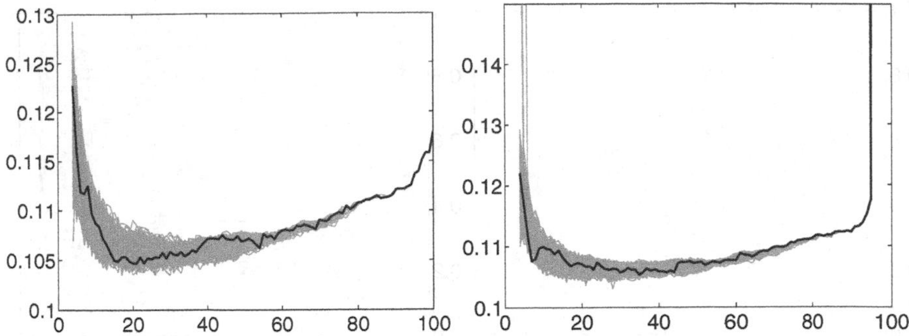

Fig. 1 500 random start forward searches for a synthetic dataset, homogeneous (left panel) and with 5 shuffled units (right panel)

but then the algorithm is immediately able to recover and to substitute the anomalous observations in the fitting subset with uncontaminated ones. In the parlance of the FS, we say that *interchanges* have occurred in the first steps of the algorithm.

4.3 EuroVoc Data

Figure 2 shows the minimum cosine dissimilarity trajectories of 500 randomly started forward searches, for two EuroVoc datasets. The left panel is about category C7, formed by 26 units and 119112 variables. The structure of this plot is very different from what we have seen in Figure 1, both for the case of

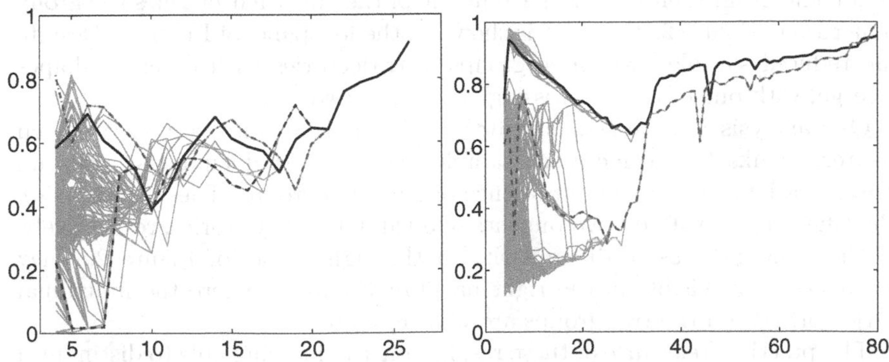

Fig. 2 500 random start forward searches for two EuroVoc datasets, classified by professional librarians to categories identified with C7 (left panel) and C174 (right panel)

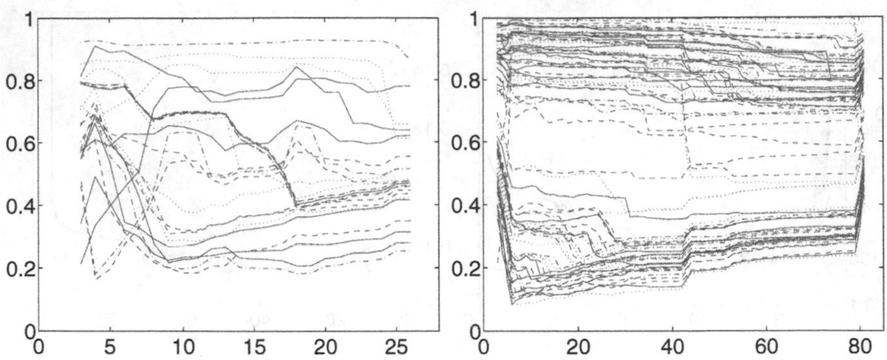

Fig. 3 Individual trajectories of the cosine dissimilarity measures for all the documents from their overall mean in one of the runs that clarifies the group structure in Figure 2. Left panel: category C7; right panel: category C174.

uncontaminated data and when outliers are present. Specifically, at step 18 there are only three groups of 418, 15 and 67 trajectories (respectively from the top to the bottom one). Each group is formed by trajectories that, starting from different initial subsets of documents, converge to the same path. This behaviour provides clear evidence of a cluster structure, because the searches that start in individual groups continue to add observations from the group until all observations in that cluster have been used in estimation. There is then a sudden change in the cosine dissimilarity measure as units from other clusters enter the subset used for estimation. We conclude that category C7 of the EuroVoc corpus cannot be considered homogeneous, but displays three different substructures. This analysis can be supplemented by a forward plot of the individual trajectories of the cosine dissimilarity measures for all the documents from their overall mean in one of the runs that clarifies the group structure in Figure 2. This plot is shown in the left panel of Figure 3. Despite the reduced sample size, three groups of trajectories with different shapes emerge, with one of them "crossing" the other two.

Our analysis is repeated for category C174. Here the pictures are even clearer, thanks to the increased sample size ($n = 81$ documents, again on 119112 variables) and to the presence of only two groups. These clusters are identified, at step 40, by two bunches of 39 and 461 trajectories, respectively, of the minimum cosine dissimilarity in the right panel of Figure 2. They are also clearly visible in the right panel of Figure 3, where the individual trajectories from the two groups are well separated.

The practical relevance of these results consists in being able to distinguish between rather homogeneous sets of documents, possibly contaminated by isolated outliers, and sets formed by different groups. Depending on the final application, outliers and subgroups can be treated differently. For instance, groups can be used to build a committee of classifiers rather than a single one for the entire dataset.

5 Summary

In this paper we have extended the Forward Search approach for robust data analysis to address some relevant issues in text mining, such as the detection of outlying documents or the identification of possible clusters in the data. This achievement has been reached by replacing the traditional Mahalanobis metric of multivariate analysis, which cannot be applied in situations where the sample size is smaller by order of magnitudes than the number of variables, with the cosine dissimilarity measure.

It is well known that when using the VSM, documents can talk about the same theme even using very different set of terms, resulting in low cosine similarity. In this case our approach would identify different groups in the set of documents. This effect can be limited by the adoption of more sophisticated text representation schemes such as the Concept VSM, where each component of the numeric vector represents a concept that is identified by a group of semantically similar terms. As the cosine similarity is a reasonable distance also for concept vectors, our Forward Search extension to text mining would be still applicable.

References

1. Atkinson, A.C., Riani, M.: Robust Diagnostic Regression Analysis. Springer, Berlin (2000)
2. Atkinson, A.C., Riani, M.: Exploratory tools for clustering multivariate data. Comput. Stat. Data Anal. 52, 272–285 (2007)
3. Atkinson, A.C., Riani, M., Cerioli, A.: Exploring Multivariate Data with the Forward Search. Springer, Berlin (2004)
4. Billhardt, H., Borrajo, D., Maojo, V.: A context vector model for information retrieval. J. Am Soc. Inf. Sci. Tec. 53, 236–249 (2002)
5. Garcia-Escudero, L., Gordaliza, A., Matran, C., Mayo-Iscar, A.: A general trimming approach to robust cluster analysis. Ann. Stat. 36, 1324–1345 (2008)
6. Huang, A.: Similarity measures for text document clustering. In: Proc. of the 6th New Zealand Computer Science Research Student Conference, Christchurch, New Zealand, pp. 49–56 (2008)
7. Hubert, M., Rousseeuw, P.J., Van Aelst, S.: High-breakdown robust multivariate methods. Stat. Sci. 23, 92–119 (2008)
8. Mao, W., Chu, W.W.: Free-text medical document retrieval via phrase-based vector space model. In: Proc. of the AMIA Symposium, p. 489 (2002)
9. Pouliquen, B., Steinberger, R., Ignat, C.: Automatic annotation of multilingual text collections with a conceptual thesaurus In: Proc. of the Workshop Ontologies and Information Extraction at the EUROLAN 2003, Bucharest, Romania (2003)

10. Riani, M., Perrotta, D., Torti, F.: FSDA: A MATLAB toolbox for robust analysis and interactive data exploration. Chemometr. Intell. Lab. 116, 17–32 (2012)
11. Salton, G., Wong, A., Yang, C.S.: A vector space model for automatic indexing. Commun. ACM 18, 613–620 (1975)
12. Salton, G., Buckley, C.: Term-weighting approaches in automatic text retrieval. Inform. Process. Manag. 24, 513–522 (1988)
13. Steinberger, R., Ebrahim, M., Turchi, M.: JRC Eurovoc Indexer JEX — A freely available multi-label categorisation tool. In: Proc. of the 8th Int. Conf. on Language Resources and Evaluation (LREC 2012), Istanbul, Turkey (2012)
14. Yates, R.B., Neto, B.R.: Modern Information Retrieval. Addison-Wesley, Reading (1999)

An Alternative Approach to the Median of a Random Interval Using an L^2 Metric

Beatriz Sinova, Gil González-Rodríguez, and Stefan Van Aelst

Abstract. Since the Aumann-type expected value of a random interval is not robust, the aim of this paper is to propose a new central tendency measure for interval-valued data. The median of a random interval has already been defined as the interval minimizing the mean distance, in terms of an L^1 metric extending the Euclidean distance, to the values of the random interval. Inspired by the spatial median, we now follow a more common approach to define the median using an L^2 metric.

Keywords: Interval-valued data, L^2 metric, median, robustness.

1 Introduction and Motivation

Interval data are usually obtained from random experiments involving intrinsically imprecise measurements. Many examples can be found in research studies (from very different fields) which pay more attention to the range of values that a variable can take along a period than to the detailed records.

To analyze the information given by random intervals (that is, interval-valued random elements) some central tendency measures based on the interval arithmetic have been proposed. Although the most often used measure, the Aumann-type expected value, possesses very good properties from both a probabilistic and a statistical point of view, its high sensitivity to data changes or the existence of great magnitude data motivates the search for a

Beatriz Sinova · Gil González-Rodríguez
Universidad de Oviedo, 33007 Oviedo, Spain
e-mail: {sinovabeatriz,gil}@uniovi.es

Stefan Van Aelst
Universiteit Gent, 9000 Gent, Belgium
e-mail: Stefan.VanAelst@UGent.be

R. Kruse et al. (Eds.): Synergies of Soft Computing and Statistics, AISC 190, pp. 273–281.
springerlink.com © Springer-Verlag Berlin Heidelberg 2013

more robust central tendency measure. Inspired by the real case, we define the median of a random interval. The definition of the median as a 'middle' position value cannot be extended because there is not any universally accepted total order criterion in the space of non-empty compact intervals. However, it can still be defined as the element of the space minimizing the mean distance to all the values that the random interval can take (w.r.t. an L^1 metric extending the Euclidean distance in $\mathbb{R}$). The two considered choices for the L^1 metric are the generalized Hausdorff metric (see Sinova *et al.* [5]) and the metric based on the 1-norm, as introduced by Vitale [9] (see Sinova and Van Aelst [6]).

The new proposal uses an L^2 metric instead of an L^1 metric in the definition of the median, similarly as in the spatial median also called mediancentre wich is a well-known generalization of the median to multivariate settings (see, for instance, Gower [4] or Milasevic and Ducharme [2]). We use the L^2 metrics introduced by Bertoluzza *et al.* [1] (as expressed by Gil *et al.* [3]; see also Trutschnig *et al.* [8] for a recent review), a wide and valuable family of metrics for interval data. Furthermore, one of its particular cases was proven to be equivalent to the well-known Vitale L^2 metric (cf. Vitale [9]). One of the advantages of this metric is that it weights squared distances between data location (mid-points/centers) and squared distances between data imprecision (spread/radius), similarly as the generalized Hausdorff metric used in [5] to define the median. In Sect. 2 the notation and basic operations and concepts in the space of interval data are recalled. The definition of the three medians and some details about the computation of the new proposal are presented in Sect. 3. In Sect. 4, the three approaches to the median of a random interval are compared by means of some simulation studies. Finally, Sect. 5 presents some conclusions and open problems.

2 The Space of Intervals $\mathcal{K}_c(\mathbb{R})$: Preliminaries

In this section, the notation used in the paper is established, as well as the basic concepts involving nonempty compact intervals and random intervals in. Each interval $K \in \mathcal{K}_c(\mathbb{R})$, where $\mathcal{K}_c(\mathbb{R})$ denotes the class of nonempty compact intervals in $\mathbb{R}$, can be characterized in terms of its infimum and supremum, $K = [\inf K, \sup K]$, or in terms of its mid-point and spread or radius, $K = [\operatorname{mid} K - \operatorname{spr} K, \operatorname{mid} K + \operatorname{spr} K]$, where

$$\operatorname{mid} K = \frac{\inf K + \sup K}{2}, \quad \operatorname{spr} K = \frac{\sup K - \inf K}{2}.$$

The usual interval arithmetic provides the two most relevant operations from a statistical point of view, the addition and the product by a scalar:

- The Minkowski *sum* of two nonempty compact intervals, $K, K' \in \mathcal{K}_c(\mathbb{R})$, is defined as the interval

$$K + K' = [\inf K + \inf K', \sup K + \sup K']$$

$$= [(\text{mid}\, K + \text{mid}\, K') - (\text{spr}\, K + \text{spr}\, K'), (\text{mid}\, K + \text{mid}\, K') + (\text{spr}\, K + \text{spr}\, K')].$$

- The *product of an interval* $K \in \mathcal{K}_c(\mathbb{R})$ *by a scalar* $\gamma \in \mathbb{R}$ is defined as the element of $\mathcal{K}_c(\mathbb{R})$ such that

$$\gamma \cdot K = \begin{cases} [\gamma \cdot \inf K, \gamma \cdot \sup K] \text{ if } \gamma \geq 0 \\ [\gamma \cdot \sup K, \gamma \cdot \inf K] \text{ otherwise} \end{cases}$$

$$= [\gamma \cdot \text{mid}\, K - |\gamma| \cdot \text{spr}\, K, \gamma \cdot \text{mid}\, K + |\gamma| \cdot \text{spr}\, K].$$

With these two operations the space is only semilinear (with a conical structure) and, to overcome the nonexistence of a difference, distances play a crucial role in statistical developments. We first recall the definition of the medians introduced earlier, based on two L^1 distances between intervals:

- The *generalized Hausdorff metric* (Sinova *et al.* [5]), which is partially inspired by the Hausdorff metric for intervals and the L^2 metrics in Trutschnig *et al.* [8], is defined as follows. Given two intervals $K, K' \in \mathcal{K}_c(\mathbb{R})$ and any $\theta \in (0, \infty)$, their generalized Hausdorff distance is defined as:

$$d_{H,\theta}(K, K') = |\text{mid}\, K - \text{mid}\, K'| + \theta \cdot |\text{spr}\, K - \text{spr}\, K'|.$$

- The *1-norm metric*, introduced by Vitale [9]. Given any two intervals $K, K' \in \mathcal{K}_c(\mathbb{R})$, their 1-norm distance is defined as:

$$\rho_1(K, K') = \frac{1}{2}|\inf K - \inf K'| + \frac{1}{2}|\sup K - \sup K'|.$$

The L^2 distance that will also be used to generalize the median is the following:

- The d_θ *metric* (by Bertoluzza *et al.* [1] and Gil *et al.* [3]) is defined as:

$$d_\theta(K, K') = \sqrt{(\text{mid}\, K - \text{mid}\, K')^2 + \theta \cdot (\text{spr}\, K - \text{spr}\, K')^2},$$

  where $K, K' \in \mathcal{K}_c(\mathbb{R})$ and $\theta \in (0, \infty)$ (it is often supposed that $\theta \leq 1$).

The generalization of the concept of random variable as the process of randomly generating elements of the space $\mathcal{K}_c(\mathbb{R})$ is the *random interval*, usually defined as a Borel measurable mapping $X : \Omega \to \mathcal{K}_c(\mathbb{R})$, where $(\Omega, \mathcal{A}, P)$ is a probability space, w.r.t. $\mathcal{A}$ and the Borel σ-field generated by the topology induced by the Hausdorff metric or any of the previous metrics, since all of them are topologically equivalent. It can also be defined in terms of real-valued random variables: X is a random interval iff both functions $\inf X : \Omega \to \mathbb{R}$ and

$\sup X : \Omega \to \mathbb{R}$ (or equivalently, $\operatorname{mid} X : \Omega \to \mathbb{R}$ and $\operatorname{spr} X : \Omega \to [0, \infty)$) are real-valued random variables.

The most common central tendency measure to summarize the information given by a random interval is the *Aumann expectation*. This mean value is indeed the Fréchet expectation with respect to the d_θ metric and admits an alternative expression as the interval whose infimum and supremum equal the expected values of inf X and sup X, respectively (and, hence, the mid-point and spread equal the expected values of midX and sprX, respectively). Although this measure inherits many very good probabilistic and statistical properties from the expectation of a real-valued random variable, it also preserves its high sensitivity to data changes or extreme data.

3 The Median of a Random Interval

As it has already been explained in Sect. 1, the idea of extending the concept of median to overcome the fact that the Aumann expectation of a random interval is not robust enough can be put into practice by defining it as the value with the smallest mean distance (w.r.t. a metric extending the Euclidean one) to the values of the random interval. Till now, the distances used were of the L^1 kind (see Sinova *et al.* [5] and Sinova and Van Aelst [6]):

Definition 1. The $d_{H,\theta}$-median (or medians) of a random interval $X : \Omega \to \mathcal{K}_c(\mathbb{R})$ is (are) defined as the interval(s) $\operatorname{Me}[X] \in \mathcal{K}_c(\mathbb{R})$ such that:

$$E(d_{H,\theta}(X, \operatorname{Me}[X])) = \min_{K \in \mathcal{K}_c(\mathbb{R})} E(d_{H,\theta}(X, K)), \tag{1}$$

if these expected values exist.

Definition 2. The ρ_1-median (or medians) of a random interval $X : \Omega \to \mathcal{K}_c(\mathbb{R})$ is (are) defined as the interval(s) $\operatorname{Med}[X] \in \mathcal{K}_c(\mathbb{R})$ such that:

$$E(\rho_1(X, \operatorname{Med}[X])) = \min_{K \in \mathcal{K}_c(\mathbb{R})} E(\rho_1(X, K)), \tag{2}$$

if these expected values exist.

The main advantage of these medians is that there exists a very practical result that guarantees their existence and simplifies their computation:

Proposition 1. *Given a probability space $(\Omega, \mathcal{A}, P)$ and an associated random interval X, the minimization problems (1) and (2) both have at least one solution, given by a nonempty compact interval such that*

(1) $\operatorname{mid} \operatorname{Me}[X] = \operatorname{Me}(\operatorname{mid} X)$, $\quad \operatorname{spr} \operatorname{Me}[X] = \operatorname{Me}(\operatorname{spr} X)$
(2) $\inf \operatorname{Med}[X] = \operatorname{Me}(\inf X)$, $\quad \sup \operatorname{Med}[X] = \operatorname{Me}(\sup X)$.

It should be pointed out that the solution for (1) does not depend on the value chosen for θ, although the mean error does. One remark is that if either Me(mid X) or Me(spr X) (which are medians of real-valued random variables) are not unique, then the $d_{H,\theta}$-median will not be unique, but any of the possible choices gives a solution to the problem. However, to guarantee that Med(X) is nonempty, it is necessary to establish a criterion in case of nonuniqueness of the medians of inf X or sup X, like the criterion consisting of choosing the mid-point of the interval of possible medians.

Both medians preserve most of the elementary operational properties of the median in real settings (see Sinova *et al.* [6]). However, inspired by the spatial median (or mediancentre) as extension of the median to higher dimensional Euclidean spaces and even Banach spaces, we now introduce a new definition of median of a random interval based on an L^2 type distance between intervals. In this paper we use the class of d_θ distances because, as mentioned in Sect. 1, it provides a wide class of metrics for interval data:

Definition 3. The d_θ-median (or medians) of a random interval $X : \Omega \to \mathcal{K}_c(\mathbb{R})$ is (are) defined as the interval(s) $\mathrm{M}[X] \in \mathcal{K}_c(\mathbb{R})$ such that:

$$E(d_\theta(X, \mathrm{M}[X])) = \min_{K \in \mathcal{K}_c(\mathbb{R})} E(d_\theta(X, K)), \tag{3}$$

if these expected values exist.

Since this is an empirical study on the behavior of the d_θ-median, its existence and uniqueness will not be proven here, but in a paper in preparation [7]. In fact, the d_θ-median will be computed for finite samples and for the time being we only use its sample version:

Definition 4. The sample d_θ-median (or medians) of a simple random sample $(X_1, \ldots, X_n)$ from a random interval $X : \Omega \to \mathcal{K}_c(\mathbb{R})$ is (are) defined as the interval(s) $\widehat{\mathrm{M}[X]} \in \mathcal{K}_c(\mathbb{R})$ which is (are) the solution(s) of the following optimization problem:

$$\min_{K \in \mathcal{K}_c(\mathbb{R})} \frac{1}{n} \sum_{i=1}^{n} d_\theta(X_i, K) \tag{4}$$

$$= \min_{(y,z) \in \mathbb{R} \times \mathbb{R}^+} \frac{1}{n} \sum_{i=1}^{n} \sqrt{(\mathrm{mid}\, X_i - y)^2 + \theta \cdot (\mathrm{spr}\, X_i - z)^2}. \tag{5}$$

We now explain the most natural algorithm to compute the sample d_θ-median. Surely, more efficient algorithms to compute the sample d_θ-median can be developed, but this is a topic for further research.

It is easy to notice that the objective function in the minimization problem (5) is differentiable at any point of the domain $\mathbb{R} \times \mathbb{R}^+$ except at the sample points $\{(\mathrm{mid}\, X_i, \mathrm{spr}\, X_i)\}_{i=1}^n$. Hence, the minimum will be reached either by a sample point or by the point in which both partial derivatives are equal to zero. That is, at the point (y_0, z_0) which satisfies:

$$y_0 = \frac{\sum_{i=1}^n \frac{\operatorname{mid} X_i}{\sqrt{(\operatorname{mid} X_i - y_0)^2 + \theta\cdot(\operatorname{spr} X_i - z_0)^2}}}{\sum_{i=1}^n \frac{1}{\sqrt{(\operatorname{mid} X_i - y_0)^2 + \theta\cdot(\operatorname{spr} X_i - z_0)^2}}}, \quad z_0 = \frac{\sum_{i=1}^n \frac{\operatorname{spr} X_i}{\sqrt{(\operatorname{mid} X_i - y_0)^2 + \theta\cdot(\operatorname{spr} X_i - z_0)^2}}}{\sum_{i=1}^n \frac{1}{\sqrt{(\operatorname{mid} X_i - y_0)^2 + \theta\cdot(\operatorname{spr} X_i - z_0)^2}}}$$

In this case the mid-point and the spread of the sample d_θ-median thus are a weighted mean of the mid-points and the spreads of the intervals in the sample, respectively. Then, the algorithm used follows these steps:

Algorithm to Compute the Sample d_θ-Median:

Step 0. If the data intervals are specified in terms of their inf/sup characterization, then first compute their mid-point and spread:

$$\operatorname{mid} X_i = \frac{\inf X_i + \sup X_i}{2}, \quad \operatorname{spr} X_i = \frac{\sup X_i - \inf X_i}{2}, \quad \text{for } i = 1, \ldots, n.$$

Step 1. Fix the maximum number of iterations, the tolerance of the approximation and set $m = 1$. Moreover, fix a seed $(y_m, z_m) \in \mathbb{R} \times \mathbb{R}^+$ and the weight $\theta > 0$, and calculate the corresponding error

$$\text{Error}_m = \frac{1}{n} \sum_{i=1}^n \sqrt{(\operatorname{mid} X_i - y_m)^2 + \theta \cdot (\operatorname{spr} X_i - z_m)^2}. \tag{6}$$

Step 2. Compute the weights and update the estimate:

$$v_i = \frac{\frac{1}{\sqrt{(\operatorname{mid} X_i - y_m)^2 + \theta\cdot(\operatorname{spr} X_i - z_m)^2}}}{\sum_{j=1}^n \frac{1}{\sqrt{(\operatorname{mid} X_j - y_m)^2 + \theta\cdot(\operatorname{spr} X_j - z_m)^2}}} \quad \text{for all } i = 1, \ldots, n$$

$$y_{m+1} = \sum_{i=1}^n v_i \cdot \operatorname{mid} X_i, \quad z_{m+1} = \sum_{i=1}^n v_i \cdot \operatorname{spr} X_i.$$

Step 3. For the new estimate (y_{m+1}, z_{m+1}), compute the corresponding error Error_{m+1} as given by (6). If the difference $\text{Error}_m - \text{Error}_{m+1}$ exceeds the specified tolerance and the number of iterations is lower than the maximum, then increase m by 1 and return to Step 2. Otherwise, go to Step 4.

Step 4. Compare the final error Error_{m+1} obtained in Step 3 with the errors $\text{Error}(X_j)$ corresponding to each sample interval X_j, where

$$\text{Error}(X_j) = \frac{1}{n} \sum_{i=1}^n \sqrt{(\operatorname{mid} X_i - \operatorname{mid} X_j)^2 + \theta \cdot (\operatorname{spr} X_i - \operatorname{spr} X_j)^2}$$

If $\text{Error}_{m+1} < \min_j \text{Error}(X_j)$ then return the solution (y_{m+1}, z_{m+1}). Otherwise, return the solution $(\operatorname{mid} X_{j_0}, \operatorname{spr} X_{j_0})$ where X_{j_0} is a solution of $\min_j \text{Error}(X_j)$.

4 Preliminary Empirical Study on the d_θ-Median

In our empirical studies we calculate the sample medians for a randomly generated sample of $n = 10000$ observations from a random interval characterized by the distribution of two real-valued random variables, $\text{mid}\, X$ and $\text{spr}\, X$. Both cases where the two random variables are independent (Case 1) and dependent (Case 2) have been considered. The sample has been split into two subsamples, one of size $n \cdot c_p$ associated with a contaminated distribution (hence c_p represents the proportion of contamination) and the other one, of size $n \cdot (1 - c_p)$, without any perturbation. A second parameter, C_D, has also been included to measure the relative distance between the distribution of the two subsamples. In detail, for different values of c_p and C_D the data for Case 1 are generated according to

- $\text{mid}\, X \rightsquigarrow \mathcal{N}(0,1)$ and $\text{spr}\, X \rightsquigarrow \chi_1^2$ for the non contaminated subsample,
- $\text{mid}\, X \rightsquigarrow \mathcal{N}(0,3) + C_D$ and $\text{spr}\, X \rightsquigarrow \chi_4^2 + C_D$ for the contaminated subsample,

while for Case 2 we use

- $\text{mid}\, X \rightsquigarrow \mathcal{N}(0,1)$ and $\text{spr}\, X \rightsquigarrow \left(\frac{1}{(\text{mid}\, X)^2+1}\right)^2 + .1 \cdot \chi_1^2$ for the non contaminated subsample,
- $\text{spr}\, X \rightsquigarrow \mathcal{N}(0,3) + C_D$ and $\text{spr}\, X \rightsquigarrow \left(\frac{1}{(\text{mid}\, X)^2+1}\right)^2 + .1 \cdot \chi_1^2 + C_D$ for the contaminated subsample.

Table 1 Monte Carlo approximation (1000 iterations) of the three medians in Case 1

c_p	c_D	d_θ-median	$d_{H,\theta}$-median	ρ_1-median
.0	0	$[-0, 6555071, 0, 6555835]$	$[-0, 4552839, 0, 4554768]$	$[-0, 7381633, 0, 7386939]$
.0	1	$[-0, 6550019, 0, 6544749]$	$[-0, 4545302, 0, 4543579]$	$[-0, 7378337, 0, 7373728]$
.0	5	$[-0, 6557143, 0, 6551791]$	$[-0, 4555520, 0, 4550171]$	$[-0, 7381773, 0, 7380322]$
.0	10	$[-0, 6553301, 0, 6554320]$	$[-0, 4547221, 0, 4549639]$	$[-0, 7382577, 0, 7386385]$
.1	0	$[-0, 6548555, 0, 6552054]$	$[-0, 4551417, 0, 4551000]$	$[-0, 7377625, 0, 7381397]$
.1	1	$[-0, 6554750, 0, 6546888]$	$[-0, 4556005, 0, 4547448]$	$[-0, 7382113, 0, 7376722]$
.1	5	$[-0, 6556373, 0, 6553330]$	$[-0, 4549310, 0, 4549547]$	$[-0, 7386306, 0, 7381864]$
.1	10	$[-0, 6555189, 0, 6548982]$	$[-0, 4553185, 0, 4546387]$	$[-0, 7379289, 0, 7379623]$
.2	0	$[-0, 6556167, 0, 6554632]$	$[-0, 4556358, 0, 4553282]$	$[-0, 7383702, 0, 7384272]$
.2	1	$[-0, 6553042, 0, 6554212]$	$[-0, 4548549, 0, 4552123]$	$[-0, 7387323, 0, 7382399]$
.2	5	$[-0, 6546356, 0, 6553408]$	$[-0, 4545946, 0, 4554151]$	$[-0, 7373326, 0, 7379313]$
.2	10	$[-0, 6553221, 0, 6559287]$	$[-0, 4548740, 0, 4553467]$	$[-0, 7380468, 0, 7393618]$
.4	0	$[-0, 6552048, 0, 6551526]$	$[-0, 4549726, 0, 4549666]$	$[-0, 7378274, 0, 7379544]$
.4	1	$[-0, 6552756, 0, 6559164]$	$[-0, 4550163, 0, 4553629]$	$[-0, 7383721, 0, 7387192]$
.4	5	$[-0, 6553724, 0, 6555041]$	$[-0, 4547553, 0, 4554061]$	$[-0, 7384876, 0, 7377935]$
.4	10	$[-0, 6554173, 0, 6556545]$	$[-0, 4550900, 0, 4554940]$	$[-0, 7380930, 0, 7384071]$
.4	100	$[-0, 6544593, 0, 6553595]$	$[-0, 4545915, 0, 4549898]$	$[-0, 7372087, 0, 7384379]$

The population medians have been approximated by a Monte Carlo approach using their sample versions. The results for Case 1 are shown in Table 1.

Then, the behavior of these medians is very similar under contamination effects, also in Case 2 (those results are not explicitly shown because of space limitations). In Sinova and Van Aelst [6], the robustness of the $d_{H,\theta}$ and the ρ_1-median had been proved with the finite sample breakdown point, so empirically it seems that the d_θ-median will be as robust as these L^1-medians.

5 Concluding Remarks about Open Problems

As this study is a preliminary contribution, there are a lot of open problems: the theoretical study of the d_θ-median and its properties (which includes the study of its robustness using tools like the finite sample breakdown point), more simulation studies comparing it with other central tendency measures like trimmed means or its extension to the fuzzy-valued case.

Acknowledgements. The research by B. Sinova and G. González-Rodríguez was partially supported by/benefited from the Spanish Ministry of Science and Innovation Grant MTM2009-09440-C02-01 and the COST Action IC0702. B. Sinova has also been granted with the Ayuda del Programa de FPU AP2009-1197 from the Spanish Ministry of Education, an Ayuda de Investigación 2011 from the Fundación Banco Herrero and three Short Term Scientific Missions associated with the COST Action IC0702. The research by S. Van Aelst was supported by a grant of the Fund for Scientific Research-Flanders (FWO-Vlaanderen). Their financial support is gratefully acknowledged.

References

1. Bertoluzza, C., Corral, N., Salas, A.: On a new class of distances between fuzzy numbers. Math. Soft Comput. 2, 71–84 (1995)
2. Milasevic, P., Ducharme, G.R.: Uniqueness of the spatial median. Ann. Stat. 15, 1332–1333 (1987)
3. Gil, M.A., Lubiano, M.A., Montenegro, M., López-García, M.T.: Least squares fitting of an affine function and strength of association for interval data. Metrika 56, 97–111 (2002)
4. Gower, J.C.: Algorithm AS 78: The mediancentre. Appl. Stat. 23, 466–470 (1974)
5. Sinova, B., Casals, M.R., Colubi, A., Gil, M.Á.: The Median of a Random Interval. In: Borgelt, C., González-Rodríguez, G., Trutschnig, W., Lubiano, M.A., Gil, M.Á., Grzegorzewski, P., Hryniewicz, O. (eds.) Combining Soft Computing and Statistical Methods in Data Analysis. AISC, vol. 77, pp. 575–583. Springer, Heidelberg (2010)

6. Sinova, B., Van Aelst, S.: Comparing the Medians of a Random Interval Defined by Means of Two Different L^1 Metrics. In: Borgelt, C., Gil, M.Á., Sousa, J.M.C., Verleysen, M. (eds.) Towards Advanced Data Analysis. STUDFUZZ, vol. 285, pp. 75–86. Springer, Heidelberg (2012)
7. Sinova, B., Gil, M.A., González-Rodríguez, G., Van Aelst, S.: An L^2 median for random intervals (forthcoming)
8. Trutschnig, W., González-Rodríguez, G., Colubi, A., Gil, M.A.: A new family of metrics for compact, convex (fuzzy) sets based on a generalized concept of mid and spread. Inf. Sci. 179, 3964–3972 (2009)
9. Vitale, R.A.: L_p metrics for compact, convex sets. J. Approx. Theory 45, 280–287 (1985)

Comparing Classical and Robust Sparse PCA

Valentin Todorov and Peter Filzmoser

Abstract. The main drawback of principal component analysis (PCA) especially for applications in high dimensions is that the extracted components are linear combinations of all input variables. To facilitate the interpretability of PCA various sparse methods have been proposed recently. However all these methods might suffer from the influence of outliers present in the data. An algorithm to compute sparse and robust PCA was recently proposed by Croux *et al.* We compare this method to standard (non-sparse) classical and robust PCA and several other sparse methods. The considered methods are illustrated on a real data example and compared in a simulation experiment. It is shown that the robust sparse method preserves the sparsity and at the same time provides protection against contamination.

Keywords: Principcal component analysis, robust statistics.

1 Introduction

Principal component analysis (PCA) is a widely used technique for dimension reduction achieved by finding a smaller number q of linear combinations of the originally observed p variables and retaining most of the variability of the data. It is important to be able to interpret these new variables, referred to

Valentin Todorov
United Nations Industrial Development Organization (UNIDO), Vienna, Austria
e-mail: v.todorov@unido.org

Peter Filzmoser
Department of Statistics and Probability Theory,
Vienna University of Technology, Vienna, Austria
e-mail: p.filzmoser@tuwien.ac.at

R. Kruse et al. (Eds.): Synergies of Soft Computing and Statistics, AISC 190, pp. 283–291.
springerlink.com © Springer-Verlag Berlin Heidelberg 2013

as *principal components*, especially when the original variables have physical meaning. The link between the original variables and the principal components is given by the so called *loadings matrix* used for transforming the data and thus it should serve as a means for interpreting the PCs. However, PCA usually tends to provide PCs which are linear combinations of all the original variables (by giving them non-zero loadings). Regarding the interpretability of the results it would be very helpful to reduce not only the dimensionality but also the number of used variables (ideally to relate each PC to only a few variables). It is not surprising that vast research effort was devoted to this issue and various proposals have been introduced in the literature. A straightforward informal method is to set to zeros those PC loadings which have absolute values below a given threshold (*simple thresholding*). In [6] SCoTLASS was proposed which applies a *lasso* penalty on the loadings in a PCA optimization problem. Recently a reformulated PCA as a regression problem has been proposed [13] that uses the *elastic net* to obtain a sparse version (SPCA).

Despite more or less successful in achieving sparsity, all these methods suffer a common drawback - all are based on the classical approach to PCA which measures the variability through the empirical variance and is essentially based on computation of eigenvalues and eigenvectors of the sample covariance or correlation matrix. Therefore the results may be very sensitive to the presence of even a few atypical observations in the data. The outliers could artificially increase the variance in an otherwise uninformative direction and this direction will be determined as a PC direction. To cope with the possible presence of outliers in the data, recently a method has been proposed [1] which is sparse and robust at the same time. It utilizes the *projection pursuit* approach where the PCs are extracted from the data by searching the directions that maximize a robust measure of variance of data projected on it. An efficient computational algorithm was proposed in [2]. Another robust sparse PCA algorithm was proposed by [9] maximizing the L1-norm variance instead of the classical variance but unfortunately no R implementation was available and in the short time we could not include it in the comparison.

The paper [13] defined the (minimal) requirements for a good sparse method as follows: (i) without any penalty constraint the method is equivalent to standard PCA; (ii) the method is computationally efficient for both large n and large p and (iii) it avoids misidentifying important variables. To these requirements we will add one more: (iv) the method should attain the properties (i) to (iii) even in the presence of outliers in the data.

The remainder of the paper is organized as follows. Section 2 presents briefly the sparse and robust methods considered. Section 3 illustrates these methods on real data examples and Section 4 compares them on simulated data sets. The final Section 5 concludes.

2 Methods and Algorithms

Consider an $n \times p$ data matrix $\boldsymbol{X}$. Without loss of generality we can assume that the column means of $\boldsymbol{X}$ are all zeros. Note that in the context of robust PCA, the centering has to be done also in a robust way - [2]. We are looking for linear combinations $\boldsymbol{t}_j$ that result from a projection of the centered data on a direction $\boldsymbol{p}_j$,

$$\boldsymbol{t}_j = \boldsymbol{X}\boldsymbol{p}_j \tag{1}$$

such that

$$\boldsymbol{p}_j = \underset{\boldsymbol{p}}{\operatorname{argmax}} \operatorname{Var}(\boldsymbol{X}\boldsymbol{p}) \tag{2}$$

subject to $\|\boldsymbol{p}_j\| = 1$ and $\operatorname{Cov}(\boldsymbol{X}\boldsymbol{p}_j, \boldsymbol{X}\boldsymbol{p}_l) = 0$ for $l < j$ and $j = 1, \ldots, q$ with $q \leq \min(n, p)$. The solutions of these maximization problems are obtained by solving a Lagrangian problem, and the result is that the principal components of $\boldsymbol{X}$ are the eigenvectors of the covariance matrix $\operatorname{Cov}(\boldsymbol{X})$, and the variances are the corresponding eigenvalues $l_j = \operatorname{Var}(\boldsymbol{X}\boldsymbol{p}_j)$. Classical PCA is obtained if the sample covariance matrix $\boldsymbol{S}$ is used. The vectors $\boldsymbol{t}_j$ are collected as columns in the $n \times q$ *scores* matrix $\boldsymbol{T}$, and the vectors $\boldsymbol{p}_j$ as columns in the *loadings* matrix $\boldsymbol{P}$. The eigenvalues l_j are arranged in the diagonal of the $q \times q$ diagonal matrix $\boldsymbol{\Lambda}$.

The most straightforward way to robustify PCA is to replace $\boldsymbol{S}$ by a robust version like for example MCD (see [12]). Another approach to robust PCA uses *projection pursuit* (PP) and calculates directly the robust estimates of the eigenvalues and eigenvectors. Directions are sought for, which maximize the variance of the data projected onto them. The advantage of this approach is that the principal components can be computed sequentially, and that one can stop after q components have been extracted. Thus, this approach is appealing for high-dimensional data, in particular for problems with $p \gg n$. Using the empirical variance in the maximization problem would lead to classical PCA, and robust scale estimators result in robust PCA. Suitable robust measures are the squared median absolute deviation (MAD) or the more efficient Q_n. A tractable algorithm in these lines was proposed in [3]. When solving the maximization problem the algorithm does not investigate all possible directions but considers only those defined by a data point and the robust center of the data. The robust variance estimate is computed for the data points projected on these n directions and the direction corresponding to the maximum of the variance is the searched approximation of the first principal component. After that the search continues in the same way in the space orthogonal to the first component. An improved version of this algorithm, being more precise especially for high-dimensional data, was proposed in [2]. The space of all possible directions is scanned more thoroughly. This is done by restricting the search for an optimal direction on a regular grid in a plane.

To introduce sparseness in PCA the authors of [1] add an L_1 constraint in the definition (2) which yields

$$\boldsymbol{p}_j = \underset{\boldsymbol{p}}{\operatorname{argmax}} \operatorname{Var}(\boldsymbol{X}\boldsymbol{p}) - \lambda_j \|\boldsymbol{p}\|_1 \tag{3}$$

where λ_j is a tuning parameter. To solve this optimization problem, again the grid algorithm [2] can be used as described in detail in [1] and implemented in the R package **pcaPP**. All versions of these algorithms (standard and sparse classical, standard and sparse robust) are provided with a unified interface in the R package **rrcovHD**. In the following we will denote these by the obvious abbreviations PCA, SPCA-grid, RPCA-grid and RSPCA-grid.

One of the most popular sparse PCA algorithms is the SPCA proposed in [13]. It relies on the fact that PCA can be rewritten as a regression-type optimization problem which is solved by the sparse *elastic net* regression. The main drawback of this algorithm is that orthogonality of the components is not guaranteed. The algorithm fits the situation when $p \gg n$ but might be computationally very expensive when requiring a large number of nonzero loadings. Therefore the authors propose a variant of the algorithm based on *soft thresholding* suitable for such cases. Both versions of the algorithm are available in the R package **elasticnet**.

3 Example

We will use a real data example to compare the standard and robust sparse methods. The `bus` data set [4] which is also available in the R package **rrcov** was used to study methods for automatic vehicle recognition [11], see also [8], page 213, Example 6.3. This data set from the Turing Institute, Glasgow, Scotland, contains measures of shape features extracted from vehicle silhouettes. The images were acquired by a camera looking downward at the model vehicle from a fixed angle of elevation. Each of the 218 rows corresponds to a view of a bus silhouette, and contains 18 attributes of the image. The median absolute deviations (MAD) of the columns vary from 0 (for variable `V9`) to 34.8. Therefore we remove `V9` from the analysis and divide each variable by the corresponding MAD. The first four classical PCs explain more than 97% of the total variance and the first four robust PCs explain more than 85%, therefore we decide to retain four components in both cases. Next we need to choose the degree of sparseness which is controlled by the regularization parameter λ. Since the sparse PCs have to provide a good trade-off between sparseness and achieved percentage of explained variance we can proceed similarly as in the selection of the number of principal components with the scree plot - we compute the sparse PCA for many different values of λ and plot the percent of explained variance against λ. We choose $\lambda = 1.64$ for classical PCA and $\lambda = 2.07$ for robust PCA, thus attaining 92 and 84 percent of explained variance, respectively, which is only an acceptable reduction compared to the non-sparse PCA. Retaining $k = 4$ principal components as above and using the selected parameters λ, we can construct the so called *diagnostic plots* which are especially useful for identifying outlying observations. The diagnostic plot is based on the *score distances* and *orthogonal distances* computed for each observation. The *diagnostic plot* shows the score versus the orthogonal distance, and indicates with a horizontal and vertical line the cut-off values that allow to distinguish regular observations (those with small score and small orthogonal distance) from the different types of outliers: *bad leverage points* with large score and large orthogonal distance, *good leverage points* with large score and small orthogonal distance and *orthogonal outliers* with small score and large orthogonal distance (for detailed description see [5]). In Figure 1 the classical and robust diagnostic plot as well as their sparse alternatives are presented. The diagnostic plots for the standard PCA reveals only several orthogonal outliers and identifies two observations as good leverage points. These two observations are identified as bad leverage points by the sparse standard PCA which is already an improvement, but only the robust methods identify a large cluster of outliers. These outliers are masked by the non-robust score and orthogonal distances and cannot be identified by the classical methods. It is important to note that the sparsity feature added to the robust PCA did not influence its ability to detect properly the outliers.

4 Simulation

In order to compare different methods for extraction of sparse features from data sets with varying degree of contamination we should be able to generate data sets with known sparseness. Most of the simulation studies in the literature follow the simulation example in [13] which use a fixed configuration with three underlying factors and three blocks of variables, each of them revealing one of the factors. In [1] a contamination model has been proposed to be superimposed on the so generated data. The most straightforward method to generate data with sparse structure in R^p is to choose the leading q, $(q < p)$, $\boldsymbol{p}_1, \ldots, \boldsymbol{p}_q$ eigenvectors of the covariance matrix $\boldsymbol{\Sigma}$, which are sparse and orthonormal [10]. The covariance matrix $\boldsymbol{\Sigma}$ is decomposed as $\boldsymbol{\Sigma} = \boldsymbol{P}\boldsymbol{D}\boldsymbol{P}^T$ where $\boldsymbol{D} = diag(d_1, \ldots, d_p)$ is a diagonal matrix containing the positive eigenvalues of $\boldsymbol{\Sigma}$ on the main diagonal sorted in decreasing order. The matrix $\boldsymbol{P}$ is the orthogonal loadings matrix. The first $q, q \leq p$ eigenvectors are chosen to be sparse and the remaining $p-q$ are arbitrary. We start by forming the full rank matrix $\boldsymbol{P}$ by randomly drawing its elements from say, $U(0, 1)$ and replacing the first q columns with the pre-specified sparse vectors $\boldsymbol{p}_1, \ldots, \boldsymbol{p}_q$. Then the matrix $\boldsymbol{P}$ is rendered orthonormal by applying Gram-Schmidt orthogonalization to it (the matrix $\boldsymbol{Q}$ of the QR-decomposition of $\boldsymbol{P}$).

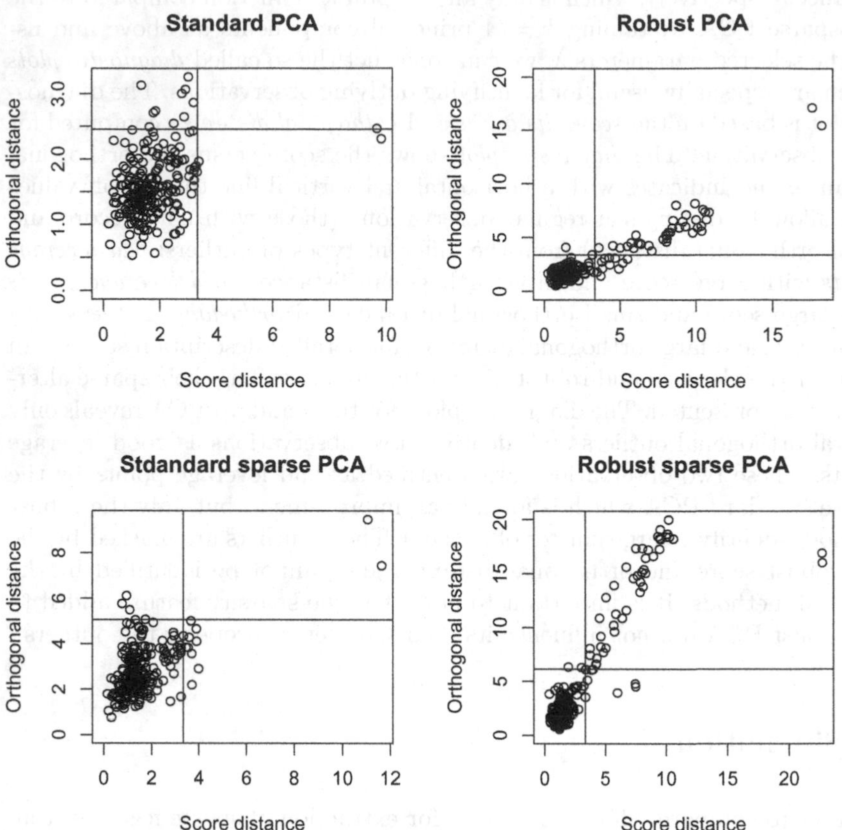

Fig. 1 Distance-distance plots for standard and sparse PCA and their robust versions for the bus data

The generated data sets $\boldsymbol{X}$ will consist of n observations drawn from a p-variate zero-mean normal distribution $\boldsymbol{X} \sim N(\boldsymbol{0}, \boldsymbol{\Sigma})$. Different proportions of contamination will be added to this data sets by replacing ε percent of the observations in $\boldsymbol{X}$, $\varepsilon = 0, 10, 20, 30$, with normally distributed p-variates $\tilde{\boldsymbol{x}} \sim N(\boldsymbol{\mu}, \sigma \boldsymbol{I}_p)$ with $\boldsymbol{\mu} = (2, 4, 2, 4, 0, -1, 1, 0, 1, \ldots, 0, 1, -1)^T$ and $\sigma = 20$.

To compare the performance of the different methods for each generated data set we estimate the first q principal components by each method and compute the angles between the estimated leading eigenvectors and the corresponding true vectors $\boldsymbol{p}_1, \ldots, \boldsymbol{p}_q$ as well as the maximal angle between the subspace spanned by the first q estimated principal components and the subspace spanned by the first q eigenvectors of $\boldsymbol{\Sigma}$. This angle between subspaces, which we will call *maxsub*, can be computed by a method proposed by Krzanowski [7]. We want that these angles are as close to zero as possible.

For our example we take $n = 150, p = 10, q = 2$, i.e. the data sets will be generated in R^{10} and two sparse leading eigenvectors will be chosen with degree of sparseness 6, as:

$$\tilde{\boldsymbol{p}}_1 = (1,1,1,1,0,0,0,0,0,0) \text{ and } \tilde{\boldsymbol{p}}_2 = (0,0,0,0,1,1,1,1,0,0).$$

The eigenvalues are $\boldsymbol{d} = (400, 200, 1, ..., 1)$ and thus the first two eigenvectors explain almost 90 percent of the total variance. With all methods we will extract the true number of principal components, $q = 2$. Since SPCA has no automatic method for selecting the degree of sparseness we will use for all sparse methods the true one and will refer to this as the oracle method. We average the computed measures by performing the complete procedure $m = 100$ times and taking the median of the corresponding angles.

Table 1 Comparison of PCA, sparse PCA and robust (sparse) PCA methods: median angle between the true and extracted subspaces (*maxsub*) and median angles between the true and extracted first two loading vectors for clean data and two levels of contamination (ε = 0, 10 and 20 percent).

	$\varepsilon = 0\%$			$\varepsilon = 10\%$			$\varepsilon = 20\%$			$\varepsilon = 30\%$
	maxsub	PC1	PC2	*maxsub*	PC1	PC2	*maxsub*	PC1	PC2	*maxsub*
PCA	0.07	0.03	0.07	0.93	0.93	1.00	0.93	0.93	1.01	0.93
RPCA-grid	0.09	0.06	0.09	0.18	0.13	0.20	0.19	0.15	0.18	0.22
SPCA	0.06	0.04	0.07	0.93	0.93	1.00	0.93	0.93	1.01	0.93
SPCA-grid	0.02	0.01	0.02	0.94	0.93	1.00	0.94	0.93	1.00	0.94
SRPCA-grid	0.06	0.04	0.05	0.16	0.12	0.12	0.20	0.14	0.22	0.21

The results are presented in Table 1. When there is no contamination all methods perform reasonably well. Since the generated data have sparse structure, all sparse methods result in lower median angles than the standard PCA and the best performer is SPCA-grid (sparse PCA computed by the grid method). The robust non-sparse method RPCA-grid performs only slightly worse than the classical PCA but the robust sparse (SRPCA-grid), while worse that SPCA-grid is still better than the classical PCA.

The picture changes drastically when we add even only 10% of contamination. All non-robust methods produce median angles close to one (the vectors, respectively the subspaces are almost perpendicular) while with the robust methods only slight change is observed. It is important to note that the advantage of using a sparse method disappears in the presence of contamination while the robust sparse method is still better than the robust non-sparse one. If we increase the contamination to 20% and even 30% almost nothing changes - *maxsub* remains below 0.22 for the robust methods and is above 0.93 for the non-robust ones.

An important characteristic of any algorithm is the computational efficiency. Due to the space restrictions we are not going to present here a comparison of the computation times of the algorithms but some observations are in order. The speed of all methods computed by the grid algorithm does not depend on the degree of sparseness. However, it depends on the selected variance estimator (the sample variance is much faster (but not robust) than the median absolute deviation (MAD), which in turn is faster (but not statistically efficient) than Qn estimator). It depends also on the number of extracted principal components, and the possibility to extract only the necessary components is a great advantage in high dimensional settings. The computation time of SPCA depends on the selected degree of sparseness but not on the selected number of principal components. It could be prohibitive to use this algorithm on high dimensional data but here the soft-thresholding version comes handy. The sparse non-robust and robust versions of the grid algorithm are faster than SPCA, when providing comparable degree of sparseness and extracting a reasonable number of components.

5 Summary and Conclusions

In this article we investigated several methods for sparse PCA in terms of their efficiency and their resistance to the presence of outliers in the data. Standard PCA, several sparse methods, robust PCA, and robust sparse PCA recently proposed in [1] are compared on a simulation experiment. The robust sparse method attains the requested sparseness and at the same time provides adequate principal components even when the data are contaminated with as much as 30%, while all the non-robust methods break down. All considered methods and data sets are available in the R package **rrcovHD**. There are various other sparse PCA methods and algorithms as well as other issues which were not investigated in this work: selecting the number of PCs, selecting the tuning parameter (the degree of sparseness) as a trade-off between explained variance and interpretability, considering more types of contamination. These could be in the focus of a more extended comparative study.

Acknowledgements. The views expressed herein are those of the authors and do not necessarily reflect the views of the United Nations Industrial Development Organization.

References

1. Croux, C., Filzmoser, P., Fritz, H.: Robust sparse principal component analysis. Reserach report sm-2011-2, Vienna University of Technology (2011)
2. Croux, C., Filzmoser, P., Oliveira, M.: Algorithms for projection-pursuit robust principal component analysis. Chemometr. Intel. Lab. 87, 218–225 (2007)

3. Croux, C., Ruiz-Gazen, A.: High breakdown estimators for principal components: The projection-pursuit approach revisited. J. Multivariate Anal. 95, 206–226 (2005)
4. Hettich, S., Bay, S.D.: The UCI KDD archive (1999), `http://kdd.ics.uci.edu`
5. Hubert, M., Rousseeuw, P., Vanden Branden, K.: ROBPCA: A new approach to robust principal component analysis. Technometrics 47, 64–79 (2005)
6. Jolliffe, I.T., Trendafilov, N.T., Uddin, M.: A modified principal component technique based on the LASSO. J. Comput. Graph. Stat. 12, 531–547 (2003)
7. Krzanowski, W.J., Marriott, F.H.C.: Multivariate Analysis, Part 2: Classification, Covariance Structure and Repeated Measurements. Arnold, London (1995)
8. Maronna, R.A., Martin, D., Yohai, V.: Robust Statistics: Theory and Methods. Wiley, New York (2006)
9. Meng, D., Zhao, Q., Xu, Z.: Improve robustness of sparse PCA by L1-norm maximization. Pattern Recogn. 45(1), 487–497 (2012)
10. Shen, H., Huang, J.Z.: Sparse principal component analysis via regularized low rank matrix approximation. J. Multivariate Anal. 99, 1015–1034 (2008)
11. Siebert, J.P.: Vehicle recognition using rule based methods. Turing Institute Research Memorandum TIRM-87-018 (1987)
12. Todorov, V., Filzmoser, P.: An object oriented framework for robust multivariate analysis. J. Stat. Softw. 32, 1–47 (2009)
13. Zou, H., Hastie, T., Tibshirani, R.: Sparse principal component analysis. J. Comput. Graph. Stat. 15, 265–286 (2006)

An Exact Algorithm for Likelihood-Based Imprecise Regression in the Case of Simple Linear Regression with Interval Data

Andrea Wiencierz and Marco E.G.V. Cattaneo

Abstract. Likelihood-based Imprecise Regression (LIR) is a recently introduced approach to regression with imprecise data. Here we consider a robust regression method derived from the general LIR approach and we establish an exact algorithm to determine the set-valued result of the LIR analysis in the special case of simple linear regression with interval data.

Keywords: Interval data, likelihood inference, robust regression.

1 Introduction

In [3], Likelihood-based Imprecise Regression (LIR) was introduced as a very general theoretical framework for regression analysis with imprecise data. Within the context of LIR, the term imprecise data refers to imprecisely observed quantities. This means that one is actually interested in analyzing the relation between precise variables, but the available data provide only the partial information that the values each lie in some subset of the observation space. In the general formulation of LIR, the imprecise observations can be arbitrary subsets of the observation space, including as special cases actually precise data (where the subset is a singleton) and missing data (where the subset is the entire observation space).

The aim of a LIR analysis is to identify plausible descriptions of the relation between the unobserved precise quantities on the basis of the imprecise observations. This is achieved by applying a general methodology for likelihood inference with imprecise data to the regression problem with imprecise

---

Andrea Wiencierz · Marco E.G.V. Cattaneo
Department of Statistics, LMU Munich, 80539 München, Germany
e-mail: {andrea.wiencierz,cattaneo}@stat.uni-muenchen.de

R. Kruse et al. (Eds.): Synergies of Soft Computing and Statistics, AISC 190, pp. 293–301.
springerlink.com © Springer-Verlag Berlin Heidelberg 2013

data as a problem of statistical inference. The mathematical details of the LIR approach are set out in [3].

In this paper, we deal with the implementation of the robust regression method derived from the general LIR approach in [3]. There, a grid search was proposed as a first implementation, which served to obtain the (approximate) result of the LIR analysis for a quadratic regression problem with interval data. Here, we consider the special case of simple linear regression with interval data and we derive an exact algorithm to determine the set-valued result of the LIR analysis in this particular situation. In the following section, we review the relevant technical details of the robust LIR method, before we establish the exact algorithm in Sect. 3.

2 LIR in the Case of Simple Linear Regression with Interval Data

In the case of simple linear regression, the relation between two real-valued variables, X and Y, shall be described by means of a linear function. Thus, the set of regression functions considered here can be written as $\mathcal{F} = \{f_{a,b} : (a,b) \in \mathbb{R}^2\}$ with $f_{a,b} : \mathbb{R} \to \mathbb{R}$, $x \mapsto a + b\,x$. Furthermore, we here focus on the particular case of interval data, where the imprecise data $V_i^* := [\underline{X}_i, \overline{X}_i] \times [\underline{Y}_i, \overline{Y}_i]$, $i = 1, \ldots, n$ are (possibly unbounded) rectangles. To keep the notation simple, throughout the paper, we write $[\underline{I}, \overline{I}]$ for the set of all real numbers z such that $\underline{I} \leq z \leq \overline{I}$. This is not the standard notation if $\underline{I} = -\infty$ or $\overline{I} = +\infty$. Figure 1 gives an example of such a data set containing 17 observations with varying amounts of imprecision.

The robust regression method we consider in this paper is based on a fully nonparametric probability model. It is only assumed that the n

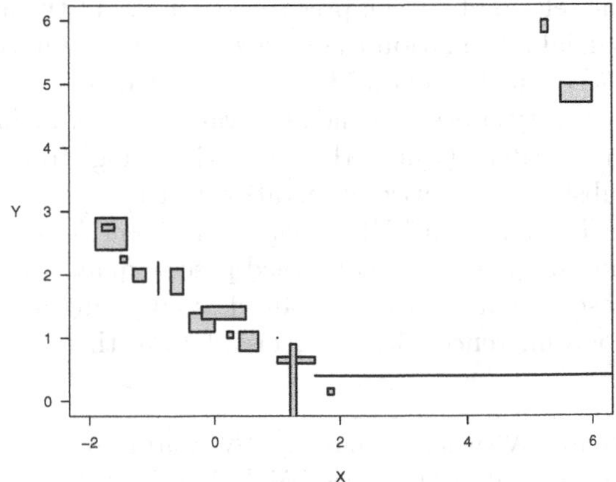

Fig. 1 Example data set containing 17 observations with varying amounts of imprecision: there is one actually precisely observed data point $V_i^* = [1,1] \times [1,1] = \{(1,1)\}$, there are two line segments (one of which is unbounded towards $+\infty$ in the X dimension), and, finally, there are 14 rectangles of different sizes and shapes (one of which is unbounded towards $-\infty$ in the Y dimension).

random objects (V_i, V_i^*), $i = 1, \ldots, n$ (where $V_i := (X_i, Y_i)$ are the unobserved precise values) are independent and identically distributed, and that $P(V_i \in V_i^*) \geq 1 - \varepsilon$, for some $\varepsilon \in [0, 1]$. If $\varepsilon > 0$, this assumption implies that an imprecise observation may not cover the precise value with probability at most ε. Apart from this assumption, there is no restriction on the set of possible distributions of the data.

The relation between X and Y shall be described by a linear function. Which linear function is a suitable description of the relation when no particular structure of the joint distribution of X and Y is assumed? The basic idea behind the robust LIR method is that a possible description f can be evaluated by the p-quantile, $p \in]0, 1[$, of the distribution of the corresponding (absolute) residual $|Y - f(X)|$. The closer to zero the p-quantile is, the better the associated function describes the relation between X and Y. Therefore, the linear function for which the p-quantile of the residual's distribution is minimal can be considered as the best description of the relation of interest. This linear function can be characterized geometrically as the central line of the thinnest band of the form $f \pm q$, $q \geq 0$, that contains (X, Y) with probability at least p.

This idea is very similar to the idea behind the robust regression method of least quantile of squares (or absolute deviations) regression, introduced in [5] as a generalization of the method of least median of squares regression (corresponding to the choice $p = 0.5$). Therefore, the LIR method can be seen as a generalization of these robust regression methods to the setting with imprecise data, where not only the optimal line is estimated, but a whole set of plausible descriptions is idenified.

To see how the robust LIR method works in detail, consider $V_1^* = A_1, \ldots, V_n^* = A_n$ as (nonempty) realizations of the imprecise data. Applying the general methodology for likelihood inference with imprecise data on which the LIR method is based, likelihood-based confidence regions for the p-quantile of the distribution of the precise residuals $R_{f,i} := |Y_i - f(X_i)|$, $i = 1, \ldots, n$, are determined for each considered regression function $f \in \mathcal{F}$. The confidence regions are obtained by cutting the (normalized) profile likelihood function for the p-quantile induced by the imprecise data at some cutoff point $\beta \in]0, 1[$. The confidence regions cover the values of the p-quantiles corresponding to all probability distributions that give at least a certain probability to the observations, i.e. whose likelihood exceeds the threshold β.

To obtain the confidence regions, for each $f \in \mathcal{F}$ lower and upper (absolute) residuals are defined as follows

$$\underline{r}_{f,i} = \min_{(x,y)\in A_i} |y - f(x)| \quad \text{and} \quad \overline{r}_{f,i} = \sup_{(x,y)\in A_i} |y - f(x)|, \qquad i = 1, \ldots, n.$$

Let $0 =: \underline{r}_{f,(0)} \leq \underline{r}_{f,(1)} \leq \cdots \leq \underline{r}_{f,(n)} \leq \underline{r}_{f,(n+1)} := +\infty$ be the ordered lower residuals and $0 =: \overline{r}_{f,(0)} \leq \overline{r}_{f,(1)} \leq \cdots \leq \overline{r}_{f,(n)} \leq \overline{r}_{f,(n+1)} := +\infty$ be the ordered upper residuals. Furthermore, define $\underline{i} = \max(\lceil (p - \varepsilon)\, n \rceil, 0)$ and $\overline{i} = \min(\lfloor (p + \varepsilon)\, n \rfloor, n) + 1$. According to Corollary 1 of [3] the profile likelihood function for the p-quantile of the distribution of the residuals corresponding to some function $f \in \mathcal{F}$ is a piecewise constant function whose points of discontinuity are given by $\underline{r}_{f,(0)}, \ldots, \underline{r}_{f,(\underline{i})}, \overline{r}_{f,(\overline{i})}, \ldots, \overline{r}_{f,(n+1)}$. To obtain the confidence region C_f it thus suffices to identify the $(\underline{k}+1)$-th ordered lower residual and the $\overline{k}$-th ordered upper residual, which correspond to the points where the profile likelihood function jumps above and below the chosen threshold β, provided the condition $(\max\{p, 1-p\} + \varepsilon)^n \leq \beta$ holds. The values of $\underline{k}$ and $\overline{k}$ are determined on the basis of the explicit formula for the profile likelihood function given in [3]. They depend on n, on the choice of p and β, as well as on ε, which is part of the assumed probability model.

Thus, if $(\max\{p, 1-p\} + \varepsilon)^n \leq \beta$ is fulfilled, for each function $f \in \mathcal{F}$ the likelihood-based confidence region is the interval $C_f := [\underline{r}_{f,(\underline{k}+1)}, \overline{r}_{f,(\overline{k})}]$ (see Corollary 2 of [3]). In order to find the best description of the relation between X and Y it is possible to follow a minimax approach and minimize the upper endpoint of the confidence interval over all considered regression functions. When there is a unique $f \in \mathcal{F}$ that minimizes $\sup C_f$, it is optimal according to the Likelihood-based Region Minimax (LRM) criterion (see [1]) and therefore called f_{LRM}. If we consider the closed bands $\overline{B}_{f,q}$ defined for each function $f \in \mathcal{F}$ and each $q \in [0, +\infty[$ by

$$\overline{B}_{f,q} = \left\{(x,y) \in \mathbb{R}^2 : |y - f(x)| \leq q\right\},$$

the function f_{LRM} can be characterized geometrically. The closed band $\overline{B}_{f_{LRM},\overline{q}_{LRM}}$ (where $\overline{q}_{LRM} := \sup C_{f_{LRM}}$) is the thinnest band of the form $\overline{B}_{f,q}$ containing at least $\overline{k}$ imprecise data, for all $f \in \mathcal{F}$ and all $q \in [0, +\infty[$. Thus, to determine the function f_{LRM} it suffices to adapt to the case of imprecise data an algorithm for the least quantile of squares regression, as we do in Sect. 3.1. Figure 2 shows f_{LRM} (solid line) for the LIR analysis of the

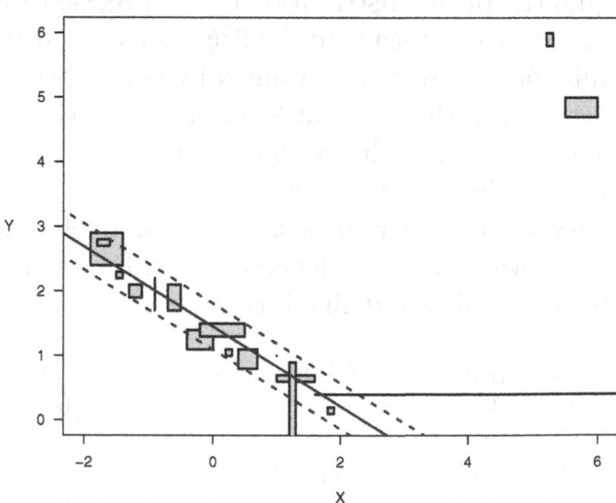

Fig. 2 Function f_{LRM} (solid line) for the LIR analysis of the example data set introduced in Fig. 1 with $p = 0.5, \beta = 0.8, \varepsilon = 0$ (implying $\underline{k} = 7$ and $\overline{k} = 10$) and band $\overline{B}_{f_{LRM},\overline{q}_{LRM}}$ (dashed lines).

example data set introduced in Fig. 1, as well as the closed band $\overline{B}_{f_{LRM},\overline{q}_{LRM}}$ of width $2\,\overline{q}_{LRM}$ (dashed lines).

However, f_{LRM} is not regarded as the final result of the LIR analysis. The aim of a LIR analysis is to describe the whole uncertainty about the relation between X and Y, including the statistical uncertainty due to the finite sample as well as the indetermination related to the fact that the quantities are only imprecisely observed. Therefore, the set of all functions that are plausible in the light of the data is considered as the set-valued result of the LIR analysis, which describes the entire uncertainty involved in the regression problem with imprecise data. A regression function $f \in \mathcal{F}$ is regarded as plausible, if the corresponding confidence interval $\mathcal{C}_f$ is not strictly dominated by another one. Thus, the result of the LIR analysis is the set

$$\{f \in \mathcal{F} : \min \mathcal{C}_f \leq \overline{q}_{LRM}\} = \{f \in \mathcal{F} : \underline{r}_{f,(\underline{k}+1)} \leq \overline{q}_{LRM}\}.$$

The undominated functions can be characterized geometrically by the fact that the corresponding closed bands $\overline{B}_{f,\overline{q}_{LRM}}$ (i.e. the bands have width $2\,\overline{q}_{LRM}$) intersect at least $\underline{k}+1$ imprecise data. This characterization is the basis of the second part of the algorithm presented in the next section.

3 An Exact Algorithm for LIR

As a first implementation of the robust LIR method, we suggested in [3] a grid search over the space of parameters identifying the considered regression functions, while we considered a random search in [2]. Here, we derive an exact algorithm to determine the result of the robust LIR analysis in the case of simple linear regression with interval data. The algorithm consists of two parts: first, we find the optimal function f_{LRM}, which is then used to identify the set of all undominated regression lines. It can be proved that the computational complexity of the algorithm is $O(n^3 \log n)$, i.e. it is of the same order as the complexity of the initial algorithm for least median of squares regression (see [6]).

3.1 Part 1: Finding the LRM Line

Analogously to what is shown in [6] for the case with precise data, it is possible to prove that, if the slope b_{LRM} of the function f_{LRM} is different from zero, the band $\overline{B}_{f_{LRM},\overline{q}_{LRM}}$ is determined by three imprecise observations V_i^* for which $\overline{r}_{f_{LRM},i} = \overline{q}_{LRM}$. Figure 3 illustrates this fact for the example of Fig. 2. From this property follows that b_{LRM} is either zero or given by the slope of the line connecting the corresponding corner points of two of the

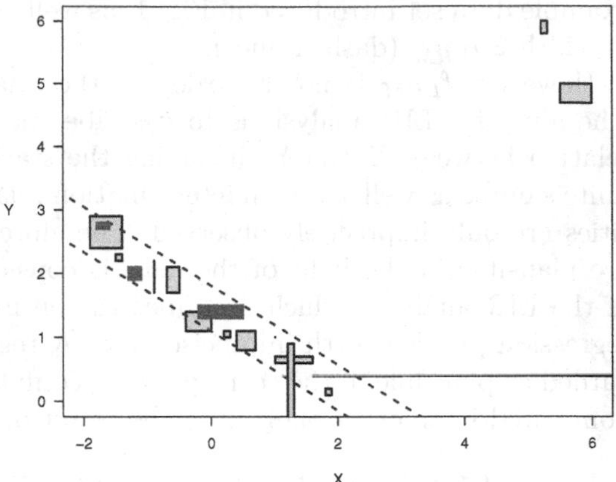

Fig. 3 Band $\overline{B}_{f_{LRM},\overline{q}_{LRM}}$ (dashed lines) for the LIR analysis considered in Fig. 2. The three imprecise data determining $\overline{B}_{f_{LRM},\overline{q}_{LRM}}$ in this case are highlighted.

observations. Thus, in order to identify candidates for b_{LRM} it suffices to consider the four slopes between the corresponding vertices of each pair of (nonidentical) bounded imprecise observations. In this way, we obtain a set of at most $4\binom{n}{2}+1$ candidates for the slope of f_{LRM}.

For a given slope b, it is easy to determine the intercept a for which the width of the resulting closed band around $f_{a,b}$ (containing at least $\overline{k}$ imprecise data) is minimal (over all linear functions with slope b). Consider the transformed data $Z_i^* := [\underline{Z}_i, \overline{Z}_i]$, $i = 1, \ldots, n$ obtained as

$$\underline{Z}_i = \begin{cases} \underline{Y}_i - b\,\overline{X}_i\,, & b > 0 \\ \underline{Y}_i - b\,\underline{X}_i\,, & b \leq 0 \end{cases} \quad \text{and} \quad \overline{Z}_i = \begin{cases} \overline{Y}_i - b\,\underline{X}_i\,, & b > 0 \\ \overline{Y}_i - b\,\overline{X}_i\,, & b \leq 0 \end{cases}.$$

Then finding the thinnest band containing (at least) $\overline{k}$ of the imprecise data V_i^* corresponds to finding the shortest interval containing (at least) $\overline{k}$ of the transformed imprecise data Z_i^*. Since the bands $\overline{B}_{f,q}$ are symmetric around f, the optimal intercept for a fixed candidate slope is given by the center of the shortest interval containing (at least) $\overline{k}$ of the transformed imprecise data Z_i^*. It can be proved that this shortest interval is one of the $n-\overline{k}+1$ intervals going from the j-th ordered lower endpoint $\underline{Z}_{(j)}$ to the $\overline{k}$-th of those ordered upper endpoints whose corresponding lower endpoints are not smaller than $\underline{Z}_{(j)}$, for $j = 1, \ldots, n-\overline{k}+1$. The interval with the shortest length provides the optimal intercept by its midpoint and the corresponding bandwidth by its length.

In this way, we obtain for each of the candidate slopes the associated optimal intercept and the resulting upper endpoint of the confidence interval, which corresponds to half of the width of the associated closed band. The function f_{LRM} is then given by the function with the minimal upper endpoint.

3.2 Part 2: Identifying the Set of All Undominated Lines

Once f_{LRM} and the associated $\overline{q}_{LRM}$ are known, the actual result of the LIR analysis is determined, which is the set of all regression lines that are not strictly dominated by f_{LRM}. For each $b \in \mathbb{R}$ there is a (possibly empty) set $\mathcal{A}_b$ consisting of all intercept values a such that the function $f_{a,b}$ is not strictly dominated by f_{LRM}.

To determine $\mathcal{A}_b$, we make use of the fact that the closed band of width $2\,\overline{q}_{LRM}$ around an undominated regression line intersects at least $\underline{k}+1$ imprecise data. Consider again the transformed data Z_i^*, then finding the centers of all bands of width $2\,\overline{q}_{LRM}$ that intersect (at least) $\underline{k}+1$ of the imprecise data V_i^* reduces to finding the centers of all intervals (of length $2\,\overline{q}_{LRM}$) that intersect (at least) $\underline{k}+1$ of the transformed imprecise data Z_i^*. Thus, for each b we look for the values a such that the intervals $[a-\overline{q}_{LRM}, a+\overline{q}_{LRM}]$ intersect at least $\underline{k}+1$ of the Z_i^*, $i = 1,\ldots,n$. For each subset of $\underline{k}+1$ transformed imprecise data, $Z_{i_1}^*,\ldots,Z_{i_{\underline{k}+1}}^*$, the set of undominated interval centers is the interval

$$\left[\max_{i\in\{i_1,\ldots,i_{\underline{k}+1}\}} \underline{Z}_i - \overline{q}_{LRM},\ \min_{i\in\{i_1,\ldots,i_{\underline{k}+1}\}} \overline{Z}_i + \overline{q}_{LRM}\right].$$

If the lower interval endpoint exceeds the upper one, the set of undominated interval centers associated with the considered subset of imprecise data is empty. This means that there is no interval of length $2\,\overline{q}_{LRM}$ intersecting all of the considered imprecise data.

Employing this idea, we can prove that for each b the set $\mathcal{A}_b$ can be obtained as the union of the intervals $[\underline{Z}_{(\underline{k}+j)} - \overline{q}_{LRM}, \overline{Z}_{(j)} + \overline{q}_{LRM}]$, $j = 1,\ldots,n-\underline{k}$, where $\underline{Z}_{(i)}$ and $\overline{Z}_{(i)}$ are the i-th ordered lower and upper

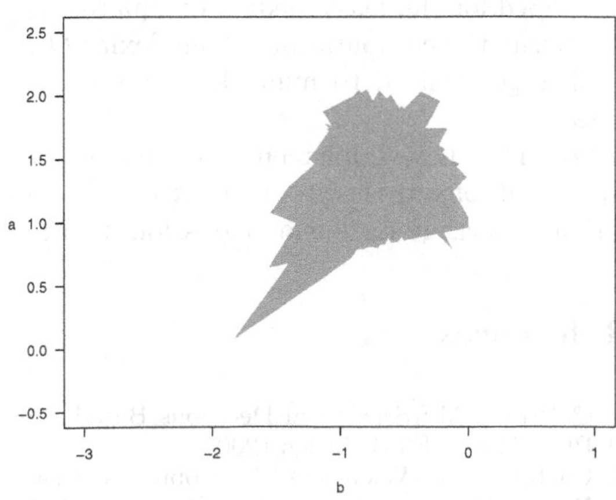

Fig. 4 Set of parameters corresponding to the set of undominated regression lines for the LIR analysis considered in Fig. 2.

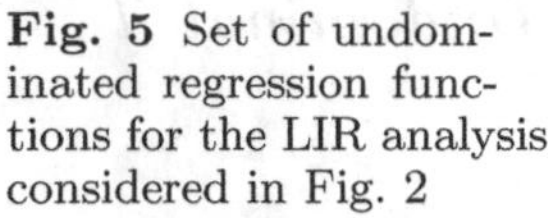

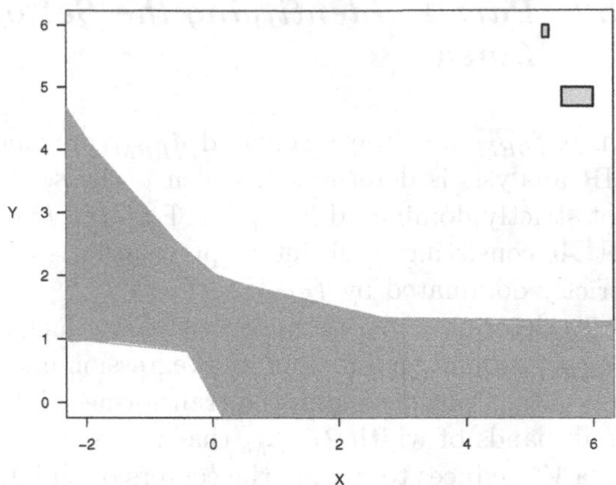

Fig. 5 Set of undominated regression functions for the LIR analysis considered in Fig. 2

endpoints of the imprecise data, respectively. Finally, the whole set of parameters (a, b) identifying the undominated functions is given by the union of the sets $\mathcal{A}_b \times \{b\}$ over all $b \in \mathbb{R}$. It can be shown that this set is polygonal, but it is not necessarily convex nor connected. Figure 4 shows the complex shape of this set in our example and in Fig. 5 the corresponding regression functions are plotted.

4 Conclusions

We presented an algorithm to determine the set-valued result of a robust LIR analysis in the case of simple linear regression with interval data. The algorithm is directly derived from the geometrical properties of the LIR results and it is exact. The proofs will be given in an extended version of this paper. The presented algorithm can be seen as a generalization of an algorithm developed for the least median of squares regression (see [5, 6]), from which it inherits the computational complexity $O(n^3 \log n)$. The algorithm can be further generalized to multiple regression and to other kinds of imprecise data.

So far, we have implemented the algorithm as a general function using the statistical software environment R (see [4]). In future work, we intend to set up an R package for linear regression with LIR.

References

1. Cattaneo, M.: Statistical Decisions Based Directly on the Likelihood Function. PhD Thesis, ETH Zurich (2007)
2. Cattaneo, M., Wiencierz, A.: Robust regression with imprecise data. Technical Report 114, Department of Statistics, LMU Munich (2011)

3. Cattaneo, M., Wiencierz, A.: Likelihood-based Imprecise Regression. Int. J. Approx. Reason. (preliminary version of the paper is available as Technical Report 116, Department of Statistics, LMU Munich) (accepted)
4. R Development Core Team: R: A Language and Environment for Statistical Computing. R Foundation for Statistical Computing (2012)
5. Rousseeuw, P.J., Leroy, A.M.: Robust Regression and Outlier Detection. Wiley (1987)
6. Steele, J.M., Steiger, W.L.: Algorithms and complexity for least median of squares regression. Discret Appl. Math. 14, 93–100 (1986)

Part IV
Mathematical Aspects

Evolution of the Dependence of Residual Lifetimes

Fabrizio Durante and Rachele Foschi

Abstract. We investigate the dependence properties of a vector of residual lifetimes by means of the copula associated with the conditional distribution function. In particular, the evolution of positive dependence properties (like quadrant dependence and total positivity) are analyzed and expressions for the evolution of measures of association are given.

Keywords: Association measures, copulas, positive dependence, residual lifetimes.

1 Introduction

In the present note, we are interested in multivariate stochastic models related to a system composed by several components, whose behaviour can be represented by a random vector $\mathbf{X} = (X_1, X_2, \ldots, X_d)$ defined on a suitable probability space and taking values in $\mathbb{R}_+^d$. Specifically, each X_i is a continuous and positive random variable having the meaning of lifetime. For instance, in reliability theory, X_i's represent the lifetimes of certain disposals working in the same system; in credit risk, X_i's may represent the times-to-default of some companies. Regardless of their specific interpretation, it has been long recognized that the behaviour of $\mathbf{X}$ depends on both the individual behaviour

Fabrizio Durante
Free University of Bozen-Bolzano, School of Economics and Management, Bolzano, Italy
e-mail: fabrizio.durante@unibz.it

Rachele Foschi
IMT Advanced Studies, Lucca, Italy
e-mail: rachele.foschi@imtlucca.it

R. Kruse et al. (Eds.): Synergies of Soft Computing and Statistics, AISC 190, pp. 305–312.
springerlink.com © Springer-Verlag Berlin Heidelberg 2013

of each component and the dependence structure of $\mathbf{X}$ as interpreted by its copula C (see [10] and the references therein).

Here we are interested in the evolution of the dependence structure when one knows that all the components of $\mathbf{X}$ have survived up to time $\tau > 0$. For such a situation, the following facts can be revealed:

- the evolution of the dependence has "no jumps", in the sense that it evolves smoothly and does not admit drastic changes (as it would be in presence of exogenous shocks);
- the evolution of the dependence is stable with respect to misspecification, in the sense that "small" errors in selecting the dependence structure at time $\tau = 0$ do not amplify;
- some "weak" positive dependence among the components of $\mathbf{X}$ may disappear when τ increases. Contrarily, a "stronger" positive dependence is preserved at any time τ.

In the following, we clarify how these facts can be described rigorously in terms of copulas and present some additional results, especially concerning association measures, and relevant examples.

2 Copulas of Residual Lifetimes

For sake of simplicity, we treat only the case $d = 2$ (most of the considerations can be easily extended to the general case). Thus, let us consider a pair (X_1, X_2) of lifetimes whose copula is given by C. We denote by $\mathcal{C}$ the class of bivariate copulas.

As known (see [4]), the dependence properties of the family of distribution functions $(F_\tau)_{\tau \geq 0}$ of

$$[X_1 - \tau, X_2 - \tau \mid X_1 > \tau, X_2 > \tau]$$

can be described by means of a suitable copula process $(C_\tau)_{\tau > 0}$. Moreover, it is convenient to reparametrize the latter copula process in terms of a parameter $t \in (0,1]$, obtaining the process $(C_t)_{t \in (0,1]}$. In other words, C_1 represents the dependence structure of F_τ when $\tau = 0$ and, as t tends to 0, C_t represents the limiting dependence structure of F_τ as τ tends to ∞.

The following result allows us to derive some analytical properties of the copula process $(C_t)_{t \in (0,1]}$.

Proposition 1 ([4]). *For every $t \in (0,1]$, C_t is completely described by the restriction of the copula C to $[0,t]^2$. Specifically, one has*

$$C_t(u,v) = \frac{C(h_t^{-1}(uh_t(t)), k_t^{-1}(vk_t(t)))}{C(t,t)}, \tag{1}$$

where $h_t(u) = C(u,t)$ and $k_t(u) = C(t,u)$ for all $t \in [0,1]$.

Formally, the transformation (1) can be described in terms of the following mapping:

$$\Psi : (0,1] \times \mathcal{C} \to \mathcal{C}, \quad \Psi(t, C) = C_t.$$

Such a Ψ has the following features:

- Let $C \in \mathcal{C}$. The mapping $\Psi(\cdot, C) : (0,1] \to \mathcal{C}$, $t \mapsto C_t$, is continuous, i.e., C_t converges uniformly to C_{t_0} when t tends to t_0. Roughly speaking, the evolution of the dependence has no jumps (see [4]).
- Let $t \in (0,1]$. The mapping $\Psi(t, \cdot) : \mathcal{C} \to \mathcal{C}$ is continuous with respect to the L^∞–norm. In other words, if the copulas C and C' are sufficiently close each other (with respect to a suitable norm), then, for any t, C_t and C'_t are sufficiently close each other (see [1]). Roughly speaking, the evolution of the dependence is stable with respect to misspecification of C_1.

Moreover, notice that Ψ can be interpreted as the action of a suitable semi–group $((0,1], *)$ on $\mathcal{C}$. In particular, for all $t, s \in (0,1]$, we have $(C_t)_s = \Psi(s, C_t) = C_{t*s}$ (see [8, 9]).

It can be easily seen that the limit of C_t, as t tends to 0, may not exists. To this end, it is enough to consider a special kind of ordinal sum of copulas or a copula with fractal support (see, for instance, [1, Remark 3.3]). However, when such a limit exists, it follows that the limiting copula is invariant under the transformation defined in Eq. (1). For example, the independence copula $\Pi(u,v) = uv$ and the comonotone copula $M(u,v) = \min(u,v)$ are invariant; moreover copulas belonging to the Clayton family of copulas $\{C_\theta^{Cl}\}$,

$$C_\theta^{Cl}(u,v) = \left(\max\left(0, u^{-\theta} + v^{-\theta} - 1\right)\right)^{-1/\theta}, \quad \theta \geq -1,\ \theta \neq 0, \tag{2}$$

are invariant (for more details, see [5, 6]).

3 Dependence of Residual Lifetimes

Now, suppose that $C \in \mathcal{C}$ satisfies some positive dependence property. Our aim is to investigate whether the positive dependence is preserved by the process $(C_t)_{t\in(0,1]}$. First we introduce some definitions (see, e.g., [11]).

Definition 1. Let $C \in \mathcal{C}$.

- C is PQD (positive quadrant dependent) if and only if $C(u,v) \geq uv$ for all $u, v \in [0,1]$.
- C is TP2 (totally positive of order 2) if and only if for all u, u', v, v' in $[0,1]$, $u \leq u'$, $v \leq v'$,

$$C(u,v)C(u',v') \geq C(u,v')C(u',v). \tag{3}$$

- C is PLR (positively likelihood ratio dependent) if and only if it is absolutely continuous and its density satisfies (3).

Notice that PLR implies TP2; moreover, if C is TP2, then C is PQD. The following result also holds.

Proposition 2 ([4]). *Let $C \in \mathcal{C}$.*

- *If C is TP2, then C_t is TP2 for all $t \in (0, 1]$.*
- *If C is PLR, then C_t is PLR for all $t \in (0, 1]$.*

If C is PQD, instead, C_t may not be PQD for some t (see e.g. [4, Example 10]). In order to guarantee that positive quadrant dependence of C is preserved by any C_t, we need some stronger conditions, as specified in the following result.

Proposition 3 ([4]). *Let $C \in \mathcal{C}$. Then C_t is PQD for all $t \in \Lambda \subseteq (0, 1]$ if and only if, for all $u, v, t \in \Lambda$, $u, v \leq t$,*

$$C(u, v)C(t, t) \geq C(u, t)C(t, v). \tag{4}$$

In particular, C is said to be *hyper-PQD* if C satisfies (4) for $\Lambda = (0, 1]$.

Another way to look at the dependence evolution of the process $(C_t)_{t \in (0,1]}$ consists in introducing a suitable way to compare the copulas at different times. To this end, we consider the following definitions.

Definition 2. Let $C_1, C_2 \in \mathcal{C}$. C_1 is smaller than C_2 in the PQD order (written $C_1 \preceq_{PQD} C_2$) if $C_1(u, v) \leq C_2(u, v)$ for all $u, v \in [0, 1]$.

Definition 3. Let $C \in \mathcal{C}$. Then $(C_t)_{t \in (0,1]}$ is increasing (in the PQD order) if $C_{t'} \preceq_{PQD} C_{t''}$ for any $t' < t''$.

The following example shows that, regardless of the specific positive dependence of $C \in \mathcal{C}$, the dependence may evolve in different ways.

Example 1. Given a continuous and increasing function $f\colon [0, 1] \to [0, 1]$ such that $f(1) = 1$ and $\frac{f(t)}{t}$ is decreasing on $(0, 1]$, consider the copula

$$C(u, v) = \min(u, v) f(\max(u, v))$$

(see [3, 7] for more details about this construction). Such a copula C is TP2 (see [3]). Therefore, as follows by condition (4), C is hyper-PQD. Furthermore, by Prop. 2,

$$C_t(u, v) = \min(u, v) \frac{f(t \max(u, v))}{f(t)}$$

is TP2 for any $t \in (0, 1]$. The evolution of the strength of the dependence, instead, is influenced by the choice of f. In particular, the following cases can be considered.

- If $f(t) = t^\alpha$, $\alpha \in [0,1]$, then C is a Cuadras–Augé copula (see [2]) and $C_t = C$ for every $t \in (0,1]$ (see also [1, Example 4.1]).
- If $f(t) = \alpha t + (1-\alpha)$, $\alpha \in [0,1]$, then C is a Fréchet copula and, for all $t_1, t_2 \in (0,1]$, $t_1 \leq t_2$, we have $C_{t_1} \succeq_{PQD} C_{t_2}$. In particular, $\lim_{t\to 0^+} C_t(u,v) = \min(u,v)$.
- If $f(t) = \min(\alpha t, 1)$, $\alpha \geq 1$, then C is an ordinal sum of the copulas $\Pi(u,v) = uv$ and $M(u,v) = \min(u,v)$ with respect to the partition $([0,1/\alpha],[1/\alpha,1])$. Thus, for all $t_1, t_2 \in (0,1]$, $t_1 \leq t_2$, we have $C_{t_1} \preceq_{PQD} C_{t_2}$. In particular, $\lim_{t\to 0^+} C_t(u,v) = uv$.

Thus, depending on f, the mapping $t \mapsto C_t$ may be constant, increasing or decreasing in the PQD order. □

4 Measures of Association of Residual Lifetimes

The analysis of the dependence properties of C_t can be sometimes complicated because of technical difficulties in computing Eq. (1). In such a case, it could be convenient to consider some suitable association measures that are related to copulas.

Here, we concentrate on the most widespread *Kendall's tau* and *Spearman's rho*, which measure the concordance between two random variables. As known (see e.g. [12]), for every copula C, they are given by:

$$\tau_K(C) = 4\int_{[0,1]^2} C(u,v)\, dC(u,v) - 1,$$

$$\rho_S(C) = 12\int_{[0,1]^2} (C(u,v) - uv)\, du\, dv.$$

To compute such measures for C_t, we assume here that C is absolutely continuous and, hence, C_t is absolutely continuous for all $t \in (0,1]$ (see [4, Proposition 17]).

Proposition 4. *Let C be an absolutely continuous copulas. For every $t \in (0,1]$, one has*

- $\tau_K(C_t) = \dfrac{4}{C(t,t)^2}\displaystyle\int_{[0,t]^2} C(x,y)\partial^2_{12}C(x,y)\, dxdy - 1;$
- $\rho_S(C_t) = \dfrac{12}{C(t,t)^4}\displaystyle\int_{[0,t]^2} (C(x,y)C(t,t) - C(x,t)C(t,y))\partial_1 C(x,t)\partial_2 C(t,y)$ $dxdy.$

Proof. By the formula for calculating Kendall's τ, one has

$$\tau_K(C_t) = 4\int_{[0,1]^2} C_t(u,v)\partial^2_{12}C_t(u,v)dudv - 1.$$

Therefore, by the change of variable $x = h_t^{-1}(uh_t(t))$ and $y = k_t^{-1}(vk_t(t))$,

$$\tau_K(C_t) = 4 \int_{[0,t]^2} \frac{C(x,y)\partial_{12}^2 C(x,y)}{\partial_1 C(x,t)\partial_2 C(t,y)} \cdot \frac{\partial_1 C(x,t)\partial_2 C(t,y)}{C(t,t)^2} dxdy - 1.$$

Analogously,

$$\begin{aligned}\rho_S(C_t) &= 12 \int_{[0,1]^2} (C_t(u,v) - uv)dudv \\ &= 12 \int_{[0,t]^2} \left(\frac{C(x,y)}{C(t,t)} - \frac{C(x,t)C(t,y)}{C(t,t)^2} \right) \frac{\partial_1 C(x,t)\partial_2 C(t,y)}{C(t,t)^2} dxdy,\end{aligned}$$

which concludes the proof. □

Notice that the measures of association of C_t only depend on the value of C on the subdomain $[0,t]^2$.

Finally, together with τ_K and ρ_S, a measure of dependence may be considered for the process $(C_t)_{t\in(0,1]}$. Here we consider the Schweizer-Wolff's index σ (see [13]), given by

$$\sigma(C) = 12 \int_0^1 \int_0^1 |C(u,v) - uv| dudv.$$

By the same arguments in the proof of Prop. 4, it straightly follows that

$$\sigma(C_t) = \frac{12}{C(t,t)^4} \int_0^t \int_0^t |C(u,v)C(t,t) - C(u,t)C(t,v)| \partial_1 C(u,t)\partial_2 C(t,v) dudv.$$

It is immediate that, if C is PQD, then $\sigma(C) = \rho_S(C)$. Therefore, when C is hyper-PQD, $\sigma(C_t) = \rho_S(C_t)$. However, in general, $\sigma(C_t)$ is not directly obtained from $\rho_S(C_t)$ (see [11, Examples 5.18, 5.19]).

Example 2. Let us consider the copula

$$C(u,v) = uv + uv(1-u)(1-v).$$

In view of Prop. 4, for every $t \in (0,1]$, we have

$$\tau_K(C_t) = \frac{2t^2}{9(2 - 2t + t^2)^2}.$$

In particular, $\tau_K(C_t) \to 0$ as $t \to 0^+$. Analogously, for every $t \in (0,1]$, we have

$$\rho_S(C_t) = \frac{t^6}{3},$$

and $\rho_S(C_t) \to 0$ as $t \to 0^+$. Intuitively, the residual lifetimes are asymptotically (as the time τ tends to infinity) uncorrelated.

A stronger conclusion can be achieved by means of the σ index. Since C is TP2 and hence hyper-PQD, $\sigma(C_t) = \rho_S(C_t)$ for any t. Thus we also have $\sigma(C_t) \to 0$ as $t \to 0^+$, implying that the residual lifetimes are asymptotically independent. □

5 Conclusions

We have considered a copula process that allows to study the dependence behaviour of a random vector of lifetimes $\mathbf{X}$, knowing that all the components are surviving up to time τ. The study of such a copula process provides a way for looking at the tail dependence of the joint distribution of the vector. Moreover, the association measures related to the process may provide another way for expressing how the residual lifetimes evolve when the time increases.

Acknowledgements. The first author acknowledges the support of Free University of Bozen-Bolzano, School of Economics and Management, via the project "Stochastic Models for Lifetimes".

References

1. Charpentier, A., Juri, A.: Limiting dependence structures for tail events, with applications to credit derivatives. J. Appl. Probab. 43(2), 563–586 (2006)
2. Cuadras, C.M., Augé, J.: A continuous general multivariate distribution and its properties. Comm. Statist. A 10(4), 339–353 (1981)
3. Durante, F.: A new class of symmetric bivariate copulas. J. Nonparametr. Stat. 18(7-8), 499–510 (2006)
4. Durante, F., Foschi, R., Spizzichino, F.: Threshold copulas and positive dependence. Stat. Probab. Lett. 78(17), 2902–2909 (2008)
5. Durante, F., Jaworski, P.: Invariant dependence structure under univariate truncation. Stat. 46, 263–267 (2012)
6. Durante, F., Jaworski, P., Mesiar, R.: Invariant dependence structures and Archimedean copulas. Stat. Probab. Lett. 81(12), 1995–2003 (2011)
7. Durante, F., Kolesárová, A., Mesiar, R., Sempi, C.: Semilinear copulas. Fuzzy Set. Syst. 159(1), 63–76 (2008)
8. Foschi, R.: Semigroups of semi-copulas and a general approach to hyper-dependence properties (under revision)
9. Foschi, R., Spizzichino, F.: Semigroups of semicopulas and evolution of dependence at increase of age. Mathware Soft. Comput. XV(1), 95–111 (2008)

10. Jaworski, P., Durante, F., Härdle, W., Rychlik, T. (eds.): Copula Theory and its Applications. Lecture Notes in Statistics, vol. 198. Springer, Heidelberg (2010)
11. Nelsen, R.B.: An introduction to copulas, 2nd edn. Springer Series in Statistics. Springer, New York (2006)
12. Schmid, F., Schmidt, R., Blumentritt, T., Gaisser, S., Ruppert, M.: Copula-based measures of multivariate association. In: Jaworski, P., Durante, F., Härdle, W., Rychlik, T. (eds.) Copula Theory and its Applications. Lecture Notes in Statistics, vol. 198, pp. 209–236. Springer, Heidelberg (2010)
13. Schweizer, B., Wolff, E.F.: On nonparametric measures of dependence for random variables. Ann. Statist. 9(4), 879–885 (1981)

A Spatial Contagion Test for Financial Markets

Fabrizio Durante, Enrico Foscolo, and Miroslav Sabo

Abstract. By using some ideas recently introduced by Durante and Jaworski, we present a test for spatial contagion among financial markets. This test is based on a comparison between threshold copulas associated with a given pair of random variables representing two financial markets. Moreover, the described methodology is used in order to check the presence of contagion among European markets in the recent financial crisis.

Keywords: Financial crisis, spatial contagion, threshold copulas.

1 Introduction

Financial contagion is usually referred to the process that describes the spread of financial difficulties from one economy to others in the same region and beyond. Nowadays, it is an ubiquitous terms in the financial literature, which has been applied to a variety of different situations. For instance, Pericoli and Sbracia [9] have listed five different definitions of contagion stressing the different perspectives that have been adopted in the literature. For a recent overview about contagion, see [7].

Following [9], we say that "*contagion is a significant increase in comovements of prices and quantities across markets, conditional on a crisis occur-*

Fabrizio Durante · Enrico Foscolo
School of Economics and Management, Free University
of Bozen–Bolzano, Bolzano, Italy
e-mail: {fabrizio.durante,enrico.foscolo}@unibz.it

Miroslav Sabo
Department of Mathematics and Constructive Geometry,
Slovak University of Technology, Bratislava, Slovakia
e-mail: sabo@math.sk

R. Kruse et al. (Eds.): Synergies of Soft Computing and Statistics, AISC 190, pp. 313–320.
springerlink.com © Springer-Verlag Berlin Heidelberg 2013

ring in one market or group of markets". In other words, financial contagion is related to a change in the positive association (interconnectedness) among markets, when a group of markets is affected by panics (i.e., noticeable disturbances).

Checking the presence of contagion among a group of markets has an impact in the estimation of the risk of a portfolio of assets. In fact, international investors may be concerned with the benefits of diversification, i.e. they aim at reducing the risk by investing in a variety of assets. However, if cross-country correlations of asset prices are significantly higher in periods of crisis, portfolio diversification may fail to deliver exactly when its benefits are needed most.

The standard approach to check for contagion is to draw a distinction between *normal comovements*, due to simple interdependence among markets, and *excessive comovements* in prices and quantities due to some structural break in the data. This issue is usually addressed by comparing cross-country correlations in tranquil and crisis periods. Specifically, contagion is said to have occurred if there is a significant increase in correlation during the crisis period. This phenomenon is also referred to as *correlation breakdown*. From this viewpoint, a test for contagion appears usually in the form:

$$\begin{array}{rll} & H_0 : \widetilde{\rho} \leq \rho & \text{no contagion} \\ \text{against} & H_1 : \widetilde{\rho} > \rho & \text{contagion} \end{array}$$

where $\widetilde{\rho}$ and ρ represent the correlation coefficient in crisis and untroubled periods, respectively.

However, these tests present (at least) two pitfalls. First, Pearson (linear) correlation does not completely describe the dependence among random variables (see, for instance, [4]); secondly, correlation breakdown can be biased by heteroscedasticity effects (see, for instance, [5]). We will see in the following how these two aspects will be considered.

In this note, we review the basic ideas about contagion described in [3] (Sect. 2). Then, we present a new test for financial contagion that aims at reducing bias introduced by heteroscedasticity of the univariate time series (Sect. 3). The procedure is used in order to check the presence of contagion among European markets in the recent financial crisis (Sect. 4).

2 Definition of Financial Contagion via Copulas

For basic definitions and properties about copulas, we refer the reader to [6].

Let (X, Y) be a pair of continuous random variables (=r.v.'s) on $(\Omega, \mathcal{F}, \mathbb{P})$. It is well known that the dependence of (X, Y) can be conveniently described by means of the copula associated with it. Actually, this idea also extends to conditional distribution functions (=d.f.'s).

Definition 1. Let B be a Borel set in $\overline{\mathbb{R}}^2$ such that $\mathbb{P}((X,Y) \in B) > 0$. We denote by H_B the conditional d.f. of (X,Y) given $(X,Y) \in B$, defined, for all $(x,y) \in B$, by

$$H_B(x,y) = \mathbb{P}\left(X \leq x, Y \leq y \mid (X,Y) \in B\right).$$

We call *threshold copula* related to (X,Y) and to the Borel set B the unique copula C_B of H_B.

Here, we are interested on the conditional d.f. of $[X,Y \mid (X,Y) \in B]$ where the set $B \subset \overline{\mathbb{R}}^2$ is of the following type:

- $T_{\alpha_1,\alpha_2} = [-\infty, q_X(\alpha_1)] \times [-\infty, q_Y(\alpha_2)]$
- $M_{\beta_1,\beta_2} = [q_X(\beta_1), q_X(1-\beta_1)] \times [q_Y(\beta_2), q_Y(1-\beta_2)]$,

for some $\alpha_1, \alpha_2 \in \left]0, \frac{1}{2}\right[$ and $\beta_1, \beta_2 \in \left[0, \frac{1}{2}\right[$. Here q_X and q_Y are the quantile functions associated with X and Y, respectively. Specifically, T_{α_1,α_2} is called *tail set*, and it is usually related to a "risky scenario". It includes the observations of X (respectively, Y) that are less than a given threshold. The set M_{β_1,β_2} is called *central set* (or *mediocre set*), and it is refereed to an "untroubled scenario". It is used when one wants to focus on observations that are not extremal.

Let X and Y be the r.v.'s representing the log–returns of two financial markets and let C be the associated copula. Spatial contagion can be introduced in terms of a suitable comparison between threshold copulas. To this end, let us introduce the following definition.

Definition 2. Let C_1 and C_2 be two bivariate copulas. we say that C_1 is less than C_2 in the positive quadrant dependence order (one writes: $C_1 \preceq C_2$) when $C_1(u,v) \leq C_2(u,v)$ for all $(u,v) \in [0,1]^2$. In particular, we write $C_1 \prec C_2$ when $C_1 \preceq C_2$ and $C_1(u,v) \neq C_2(u,v)$ for at least one $(u,v) \in [0,1]^2$.

Let $\alpha, \beta \in \left]0, \frac{1}{2}\right[$, $\alpha \leq \beta$. Following [3], the following definition of (asymmetric) contagion can be given.

Definition 3. Let X and Y be the r.v.'s representing the returns of two financial markets. There is *contagion from market X to market Y* with respect to $T_{\alpha,1}$ and $M_{\beta,0}$ if

$$C_{M_{\beta,0}} \prec C_{T_{\alpha,1}}$$

where $M_{\beta,0} = [q_X(\beta), q_X(1-\beta)] \times \overline{\mathbb{R}}$, $T_{\alpha,1} = [-\infty, q_X(\alpha)] \times \overline{\mathbb{R}}$.

Therefore, following [1], contagion is based on the comparison between the dependence in a tail region and in a central region of the joint d.f. of (X,Y). For such a reason, it is usually denoted as *spatial contagion*.

Some preliminary comments are needed here: first, contagion is based on copulas and, as such, it is more informative than other methods based

Pearson's correlation coefficient or tail dependence coefficients; secondly, contagion appears when there is a strict order between the copulas. For instance, two markets that are perfectly comonotone (i.e., one market is an increasing function of the other) exhibit no contagion, since their dependence does not change at any time. These markets are simply interdependent (compare with [5]).

Most noticeably, contagion depends on some suitable thresholds α and β, which should be chosen taking into account suitable definitions of tranquil/panic periods (usually, $\alpha = \beta$). In our approach we restrict to such situations:

- $\alpha = \beta = 0.05$: extreme contagion
- $\alpha = \beta = 0.10$: moderate contagion
- $\alpha = \beta = 0.25$: weak contagion

To clarify the notation, "extreme contagion" refers to the fact that the dependence between markets shifts upward only in presence of very extreme losses in one market. Weak contagion, instead, means that small losses in one market are sufficient to strength the comovements between the markets.

3 Test for Financial Contagion

A test for contagion based directly on a comparison of copulas as in Def. 3 could be difficult to implement. In fact, in general, the calculation of threshold copulas is not easily manageable. Moreover, it is always difficult to find an appropriate copula model to the data at disposal. In order to avoid such troubles, a non-parametric procedure should be preferred.

The idea is based on the following fact. Given two copulas C and D, if $C \prec D$, then $\rho(C) \leq \rho(D)$, where ρ is the Spearman's rank-correlation coefficient, given, for any copula C, by

$$\rho(C) = 12 \int_{[0,1]^2} C(u,v) \mathrm{d}u \, \mathrm{d}v - 3$$

(for more details, see [10]). Then, we might check the absence of contagion by comparing not the copulas, but the values of the associated Spearman's rank-correlation.

In order to describe the test for contagion, for $i \in \{1,2\}$ let P_t^i be a time series from a stock market index. Let L_t^i be the time series of the log-returns defined as $\log(P_t^i / P_{t-1}^i)$. The joint model of (L_t^1, L_t^2) may be determined in two steps (see, for instance, [2, 8]): first, we select a model for the marginal times series; then, we choose a suitable copula between them. Here we suppose that, for $i \in \{1,2\}$, L_t^i is modelled by a AR(1)–GARCH(1,1) process with innovation distribution being Student distribution; while the innovations are

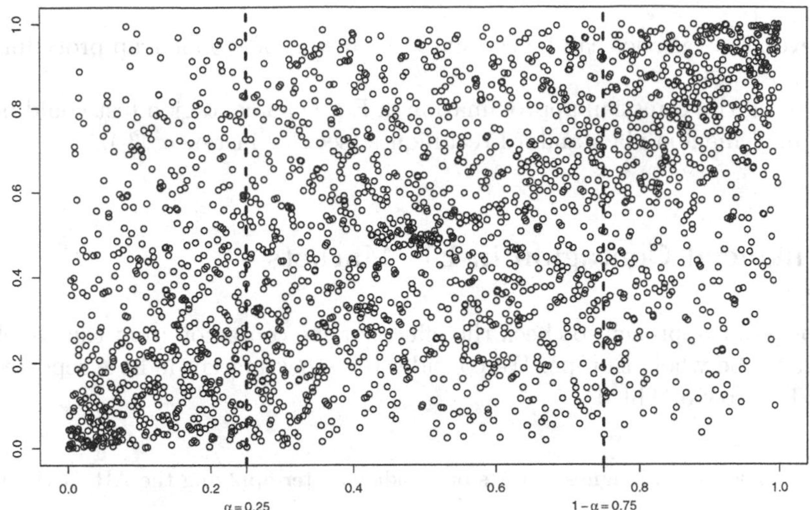

Fig. 1 Illustration of the procedure for selecting tail and central regions

assumed to be independent and identically distributed with a common copula C. Now, the following hypothesis test can be performed:

$$\begin{aligned} & H_0\colon\ \rho(C_T) \le \rho(C_M) \quad \text{(no contagion)} \\ \text{against} \quad & H_1\colon\ \rho(C_T) > \rho(C_M) \end{aligned}$$

where C_T and C_M are the copulas associated with C (the copula associated with the residuals) and suitable tail and central sets T and M, respectively.

We would like to stress that we apply a contagion test to the copula linking the innovations of the time series, because we would like to remove the effects of heteroscedasticity on the univariate time series and avoid possible bias in the test (see [5] for a discussion about this latter point).

Practically, after applying the AR–GARCH filter to each univariate time series, the obtained residuals (R_t^1, R_t^2) will be rescaled to $[0,1]^2$ by obtaining the sample (S_t^1, S_t^2). Such a rescaling has no effect in the calculation of Spearman's ρ that is invariant under rank transformation. Therefore, we may calculate:

- the empirical version $\widehat{\rho}(C_T)$ by using the points (S_t^1, S_t^2) such that $S_t^1 \le \alpha$;
- the empirical version $\widehat{\rho}(C_M)$ by using the points (S_t^1, S_t^2) when $\alpha \le S_t^1 \le 1 - \alpha$.

For an illustration see Fig. 1.

Now, we define $\Delta\rho = \rho(C_T) - \rho(C_M)$ and $\Delta\widehat{\rho} = \widehat{\rho}(C_T) - \widehat{\rho}(C_M)$ its empirical estimation. Under some technical assumptions on the involved threshold copulas, as the length of the time series $N \to +\infty$,

$$\sqrt{N}\,(\Delta\widehat{\rho} - \Delta\rho) \xrightarrow{d} \mathcal{N}(0, \sigma^2_{T,M}). \tag{1}$$

Moreover, the variance can be estimated by means of a bootstrap procedure as shown in [3].

By using the Gaussian approximation in Eq. (1) a contagion test could be easily implemented via standard techniques (see, for instance, [3]).

4 Financial Contagion in EU Markets

We apply our approach to check the effects of the contagion from the Greek market to the whole Europe. We consider the stock market indices reported in the first row of Tbl. 1.

Table 1 p–values of diagnostic tests on residuals after applying the AR–GARCH filter

	ATHEX	FTSE MIB	DAX	IBEX 35	Euro Stoxx 50	CAC 40	FTSE 100
AR(5)	0.33	0.61	0.44	0.70	0.48	0.39	0.35
ARCH(5)	0.23	0.09	0.08	0.00	0.11	0.10	0.43
KS (Student)	0.18	0.00	0.00	0.01	0.02	0.02	0.01

We focus on daily log-returns of the indices in the period 2003–2011 (data from Datastream). As described, to each univariate time series an AR–GARCH model has been applied. The adequacy of the fit has been checked via standard tests in order to check the residual uncorrelatedness and homoscedasticity (see Tbl. 1).

Then, we test an asymmetric contagion originated from the Greek stock market towards the other markets listed above. As for the choice of the thresholds, we consider $\alpha = \beta \in \{0.05, 0.10.0.25\}$. An example of the different values assumed by the Spearman's ρ in the tail and central set is given in Fig. 2.

Tables 2, 3 and 4 present the results of the performed tests. There is some evidence of weak contagion (at a significance level of 5%). Thus, the comovements between Greek market and the other European markets tend to appear more frequently when Greece is doing moderately badly. In other words, the link between the Greek stock exchange and the other stock exchange shifts upwards already when Greek market is having weak losses.

On the other hand, the absence of contagion is not rejected when we restrict to moderate or extreme contagion. The linkage between the two markets under consideration does not shift upwards when one market (the Greek stock markets) is having severe losses. Notice that this does not mean that the markets are weakly connected, but simply that their relation does not jump when we know that one market is doing quite badly.

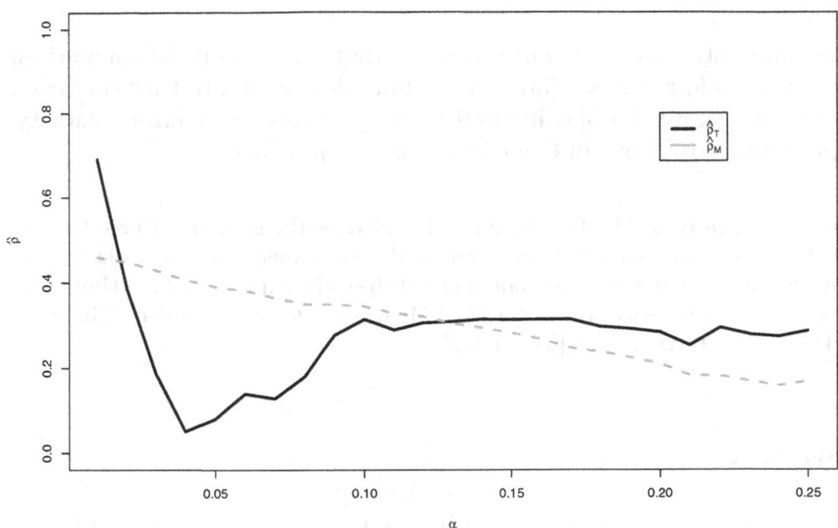

Fig. 2 Spearman's rho of the tail and the central set (for different threshold α) related to FTSE ATHEX 140 and EURO STOXX 50, respectively

Table 2 p–value of the asymmetric contagion test ($\alpha = 0.05$) from ATHEX to the other markets

FTSE MIB	DAX 30	IBEX 35	Euro Stoxx 50	CAC 40	FTSE 100
0.9989	0.9992	0.9996	0.9999	0.9987	0.9834

Table 3 p–value of the asymmetric contagion test ($\alpha = 0.10$) from ATHEX to the other markets

FTSE MIB	DAX 30	IBEX 35	Euro Stoxx 50	CAC 40	FTSE 100
0.5942	0.6201	0.5077	0.6994	0.7641	0.7216

Table 4 p–value of the asymmetric contagion test ($\alpha = 0.25$) from ATHEX to the other markets

FTSE MIB	DAX 30	IBEX 35	Euro Stoxx 50	CAC 40	FTSE 100
0.0036	0.0007	0.0114	0.0031	0.0070	0.0115

5 Conclusions

We have presented the notion of spatial contagion based on the comparison between threshold copulas. Moreover, we have described a test for contagion that tries to remove the bias induced by the presence of heteroscedasticity. An application to European financial markets is provided.

Acknowledgements. The first author acknowledges the support of Free University of Bozen-Bolzano, School of Economics and Management, via the project "Risk and Dependence". The second author acknowledges the support of Free University of Bozen-Bolzano, School of Economics and Management, via the project "Handling High–Dimensional Systems in Economics".

References

1. Bradley, B., Taqqu, M.: Framework for analyzing spatial contagion between financial markets. Finance Letters 2(6), 8–16 (2004)
2. Chen, X., Fan, Y.: Estimation and model selection of semiparametric copula-based multivariate dynamic models under copula misspecification. Journal of Econometrics 135(1-2), 125–154 (2006)
3. Durante, F., Jaworski, P.: Spatial contagion between financial markets: a copula-based approach. Appl. Stoch. Models Bus Ind. 26(5), 551–564 (2010)
4. Embrechts, P., McNeil, A.J., Straumann, D.: Correlation and dependence in risk management: properties and pitfalls. In: Dempster, M. (ed.) Risk Management: Value at Risk and Beyond, pp. 176–223. Cambridge University Press, Cambridge (2002)
5. Forbes, K.J., Rigobon, R.: No contagion, only interdependence: measuring stock market comovements. The Journal of Finance 57(5), 2223–2261 (2002)
6. Jaworski, P., Durante, F., Härdle, W., Rychlik, T. (eds.): Lecture Notes in Statistics - Proceedings. Lecture Notes in Statistics - Proceedings, vol. 198. Springer, Heidelberg (2010)
7. Kolb, R. (ed.): Financial contagion: the viral threat to the wealth of nations. Wiley, Hoboken (2011)
8. Nikoloulopoulos, A., Joe, H., Li, H.: Vine copulas with asymmetric tail dependence and applications to financial return data. Computational Statistics & Data Analysis (in press, 2012)
9. Pericoli, M., Sbracia, M.: A primer on financial contagion. Journal of Economic Surveys 17(4), 571–608 (2003)
10. Schmid, F., Schmidt, R., Blumentritt, T., Gaisser, S., Ruppert, M.: Copula-based measures of multivariate association. In: Jaworski, P., Durante, F., Härdle, W., Rychlik, T. (eds.) Copula Theory and its Applications. Lecture Notes in Statistics - Proceedings, vol. 198, pp. 209–236. Springer, Heidelberg (2010)

On Copulas and Differential Inclusions

Piotr Jaworski

Abstract. We construct a class of differential inclusions such that their solutions are horizontal sections of copulas. Furthermore we show that the horizontal sections of any copula can be obtained in such a way.

Keywords: Copulas, discontinuous differential equations, horizontal sections of copulas.

1 Introduction

The objective of this paper is to study the *generalized* solutions of first order ordinary differential equations with discontinuous right side

$$z' = F(x, z). \tag{1}$$

Namely, we are dealing with absolutely continuous functions $z(x)$, such that almost everywhere in their domain they fulfill a differential inclusion

$$z'(x) \in [\liminf_{y \to z} F(x, y), \limsup_{y \to z} F(x, y)]. \tag{2}$$

Our aim is to show that for a certain class of such equations the general solution consists of horizontal sections of some copula. Since, vice versa, every copula can be obtained in such a way we get a kind of a correspondence.

The paper is organized as follows. Section 2 is important from the point of view of self consistency of the paper. We present here the basic facts concerning the notion of an absolutely continuous solution of ordinary differential

Piotr Jaworski
Institute of Mathematics, University of Warsaw, 02-097 Warszawa, Poland
e-mail: p.jaworski@mimuw.edu.pl

R. Kruse et al. (Eds.): Synergies of Soft Computing and Statistics, AISC 190, pp. 321–329.
springerlink.com © Springer-Verlag Berlin Heidelberg 2013

equations and differential inclusions. In Section 3 we state the main theorem. The proofs and auxiliary results are collected in Section 4.

2 Notation

We are going to deal with differential equations with discontinuous right side. Therefore we apply the notion of an absolutely continuous solution, known in the literature under the name *Filippov solution* ([11]).

Definition 1. Let f be a real valued function (may be discontinuous)

$$f : \mathbb{I} \times [y_1, y_2] \longrightarrow \mathbb{R},$$

where $\mathbb{I}$ is an interval (closed, half-closed or open). We say that a function $g(x)$ defined on the interval $\mathbb{I}$ with values in $[y_1, y_2]$ is a generalized solution of the differential equation

$$y' = f(x, y),$$

if g is absolutely continuous and almost everywhere satisfies the differential inclusion

$$g'(x) \in [\liminf_{y \to g(x)} f(x, y), \limsup_{y \to g(x)} f(x, y)].$$

The basic facts about the differential inclusions and their solutions the reader may find for example in [11]. Note, that the absolute continuity of g implies that its derivative exists almost everywhere. Moreover

$$\int_a^b g'(t)dt = g(b) - g(a)$$

for any $a < b$ from $\mathbb{I}$ ([8] Theorem 7.4.4). Furthermore, if h is any integrable function on $[a, b]$ then the function

$$g(x) = \int_a^x h(t)dt$$

is absolutely continuous on $[a, b]$ and ([8] Theorem 7.1.7 and formula 7.1.14 below)

$$g'(x) = h(x) \quad a.e.$$

As a consequence, the generalized solutions are closely related with so called weak solutions in the Sobolev space $W^{1,1}(\mathbb{I})$ (compare [1] Theorem 8.2).

3 Main Results

Definition 2. Let Δ be a triangle $\Delta = \{(x,z) \in [0,1]^2 : z \leq x\}$. By $\mathbb{F}$ we denote the set of real valued functions F, $F : \Delta \longrightarrow \mathbb{R}$, such that:
F1. For fixed $z \in [0,1]$ the functions $F_z = F(\cdot, z) : [z,1] \to \mathbb{R}$ are measurable;
F2. F fulfill the boundary conditions

$$\forall x \in (0,1] \quad F(x,0) = 0, \quad F(x,x) = 1;$$

F3. The functions $F(x,z)$ are nondecreasing in the second variable

$$\forall x \in (0,1] \quad \forall 0 \leq z_1 \leq z_2 \leq x \quad F(x,z_1) \leq F(x,z_2).$$

Note that the conditions F1 and F3 imply that each $F \in \mathbb{F}$ is a measurable function of two variables (see [12] Theorem 11). For notational simplicity for $x \in [0,1]$ we put $F(x,z) = 0$ for $z < 0$ and $F(x,z) = 1$ for $z > x$.

Theorem 1. *For every $F \in \mathbb{F}$, there exists a unique function $C : [0,1]^2 \to [0,1]$ such that for fixed $y \in [0,1]$:*
1. $C_y(x) = C(x,y)$ is absolutely continuous and $C(x,y) \leq x$;
2. $C(1,y) = y$ and

$$\frac{\partial C(x,y)}{\partial x} \in [F(x, C(x,y)^-), F(x, C(x,y)^+)] \ a.e.$$

Furthermore:
3. The function $C(x,y)$ is a copula;
4. If $F_1, F_2 \in \mathbb{F}$ are equal almost everywhere to each other, then the corresponding solutions C_1 and C_2 coincide

$$\forall x,y \in [0,1] \quad C_1(x,y) = C_2(x,y);$$

5. If $F_1, F_2 \in \mathbb{F}$ and $F_1(x,z) \geq F_2(x,z)$ almost everywhere, then C_2 is bigger than C_1

$$\forall x,y \in [0,1] \quad C_1(x,y) \leq C_2(x,y).$$

We recall that a bivariate *copula* is a restriction to $[0,1]^2$ of a distribution function whose univariate margins are uniformly distributed on $[0,1]$. Specifically, $C\colon [0,1]^2 \to [0,1]$ is a copula if it satisfies the following properties:

(C1) $C(x,0) = C(0,x) = 0$ for every $x \in [0,1]$, i.e. C is *grounded*,
(C2) $C(x,1) = C(1,x) = x$ for every $x \in [0,1]$,
(C3) C is 2*–increasing*, that is, for every $x_1, y_1, x_2, y_2 \in [0,1]$, $x_1 \leq x_2$ and $y_1 \leq y_2$, it holds

$$C(x_1,y_1) + C(x_2,y_2) \geq C(x_1,y_2) + C(x_2,y_1).$$

For more details about the copula theory we refer to [10, 7, 6].

Example 1. Let for $x \in (0,1]$

$$F(x,z) = \begin{cases} 0 & \text{for } 0 \le z < \frac{x}{2} \\ 0.3 & \text{for } \quad z = \frac{x}{2} \\ 1 & \text{for } \frac{x}{2} < z \le x \end{cases}$$

The extended general solution C is piecewise linear, for $x \in [0,1]$

$$C(x,y) = \begin{cases} y & \text{for} \quad 0 \le y \le \frac{1}{2}x, \\ \frac{1}{2}x & \text{for} \quad \frac{1}{2}x < y < \frac{1}{2}(2-x), \\ y + x - 1 & \text{for} \quad \frac{1}{2}(2-x) \le y \le 1. \end{cases}$$

We obtain a copula, which describes the singular probability distribution, where the probability mass is uniformly distributed on two segments $Conv\{(0,0),(1,0.5)\}$ and $Conv\{(0,1),(1,0.5)\}$ (the rotated letter V, compare [10] Example 3.3).

Theorem 1 describes all horizontal sections of bivariate copulas. Indeed:

Theorem 2. *For any bivariate copula C there exists $F_C \in \mathbb{F}$ such that for every fixed $y \in [0,1]$*

$$\frac{\partial C(x,y)}{\partial x} = F_C(x, C(x,y)) \quad a.e.$$

4 Proofs and Auxiliary Results

The existence of the solutions of the Cauchy problem

$$z' \in [F(x,z^-), F(x,z^+)], \quad z(1) = y, \tag{3}$$

follows from Theorem 4.7 in [11]. Note that for $x \in (0,1]$ $F(x,0) = 0$ and $F(x,x) = 1$ hence there is a constant minimal solution

$$g_{min}(x) = 0, \quad x \in [0,1].$$

and a linear maximal solution

$$g_{max}(x) = x, \quad x \in [0,1].$$

Therefore, if g is any solution of (3), then g is defined on the interval [0,1] and

$$\forall x \in [0,1] \quad 0 = g_{min}(x) \le g(x) \le g_{max}(x) = x. \tag{4}$$

The uniqueness of the solutions is due to the monotonicity of F in the second variable. Indeed:

Lemma 1. *Let $F_1, F_2 \in \mathbb{F}$ and $F_1(x,z) \geq F_2(x,z)$ almost everywhere on Δ. If for some point $x_0 \in (0,1)$ and any two generalized solutions g_1 and g_2, respectively of $z' = F_1(x,z)$ and $z' = F_2(x,z)$,*

$$g_1(x_0) > g_2(x_0),$$

then

$$\forall x \in [x_0, 1] \quad g_1(x) - g_2(x) \geq g_1(x_0) - g_2(x_0).$$

Proof
F_1 and F_2 are measurable functions and $F_1(x,z) \geq F_2(x,z)$ almost everywhere in Δ. Therefore for x almost everywhere in $(0,1]$, for every two points $z_1, z_2 \in [0,x]$, $z_1 > z_2$, we can select a point $z^* \in (z_2, z_1)$ such that

$$F_1(x, z^*) \geq F_2(x, z^*).$$

Due to the monotonicity of the functions $F_i(x,z)$ in z we get

$$F_2(x, z_2^+) \leq F_2(x, z^*) \leq F_1(x, z^*) \leq F_1(x, z_1^-).$$

Let

$$x_1 = \inf\{x \in [x_0, 1] : x = 1 \vee g_1(x) \leq g_2(x)\}.$$

Since g_i are continuous, $x_1 > x_0$ and on $[x_0, x_1]$ g_1 is greater or equal to g_2. Hence we have for $t \in [x_0, x_1]$

$$g_2'(t) \leq F_2(t, g_2(t)^+) \leq F_1(t, g_1(t)^-) \leq g_1'(t).$$

$$\begin{aligned} g_1(x) - g_2(x) &= g_1(x_0) - g_2(x_0) + \int_{x_0}^{x} g_1'(t)dt - \int_{x_0}^{x} g_2'(t)dt \\ &\geq g_1(x_0) - g_2(x_0) > 0. \end{aligned}$$

Therefore $x_1 = 1$, which proves the thesis of Lemma. □

Corollary 1. *Let $F_1(x,z) \geq F_2(x,z)$ almost everywhere on Δ. If for any two generalized solutions g_1 and g_2, respectively of $z' = F_1(x,z)$ and $z' = F_2(x,z)$, $g_1(1) = g_2(1)$, then $g_1(t) \leq g_2(t)$ for $t \in [0,1]$.*

Putting $F_1(x,z) = F_2(x,z) = F(x,z)$, we obtain the uniqueness of solutions.

Corollary 2. *If for any two generalized solutions g_1 and g_2 of $z' = F(x,z)$ $g_1(1) = g_2(1)$, then $g_1 = g_2$.*

Putting together the existence and uniqueness of solutions we obtain the existence and uniqueness of the function C fulfilling the first two points of Theorem 1.

Corollary 3. *There is a unique function* $C : [0,1]^2 \to [0,1]$ *such that*

$$\forall x \in [0,1] \quad C(x,y) = g(x),$$

where g *is a generalized solution of the Cauchy problem*

$$z' = F(x,z), \quad z(1) = y.$$

Proposition 1. *The function* $C(x,y)$ *defined in Corollary 3 is a copula.*

Proof
First, we show that C fulfills the boundary conditions C1 and C2.
Directly from the definition and formula (4) we have that

$$C(0,y) = 0 \quad \text{and} \quad C(1,y) = y.$$

To prove the other two equalities we observe that g_{min} and g_{max}

$$g_{min}(x) = 0, \quad g_{max}(x) = x, \quad x \in [0,1],$$

are solutions of (3) for respectively $y = 0$ and $y = 1$. Hence

$$C(x,0) = g_{min}(x) = 0 \quad \text{and} \quad C(x,1) = g_{max}(x) = x.$$

The last condition – C3 is a direct corollary of Lemma 1 applied for $F_1(x,z) = F_2(x,z) = F(x,z)$ and $g_2(1) = y_1 \leq y_2 = g_1(1)$. Indeed, then also $g_2(x_1) \leq g_1(x_1)$ and for $x_2 \geq x_1$ we get from the same Lemma applied for $x_0 = x_1$ and $x = x_2$

$$\begin{aligned} &C(x_1,y_1) + C(x_2,y_2) - C(x_1,y_2) - C(x_2,y_1) \\ &= g_2(x_1) + g_1(x_2) - g_1(x_1) - g_2(x_2) \geq 0. \end{aligned}$$

□

To accomplish the proof of Theorem 1 it is enough to observe that Corollary 1 implies the remaining two points (4 and 5) of the thesis.

To prove Theorem 2 we construct F_C basing on the conditional copulas and so called Dini derivatives.

If C is the copula of the pair (X,Y), then the copula of the conditional distribution function of (X,Y) with respect to the condition $X \leq x$, where $P\{X \leq x\} = \alpha \in (0,1]$ is called a conditional copula and denoted by $C_{[\alpha]}$ (compare [9, 3]). We recall (for a proof see [4], [5] or [9]):

Proposition 2. *Let* $C(x,y)$ *be a copula. The functional equation*

$$H\left(\alpha, \xi, \frac{C(\alpha,y)}{\alpha}\right) = \frac{C(\alpha\xi,y)}{\alpha}, \quad \xi, y \in [0,1], \ \alpha \in (0,1] \tag{5}$$

has a unique solution $H(\alpha,\xi,z)$, *which for fixed* α *is equal to the conditional copula* $C_{[\alpha]}(\xi,z)$.

Next step is to differentiate formula (5) at $\xi = 1$. Since H may not be differentiable, we apply the left-side Dini upper derivative. We recall that the concept of *Dini derivative* (or Dini derivate) generalizes the classical notion of the derivative of a real-valued function. Let $a, b \in \mathbb{R}$, $a < b$, and let $f : (a, b] \to \mathbb{R}$ be a continuous function. Let x be a point in $(a, b]$. The limit

$$D^- f(x) = \limsup_{h \to 0^+} \frac{f(x) - f(x-h)}{h}. \tag{6}$$

is called *left-side upper Dini derivative* of f at x. For more details, we refer to [8] and [2]. Note that if $f_y(x)$ is a continuous family of continuous functions then the Dini derivative may not be continuous but nevertheless it is a Borel function.

Definition 3. For a copula C, we define F_C as a partial Dini derivative of $H(x, t, z/x)$ with respect to t at $t = 1$, where H is the family of copulas introduced in Proposition 2.

$$F_C(x, z) = D_t^- H(x, t, \frac{z}{x})_{|t=1} = \limsup_{t \to 1^-} \frac{H\left(x, 1, \frac{z}{x}\right) - H\left(x, t, \frac{z}{x}\right)}{1 - t}, \tag{7}$$

where $(x, z) \in \Delta$.

Proposition 3. *F_C has the following properties:*
1. For fixed y $\frac{\partial C(x,y)}{\partial x} = F_C(x, C(x, y))$ almost everywhere.
2. F_C belongs to $\mathbb{F}$.

Proof.
To prove the first point we substitute in equation (5) $\xi = t$ and $\alpha = x$. We fix x and y and compute the Dini derivative at $t = 1$

$$\begin{aligned} &\limsup_{t \to 1^-} \frac{H\left(x, 1, \frac{C(x,y)}{x}\right) - H\left(x, t, \frac{C(x,y)}{x}\right)}{1 - t} \\ &= \limsup_{t \to 1^-} \frac{C(x, y) - C(tx, y)}{x(1 - t)} = \limsup_{\xi \to x^-} \frac{C(x, y) - C(\xi, y)}{x - \xi}. \end{aligned} \tag{8}$$

Since C is Lipschitz, its Dini derivative is equal almost everywhere to the usual derivative. Therefore for fixed y

$$F_C(x, C(x, y)) = \frac{\partial C(x, y)}{\partial x} \quad \text{a.e.}$$

To prove point (2) we observe that H is a continuous family of copulas. Therefore Dini derivatives are Borel functions and (F1) is valid. Moreover, since $H(x, t, 0) = 0$ and $H(x, t, 1) = t$, we get the boundary conditions (F2).

The monotonicity in z (condition (F3)) follows from the two-nondecreasingness of copulas – condition (C2). Indeed, let $0 \le z_1 \le z_2 \le x$. Since copulas are two-nondecreasing, we have for $1 > t > 0$

$$H\left(x,1,\frac{z_1}{x}\right) - H\left(x,t,\frac{z_1}{x}\right) \le H\left(x,1,\frac{z_2}{x}\right) - H\left(x,t,\frac{z_2}{x}\right).$$

Therefore, having Dini-differentiated H with respect to the second variable we get

$$F_C(x,z_1) = D_t^- H\left(x,t,\frac{z_1}{x}\right)_{|t=1} \le D_t^- H\left(x,t,\frac{z_2}{x}\right)_{|t=1} = F_C(x,z_2).$$

□

Formula (8) implies the following characterization of F_C:

Corollary 4. *The function F_C is equal to a composition of the partial left-side upper Dini derivative of copula C with respect to the first variable and any family of right-inverses of the vertical sections of C*

$$F_C(x,z) = D_\xi^- C(\xi, \psi_x(z))_{|\xi=x}, \quad C(x, \psi_x(z)) = z.$$

Acknowledgements. The author would like to express his gratitude to the anonymous referees for their valuable comments. He also acknowledges the support of Polish Ministry of Science and Higher Education, via the grant N N201 547838.

References

1. Brezis, H.: Functional analysis, Sobolev spaces and partial differential equations. Springer, New York (2011)
2. Durante, F., Jaworski, P.: A new characterization of bivariate copulas. Comm. Stat. Theory Methods 39, 2901–2912 (2010)
3. Durante, F., Jaworski, P.: Spatial contagion between financial markets: a copula-based approach. Applied Stochastic Models in Business and Industry 26, 551–564 (2010)
4. Durante, F., Jaworski, P.: Invariant dependence structure under univariate truncation. Statistics 46, 263–277 (2012)
5. Jaworski, P.: Invariant dependence structure under univariate truncation: the high-dimensional case. Statistics (2012), doi:10.1080/02331888.2012.664143
6. Jaworski, P., Durante, F., Härdle, W., Rychlik, T. (eds.): Copula Theory and Its Applications, Proc. of the Workshop held in Warsaw, Poland, September 25-26. Lecture Notes in Statistics, vol. 198. Springer (2010)
7. Joe, H.: Multivariate models and dependence concepts. Monographs on Statistics and Applied Probability, vol. 73. Chapman & Hall, London (1997)
8. Łojasiewicz, S.: An introduction to the theory of real functions, 3rd edn. John Wiley & Sons, Ltd., Chichester (1988)

9. Mesiar, R., Jágr, V., Juráňová, M., Komorníková, M.: Univariate conditioning of copulas. Kybernetika 44, 807–816 (2008)
10. Nelsen, R.B.: An introduction to copulas, 2nd edn. Springer Series in Statistics. Springer, New York (2006)
11. Smirnov, G.V.: Introductory to the Theory of Differential Inclusions. American Mathematical Society, Providence (2002)
12. Ursell, H.D.: Some methods of proving measurability. Fund. Math. 32, 311–330 (1939)

An Analytical Characterization of the Exchangeable Wide-Sense Geometric Law

Jan-Frederik Mai, Matthias Scherer, and Natalia Shenkman

Abstract. The exchangeable d-variate wide-sense geometric law is uniquely characterized by $(d+1)$-monotone sequences of parameters in [3]. The proof of sufficiency in [3] requires a probabilistic model. We provide an alternative, purely analytical proof of sufficiency of the $(d+1)$-monotonicity of a sequence to define admissible parameters of a d-variate wide-sense geometric law.

Keywords: d-monotone sequences, exchangeability, lack of memory, multivariate geometric law, rectangular inequalities.

1 Introduction

It is well known that the univariate geometric distribution allows for a unique characterization by means of the discrete lack-of-memory (LM) property, i.e. a random variable τ with values in $\mathbb{N}$ follows a geometric distribution if and only if $\mathbb{P}(\tau > n + m \,|\, \tau > m) = \mathbb{P}(\tau > n)$ for all $n, m \in \mathbb{N}_0$. A possible d-variate extension of the univariate discrete LM property for a random vector $(\tau_1, \ldots, \tau_d)$ is the local discrete LM property, given by the functional equation

$$\mathbb{P}(\tau_{i_1}>n_{i_1}+m, \ldots, \tau_{i_k}>n_{i_k}+m \,|\, \tau_{i_1}>m, \ldots, \tau_{i_k}>m) = \mathbb{P}(\tau_{i_1}>n_{i_1}, \ldots, \tau_{i_k}>n_{i_k}),$$

$$\text{for } k = 1, \ldots, d,\ 1 \leq i_1 < \ldots < i_k \leq d,\ m, n_{i_1}, \ldots, n_{i_k} \in \mathbb{N}_0. \quad (1)$$

Jan-Frederik Mai
Assenagon Credit Managemant GmbH, 80339 München, Germany
e-mail: `jan-frederik.mai@assenagon.com`

Matthias Scherer · Natalia Shenkman
TU Munich, 85748 Garching bei München, Germany
e-mail: `{scherer,shenkman}@tum.de`

R. Kruse et al. (Eds.): Synergies of Soft Computing and Statistics, AISC 190, pp. 331–340.
springerlink.com © Springer-Verlag Berlin Heidelberg 2013

Recently, it was shown in [3] that (1) uniquely characterizes the parametric family of wide-sense geometric distributions, see Theorem 1. The definition of the wide-sense geometric distribution in terms of its survival function is provided in Definition 1.

Definition 1 (Wide-sense geometric distribution). Let $(\Omega, \mathcal{F}, \mathbb{P})$ be a probability space supporting an $\mathbb{N}^d$-valued random vector $(\tau_1, \ldots, \tau_d)$. The distribution of $(\tau_1, \ldots, \tau_d)$ is called *d-variate wide-sense geometric* with parameters $\{p_I\}_{I \subseteq \{1,\ldots,d\}}$ if and only if its discrete survival function $\bar{F}^{\mathcal{W}}$ is given by

$$\bar{F}^{\mathcal{W}}(n_1, \ldots, n_d) := \mathbb{P}(\tau_1 > n_1, \ldots, \tau_d > n_d) = \prod_{k=1}^{d} \Big(\sum_{\substack{I \subseteq \{1,\ldots,d\} \\ \pi_{\mathbf{n}}(i) \notin I \; \forall \, i=k,\ldots,d}} p_I \Big)^{n_{(k)} - n_{(k-1)}}, \tag{2}$$

for some parameters $p_I \in [0,1]$, $I \subseteq \{1, \ldots, d\}$, with $\sum_I p_I = 1$, $\sum_{I: k \notin I} p_I < 1$ for $k = 1, \ldots, d$, and where $n_{(0)} := 0$ and $\pi_{\boldsymbol{n}}\colon \{1, \ldots, d\} \to \{1, \ldots, d\}$ is a permutation depending on $\boldsymbol{n} := (n_1, \ldots, n_d)$ such that $n_{\pi_n(1)} \leq n_{\pi_n(2)} \leq \ldots \leq n_{\pi_n(d)}$ and $n_{(1)} \leq \ldots \leq n_{(d)}$ denotes the ordered list of $n_1, \ldots, n_d$.

In (2), the permutation $\pi_{\boldsymbol{n}}$ is independent of k and sorts the indices of the vector $(n_1, \ldots, n_d)$ such that $n_{\pi_n(1)} \leq n_{\pi_n(2)} \leq \ldots \leq n_{\pi_n(d)}$. More intuition behind Equation (2) can be gained through studying the stochastic model of the multivariate wide-sense geometric law. A probabilistic construction based on waiting times for outcomes in a sequence of multinomial trials was first introduced in [1] and is revisited in [3], where an alternative characterization in terms of the survival function is provided.

Theorem 1 (Local LM of the wide-sense geometric distribution). *The distribution of an $\mathbb{N}^d$-valued random vector $(\tau_1, \ldots, \tau_d)$ satisfies the local discrete LM property (1) if and only if it is a d-variate wide-sense geometric distribution.*

Proof. See [3]. □

The condition in (1) requires each subvector $(\tau_{i_1}, \ldots, \tau_{i_k})$ of $(\tau_1, \ldots, \tau_d)$ of length $k \in \{1, \ldots, d\}$ to satisfy the local discrete LM property. By Theorem 1, this suffices to uniquely characterize the d-variate wide-sense geometric law. An alternative requirement is to postulate iteratively for all lower-dimensional marginal distributions of $(\tau_1, \ldots, \tau_d)$ to be geometric in the wide sense together with

$$\mathbb{P}(\tau_1 > n_1 + m, \ldots, \tau_d > n_d + m \,|\, \tau_1 > m, \ldots, \tau_d > m) = \mathbb{P}(\tau_1 > n_1, \ldots, \tau_d > n_d)$$

for all $m, n_1, \ldots, n_d \in \mathbb{N}_0$.

Restriction to *exchangeable* wide-sense geometric distributions, i.e. distributions invariant under all permutations of their components, allows for an interesting and very convenient analytical characterization by d-monotone sequences. A simple exchangeability criterion for the wide-sense geometric survival functions is given in [3]: the survival function (2) (and, hence, the corresponding distribution) is exchangeable if and only if its parameters $\{p_I\}_{I\subseteq\{1,\dots,d\}}$ satisfy the condition $p_{I_1} = p_{I_2}$ for all sets $I_1, I_2 \subseteq \{1,\dots,d\}$ with the same cardinality $|I_1| = |I_2|$. An important conclusion is that exchangeable wide-sense geometric distributions are parameterized by d parameters $\{\beta_k\}_{k=1}^d$ only. Indeed, the survival function simplifies to

$$\bar{F}^{\mathcal{W}}(n_1,\dots,n_d) = \prod_{k=1}^{d} \beta_k^{n_{(d-k+1)}-n_{(d-k)}}, \tag{3}$$

$$\beta_k = \sum_{i=1}^{d-k+1} \binom{d-k}{i-1} p_i, \tag{4}$$

where $p_i := p_{\{1,\dots,i-1\}}$ for $i = 2,\dots,d$ and $p_1 := p_\emptyset$.

A full characterization of finite exchangeability is only known for distributions on $\{0,1\}^d$ and is discussed in [4], where a new characterization based on d-monotone functions on $\{0,1,\dots,d\}$ is provided.

2 Characterization by d-monotonicity

The exchangeable d-variate wide-sense geometric distribution can be uniquely characterized by $(d+1)$-monotone sequences of parameters. This result was first established by [3], who show that $(d+1)$-monotonicity of the sequence $(1,\beta_1,\beta_2,\dots,\beta_d)$ in (4) is a *necessary and sufficient* condition for $\bar{F}^{\mathcal{W}}$ in (3) to define a survival function. Their proof of sufficiency is based on a probabilistic construction due to [1]. We provide an alternative analytical proof of sufficiency using the definition of a multivariate distribution function via its analytical properties. More precisely, we exploit the specific form of $\{\bar{F}^{\mathcal{W}}(n_1,\dots,n_d)\}_{n_1,\dots,n_d\in\mathbb{N}_0}$ related to exchangeability and the local LM property to reduce the necessary and sufficient conditions (in particular, rectangular inequalities) for $\{\bar{F}^{\mathcal{W}}(n_1,\dots,n_d)\}_{n_1,\dots,n_d\in\mathbb{N}_0}$ to define a survival function to $(d+1)$-monotonicity conditions on the sequence $(1,\beta_1,\beta_2,\dots,\beta_d)$. This is interesting, since verifying that a multivariate function is a survival function via the analytical properties is usually very difficult or even impossible. The proof of Theorem 2 illustrates how $(d+1)$-monotonicity arises naturally in order to verify rectangular inequalities for the survival function.

Lemma 1 gives a characterization of an arbitrary d-variate survival function of an $\mathbb{N}^d$-valued random vector by its analytical properties. The result of Lemma 1 is only secondary in this article. However, not knowing a reference for this result, we include the proof for the sake of completeness.

Lemma 1 (Analytical characterization of a survival function). *The multi-indexed sequence* $\{\bar{F}(n_1,\dots,n_d)\}_{n_1,\dots,n_d\in\mathbb{N}_0}$ *is a d-dimensional survival function of an* $\mathbb{N}^d$*-valued random vector if and only if*

(i) $\bar{F}(0,\dots,0)=1$;
(ii) $\lim_{n_k\to\infty}\bar{F}(n_1,\dots,n_d)=0,\quad k=1,\dots,d$;
(iii) rectangular inequalities: for all $(\tilde{n}_1,\dots,\tilde{n}_d)$, $(n_1,\dots,n_d)\in\mathbb{N}_0^d$ *with* $\tilde{n}_k<n_k$, $k=1,\dots,d$, *it holds that*

$$\sum_{i_1=1}^{2}\cdots\sum_{i_d=1}^{2}(-1)^{i_1+\ldots+i_d}\bar{F}(x_{1,i_1},\dots,x_{d,i_d})\geq 0,$$

where $x_{j,1}=n_j$, $x_{j,2}=\tilde{n}_j$.

Proof. We establish the claim by making use of a characterization of a multivariate discrete distribution function by its analytical properties. Appealing to Section 1.4.2 in [2], the necessary and sufficient conditions for a discrete function $\{F(n_1,\dots,n_d)\}_{n_1,\dots,n_d\in\mathbb{N}_0}$ to define a distribution function of an $\mathbb{N}^d$-valued random vector are:

(i′) $F(0,n_2,\dots,n_d)=F(n_1,0,\dots,n_d)=\ldots=F(n_1,\dots,n_{d-1},0)=0$;
(ii′) $\lim_{n_k\to\infty\,\forall k}F(n_1,\dots,n_d)=1$;
(iii′) rectangular inequalities: for all $(\tilde{n}_1,\dots,\tilde{n}_d)$, $(n_1,\dots,n_d)\in\mathbb{N}_0^d$ with $\tilde{n}_k<n_k$ for $k=1,\dots,d$, it holds that

$$\sum_{i_1=1}^{2}\cdots\sum_{i_d=1}^{2}(-1)^{i_1+\ldots+i_d}F(x_{1,i_1},\dots,x_{d,i_d})\geq 0,$$

where $x_{j,1}=\tilde{n}_j$, $x_{j,2}=n_j$.

We show that these conditions can be rewritten in terms of the corresponding survival function $\bar{F}$ as $(i)-(iii)$. To this end, let $(\tau_1,\dots,\tau_d)\sim F$. (i) and (i') are both equivalent to $(\tau_1,\dots,\tau_d)$ having components greater than or equal to 1. (ii) and (ii') are both equivalent to $(\tau_1,\dots,\tau_d)$ being almost surely finite. Denoting by $\bar{F}_I$ the I-marginal of $\bar{F}$, i.e. $\bar{F}_I(x_{j,i_j},j\in I)=\mathbb{P}(\tau_j>x_{j,i_j},j\in I)$, the equivalence of (iii) and (iii') follows from

$$\begin{aligned}&\sum_{i_1=1}^{2}\cdots\sum_{i_d=1}^{2}(-1)^{i_1+\ldots+i_d}F(x_{1,i_1},\dots,x_{d,i_d})\\&=\sum_{i_1=1}^{2}\cdots\sum_{i_d=1}^{2}(-1)^{i_1+\ldots+i_d}\Big(\sum_{k=1}^{d}\sum_{\substack{\emptyset\neq I\subseteq\{1,\dots,d\}\\|I|=k}}(-1)^k\bar{F}_I(x_{j,i_j},j\in I)\Big)\\&=\sum_{i_1=1}^{2}\cdots\sum_{i_d=1}^{2}(-1)^{i_1+\ldots+i_d}(-1)^d\bar{F}(x_{1,i_1},\dots,x_{d,i_d}),\end{aligned}\tag{5}$$

where the last equality is due to

$$\sum_{k=1}^{d-1} \sum_{\substack{\emptyset \neq I \subseteq \{1,\ldots,d\} \\ |I|=k}} \sum_{i_1=1}^{2} \cdots \sum_{i_d=1}^{2} (-1)^{i_1+\ldots+i_d} (-1)^k \bar{F}_I(x_{j,i_j}, j \in I) = 0.$$

More precisely, let $k \in \{1, \ldots, d-1\}$, $\emptyset \neq I \subseteq \{1, \ldots, d\}$ with $|I| = k$. Then $\exists l \in \{1, \ldots, d\} \cap I^c$ and, hence,

$$\sum_{i_1=1}^{2} \cdots \sum_{i_d=1}^{2} (-1)^{i_1+\ldots+i_d} (-1)^k \bar{F}_I(x_{j,i_j}, j \in I) = \sum_{i_l=1}^{2} (-1)^{i_l} \cdot const = 0,$$

where *const* is independent of i_l. Finally, note that multiplication with the coefficient $(-1)^d$ in (5) is equivalent to setting $x_{j,1} = n_j$ and $x_{j,2} = \tilde{n}_j$ in (iii). □

We are now in a position to analytically characterize the subclass of exchangeable wide-sense geometric distributions by d-monotonicity of their parameter sequences. Definition 2 recalls the notion of a d-monotone sequence.

Definition 2 (d-monotone sequence). A finite sequence $\{x_k\}_{k=0}^{d-1} \in \mathbb{R}^d$ is said to be d-monotone if it satisfies

$$\nabla^j x_k := \sum_{i=0}^{j} (-1)^i \binom{j}{i} x_{k+i} \geq 0,$$

for $k = 0, 1, \ldots, d-1$, $j = 1, 2, \ldots, d-k-1$. The set of d-monotone sequences $\{x_k\}_{k=0}^{d-1}$ starting with $x_0 = 1$, $x_1 < 1$ is denoted by $\mathcal{M}_d$.

Theorem 2 shows that $(d+1)$-monotonicity of the sequence $(1, \beta_1, \ldots, \beta_d)$ in (4) is sufficient for $\{\bar{F}^{\mathcal{W}}(n_1, \ldots, n_d)\}_{n_1,\ldots,n_d \in \mathbb{N}_0}$ in (3) to define a survival function. The proof presented here is purely analytic and involves interesting combinatorial ideas. Furthermore, it illustrates how $(d+1)$-monotonicity arises naturally in order to verify the rectangular inequalities in Lemma 1 (iii).

Theorem 2 (Exchangeable wide-sense geometric distribution). *The multi-indexed sequence* $\{\bar{F}(n_1, \ldots, n_d)\}_{n_1,\ldots,n_d \in \mathbb{N}_0}$*, given by*

$$\bar{F}(n_1, \ldots, n_d) = \prod_{k=1}^{d} \beta_k^{n_{(d-k+1)} - n_{(d-k)}}, \quad n_1, \ldots, n_d \in \mathbb{N}_0,$$

is a d-dimensional survival function if $(1, \beta_1, \ldots, \beta_d) \in \mathcal{M}_{d+1}$. *In this case,* $\{\bar{F}(n_1, \ldots, n_d)\}_{n_1,\ldots,n_d \in \mathbb{N}_0}$ *defines a survival function of an exchangeable wide-sense geometric distribution.*

Proof. We show that $(1, \beta_1, \ldots, \beta_d) \in \mathcal{M}_{d+1}$ is sufficient for (i) - (iii) in Lemma 1. To this end, denote $\beta_0 := 1$ and assume that $\{\beta_k\}_{k=0}^{d} \in \mathcal{M}_{d+1}$. Trivially, (i) holds. (ii) is an immediate consequence of $\beta_1 < 1$ and $\nabla^1 \beta_k \geq 0$ for $k = 1, \ldots, d-1$. The condition in (iii) can be simplified by decomposing a d-dimensional cuboid into d-dimensional cubes with unit edge length, i.e.

$$\sum_{i_1=1}^{2} \cdots \sum_{i_d=1}^{2} (-1)^{i_1+\ldots+i_d} \bar{F}(x_{1,i_1}, \ldots, x_{d,i_d})$$
$$= \sum_{j_1=\tilde{n}_1}^{n_1-1} \cdots \sum_{j_d=\tilde{n}_d}^{n_d-1} \Big(\sum_{i_1=1}^{2} \cdots \sum_{i_d=1}^{2} (-1)^{i_1+\ldots+i_d} \bar{F}(j_1 + \tilde{x}_{1,i_1}, \ldots, j_d + \tilde{x}_{d,i_d}) \Big),$$

where $\tilde{x}_{k,1} = 1$, $\tilde{x}_{k,2} = 0$. Therefore, for exchangeable functions $\bar{F}$, (iii) is equivalent to proving that

$$\sum_{i_1=1}^{2} \cdots \sum_{i_d=1}^{2} (-1)^{i_1+\ldots+i_d} \bar{F}(j_1 + \tilde{x}_{1,i_1}, \ldots, j_d + \tilde{x}_{d,i_d}) \geq 0,$$

for all $j_k \in \mathbb{N}_0$ with $j_1 \geq j_2 \geq \ldots \geq j_d$, and where $\tilde{x}_{k,1} = 1$, $\tilde{x}_{k,2} = 0$. This is shown in three steps. First, consider the case $j_1 = j_2 = \ldots = j_d =: j$. Using (3) and the exchangeability of $\bar{F}$, we obtain with $\tilde{x}_{k,1} = 1$, $\tilde{x}_{k,2} = 0$,

$$\sum_{i_1=1}^{2} \cdots \sum_{i_d=1}^{2} (-1)^{i_1+\ldots+i_d} \bar{F}(j + \tilde{x}_{1,i_1}, \ldots, j + \tilde{x}_{d,i_d})$$
$$= \binom{d}{0} (-1)^d \bar{F}(j+1, \ldots, j+1) + \binom{d}{1} (-1)^{d-1} \bar{F}(j+1, \ldots, j+1, j)$$
$$+ \ldots + \binom{d}{d-1} (-1)^1 \bar{F}(j+1, j, \ldots, j) + \binom{d}{d} (-1)^0 \bar{F}(j, \ldots, j)$$
$$= \beta_d^j \Big(\binom{d}{0} (-1)^d \beta_d + \binom{d}{1} (-1)^{d-1} \beta_{d-1} + \ldots + \binom{d}{d-1} (-1)^1 \beta_1 + \binom{d}{d} (-1)^0 \beta_0 \Big)$$
$$= \beta_d^j \sum_{i=0}^{d} (-1)^i \binom{d}{i} \beta_i = \beta_d^j \nabla^d \beta_0 \geq 0.$$

Second, consider the case $j_1 > j_2 > \ldots > j_d$. Then, for $\tilde{x}_{k,1} = 1$ and $\tilde{x}_{k,2} = 0$, $j_1 + \tilde{x}_{1,i_1} \geq j_2 + \tilde{x}_{2,i_2} \geq \ldots \geq j_d + \tilde{x}_{d,i_d}$, and, hence,

$$\sum_{i_1=1}^{2}\cdots\sum_{i_d=1}^{2}(-1)^{i_1+\ldots+i_d}\bar{F}(j_1+\tilde{x}_{1,i_1},\ldots,j_d+\tilde{x}_{d,i_d})$$
$$=\sum_{i_1=1}^{2}\cdots\sum_{i_d=1}^{2}(-1)^{i_1+\ldots+i_d}\Big(\beta_1^{j_1+\tilde{x}_{1,i_1}-j_2-\tilde{x}_{2,i_2}}\cdots\beta_{d-1}^{j_{d-1}+\tilde{x}_{d-1,i_{d-1}}-j_d-\tilde{x}_{d,i_d}}\beta_d^{j_d+\tilde{x}_{d,i_d}}\Big)$$
$$=(\beta_1^{j_1-j_2-1}\cdots\beta_{d-1}^{j_{d-1}-j_d-1}\beta_d^{j_d})\times$$
$$\times\sum_{i_1=1}^{2}\cdots\sum_{i_d=1}^{2}(-1)^{i_1+\ldots+i_d}\Big(\beta_1^{\tilde{x}_{1,i_1}-\tilde{x}_{2,i_2}+1}\cdots\beta_{d-1}^{\tilde{x}_{d-1,i_{d-1}}-\tilde{x}_{d,i_d}+1}\beta_d^{\tilde{x}_{d,i_d}}\Big)$$

$$=(\beta_1^{j_1-j_2-1}\cdots\beta_{d-1}^{j_{d-1}-j_d-1}\beta_d^{j_d})\prod_{i=1}^{d}\nabla\beta_{i-1}\geq 0,$$

where the last equality is proved by induction. More precisely, to show that

$$\sum_{i_1=1}^{2}\cdots\sum_{i_d=1}^{2}(-1)^{i_1+\ldots+i_d}\Big(\beta_1^{\tilde{x}_{1,i_1}-\tilde{x}_{2,i_2}+1}\cdots\beta_{d-1}^{\tilde{x}_{d-1,i_{d-1}}-\tilde{x}_{d,i_d}+1}\beta_d^{\tilde{x}_{d,i_d}}\Big)=\prod_{i=1}^{d}\nabla\beta_{i-1}$$

holds in low dimensions ($d\leq 3$) is an easy exercise, see also Remark 1. The induction step is carried out as follows:

$$\sum_{i_1=1}^{2}\cdots\sum_{i_d=1}^{2}(-1)^{i_1+\ldots+i_d}\Big(\beta_1^{\tilde{x}_{1,i_1}-\tilde{x}_{2,i_2}+1}\cdots\beta_{d-1}^{\tilde{x}_{d-1,i_{d-1}}-\tilde{x}_{d,i_d}+1}\beta_d^{\tilde{x}_{d,i_d}}\Big)$$
$$=\sum_{i_d=1}^{2}(-1)^{i_d}\beta_{d-1}^{1-\tilde{x}_{d,i_d}}\beta_d^{\tilde{x}_{d,i_d}}\sum_{i_1=1}^{2}\cdots\sum_{i_{d-1}=1}^{2}(-1)^{i_1+\ldots+i_{d-1}}\Big(\beta_1^{\tilde{x}_{1,i_1}-\tilde{x}_{2,i_2}+1}\cdots\beta_{d-1}^{\tilde{x}_{d-1,i_{d-1}}}\Big)$$
$$=\sum_{i_d=1}^{2}(-1)^{i_d}\beta_{d-1}^{1-\tilde{x}_{d,i_d}}\beta_d^{\tilde{x}_{d,i_d}}\Big(\prod_{i=1}^{d-1}\nabla\beta_{i-1}\Big)=(\beta_{d-1}-\beta_d)\prod_{i=1}^{d-1}\nabla\beta_{i-1}=\prod_{i=1}^{d}\nabla\beta_{i-1}.$$

Finally, consider the most general case $j_1=\ldots=j_{k_1}>j_{k_1+1}=\ldots=j_{k_2}>\ldots>j_{k_{n-1}+1}=\ldots=j_{k_n}$, where $k_n=d$. Let $k_0:=0$. For $l\in\{1,\ldots,n\}$, denote by $\tilde{x}_{[k_{l-1}+1,i_{k_{l-1}+1}]}\geq\tilde{x}_{[k_{l-1}+2,i_{k_{l-1}+2}]}\geq\cdots\geq\tilde{x}_{[k_l,i_{k_l}]}$ the descending ordered list of $\tilde{x}_{k_{l-1}+1,i_{k_{l-1}+1}},\tilde{x}_{k_{l-1}+2,i_{k_{l-1}+2}},\ldots,\tilde{x}_{k_l,i_{k_l}}$. Then

$$\sum_{i_1=1}^{2}\cdots\sum_{i_d=1}^{2}(-1)^{i_1+\ldots+i_d}\bar{F}(j_1+\tilde{x}_{1,i_1},\ldots,j_d+\tilde{x}_{d,i_d})$$

$$=\sum_{i_1=1}^{2}\cdots\sum_{i_d=1}^{2}(-1)^{i_1+\ldots+i_d}\times$$

$$\times\beta_1^{\tilde{x}[1,i_1]-\tilde{x}[2,i_2]}\cdots\beta_{k_1-1}^{\tilde{x}[k_1-1,i_{k_1-1}]-\tilde{x}[k_1,i_{k_1}]}\beta_{k_1}^{j_{k_1}+\tilde{x}[k_1,i_{k_1}]-j_{k_1+1}-\tilde{x}[k_1+1,i_{k_1+1}]}$$

$$\times\beta_{k_1+1}^{\tilde{x}[k_1+1,i_{k_1+1}]-\tilde{x}[k_1+2,i_{k_1+2}]}\cdots\beta_{k_2-1}^{\tilde{x}[k_2-1,i_{k_2-1}]-\tilde{x}[k_2,i_{k_2}]}$$

$$\times\beta_{k_2}^{j_{k_2}+\tilde{x}[k_2,i_{k_2}]-j_{k_2+1}-\tilde{x}[k_2+1,i_{k_2+1}]}$$

$$\times\ldots\times\beta_{k_{n-2}+1}^{\tilde{x}[k_{n-2}+1,i_{k_{n-2}+1}]-\tilde{x}[k_{n-2}+2,i_{k_{n-2}+2}]}\cdots\beta_{k_{n-1}-1}^{\tilde{x}[k_{n-1}-1,i_{k_{n-1}-1}]-\tilde{x}[k_{n-1},i_{k_{n-1}}]}$$

$$\times\beta_{k_{n-1}}^{j_{k_{n-1}}+\tilde{x}[k_{n-1},i_{k_{n-1}}]-j_{k_{n-1}+1}-\tilde{x}[k_{n-1}+1,i_{k_{n-1}+1}]}$$

$$\times\beta_{k_{n-1}+1}^{\tilde{x}[k_{n-1}+1,i_{k_{n-1}+1}]-\tilde{x}[k_{n-1}+2,i_{k_{n-1}+2}]}\cdots\beta_{k_n-1}^{\tilde{x}[k_n-1,i_{k_n-1}]-\tilde{x}[k_n,i_{k_n}]}\beta_{k_n}^{j_{k_n}+\tilde{x}[k_n,i_{k_n}]}$$

$$=(\beta_{k_1}^{j_{k_1}-j_{k_1+1}-1}\cdots\beta_{k_{n-1}}^{j_{k_{n-1}}-j_{k_{n-1}+1}-1}\beta_{k_n}^{j_{k_n}})\times$$

$$\times\sum_{i_1=1}^{2}\cdots\sum_{i_{k_1}=1}^{2}(-1)^{i_1+\ldots+i_{k_1}}\beta_1^{\tilde{x}[1,i_1]-\tilde{x}[2,i_2]}\cdots\beta_{k_1-1}^{\tilde{x}[k_1-1,i_{k_1-1}]-\tilde{x}[k_1,i_{k_1}]}\beta_{k_1}^{\tilde{x}[k_1,i_{k_1}]}$$

$$\times\sum_{i_{k_1+1}=1}^{2}\cdots\sum_{i_{k_2}=1}^{2}(-1)^{i_{k_1+1}+\ldots+i_{k_2}}\beta_{k_1}^{1-\tilde{x}[k_1+1,i_{k_1+1}]}$$

$$\times\beta_{k_1+1}^{\tilde{x}[k_1+1,i_{k_1+1}]-\tilde{x}[k_1+2,i_{k_1+2}]}\cdots\beta_{k_2-1}^{\tilde{x}[k_2-1,i_{k_2-1}]-\tilde{x}[k_2,i_{k_2}]}\beta_{k_2}^{\tilde{x}[k_2,i_{k_2}]}$$

$$\times\ldots\times\sum_{i_{k_{n-1}+1}=1}^{2}\cdots\sum_{i_{k_n}=1}^{2}(-1)^{i_{k_{n-1}+1}+\ldots+i_{k_n}}\beta_{k_{n-1}}^{1-\tilde{x}[k_{n-1}+1,i_{k_{n-1}+1}]}$$

$$\times\beta_{k_{n-1}+1}^{\tilde{x}[k_{n-1}+1,i_{k_{n-1}+1}]-\tilde{x}[k_{n-1}+2,i_{k_{n-1}+2}]}\cdots\beta_{k_n-1}^{\tilde{x}[k_n-1,i_{k_n-1}]-\tilde{x}[k_n,i_{k_n}]}\beta_{k_n}^{\tilde{x}[k_n,i_{k_n}]}.$$

The first sum

$$\sum_{i_1=1}^{2}\cdots\sum_{i_{k_1}=1}^{2}(-1)^{i_1+\ldots+i_{k_1}}\beta_1^{\tilde{x}[1,i_1]-\tilde{x}[2,i_2]}\cdots\beta_{k_1-1}^{\tilde{x}[k_1-1,i_{k_1-1}]-\tilde{x}[k_1,i_{k_1}]}\beta_{k_1}^{\tilde{x}[k_1,i_{k_1}]}=\nabla^{k_1}\beta_0$$

has already been computed in step 1. The remaining sums can be simplified to

$$\sum_{i_{k_{l-1}+1}=1}^{2} \cdots \sum_{i_{k_l}=1}^{2} (-1)^{i_{k_{l-1}+1}+\ldots+i_{k_l}} \beta_{k_{l-1}}^{1-\tilde{x}_{[k_{l-1}+1, i_{k_{l-1}+1}]}}$$

$$\times \beta_{k_{l-1}+1}^{\tilde{x}_{[k_{l-1}+1, i_{k_{l-1}+1}]}-\tilde{x}_{[k_{l-1}+2, i_{k_{l-1}+2}]}} \cdots \beta_{k_l-1}^{\tilde{x}_{[k_l-1, i_{k_l-1}]}-\tilde{x}_{[k_l, i_{k_l}]}} \beta_{k_l}^{\tilde{x}_{[k_l, i_{k_l}]}}$$

$$= \binom{k_l - k_{l-1}}{0}(-1)^{k_l-k_{l-1}}\beta_{k_l} + \binom{k_l - k_{l-1}}{1}(-1)^{k_l-k_{l-1}-1}\beta_{k_l-1}$$

$$+ \ldots + \binom{k_l - k_{l-1}}{k_l - k_{l-1} - 1}(-1)^1\beta_{k_{l-1}+1} + \binom{k_l - k_{l-1}}{k_l - k_{l-1}}(-1)^0\beta_{k_{l-1}}$$

$$= \sum_{i=0}^{k_l-k_{l-1}} (-1)^i \binom{k_l - k_{l-1}}{i} \beta_{k_{l-1}+i} = \nabla^{k_l-k_{l-1}}\beta_{k_{l-1}}$$

for $l = 2, \ldots, n$. Overall, this gives

$$\sum_{i_1=1}^{2} \cdots \sum_{i_d=1}^{2} (-1)^{i_1+\ldots+i_d} \bar{F}(j_1 + \tilde{x}_{1,i_1}, \ldots, j_d + \tilde{x}_{d,i_d})$$

$$= (\beta_{k_1}^{j_{k_1}-j_{k_1+1}-1} \cdots \beta_{k_{n-1}}^{j_{k_{n-1}}-j_{k_{n-1}+1}-1} \beta_{k_n}^{j_{k_n}}) \prod_{i=1}^{n} \nabla^{k_i-k_{i-1}}\beta_{k_{i-1}} \geq 0.$$

Hence, $(1, \beta_1, \ldots, \beta_d) \in \mathcal{M}_{d+1}$ is sufficient for $\bar{F}$ to define a survival function. Finally, it is easy to see that if $\bar{F}$ defines a survival function, then the corresponding distribution has geometric marginals and satisfies the local discrete LM property. Hence, it is an exchangeable wide-sense geometric distribution, and the theorem follows. □

To provide more intuition behind Theorem 2, we illustrate the proof in the bivariate case.

Remark 1 (Theorem 2 in $d = 2$). The major concern in the proof of Theorem 2 is to show that $(1, \beta_1, \ldots, \beta_d) \in \mathcal{M}_{d+1}$ suffices to guarantee the triangular inequalities in *(iii)*. In dimension $d = 2$, these are

$$\bar{F}(n_1, n_2) - \bar{F}(\tilde{n}_1, n_2) - \bar{F}(n_1, \tilde{n}_2) + \bar{F}(\tilde{n}_1, \tilde{n}_2) \geq 0 \tag{6}$$

for all (n_1, n_2), $(\tilde{n}_1, \tilde{n}_2) \in \mathbb{N}_0^2$ with $\tilde{n}_1 < n_1$, $\tilde{n}_2 < n_2$.

The first idea of the proof is to *resolve the rectangular* with coordinates $(\tilde{n}_1, \tilde{n}_2)$, $(n_1, \tilde{n}_2)$, $(\tilde{n}_1, n_2)$, (n_1, n_2) into $(n_1 - \tilde{n}_1)(n_2 - \tilde{n}_2)$ squares with unit edge length. Then, (6) is equivalent to showing that for each unit square with coordinates (j_1, j_2), $(j_1 + 1, j_2)$, $(j_1, j_2 + 1)$, $(j_1 + 1, j_2 + 1)$, $j_1, j_2 \in \mathbb{N}_0$,

$$\bar{F}(j_1 + 1, j_2 + 1) - \bar{F}(j_1, j_2 + 1) - \bar{F}(j_1 + 1, j_2) + \bar{F}(j_1, j_2) \geq 0. \tag{7}$$

The ansatz is intuitive and is justified by the σ-additivity of a probability measure:

$$
\begin{aligned}
&\bar{F}(n_1,n_2)-\bar{F}(\tilde{n}_1,n_2)-\bar{F}(n_1,\tilde{n}_2)+\bar{F}(\tilde{n}_1,\tilde{n}_2)\\
&=\mathbb{P}(\tilde{n}_1<\tau_1\leq n_1,\tilde{n}_2<\tau_2\leq n_2)=\sum_{j_1=\tilde{n}_1}^{n_1-1}\sum_{j_2=\tilde{n}_2}^{n_2-1}\mathbb{P}(j_1<\tau_1\leq j_1+1,j_2<\tau_2\leq j_2+1)\\
&=\sum_{j_1=\tilde{n}_1}^{n_1-1}\sum_{j_2=\tilde{n}_2}^{n_2-1}\big(\bar{F}(j_1+1,j_2+1)-\bar{F}(j_1,j_2+1)-\bar{F}(j_1+1,j_2)+\bar{F}(j_1,j_2)\big).
\end{aligned}
$$

The second idea of the proof is to make use of *exchangeability* of $\bar{F}$, i.e. $\bar{F}(j_1,j_2)=\bar{F}(j_2,j_1)$ for all $j_1,j_2\in\mathbb{N}_0$, to reduce the inequalities in (7) to the case $j_1\geq j_2$, $j_1,j_2\in\mathbb{N}_0$. Thus, the inequalities in (7) have to hold only for the unit squares below the identity line.

The third idea of the proof is to exploit the *local lack-of-memory* of $\bar{F}$, i.e. $\bar{F}(j_1+m,j_2+m)=\bar{F}(j_1,j_2)\bar{F}(m,m)$ for all $j_1,j_2,m\in\mathbb{N}_0$. Then, in the case $j_1=j_2$, it suffices to verify (7) just for the unit square with coordinates $(0,0)$, $(1,0)$, $(0,1)$, $(1,1)$:

$$
\begin{aligned}
&\bar{F}(j_1{+}1,j_1{+}1)-\bar{F}(j_1,j_1{+}1)-\bar{F}(j_1{+}1,j_1)+\bar{F}(j_1,j_1)\\
&\quad=\bar{F}(j_1,j_1)\big(\bar{F}(1,1)-2\bar{F}(1,0)+\bar{F}(0,0)\big)=\beta_2^{j_1}(\beta_2-2\beta_1+1)=\beta_2^{j_1}\nabla^2\beta_0.
\end{aligned}
$$

Finally, in the case $j_1>j_2$, the problem of verifying rectangular inequalities reduces to proving (7) only for the unit square with coordinates $(1,0)$, $(1,1)$, $(2,0)$, $(2,1)$:

$$
\begin{aligned}
&\bar{F}(j_1{+}1,j_2{+}1)-\bar{F}(j_1,j_2{+}1)-\bar{F}(j_1{+}1,j_2)+\bar{F}(j_1,j_2)\\
&\quad=\bar{F}(j_2,j_2)\big(\bar{F}(j_1{-}j_2{+}1,1)-\bar{F}(j_1{-}j_2,1)-\bar{F}(j_1{-}j_2{+}1,0)+\bar{F}(j_1{-}j_2,0)\big)\\
&\quad=\beta_2^{j_2}\beta_1^{j_1-j_2-1}\big(\nabla\beta_0\nabla\beta_1\big).
\end{aligned}
$$

References

1. Arnold, B.C.: A characterization of the exponential distribution by multivariate geometric compounding. Indian J. Stat. 37(1), 164–173 (1975)
2. Joe, H.: Multivariate models and dependence concepts. Chapman & Hall (1997)
3. Mai, J.-F., Scherer, M., Shenkman, N.: Multivariate geometric distributions (logarithmically) monotone sequences, and infinitely divisible laws (working paper)
4. Ressel, P.: Monotonicity properties of multivariate distribution and survival functions with an application to Lévy-frailty copulas. J. Multivar. Anal. 102(3), 393–404 (2011)

HMM and HAC

Weining Wang, Ostap Okhrin, and Wolfgang Karl Härdle

Abstract. Understanding the dynamics of a high dimensional non-normal dependency structure is a challenging task. This research aims at attacking this problem by building up a hidden Markov model (HMM) for Hierarchical Archimedean Copulae (HAC). The HAC constitute a wide class of models for high dimensional dependencies, and HMM is a statistical technique for describing time varying dynamics. HMM applied to HAC flexibly models high dimensional non-Gaussian time series. Consistency results for both parameters and HAC structures are established in an HMM framework. The model is calibrated to exchange rate data with a VaR application, and the model's performance is compared to other dynamic models.

Keywords: Hidden Markov model, hierarchical Archimedean copulae, multivariate distribution.

1 Introduction

Modelling high-dimensional time series is an often underestimated exercise of routine econometrical and statistical work. This slightly pejorative attitude towards day to day statistical analysis is unjustified since actually the

Weining Wang · Ostap Okhrin
Ladislaus von Bortkiewicz Chair of Statistics, Humboldt-Universität zu Berlin, 10178 Berlin, Germany
e-mail: wangwein@cms.hu-berlin.de, ostap.okhrin@wiwi.hu-berlin.de

Wolfgang Karl Härdle
Center for Applied Statistics and Economics, Humboldt-Universität zu Berlin, 10178 Berlin, Germany
e-mail: haerdle@wiwi.hu-berlin.de

R. Kruse et al. (Eds.): Synergies of Soft Computing and Statistics, AISC 190, pp. 341–348.
springerlink.com © Springer-Verlag Berlin Heidelberg 2013

calibration of time series models in high dimensions for standard data sizes is not only difficult on the numerical side but also on the mathematical side. Computationally speaking, integrated models for high dimensional time series become more involved when the parameter space is too large. An example is the multivariate GARCH(1,1) Baba-Engle-Kraft-Kroner (BEKK) model developed in [6], that for even two dimensions has an associated parameter space of dimension 12. For moderate sample sizes, the parameter space dimension might well be in the range of the sample size or even bigger. This data situation has evoked a new strand of literature on dimension reduction via penalty methods.

In this paper we take a different route, by calibrating an integrated dynamic model with unknown dependency structure among the d dimensional time series variables. More precisely, the unknown dependency structure may vary within a set of given dependencies. The specific dependence at each time t is unknown to the data analyst, but depends on the dependency pattern at time $t-1$. Therefore, HMM naturally come into play. This leaves us with the problem of specifying the set of dependencies.

An approach based on assuming a multivariate Gaussian or mixed normal is handicapped in capturing important types of data features such as heavy tails, asymmetry, and nonlinear dependencies. Such a simplification might in practice be too restrictive an assumption and might lead to biased results. Copulae are one possible approach to solving these problems, see [12]. Moreover, copulae allow us to separate the marginal distributions and the dependency model, see [17]. In recent decades, copula-based models have gained popularity in various fields like finance, insurance, biology, hydrology, etc. Nevertheless, many basic multivariate copulae are still too restrictive and a simple extension by putting in more parameters would lead to the extreme of a totally nonparametric approach that runs into the problem of the curse of dimensionality. A natural compromise is the class of hierarchical Archimedean copulae (HAC). An HAC allows a rich copula structure with a finite number of parameters. Recent works which have shown their flexibility are [11, 13, 18].

Many attempts have been made to obtain insights into the dynamics of the copulae: [3] assumes the underlying sequence is Markovian; [14] considers an asset-allocation problem with a time-varying parameter of bivariate copulae; [16] studies financial contagion using switching-parameter bivariate copulae. A likelihood based local adaptive method is an alternative approach for understanding the time evolution, see [8, 9]. This suggests to us a different path of modeling the dynamics: instead of taking a local point of view, we adopt a global dynamic model (HMM) for the change of both the tree structure and the parameters of the HAC along the time horizon. Under HMM, a stochastic process Y with a not directly observable underlying Markov process X is needed to determine the state of distributions of Y. This has been widely applied to speech recognition, see [15], molecular biology, and digital communications over unknown channels. For estimation and inference issues in HMM, see [1, 7], among others.

In this paper, we propose a new type of dynamic model, called HMMHAC, by incorporating HAC into an HMM framework. The theoretical problems such as parameter consistency and structure consistency are solved. The expectation maximization (EM) algorithm is developed in this framework for parameter estimation. See Sect. 2 for the model description, Sect. 3 for theorems about consistency. EM algorithm and computation issues are in Sect. 4. Section 5 is for applications.

2 Model Description

An HMM is a parameterized Markov random walk with an underlying Markov chain viewed as missing data, as in [10, 1]. Specifically, in our HMM HAC framework, let $\{X_t, t \geq 0\}$ be a stationary Markov chain on a finite state space $D = \{1, 2, \ldots, M\}$, with transition probability matrix $P = \{p_{ij}\}_{i,j=1,\ldots,M}$ and initial distribution $\pi = \{\pi_i\}_{i=1,\ldots,M}$.

$$P(X_0 = i) = \pi_i, \tag{1}$$

$$\begin{aligned} P(X_t = j | X_{t-1} = i) &= p_{ij} \\ &= P(X_t = j | X_{t-1} = i, X_{t-2} = x_{t-2}, \ldots, X_0 = x_0), \end{aligned} \tag{2}$$

for $i, j = 1, \ldots, M$. Let $\{Y_t, t \geq 0\}$ be the associated observations, and they are adjoined with $\{X_t, t \geq 0\}$ in such a way that given $X_t = i, i = 1, \ldots, M$, the distribution of Y_t is fixed:

$$P(X_t | X_{1:(t-1)}, Y_{1:(t-1)}) = P(X_t | X_{t-1}) \tag{3}$$

$$P(Y_t | Y_{1:(t-1)}, X_{(1:t)}) = P(Y_t | X_t), \tag{4}$$

where $Y_{1:(t-1)}$ stands for $\{Y_1, \ldots, Y_{t-1}\}$, $t < T$. As can be seen for simplicity we consider only order one models.

Let $f_j\{\cdot; \boldsymbol{\theta}^{(j)}, s^{(j)}\}$ be the conditional density of Y_t given X_{t-1}, $X_t = j$ with $\boldsymbol{\theta} \in \Theta, \mathbf{s} \in S$, $j = 1, \ldots, M$ being the unknown parameters. That is, $\{X_t, t \geq 0\}$ is a Markov chain, given $X_0, X_1, \ldots, X_T$, with $Y_0, Y_1, \ldots, Y_T$ being independent. Note that $\boldsymbol{\theta} = (\boldsymbol{\theta}^{(1)}, \ldots, \boldsymbol{\theta}^{(M)}) \in \mathbb{R}^{dM}$ are the unknown dependency parameters, $\mathbf{s} = (s^{(1)}, \ldots, s^{(M)})$ are the unknown HAC structure parameters, and its true value is denoted by $\boldsymbol{\theta}^*$ and $\mathbf{s}^*$.

For given d dimensional time series $y_t \in \mathbb{R}^d$ with $t = 1, \ldots, T$ and $y_t = (y_{1t}, y_{2t} \ldots, y_{dt})^T$ connected with unobservable (or missing) $x_1, \ldots, x_T$ from the given HMM, define π_{x_t} as the π_i for $x_0 = i$ $(i = 1, \ldots, M)$, and $p_{x_{t-1}x_t} = p_{ji}$ for $x_{t-1} = j$ and $x_t = i$. The full likelihood function given one realization of $\{x_t, y_t\}_{t=1}^T$ is

$$p_T(y_1,\cdots,y_T;x_1,\ldots,x_T)=\pi_{x_0}\prod_{t=1}^{T}p_{x_{t-1}x_t}f_{x_t}(y_t;\boldsymbol{\theta}^{(x_t)},s^{(x_t)}) \tag{5}$$

The novelty of our approach lies in a special parametrization of $f_i(.)$, which helps to properly understand the dynamics of a multivariate distribution. Up to now, typical parameterizations have been mixtures of log-concave or elliptical symmetric densities, such as those from Gamma or Poisson families, which are not flexible enough to model high dimensional time series. The advantage of the copula is that it splits the multivariate distribution into its margins and a pure dependency component. In other words, it captures the dependency between variables eliminating the impact of the marginal distributions.

Furthermore, we incorporate this procedure into the HMM framework. We denote the underlying Markov variable X_t as a dependency type variable. If $x_t = i$, the parameters $(\boldsymbol{\theta}^{(i)}, s^{(i)})$ determined by state $i = 1,\ldots,M$ take values on $S\times\Theta$, where S is a set of discrete candidate states corresponding to different dependency structures of the HAC, and Θ is a compact set in $\mathbb{R}^{d-1}$ wherein the HAC parameters take their values. Therefore,

$$f_i(\cdot)=c\{F_1^{\mathrm{m}}(y_1),F_2^{\mathrm{m}}(y_2),\ldots,F_d^{\mathrm{m}}(y_d),\boldsymbol{\theta}^{(i)},s^{(i)}\}f_1^{\mathrm{m}}(y_1)f_2^{\mathrm{m}}(y_2)\cdots f_d^{\mathrm{m}}(y_d), \tag{6}$$

with $f_i^{\mathrm{m}}(y_i)$ the marginal densities, $F_i^{\mathrm{m}}(y_i)$ the marginal cdf, $c(\cdot)$ the copula density.

Let $\boldsymbol{\theta}^{(i)} = (\theta_{i1},\ldots,\theta_{i,d-1})^T$ be the dependency parameters of the copulae starting from the lowest up to the highest level connected with a fixed state $x_t = i$ and the $f_i(.)$. The multistage maximum likelihood estimator is developed and discussed in [13]. The marginal densities $\hat{f}_m^{\mathrm{m}}(\cdot)$ are estimated according to the cdfs, and w_{it} is the weight associated with state i and time t, see (9). [4, 13] provide the asymptotic behavior of the estimates.

2.1 Likelihood Estimation

For the estimation of the HMM HAC model, we adopt the EM algorithm [5]. In the context of HMM, the EM algorithm is also known as the Baum–Welch algorithm which suggests estimating a sequence of parameters $\mathfrak{g}_{(i)} \equiv (P_{(i)}, \mathsf{s}_{(i)}, \boldsymbol{\theta}_{(i)})$ (for the ith iteration) by iterative maximization of $\mathcal{Q}(\mathfrak{g};\mathfrak{g}_{(i)}) \equiv \mathsf{E}_{\mathfrak{g}_{(i)}}\{\log p_T(Y_{1:T};X_{1:T})|Y_{1:T}\}$. Namely, one carries out the following two steps:

- (a) E-step: compute $\mathcal{Q}(\mathfrak{g};\mathfrak{g}_{(i)})$,
- (b) M-step: choose the update parameters $\mathfrak{g}_{(i+1)} = \arg\max_{\mathfrak{g}}\mathcal{Q}(\mathfrak{g};\mathfrak{g}_{(i)})$.

The essence of the EM algorithm is that $\mathcal{Q}(\mathfrak{g};\mathfrak{g}_{(i)})$ can be used as a surrogate for $\log p_T(y_1,\ldots,y_T;x_1,\ldots,x_T;\theta)$, see [2].

In our setting, we may write $\mathcal{Q}(\mathfrak{g};\mathfrak{g}_{(i)})$ as:

$$\mathcal{Q}(\mathfrak{g};\mathfrak{g}_{(i)}) = \sum_{i=1}^{M} P_{\mathfrak{g}_{(i)}}(X_0 = i|Y_{1:T})\log\{\pi_i f_i(y_0)\}$$
$$+ \sum_{t=1}^{T}\sum_{i=1}^{M} P_{\mathfrak{g}_{(i)}}(X_t = i|Y_{1:T})\log f_i(y_t)$$
$$+ \sum_{t=1}^{T}\sum_{i=1}^{M}\sum_{j=1}^{M} P_{\mathfrak{g}_{(i)}}(X_{t-1} = i, X_t = j|Y_{1:T})\log\{p_{ij}\}, \quad (7)$$

where $f_i(\cdot)$ is as in (6). The E-step, in which $P_{\mathfrak{g}_{(i)}}(X_t = i|Y_{1:T}), P_{\mathfrak{g}_{(i)}}(X_{t-1} = i, X_t = j|Y_{1:T})$ are evaluated, is carried out by the forward-backward algorithm and the M-step is explicit in the p_{ij} and the π_i. Adding constraints to (7) yields

$$\mathfrak{L}(\mathfrak{g},\lambda;\mathfrak{g}') = \mathcal{Q}(\mathfrak{g};\mathfrak{g}') + \sum_{i=1}^{M}\lambda_i(1 - \sum_{j=1}^{M} p_{ij}) \quad (8)$$

For the M-step, we need to take the first order partial derivative, and plug into (8). So, the dependency parameters $\boldsymbol{\theta}$ and the structure parameters $\mathbf{s}$ need to be estimated iteratively, for $\boldsymbol{\theta}^{(i)}$:

$$\frac{\partial\mathfrak{L}(\mathfrak{g},\lambda;\mathfrak{g}')}{\partial\theta_{ij}} = \sum_{t=1}^{T} P(X_t = i|Y_{1:T})\partial\log f_i(y_t)/\partial\theta_{ij}, \quad (9)$$

where, $j = 1,\ldots,d-1$. To simplify the procedure, we adopt the HAC estimation method from [13] with weights in terms of $w_{it} \equiv P(X_t = i|Y_{1:T})$. We also fix $\pi_i, i = 1,\ldots,M$ as it influences only the first observation X_0 which may be considered also as given and fixed. The estimation of the transition probabilities p_{ij} follows:

$$\frac{\partial\mathfrak{L}(\mathfrak{g},\lambda;\mathfrak{g}')}{\partial p_{ij}} = \sum_{t=1}^{T}\frac{P(X_{t-1} = i, X_t = j|Y_{1:T})}{p_{ij}} - \lambda_i \quad (10)$$

$$\frac{\partial\mathfrak{L}(\mathfrak{g},\lambda;\mathfrak{g}')}{\partial\lambda_i} = 1 - \sum_{j=1}^{M} p_{ij}. \quad (11)$$

Equating (10) and (11) yields:

$$\hat{p}_{i,j} = \frac{\sum_{t=1}^{n} P(X_{t-1} = i, X_t = j|Y_{1:T})}{\sum_{t=1}^{n}\sum_{j=1}^{M} P(X_{t-1} = i, X_t = j|Y_{1:T})} \quad (12)$$

Theorem 1. *Under regularity conditions, we find the corresponding structure:*

$$\lim_{T\to\infty} \max_{i\in 1,\dots,M} P(\hat{s}^{(i)} = s^{*(i)}) = 1, \forall i. \tag{13}$$

Moreover, Assuming $\{Y_t\}_{t=1}^T$ being i.i.d and generated from an HAC HMM model with parameters $\{s^{*(i)}, \theta^{*(i)}, \pi^*, \{p^*_{ij}\}_{i,j}\}$. The parameter $\hat{\boldsymbol{\theta}}^{(i)}$ satisfies, $\forall \varepsilon > 0$:

$$\lim_{T\to\infty} \min_{i\in 1,\dots,M} P(|\hat{\boldsymbol{\theta}}^{(i)} - \boldsymbol{\theta}^{*(i)}| > \varepsilon | \hat{s}^{(i)} = s^{*(i)}) = 0. \tag{14}$$

The proof can be sent upon request.

3 Applications

To see how HMM HAC performs on a real data set, application to financial data is offered. A good model for the dynamics of exchange rates gives insights into exogenous economic conditions, such as the business cycle. We demonstrate the forecast performance of the proposed technique by estimating the VaR of the portfolio and compare it with multivariate GARCH model such as DCC. The backtesting results show that the VaR calculated from HMMHAC performs significantly better.

The data set consists of the daily values for the exchange rates JPY/EUR, GBP/EUR and USD/EUR. The covered period is [4.1.1999; 14.8.2009], resulting in 2771 observations [9].

To eliminate intertemporal conditional heteroscedasticity, we fit to each marginal time series of log-returns a univariate GARCH(1,1) process. The residuals exhibit the typical behavior: they are not normally distributed, which motivates nonparametric estimation of the margins. From the results of the Box–Ljung test, whose p-values are $0.73, 0.01$, and 0.87 for JPY/EUR, GBP/EUR and USD/EUR, we conclude that the autocorrelation of the residuals is strongly significant only for the GBP/EUR rate. After this intertemporal correction, we work only with the residuals.

A VaR estimation example is to show the good performance of HMMHAC. We generate $N = 10^4$ paths with $T = 2219$ observations, and $|W| = 1000$ combinations of different portfolios, where $W = \{(1/3, 1/3, 1/3) \bigcup [\mathbf{w} = (w_1, w_2, w_3)]\}$, with $w_i = w'_i / \sum_{i=1}^3 w'_i$, $w'_i \in U(0,1)$. The Profit Loss (P&L) function of a weighted portfolio based on assets y_{td} is $L_{t+1} \equiv \sum_{d=1}^3 w_i (y_{t+1d} - y_{td})$, with weights $\mathbf{w} = (w_1, w_2, w_3) \in W$. The VaR of a particular portfolio at level $0 < \alpha < 1$ is defined as $VaR(\alpha) \equiv F_L^{-1}(\alpha)$, where the $\hat{\alpha}_{\mathbf{w}}$ is estimated as a relative fraction of violations $\hat{\alpha}_{\mathbf{w}} \equiv T^{-1} \sum_{t=1}^T \mathbf{I}\{L_t < \widehat{VaR}_t(\alpha)\}$, and the distance between $\hat{\alpha}_{\mathbf{w}}$ and α is $e_{\mathbf{w}} \equiv (\hat{\alpha}_{\mathbf{w}} - \alpha)/\alpha$. If the portfolio distribution is i.i.d., and a well calibrated model is properly mimicking the true underlying asset process, $\hat{\alpha}_{\mathbf{w}}$ is close to its nominal level α.

We considered four main models: HMMHAC for 500 observation windows for Gumbel and rotated Gumbel; multiple rolling window with 250 observations windows; locap change point detection, see [9](LCP) with $m_0 = 20$ and $m_0 = 40$ with Gumbel copulae; and DCC based on 500 observation windows. For all the models we made an out of sample forecast. To better evaluate the performance, we calculated the average and SD of e_W as $A_W = \frac{1}{|W|}\sum_{\mathbf{w}\in W} e_{\mathbf{w}}$ and $D_W = \left\{\frac{1}{|W|}\sum_{\mathbf{w}\in W}(e_{\mathbf{w}} - A_W)^2\right\}^{1/2}$.

Table 1 show the backtesting performance for the described models. One concludes that HMMHAC performs better than the concurring moving window, LCP, or DCC, as A_w and D_w are typically smaller.

Table 1 Robustness relative to $A_W(D_W)$

	Window\ α	0.1	0.05	0.01
HMM, RGum	500	-0.0204 (0.013)	**0.0147** (0.012)	**0.2827** (0.064)
HMM, Gum	500	**-0.0191** (0.008)	0.0233 (0.018)	0.3521 (0.029)
Rolwin, RGum	250	0.0375 (0.009)	0.0576 (0.012)	0.5076 (0.074)
Rolwin, Gum	250	0.0426 (0.009)	0.0772 (0.030)	0.6210 (0.043)
LCP, $m_0 = 40$	468	-0.0270 (0.010)	0.0391 (0.018)	0.4553 (0.037)
LCP, $m_0 = 20$	235	0.0344 (0.009)	0.0735 (0.026)	0.6888 (0.050)
DCC	500	-0.2573 (0.015)	-0.2140 (0.015)	0.6346 (0.091)

4 Conclusion

In this project, we propose a dynamic model for multivariate time series with non-Gaussian dependency. The idea has an easy extension to HMM for general copula models, and leads to a rich field for further work on dynamic models with dependency structures. This method is helpful in studying financial contagion at an extreme level over time, and naturally it can help in deriving conditional risk measures. As we have shown, dynamic copula models are good enough to mimic financial markets as well as nature.

Acknowledgements. The financial support from the Deutsche Forschungsgemeinschaft via SFB 649 Ökonomisches Risikö, Humboldt-Universität zu Berlin is gratefully acknowledged. We thank Prof. Cheng-Der Fuh for his comments.

References

1. Bickel, P.J., Ritov, Y., Rydén, T.: Asymptotic normality of the maximum-likelihood estimator for general hidden markov models. Annals of Statistics 26(4), 1614–1635 (1998)
2. Cappé, O., Moulines, E., Rydén, T.: Inference in Hidden Markov Models. Springer (2005)

3. Chen, X., Fan, Y.: Estimation of copula-based semiparametric time series models. Journal of Econometrics 130(2), 307–335 (2005)
4. Chen, X., Fan, Y.: Estimation and model selection of semiparametric copula-based multivariate dynamic models under copula misspesification. Journal of Econometrics 135, 125–154 (2006)
5. Dempster, A., Laird, N., Rubin, D.: Maximum likelihood from incomplete data via the em algorithm (with discussion). J. Roy. Statistical Society B 39, 1–38 (1997)
6. Engle, R.F., Kroner, K.F.: Multivariate simultaneous generalized arch. Econometric Theory 11, 122–150 (1995)
7. Fuh, C.D.: SPRT and CUSUM in hidden Markov Models. Ann. Statist. 31(3), 942–977 (2003)
8. Giacomini, E., Härdle, W.K., Spokoiny, V.: Inhomogeneous dependence modeling with time-varying copulae. Journal of Business and Economic Statistics 27(2), 224–234 (2009)
9. Härdle, W.K., Okhrin, O., Okhrin, Y.: Time varying hierarchical archimedean copulae (2011) (submitted for publication)
10. Leroux, B.G.: Maximum-likelihood estimation for hidden markov models. Stochastic Processes and their Applications 40, 127–143 (1992)
11. McNeil, A.J., Nešlehová, J.: Multivariate Archimedean copulas, d-monotone functions and l_1 norm symmetric distributions. Annals of Statistics 37(5b), 3059–3097 (2009)
12. Nelsen, R.B.: An Introduction to Copulas. Springer, New York (2006)
13. Okhrin, O., Okhrin, Y., Schmid, W.: On the structure and estimation of hierarchical archimedean copulas. Under Revision of Journal of Econometrics (2009)
14. Patton, A.J.: On the out-of-sample importance of skewness and asymmetric dependence for asset allocation. Journal of Financial Econometrics 2, 130–168 (2004)
15. Rabiner, L.R.: A tutorial on Hidden Markov Models and selected applications in speech recognition. Proceedings of IEEE 77(2) (1989)
16. Rodriguez, J.C.: Measuring financial contagion: a copula approach. Journal of Empirical Finance 14, 401–423 (2007)
17. Sklar, A.: Fonctions dé repartition á n dimension et leurs marges. Publ. Inst. Stat. Univ. Paris 8, 299–231 (1959)
18. Whelan, N.: Sampling from Archimedean copulas. Quantitative Finance 4, 339–352 (2004)

Some Smoothing Properties of the Star Product of Copulas

Wolfgang Trutschnig

Abstract. Three indications for the fact that the star product of copulas is smoothing are given. Firstly, it is shown that for every absolutely continuous copula A and every copula B both $A*B$ and $B*A$ are absolutely continuous. Secondly, an example of a singular copula A such that the absolutely continuous component of $A*A$ has support $[0,1]^2$ and mass at least $1/4$ is given. Finally, it is shown that for every copula B of the form $B=(1-\alpha)A+\alpha S$, whereby A is an absolutely continuous copula, S is a singular copula and $\alpha \in [0,1)$, there exists an absolutely continuous idempotent copula $\widehat{B}$ such that $\widehat{B}$ is the Cesáro limit of the sequence $(B^{*n})_{n\in\mathbb{N}}$ of iterates of the star product of B with respect to the metric D_1 introduced in [15].

Keywords: Copula, star product.

1 Introduction

Since its introduction by Darsow et al. in 1992 (see [1]) the so-called star product of copulas has been studied in various papers. In 1996 Olsen et al. showed that the space $(\mathcal{C},*)$ of (two-dimensional) copulas with the star product as binary operation and the space $(\mathcal{M},\circ)$ of Markov operators with the composition as binary operation are isomorphic (see [10] and Section 2) and that every copula $A\in\mathcal{C}$ can be written in the form $A=B^t*C$ whereby B,C are so-called completely dependent (or, equivalently, left invertible) copulas (see [10]) and B^t denotes the transpose of B. Using the above mentioned isomorphism Sempi (see [13]) showed in 2002 that there is a one-to-one

Wolfgang Trutschnig
Research Unit for Intelligent Data Analysis, European Centre for Soft Computing, 33600 Mieres (Asturias), Spain
e-mail: wolfgang.trutschnig@softcomputing.es

R. Kruse et al. (Eds.): Synergies of Soft Computing and Statistics, AISC 190, pp. 349–357.
springerlink.com © Springer-Verlag Berlin Heidelberg 2013

correspondence between the class of $*$-idempotent copulas $\mathcal{C}_{ip}$ (i.e. copulas with $A * A = A$) and the subclass of $\mathcal{M}$ consisting of conditional expectations. In 2007 Durante et al. (see [4]) studied two product-like constructions for copulas, one being a generalized version of the star product. In 2010 Darsow et al. (see [2]) classified idempotent copulas in non-atomic, atomic and totally atomic ones and, for each of the three types, gave a complete characterization of all of its members. Furthermore, based on these characterizations, they answered the question posed in [1] whether idempotent copulas are necessarily symmetric with 'yes'. In the current paper we give three indications for the fact the star product can be considered as *smoothing*: *Firstly* we will prove that for every absolutely continuous copula $A \in \mathcal{C}$ and every copula B both $A*B$ and $B*A$ are absolutely continuous. *Secondly* we will construct an example of a singular copula A for which $A * A$ has an absolutely continuous component with support full $[0,1]^2$. And *thirdly* we will show that for every copula B of the form $B = (1-\alpha)A + \alpha S$, whereby A is an absolutely continuous copula, S is a singular copula and $\alpha \in [0,1)$, there exists an absolutely continuous idempotent copula $\widehat{B}$ such that

$$\lim_{n\to\infty} D_1(s_{*n}(B), \widehat{B}) = 0,$$

holds, whereby $s_{*n}(B) := \frac{1}{n}\sum_{i=1}^{n} B^{*i}$ for every $n \in \mathbb{N}$, B^{*i} is the i-times star product of B with itself, i.e. $B^{*1} = B$, $B^{*2} = B * B$, and $B^{*(n+1)} = B * B^{*n}$ for every $n \geq 2$, and D_1 is the metric introduced in [15]. Moreover, it will be proved that in case the density k_A of A is strictly positive on $[0,1]^2$ $\widehat{B}$ has to coincide with the product copula Π.

The rest of this short paper is organized as follows: Section 2 gathers some preliminaries and notations that will be used later on. Section 3 contains the above mentioned results. Finally, Section 4 discusses open points and possible future work.

2 Notation and Preliminaries

As already mentioned before $\mathcal{C}$ will denote the family of all (two-dimensional) *copulas*, i.e. distribution functions on $[0,1]^2$ with uniform marginals, see [5], [9], [14]. For every $A \in \mathcal{C}$, μ_A will denote the corresponding *doubly stochastic measure* and $\mathcal{P}_{\mathcal{C}}$ the class of all these doubly stochastic measures. λ and λ_2 will denote the Lebesgue measure on $[0,1]$ and $[0,1]^2$ respectively, $\mathcal{B}([0,1])$ and $\mathcal{B}([0,1]^2)$ the Borel σ-fields in $[0,1]$ and $[0,1]^2$. A *Markov kernel* from $\mathbb{R}$ to $\mathcal{B}(\mathbb{R})$ is a mapping $K : \mathbb{R}\times\mathcal{B}(\mathbb{R}) \to [0,1]$ such that $x \mapsto K(x,B)$ is measurable for every fixed $B \in \mathcal{B}(\mathbb{R})$ and $B \mapsto K(x,B)$ is a probability measure for every fixed $x \in \mathbb{R}$. Suppose that X, Y are real-valued random variables on a probability space $(\Omega, \mathcal{A}, \mathcal{P})$, then a Markov kernel $K : \mathbb{R} \times \mathcal{B}(\mathbb{R}) \to [0,1]$ is called *regular conditional distribution of* Y *given* X if for every $B \in \mathcal{B}(\mathbb{R})$

$$K(X(\omega), B) = \mathbb{E}(\mathbf{1}_B \circ Y|X)(\omega) \tag{1}$$

holds $\mathcal{P}$-a.s. It is well known that for each pair (X, Y) of real-valued random variables a regular conditional distribution $K(\cdot,\cdot)$ of Y given X exists, that $K(\cdot,\cdot)$ is unique $\mathcal{P}^X$-a.s. (i.e. unique for $\mathcal{P}^X$-almost all $x \in \mathbb{R}$) and that $K(\cdot,\cdot)$ only depends on $\mathcal{P}^{X\otimes Y}$. Hence, given $A \in \mathcal{C}$ we will denote (a version of) the regular conditional distribution of Y given X by $K_A(\cdot,\cdot)$ and refer to $K_A(\cdot,\cdot)$ simply as *regular conditional distribution of* A or as *the Markov kernel of* A. Note that for every $A \in \mathcal{C}$, its conditional regular distribution $K_A(\cdot,\cdot)$, and every Borel set $G \in \mathcal{B}([0,1]^2)$ we have ($G_x := \{y \in [0,1] : (x,y) \in G\}$ denoting the x-section of G for every $x \in [0,1]$)

$$\int_{[0,1]} K_A(x, G_x)\, d\lambda(x) = \mu_A(G), \tag{2}$$

so in particular

$$\int_{[0,1]} K_A(x, F)\, d\lambda(x) = \lambda(F) \tag{3}$$

for every $F \in \mathcal{B}([0,1])$. On the other hand, every Markov kernel $K : [0,1] \times \mathcal{B}([0,1]) \to [0,1]$ fulfilling (3) induces a unique element $\mu \in \mathcal{P}_{\mathcal{C}}([0,1]^2)$ via (2). For more details and properties of conditional expectation, regular conditional distributions, and disintegration see [6] and [7].

A linear operator T on $L^1([0,1]) := L^1([0,1], \mathcal{B}([0,1]), \lambda)$ is called *Markov operator* (see [1] and [10]) if it fulfills the following three properties:

1. T is positive, i.e. $T(f) \geq 0$ whenever $f \geq 0$
2. $T(\mathbf{1}_{[0,1]}) = \mathbf{1}_{[0,1]}$
3. $\int_{[0,1]} (Tf)(x) d\lambda(x) = \int_{[0,1]} f(x) d\lambda(x)$

As mentioned in the introduction $\mathcal{M}$ will denote the class of all Markov operators on $L^1([0,1])$. It is straightforward to see that the operator norm of T is one, i.e. $\|T\| := \sup\{\|Tf\|_1 : \|f\|_1 \leq 1\} = 1$ holds. According to [1] and [10] *there is a one-to-one correspondence between* $\mathcal{C}$ *and* $\mathcal{M}$ - in fact, the mappings $\Phi : \mathcal{C} \to \mathcal{M}$ and $\Psi : \mathcal{M} \to \mathcal{C}$, defined by

$$\begin{aligned} \Phi(A)(f)(x) :&= (T_A f)(x) := \frac{d}{dx} \int_{[0,1]} A_{,2}(x,t) f(t) d\lambda(t), \\ \Psi(T)(x,y) :&= A_T(x,y) := \int_{[0,x]} (T\mathbf{1}_{[0,y]})(t) d\lambda(t) \end{aligned} \tag{4}$$

for every $f \in L^1([0,1])$ and $(x,y) \in [0,1]^2$ ($A_{,2}$ denoting the partial derivative w.r.t. y), fulfill $\Psi \circ \Phi = id_{\mathcal{C}}$ and $\Phi \circ \Psi = id_{\mathcal{M}}$. Note that in case of $f := \mathbf{1}_{[0,y]}$ we have $(T_A \mathbf{1}_{[0,y]})(x) = A_{,1}(x,y)$ λ-a.s. According to [15] the first equality in (4) can be simplified to

$$(T_A f)(x) = \mathbb{E}(f \circ Y | X = x) = \int_{[0,1]} f(y) K_A(x, dy) \qquad \lambda\text{-a.s.} \tag{5}$$

Expressing copulas in terms of their corresponding regular conditional distributions the metric D_1 on $\mathcal{C}$ can be defined as follows:

$$D_1(A,B) := \int_{[0,1]}\int_{[0,1]} \left|K_A(x,[0,y]) - K_B(x,[0,y])\right| d\lambda(x)\, d\lambda(y) \tag{6}$$

It can be shown that $(\mathcal{C}, D_1)$ is a complete metric space and that, given copulas $A, A_1, A_2 \ldots$ and their corresponding Markov operators $T_A, T_{A_1}, T_{A_2} \ldots$, the following two conditions are equivalent:

(a)$\lim_{n\to\infty} D_1(A_n, A) = 0$
(b)$\lim_{n\to\infty} \|T_{A_n} f - T_A f\|_1 = 0$ for every $f \in L^1([0,1])$,

i.e. D_1 is a metrization of the strong operator topology on $\mathcal{M}$ (see [15]). Given $A, B \in \mathcal{C}$ the *star product* $A * B \in \mathcal{C}$ is defined by (see [1], [4])

$$(A*B)(x,y) := \int_{[0,1]} A_{,2}(x,t) B_{,1}(t,y) d\lambda(t) \tag{7}$$

and fulfills

$$T_{A*B} = \Phi_{A*B} = \Phi(A) \circ \Phi(B) = T_A \circ T_B, \tag{8}$$

so the mapping Φ in (4) actually is an isomorphism (see [10]). A copula $A \in \mathcal{C}$ is called *idempotent* if $A * A = A$, the family of all idempotent copulas will be denoted by $\mathcal{C}_{ip}$. $A \in \mathcal{C}$ is called *symmetric* if $A^t = A$ whereby $A^t(x,y) := A(y,x)$ for all $x, y \in [0,1]$. The following result, stating that the Markov kernel of $A * B$ is just the standard composition of the Markov kernels of A and B, will prove useful in the sequel:

Lemma 1 ([17]). *Suppose that $A, B \in \mathcal{C}$ and let K_A, K_B denote regular conditional distributions of A and B. Then the Markov kernel $K_A \circ K_B$, defined by*

$$(K_A \circ K_B)(x,F) := \int_{[0,1]} K_B(y,F) K_A(x,dy), \tag{9}$$

is a regular conditional distribution of $A * B$.

As mentioned in the introduction, for every copula A and every $n \in \mathbb{N}$ we set

$$s_{*n}(A) = \frac{1}{n}\sum_{i=1}^{n} A^{*i}. \tag{10}$$

According to the following theorem $s_{*n}(A)$ is always convergent w.r.t. D_1:

Theorem 1 ([16]). *For every copula A there exists a copula $\widehat{A}$ such that*

$$\lim_{n\to\infty} D_1\big(s_{*n}(A), \widehat{A}\big) = 0. \tag{11}$$

This copula $\widehat{A}$ is idempotent, symmetric, and fulfills $\widehat{A} * A = A * \widehat{A} = \widehat{A}$.

3 Three Simple Results Indicating That the Star Product Is Smoothing

Lemma 2. *Suppose that $A \in \mathcal{C}$ is absolutely continuous. Then $A * B$ as well as $B * A$ are absolutely continuous for every copula $B \in \mathcal{C}$.*

Proof: W.l.o.g. we may assume that the probability density k_A of μ_A fulfills

$$\int_{[0,1]} k_A(x,y)d\lambda(x) = 1 \ \forall y \in [0,1] \text{ and } \int_{[0,1]} k_A(x,y)d\lambda(y) = 1 \ \forall x \in [0,1]. \tag{12}$$

Hence $K_A(x,E) := \int_E k_A(x,y)d\lambda(y)$ ($x \in [0,1]$ and $E \in \mathcal{B}([0,1])$) is a regular conditional distribution of A.

Fix an arbitrary $B \in \mathcal{C}$. For every $N \in \mathcal{B}([0,1]^2)$ with $\lambda_2(N) = 0$ there exists a Borel set $\Lambda \in \mathcal{B}([0,1])$ with $\lambda(\Lambda) = 1$ such that

$$0 = \lambda(N_x) = \int_{[0,1]} K_B(z,N_x)d\lambda(z)$$

for every $x \in \Lambda$. Consequently, for each $x \in \Lambda$, we have $K_B(z,N_x) = 0$ for λ-almost every $z \in [0,1]$, which implies

$$\begin{aligned}\mu_{A*B}(N) &= \int_{[0,1]} K_{A*B}(x,N_x)d\lambda(x) = \int_{[0,1]}\int_{[0,1]} K_B(z,N_x)K_A(x,dz)d\lambda(x) \\ &= \int_{[0,1]}\int_{[0,1]} K_B(z,N_x)k_A(x,z)d\lambda(z)d\lambda(x) \\ &= \int_{\Lambda}\int_{[0,1]} K_B(z,N_x)k_A(x,z)d\lambda(z)d\lambda(x) = 0.\end{aligned}$$

Furthermore, for every $x \in \Lambda$, we have

$$K_A(z,N_x) = \int_{N_x} k_A(z,y)d\lambda(y) = 0$$

for every $z \in [0,1]$, so

$$\begin{aligned}\mu_{B*A}(N) &= \int_{[0,1]}\int_{[0,1]} K_A(z,N_x)K_B(x,dz)d\lambda(x) \\ &= \int_{\Lambda}\int_{[0,1]} K_A(z,N_x)K_B(x,dz)d\lambda(x) = 0.\end{aligned}$$

Since N was arbitrary absolute continuity of $A * B$ and $B * A$ follows. ■

Lemma 2 has the following immediate consequence:

Theorem 2. *Suppose that A is an absolutely continuous copula, that S is a singular copula, and that $\alpha \in [0, 1)$. Then for the copula $B := (1-\alpha)A + \alpha S$ the D_1-limit $\widehat{B}$ of $s_{*n}(B)$ is absolutely continuous.*

Proof: Suppose that $B_i := (1-\alpha)A_i + \alpha_i S_i$, $i \in \{1, 2\}$, whereby S_i is singular, A_i is absolutely continuous, and $\alpha_i \in [0, 1]$ for every $i \in \{1, 2\}$. Then it follows directly from Lemma 2 that the singular component of $B_1 * B_2$ has at most mass $\alpha_1\alpha_2$. Suppose now that B satisfies the assumption of Theorem 2, then it follows immediately that the singular component of B^{*n} has at most mass α^n for every $n \in \mathbb{N}$. Consequently the singular component of $s_n^*(B)$ has at most mass $m_n := \frac{1}{n}\sum_{i=1}^{n} \alpha^i$. According to Theorem 1

$$s_{*n}(B) * \widehat{B} = \widehat{B}$$

holds for every $n \in \mathbb{N}$, so, using Lemma 2, the singular component of $\widehat{B}$ has at most mass m_n. Since n was arbitrary and $\lim_{n\to\infty} m_n = 0$ absolute continuity of $\widehat{B}$ follows. ■

In case the density k_A of A in Theorem 2 is strictly positive on $[0, 1]^2$ it can be shown that the limit copula $\widehat{B}$ has to be Π. To do so we will use the following lemma, in which $\mathcal{D}([0, 1])$ denotes the family of all probability densities in $L^1([0, 1])$:

Lemma 3. *Suppose that A is absolutely continuous and that $k_A > 0$ everywhere on $[0, 1]^2$, then:*

1. *For all densities $f_1, f_2 \in \mathcal{D}([0, 1])$ with $\|f_1 - f_2\|_1 > 0$ we have*

$$\|T_A f_1 - T_A f_2\|_1 < \|f_1 - f_2\|_1.$$

2. *If $B_1, B_2 \in \mathcal{C}$ with $B_1 \neq B_2$ then $D_1(A * B_1, A * B_2) < D_1(B_1, B_2)$.*

Proof: Suppose that $f_1, f_2 \in \mathcal{D}([0, 1])$ and that $\|f_1 - f_2\|_1 > 0$ as well as $\|T_A f_1 - T_A f_2\|_1 > 0$ holds. Then, using Scheffé's theorem (see [12]) and setting $G := \{x \in [0, 1] : T_A f_1(x) > T_A f_2(x)\}$, we have

$$\begin{aligned}
\|T_A f_1 - T_A f_2\|_1 &= 2\int_G (T_A f_1(x) - T_A f_2(x))\, d\lambda(x) \\
&= 2\int_G \left(\int_{[0,1]} (f_1(y) - f_2(y))\, k_A(x, y) d\lambda(y)\right) d\lambda(x) \\
&= 2\int_{[0,1]} \left((f_1(y) - f_2(y)) \int_G k_A(x, y) d\lambda(x)\right) d\lambda(y) =: I
\end{aligned}$$

and, using the fact that, by assumption $k_A(x, y) > 0$ for all $x, y \in [0, 1]$

$$I = 2\int_{\{f_1-f_2>0\}} \Big((f_1(y)-f_2(y)) \underbrace{\int_G k_A(x,y)d\lambda(x)}_{\in(0,1) \text{ for every } y\in[0,1]} \Big) d\lambda(y)$$

$$-2\int_{\{f_1-f_2<0\}} \Big((f_2(y)-f_1(y)) \underbrace{\int_G k_A(x,y)d\lambda(x)}_{\in(0,1) \text{ for every } y\in[0,1]} \Big) d\lambda(y)$$

$$< 2\int_{\{f_1-f_2>0\}} (f_1(y)-f_2(y))d\lambda(y) = \|f_1-f_2\|_1.$$

follows. Setting $f_i := T_{B_i}\mathbf{1}_{[0,y]}$ for $i \in \{1,2\}$ and $y \in (0,1)$, and assuming $B_1 \neq B_2$ this implies (both functions have the same integral so, up to a common scalar, they are densities)

$$\begin{aligned} \Phi_{A*B_1,A*B_2}(y) &:= \int_{[0,1]} \big|K_{A*B_1}(x,[0,y]) - K_{A*B_2}(x,[0,y])\big| d\lambda(x) \\ &= \|T_A f_1 - T_A f_2\|_1 < \|f_1-f_2\|_1 \\ &= \|T_{B_1}\mathbf{1}_{[0,y]} - T_{B_2}\mathbf{1}_{[0,y]}\|_1 =: \Phi_{B_1,B_2}(y). \end{aligned}$$

Hence (see [15]) $D_1(A*B_1, A*B_2) < D_1(B_1,B_2)$ follows. ∎

Theorem 3. *Suppose that A is an absolutely continuous copula, that there exists an index $j \in \mathbb{N}$ such that the density $k_{A^{*j}}$ of A^{*j} is strictly positive, that S is a singular copula, and that $\alpha \in [0,1)$. Then for the copula $B := (1-\alpha)A + \alpha S$ the D_1-limit $\widehat{B}$ of $s_{*n}(B)$ is Π. Furthermore Π is the only idempotent absolutely continuous copula with strictly positive density.*

Proof: It follows from Theorem 2 that $\widehat{B}$ is absolutely continuous. W.l.o.g we may assume that the density $k_{\widehat{B}}$ of $\widehat{B}$ fulfills (12). Using the fact that the density $k_{A^{*j}*\widehat{B}}$ of $A^{*j} * \widehat{B}$ satisfies

$$k_{A^{*j}*\widehat{B}}(x,y) = \int_{[0,1]} k_{\widehat{B}}(x,z)k_{A^{*j}}(z,y)d\lambda(z) > 0$$

for all $x,y \in [0,1]$ and the fact that, according to Theorem 1, $B^{*j} * \widehat{B} = \widehat{B}$ holds, it follows that $k_{\widehat{B}}(x,y) > 0$ for all $x,y \in [0,1]$. If $D_1(\widehat{B},\Pi) > 0$, then applying Lemma 3 yields $D_1(\widehat{B},\Pi) = D_1(\widehat{B}*\widehat{B}, \widehat{B}*\Pi) < D_1(\widehat{B},\Pi)$, so $\widehat{B} = \Pi$. The second assertion of the Theorem easily follows. ∎

In the following we construct the example mentioned in the introduction.

Example 1. According to [2] and [3] (also see [8]) we can find λ-preserving functions $f, g : [0,1] \to [0,1]$ such that (same notation as in [2], [3])

$$\Pi(x,y) = \lambda\big(f^{-1}([0,x]) \cap g^{-1}([0,y])\big) := A_{f,g}(x,y)$$

holds for all $x, y \in [0,1]$. Set $B := \frac{1}{2}(A_{f,id} + A_{id,g})$, then B is as convex combination of two singular copulas singular too (in fact B lives on the graphs of two measurable functions, see [15]). Since $B * B$ is the following convex combination of four copulas

$$B * B = \frac{1}{4}\big(A_{f,id}^{*2} + A_{f,g} + A_{id,g} * A_{f,id} + A_{id,g}^{*2}\big) \tag{13}$$

$B * B$ has an absolutely continuous component with mass at least $\frac{1}{4}$. It is not difficult to verify that $A_{f,g}$ is the only absolute continuous summand in (13), so the absolutely continuous component of $B * B$ has exactly mass $1/4$.

4 Future Work

Some first smoothing properties of the star product have been mentioned. It seems interesting to find more general conditions under which the limit copula $\widehat{B}$ in Theorem 1 is absolutely continuous or even coincides with Π.

References

1. Darsow, W.F., Nguyen, B., Olsen, E.T.: Copulas and Markov processes. Illinois J. Math. 36(4), 600–642 (1992)
2. Darsow, W.F., Olsen, E.T.: Characterization of idempotent 2-copulas. Note Math. 30(1), 147–177 (2010)
3. de Amo, E., Díaz Carrillo, M., Fernández-Sánchez, J.: Measure-preserving functions and the independence copula. Mediterr. J. Math. 8(3), 431–450 (2011)
4. Durante, F., Klement, E.P., Quesada-Molina, J., Sarkoci, J.: Remarks on two product-like constructions for copulas. Kybernetika 43(2), 235–244 (2007)
5. Durante, F., Sempi, C.: Copula theory: an introduction. In: Jaworski, P., Durante, F., Härdle, W., Rychlik, T. (eds.) Copula theory and Its Applications. Lecture Notes in Statistics, vol. 198, pp. 1–31. Springer, Berlin (2010)
6. Kallenberg, O.: Foundations of Modern Probability. Springer, Berlin (1997)
7. Klenke, A.: Probability Theory – A Comprehensive Course. Springer, Berlin (2007)
8. Kolesárová, A., Mesiar, R., Sempi, C.: Measure-preserving transformations, copulae and compatibility. Mediterr. J. Math. 5, 325–339 (2008)
9. Nelsen, R.B.: An Introduction to Copulas. Springer, New York (2006)
10. Olsen, E.T., Darsow, W.F., Nguyen, B.: Copulas and Markov operators. In: Proc. of the Conf. on Distributions with Fixed Marginals and Related Topics. IMS Lecture Notes, Monograph Series, vol. 28, pp. 244–259 (1996)
11. Parry, W.: Topics in Ergodic Theory. Cambridge University Press (1981)
12. Scheffé, H.: A useful convergence theorem for probability distributions. Ann. Math. Statist. 18, 434–438 (1947)
13. Sempi, C.: Conditional expectations and idempotent copulae. In: Cuadras, C.M., et al. (eds.) Distributions with Given Marginals and Statistical Modelling, pp. 223–228. Kluwer, Netherlands (2002)

14. Sempi, C.: Copulae: Some mathematical aspects. Appl. Stoch. Model. Bus 27, 37–50 (2011)
15. Trutschnig, W.: On a strong metric on the space of copulas and its induced dependence measure. J. Math. Anal. Appl. 384, 690–705 (2011)
16. Trutschnig, W.: A note on Cesáro convergence of iterates of the star product and a short proof that idempotent copulas are symmetric (submitted)
17. Trutschnig, W., Fernández Sánchez, J.: Idempotent and multivariate copulas with fractal support (submitted)

Statistical Hypothesis Test for the Difference between Hirsch Indices of Two Pareto-Distributed Random Samples

Marek Gagolewski

Abstract. In this paper we discuss the construction of a new parametric statistical hypothesis test for the equality of probability distributions. The test bases on the difference between Hirsch's h-indices of two equal-length i.i.d. random samples. For the sake of illustration, we analyze its power in case of Pareto-distributed input data. It turns out that the test is very conservative and has wide acceptance regions, which puts in question the appropriateness of the h-index usage in scientific quality control and decision making.

Keywords: Aggregation operators, Hirsch index, hypotheses testing, scientometrics.

1 Introduction

The process of data aggregation [7] consists in a proper synthesis of many numerical values into a single one, representative for the whole input in some sense. It plays a key role in many theoretical and practical domains, such as statistics, decision making, computer science, operational research, and management.

Particularly, in scientific quality control and research policy one often combines citation numbers in order to assess or just rank scientists, institutes, etc. Among the most notable and popular citation indices we have the Hirsch's

Marek Gagolewski
Systems Research Institute, Polish Academy of Sciences,
01-447 Warsaw, Poland
e-mail: `gagolews@ibspan.waw.pl`

Marek Gagolewski
Faculty of Mathematics and Information Science, Warsaw University of Technology, 00-661 Warsaw, Poland

R. Kruse et al. (Eds.): Synergies of Soft Computing and Statistics, AISC 190, pp. 359–367.
springerlink.com © Springer-Verlag Berlin Heidelberg 2013

h-index, which continues to be a subject of intensive and interesting debate since its introduction in 2005. Of course, the usage of the h-index is not solely limited to this particular domain of interest [6].

In this paper we deal with a highly important problem of comparing h-index values of two equal-length inputs and determining whether they differ significantly. We propose and analyze a statistical hypothesis test that may give us more insight into the very nature of the h-index.

2 The *h*-index and Its Distribution

Let us first recall the definition of Hirsch's h-index [8].

Definition 1. Let $n \in \mathbb{N}$. The ***h*-index** is a function $\mathsf{H} : \mathbb{R}_{0+}^n \to \{0, 1, \ldots, n\}$ such that

$$\mathsf{H}(\mathbf{x}) = \begin{cases} \max\{h = 1, \ldots, n : x_{(n-h+1)} \geq h\} & \text{if } x_{(n)} \geq 1, \\ 0 & \text{otherwise,} \end{cases} \tag{1}$$

where $\mathbf{x} = (x_1, \ldots, x_n) \in \mathbb{R}_{0+}^n$ and $x_{(i)}$ denotes the ith order statistic, i.e. the ith smallest value in $\mathbf{x}$.

Interestingly, the h-index is a symmetric maxitive aggregation operator [6]. It is because (1) may be equivalently written as $\mathsf{H}(\mathbf{x}) = \bigvee_{i=1}^n \lfloor x_{(n-i+1)} \rfloor \wedge i$. Therefore, if $\mathbf{x} \in \mathbb{N}_0^n$ then H reduces itself to an ordered weighted maximum (OWMax) operator [2, 7], which in turn is equivalent to Sugeno integral of $\mathbf{x}$ w.r.t. some fuzzy (nonadditive) measure; see [4] for the proof. Please note that basic statistical properties of OWMax operators have already been examined in [5]: it turns out that they are asymptotically normally distributed and they are strongly consistent estimators of a distribution's parameter of location.

The exact distribution of H is given by the following theorem.

Theorem 1. *Let $\mathbf{X} = (X_1, \ldots, X_n)$ be a sequence of i.i.d. random variables with a continuous c.d.f. F defined on $\mathbb{R}_{0+}$. Then the c.d.f. of $\mathsf{H}(\mathbf{X})$ for $x \in [0, n)$ is given by $D_n(x) = \mathcal{I}\left(F\left(\lfloor x+1 \rfloor^{-0}\right); n - \lfloor x \rfloor, \lfloor x \rfloor + 1\right)$, where $\mathcal{I}(p; a, b)$ is the regularized incomplete beta function.*

Proof. For $i = 1, 2, \ldots, n$ the c.d.f. of the ith order statistic, $X_{(i)}$, is given by $F_{(i)}(x) = \Pr(X_{(n)} \leq x) = \mathcal{I}(F(x); i, n-i+1)$ (cf. [1]). Note that $\operatorname{supp} \mathsf{H}(\mathbf{X}) \subseteq \{0, 1, \ldots, n\}$. Hence, $D_n(x) = 1$ for $x \geq n$. By (1) we have:

$$\begin{aligned}
\Pr(\mathsf{H}(\mathbf{X}) < 1) &= \Pr(X_{(n)} < 1) &&= \mathcal{I}(F(1^{-0}); n, 1), \\
\Pr(\mathsf{H}(\mathbf{X}) < 2) &= \Pr(X_{(n-1)} < 2) &&= \mathcal{I}(F(2^{-0}); n-1, 2), \\
&\cdots &&\cdots \\
\Pr(\mathsf{H}(\mathbf{X}) < n) &= \Pr(X_{(1)} < n) &&= \mathcal{I}(F(n^{-0}); 1, n), \quad \text{QED.}
\end{aligned}$$

□

As a consequence, for all $h = 0, \dots, n-1$ it holds $D_n(h) = \Pr(Z \le h)$, where $Z \sim \mathrm{Bin}(n, 1 - F(h + 1^{-0}))$. We see that the values of the c.d.f. and the p.m.f. of the h-index in most cases may only be determined numerically. For convenience, they have been implemented in the `CITAN` package [3] for `R`.

3 Test for the Difference between Two *h*-indices

Given two equal-length vectors of observations, one may be interested whether their Hirsch's indices differ significantly. More formally, let $\Theta = (0, n)$ be a parameter space that induces an identifiable statistical model $(\mathbb{R}_{0+}, \{\Pr_\theta : \theta \in \Theta\})^n$ in which $\mathbb{E}_\theta \mathsf{H} = \theta$ for all $\theta \in \Theta$, and $\Pr_\theta(\mathsf{H} = i)$ is a continuous function of θ for all i. Moreover, let $\mathbf{X} = (X_1, \dots, X_n)$ i.i.d $\Pr_{\theta_x}$ and $\mathbf{Y} = (Y_1, \dots, Y_n)$ i.i.d $\Pr_{\theta_y}$, where $\theta_x, \theta_y \in \Theta$. We would like to construct a statistical test φ which verifies at given significance level α the null hypothesis $H_0 : \theta_x = \theta_y$ against the alternative $H_1 : \theta_x \neq \theta_y$.

The most natural test statistic is of course $T(\mathbf{X}, \mathbf{Y}) = \mathsf{H}(\mathbf{Y}) - \mathsf{H}(\mathbf{X})$. Obviously, under H_0 the distribution of T is symmetric around 0. Unfortunately, it may not be independent of the values of unknown parameters $\theta_x = \theta_y$. We therefore expect that by setting an acceptance region with bounds determined by functions of only α and n (an approach traditionally used in mathematical statistics) we will not obtain test of satisfactory power in result.

Denote by $\mathcal{B}$ the set of all 0–1 symmetric square matrices $B = (b_{ij})$, $i, j \in \{0, \dots, n\}$, such that (i) $b_{ii} = 0$ for all i, and (ii) $b_{ij} = 1 \Longrightarrow b_{i,j+1} = 1$ for $i < j < n$. Each $B \in \mathcal{B}$ generates a statistical hypothesis test

$$\varphi_B(\mathbf{X}, \mathbf{Y}) = b_{\mathsf{H}(\mathbf{X}), \mathsf{H}(\mathbf{Y})}. \tag{2}$$

Such test bases on the test statistic T and has acceptance regions that depend on the value of the h-index in one of the samples. E.g. if we observed $\mathsf{H}(\mathbf{x}) = i$ and $\mathsf{H}(\mathbf{y}) = j$ then $b_{ij} = 1$ would indicate that H_0 should be rejected.

Please note that there is a bijection between $\mathcal{B}$ and the set of integer-valued sequences $\{(v_0, \dots, v_n) : (\forall i)\ 0 \le v_i \le n - i\}$, as we may set $v_i = \sum_{j=i}^{n} b_{ij}$ for $i = 0, \dots, n$ and $b_{ij} = b_{ji} = \mathbf{I}(j - i \le v_i)$ for $0 \le i \le j \le n$. Therefore the acceptance region of T is given by $[-v_{\mathsf{H}(\mathbf{X}) \wedge \mathsf{H}(\mathbf{Y})}; v_{\mathsf{H}(\mathbf{X}) \wedge \mathsf{H}(\mathbf{Y})}]$. Additionally, we have $|\mathcal{B}| = (n + 1)!$

The power function of a test φ_B, reflecting the probability of rejecting H_0 for given $\theta_x, \theta_y \in \Theta$, is given by

$$\pi_B(\theta_x, \theta_y) = \sum_{i=0}^{n} \sum_{j=0}^{n} b_{ij} \Pr_{\theta_x}(\mathsf{H} = i) \Pr_{\theta_y}(\mathsf{H} = j). \tag{3}$$

Let $\mathcal{B}_\alpha = \{B \in \mathcal{B} : \sup_{\theta \in \Theta} \pi_B(\theta, \theta) \le \alpha\}$ denote the set of all matrices which generate tests at significance level α. Our main task may be formulated formally as an optimization problem. We would like to find the matrix $B^* \in \mathcal{B}_\alpha$ which minimizes expected probability of committing Type II error, i.e.

$$B^* := \arg\min_{B\in\mathcal{B}_\alpha} \mathbb{E}\,\mathcal{L}(B) = \arg\min_{B\in\mathcal{B}_\alpha} \iint_{\Theta^2} (1 - \pi_B(\theta_x, \theta_y))\, w(\theta_x, \theta_y)\, d\theta_x\, d\theta_y, \quad (4)$$

where w is a prior distribution. If prior w is uniform (assumed by default when we have no knowledge of or preference for the underlying distribution parameters) then it may be shown that it holds:

$$\mathbb{E}\,\mathcal{L}(B) = \sum_{i=0}^{n}\sum_{j=0}^{n}(1 - b_{ij}) \int_\Theta \Pr_\theta(\mathsf{H} = i)\, d\theta \int_\Theta \Pr_\theta(\mathsf{H} = j)\, d\theta. \quad (5)$$

Note that if the uniformly most powerful (UMP) test (in this class of tests) $\varphi_{B^{**}}$ exists then $\varphi_{B^{**}} = \varphi_{B^*}$ for any w such that $\operatorname{supp} w = \Theta^2$. Unfortunately, as the whole search space is $O(n!)$, in practice we may only seek for an *approximate* solution of (4), B^+, which may be computed in a sensible amount of time.

Let us introduce the following strict partial ordering relation over $\mathcal{B}$. We write $B \prec B'$ iff $B \neq B'$, $(\forall i,j)\ b'_{ij} = 0 \Longrightarrow b_{ij} = 0$, and $b_{ij} = 1 \Longrightarrow b'_{ij} = 1$. Intuitively, if $B \prec B'$ then B' may be obtained from B by substituting some "1"s for "0"s. In such case eq. (3) implies that $(\forall \theta_x, \theta_y \in \Theta)\ \pi_B(\theta_x, \theta_y) \leq \pi_{B'}(\theta_x, \theta_y)$.

For brevity, we will also write $B \prec^1 B'$ iff $B \prec B'$ and $\sum_i \sum_{j\geq i} b_{ij} = \sum_i \sum_{j\geq i} b'_{ij} - 1$. We propose the following algorithm for obtaining an approximation of B^*.

1. Calculate upper bound matrix $B^{(0)}$: For given $i < j$ we set $b_{ij}^{(0)} = 0$ iff $\max_\theta \sum_{k=j}^{n} \Pr_\theta(\mathsf{H} = i) \Pr_\theta(\mathsf{H} = k) > \alpha/2$, as surely is such case rejection of H_0 would lead to violation of given significance level.
2. If $B^{(0)} \in \mathcal{B}_\alpha$ then return $B^* := B^{(0)}$ as result (it is easily seen that $B^{(0)}$ is UMP).
3. Otherwise we generate a sequence $B^{(0)} \succ^1 B^{(1)} \succ^1 \cdots \succ^1 B^{(k)}$ such that $B^{(k-1)} \notin \mathcal{B}_\alpha$ and $B^{(k)} \in \mathcal{B}_\alpha$ by applying:

   for $k := 1, 2, \ldots$ **do**
       if $(\exists B \prec^1 B^{(k-1)} : B \in \mathcal{B}_\alpha)$
           $B^{(k)} := \arg\min_{B\in\mathcal{B}_\alpha, B\prec^1 B^{(k-1)}} \mathbb{E}\,\mathcal{L}(B);$
           proceed to Step #4;
       else
           $B^{(k)} := \arg\min_{B\in\mathcal{B}, B\prec^1 B^{(k-1)}} \int_\Theta \pi_B(\theta,\theta)\, \mathbf{I}(\pi_B(\theta,\theta) > \alpha)\, d\theta;$

4. Improve $B^{(k)}$: Find $B^+ \succeq B^{(k)}$ such that $B^+ \in \mathcal{B}_\alpha$ and $(\forall B \succ B^+)$ $B \notin \mathcal{B}_\alpha$ by applying:

$B^+ := B^{(k)}$;
while $(\exists B \succ^1 B^+ : B \in \mathcal{B}_\alpha)$ **do**
$\quad B^+ := \operatorname*{arg\,min}_{B \in \mathcal{B}_\alpha, B \succ^1 B^+} \mathbb{E}\,\mathcal{L}(B)$;
return B^+ **as result**;

This procedure successively substitutes "1"s for "0"s in the initial upper bound matrix $B^{(0)}$ at positions which result in the greatest overall reduction of "oversized" power, down to the desired value α. This greedy approach — although quite fast to compute (we approximate the integrals by probing the power function at sufficiently many points in Θ) — does not of course guarantee convergence to optimal solution. However, the numerical results presented in the next section suggest that, at least in the considered cases, the solutions are close to optimal in terms of loss. The problem of finding accurate approximation of $\mathbb{E}\,\mathcal{L}(B^*)$ is left for further research.

4 Numerical Results

We say that a random variable X follows a Pareto distribution with shape parameter $k > 0$, denoted $X \sim \mathrm{Par}(k)$, if its cumulative distribution function is given by $F(x) = 1 - 1/(1+x)^k$ for $x \geq 0$. Although F is continuous, it is quite often used by bibliometricians to model citation distribution (or different non integer-valued paper quality metrics). Note we have $\mathsf{H}(\mathbf{X}) = \mathsf{H}(\lfloor \mathbf{X} \rfloor)$. For any n, we apply a reparametrization of the shape parameter and set $\theta_n(k) := \mathbb{E}_k \mathsf{H}(X_1, \ldots, X_n)$ (it is a decreasing bijection). It may be shown that in result we obtain a statistical model that fulfills the assumptions stated in Sec. 3.

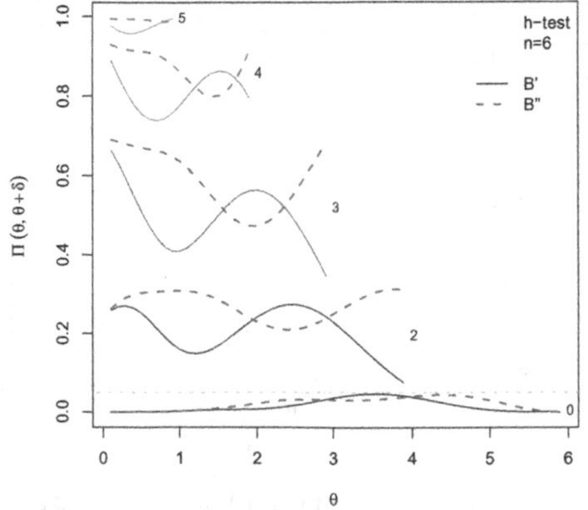

i	0 1 2 3 4 5 6
v'_i	2 **3** 2 **3** 2 1 0
i	0 1 2 3 4 5 6
v''_i	2 2 **3** 2 2 1 0
i	0 1 2 3 4 5 6
$v^{(0)}_i$	2 2 2 2 2 1 0

Fig. 1 Power functions of two optimal h-tests $\varphi_{B'}, \varphi_{B''}$ for $n = 6$ and $\alpha = 0.05$; shift value δ is printed on the right of each curve

Table 1 Computed acceptance region bounds; $n = 25$, $\alpha = 0.05$

i	0	1	2	3	4	5	6	7	8	9	10	11	12	13	14	15	16	17	18	19	20	21	22	23	24	25
v_i^+	1	2	2	3	3	4	5	4	5	6	5	5	6	5	6	5	5	5	5	4	4	3	3	2	1	0

Table 2 Computed acceptance region bounds; $n = 50$, $\alpha = 0.05$. Values improved in Step #4 of the algorithm are marked in bold.

i	0	1	2	3	4	5	6	7	8	9	10	11	12	13	14	15	16	17	18	19	20	21	22	23	24	25
v_i^+	1	2	2	3	3	4	4	**4**	5	5	6	6	6	**6**	7	7	7	7	**7**	8	8	8	**7**	8	8	8

i	26	27	28	29	30	31	32	33	34	35	36	37	38	39	40	41	42	43	44	45	46	47	48	49	50
v_i^+	9	8	8	8	8	8	7	8	7	8	7	**6**	7	7	6	6	5	5	5	4	3	3	2	1	0

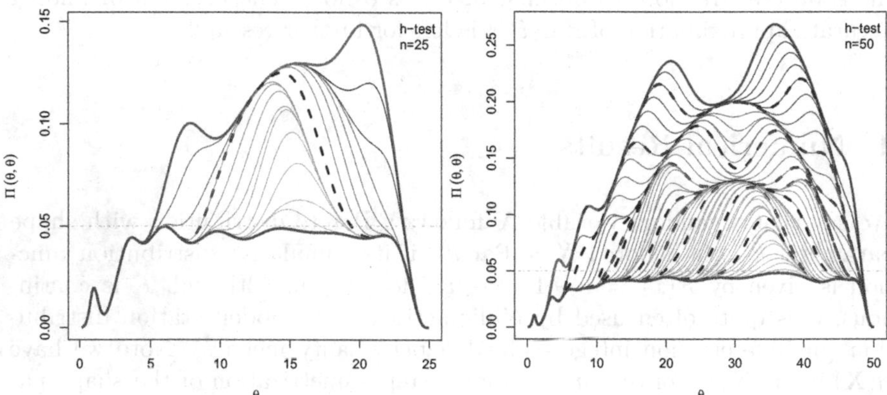

Fig. 2 Probabilities of committing Type I error at consecutive iterations of Step #3 of the algorithm; every 10th curve is dashed and marked in bold

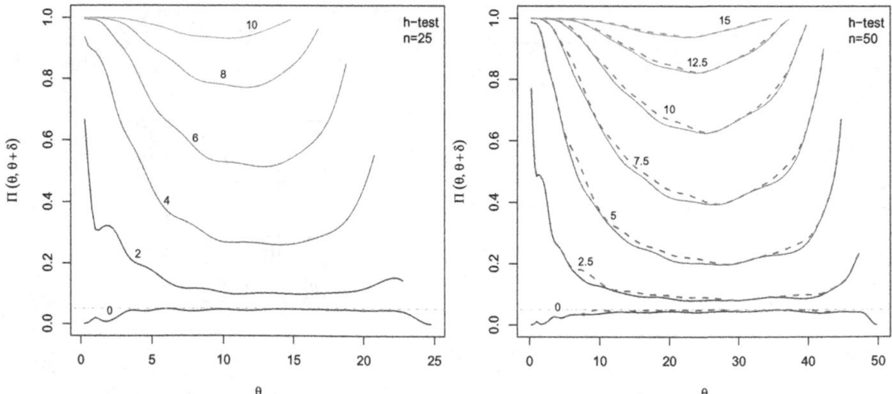

Fig. 3 Power functions of $\varphi_{B^{(20)}} = \varphi_{B^+}$ ($n = 25$, left plot), and $\varphi_{B^{(110)}} \prec \varphi_{B^+}$ ($n = 50$, right plot); shift value δ is printed above each curve

From now on, let us fix $\alpha = 0.05$. For $n \leq 5$ it holds $B^{(0)} \in \mathcal{B}_\alpha$, therefore $\varphi_{B^{(0)}}$ is uniformly most powerful in this class of tests ($\mathbb{E}\,\mathcal{L}(B^{(0)}) = 0.677$). However, e.g. for $n = 6$ we have $B^{(0)} \notin \mathcal{B}_\alpha$. In this case there are two maximal tests $\varphi_{B'}$ and $\varphi_{B''}$ (the latter is outputted by the above algorithm) in the sense that it holds $\neg(B' \prec B'')$, $\neg(B'' \prec B')$, $(B \succ B') \vee (B \succ B'') \Rightarrow B \notin \mathcal{B}_\alpha$, and $B \in \mathcal{B}_\alpha \Rightarrow (B \preceq B') \vee (B \preceq B'')$. As a consequence, the UMP test in this class does not exist (cf. Fig. 1). We have $\mathbb{E}\,\mathcal{L}(B') = 0.691$ and $\mathbb{E}\,\mathcal{L}(B'') = 0.647$. Obviously, if we assume no prior knowledge of θ then $\varphi_{B''}$ is the preferred choice for practical purposes.

We will study more deeply the two following cases. For $n = 25$ we get $k = 20$ and $\mathbb{E}\,\mathcal{L}(B^{(k)}) = \mathbb{E}\,\mathcal{L}(B^{+}) = 0.336$ (see Tab. 1 for the resulting acceptance region bounds), On the other hand, for $n = 50$ we have $k = 110$, $\mathbb{E}\,\mathcal{L}(B^{(k)}) = 0.251$, and $\mathbb{E}\,\mathcal{L}(B^{+}) = 0.247$ (cf. Tab. 2). Fig. 2 shows the plots of $\pi_{B^{(i)}}(\theta, \theta)$ for $i = 0, \ldots, k$ (cf. Step #3 of the algorithm). Additionally, in Fig. 3 we depict the plot of $\pi_{B^{(i)}}(\theta, \theta + \delta)$ and $\pi_{B^{(+)}}(\theta, \theta + \delta)$ for different values of δ. We see that the improvement of $B^{(110)}$ for $n = 50$ does not result in a drastic decrease of expected loss.

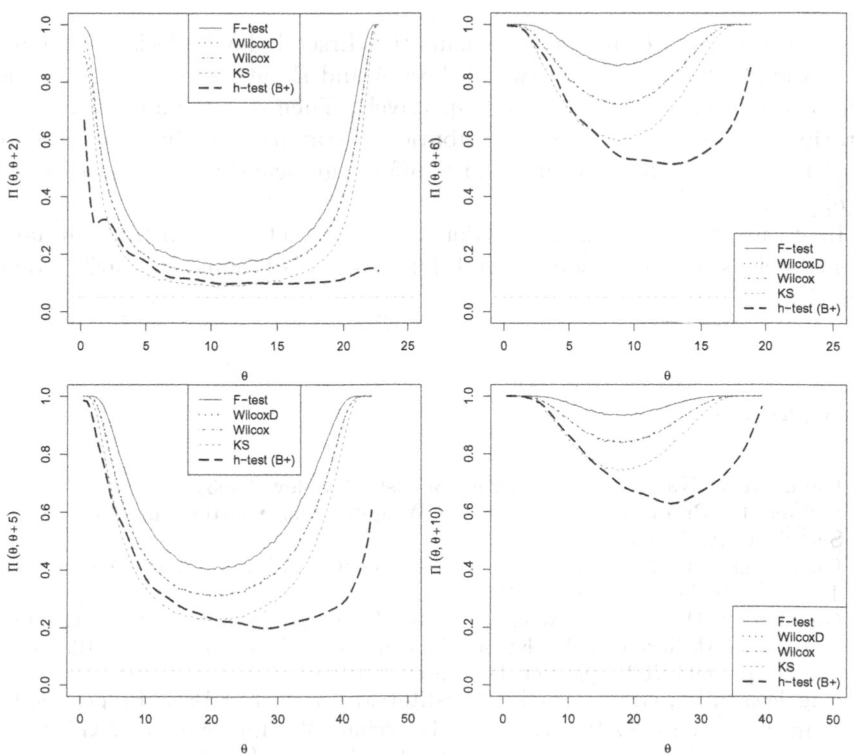

Fig. 4 Power functions of different two-sample tests; $n = 25$ (above) and $n = 50$ (below), $\alpha = 0.05$, $M = 25000$ MC iterations

Let us compare the power of the computed h-tests with some other tests for equality of distribution parameters. The parametric F-test bases on a test statistic $T(\mathbf{X}, \mathbf{Y}) = \sum_{i=1}^{n} \log(1+X_i)/\sum_{i=1}^{n} \log(1+Y_i)$ which, under H_0, has Snedecor's F distribution with $(2n, 2n)$ degrees of freedom. We also consider 3 non-parametric tools: the Wilcoxon rank sum test, the discretized Wilcoxon test (computed on $\lfloor\mathbf{X}\rfloor$ and $\lfloor\mathbf{Y}\rfloor$), and the Kolmogorov-Smirnov test.

The plots of the examined tests' estimated power functions values, generated using $M = 25000$ Monte Carlo samples, are depicted in Fig. 4. The constructed h-tests are outperformed by the F-test and Wilcoxon's test, and often by the KS test. We also observe that their power is quite small for $\theta \simeq n$, which is due to the property $\mathsf{H}(\mathbf{X} \wedge n) = \mathsf{H}(\mathbf{X})$: here the h-index "ignores" some important information. What is more, we see that in the considered cases discretization of observations did not result in a significant reduction of power of the Wilcoxon test.

5 Conclusion

We should be very cautious while using the Hirsch index in decision making. For example, let us consider two authors A and B with 25 papers each, and whose h-indices are 12 and 16, respectively. Then — assuming that their citation counts follow Pareto distributions — at 0.05 significance level we cannot state that their output quality differs significantly, as $T = (16-12) \in [-v_{12}^{+}; v_{12}^{+}] = [-6; 6]$.

In future work we will consider the construction of h-tests in non-identifiable statistical models, and for samples of non necessarily equal lengths.

References

1. David, H.A., Nagaraja, H.N.: Order Statistics. Wiley (2003)
2. Dubois, D., Prade, H., Testemale, C.: Weighted fuzzy pattern matching. Fuzzy Set. Syst. 28, 313–331 (1988)
3. Gagolewski, M.: Bibliometric impact assessment with R and the CITAN package. J. Informetr. 5(4), 678–692 (2011)
4. Gagolewski, M., Grzegorzewski, P.: Arity-Monotonic Extended Aggregation Operators. In: Hüllermeier, E., Kruse, R., Hoffmann, F. (eds.) IPMU 2010. CCIS, vol. 80, pp. 693–702. Springer, Heidelberg (2010a)
5. Gagolewski, M., Grzegorzewski, P.: S-Statistics and Their Basic Properties. In: Borgelt, C., González-Rodríguez, G., Trutschnig, W., Lubiano, M.A., Gil, M.Á., Grzegorzewski, P., Hryniewicz, O. (eds.) Combining Soft Computing and Statistical Methods in Data Analysis. AISC, vol. 77, pp. 281–288. Springer, Heidelberg (2010b)

6. Gagolewski, M., Grzegorzewski, P.: Axiomatic characterizations of (quasi-) L-statistics and S-statistics and the producer assessment problem. In: Galichet, S., Montero, J., Mauris, G. (eds.) Proc. Eusflat/LFA 2011, pp. 53–58 (2011)
7. Grabisch, M., Pap, E., Marichal, J.L., Mesiar, R.: Aggregation Functions. Cambridge Univeristy Press (2009)
8. Hirsch, J.E.: An index to quantify individual's scientific research output. P. Natl. Acad. Sci. USA 102(46), 16569–16572 (2005)

6. Gagolewski, M., Grzegorzewski, P.: Axiomatic characterizations of (quasi-) L-statistics and S-statistics and the producer assessment problem. In: Galichet, S., Montero, J., Mauris, G. (eds.) Proc. Eusflat/LFA 2011, pp. 53–58 (2011)
7. Grabisch, M., Pap, E., Marichal, J.L., Mesiar, R.: Aggregation functions. Cambridge University Press (2009)
8. Hirsch, J.E.: An index to quantify individual's scientific research output. Proc. Natl. Acad. Sci. USA 102(46), 16569–16572 (2005)

How to Introduce Fuzzy Rationality Measures and Fuzzy Revealed Preferences into a Discrete Choice Model

Davide Martinetti, Antonio Lucadamo, and Susana Montes

Abstract. The work presents a novel approach to discrete choice models. Random utility models are used to describe the utility that every individual associates to each alternative and are usually decomposed into a deterministic part and a stochastic component. Our proposal is to include in the random utility model the observations of past choices through revealed fuzzy preferences and to modify the effects of the stochastic component, by adding some measure of rationality.

Keywords: Choice function, discrete choice model, rationality measure, revealed preference.

1 Introduction

Discrete choice models are statistical tools used to model and forecast the behavior of a decision maker over a discrete bundle of alternatives. They have been used to examine the choice of mode to work [3] or the choice of the residential location [9], among numerous other applications. These models relate the observable choices made by the decision-maker to the attributes of the alternatives and of the decision-maker itself and estimate the probability that an alternative will be chosen. The usual way of representing the tastes

Davide Martinetti · Susana Montes
Departement of Statistics and Operative Research,
University of Oviedo, Oviedo, Spain
e-mail: {martinettidavide.uo,montes}@uniovi.es

Antonio Lucadamo
Department of Economic and Social System Analysis,
Universita degli Studi del Sannio, Benevento, Italy
e-mail: antonio.lucadamo@unisannio.it

R. Kruse et al. (Eds.): Synergies of Soft Computing and Statistics, AISC 190, pp. 369–377.
springerlink.com © Springer-Verlag Berlin Heidelberg 2013

over the alternatives is to assume that the decision-maker is able to assign to any alternative in the choice set an utility and then the choice is performed by just taking the alternative with the highest utility. This utility is often represented by using random utility models, called in this way due to the fact that they are composed of two components: a deterministic part and a stochastic component. The deterministic part is a function of all observable variables, while all discrepancies of the predicted model from the observed utility are explained with a random variable. Our proposal is to modify both the deterministic and the stochastic part. For the first intervention, we will make use of revealed preference theory, in its fuzzy version, recently proposed in [6]. The idea is that the preferences of the decision-maker can be inferred by observing his past choices over a family of subsets of the bundle of alternatives. The result is a fuzzy preference relation (a multi-valued version of the classic preference), that contains valuable information about the preferences of the decision-maker over every pair of alternatives. Furthermore, for the intervention on the stochastic part of the random utility model, we will refer to some works on rationality of fuzzy preference relations [5, 6, 7]. In fact, one of the basic assumption of random utility models is that the decision-maker is perfectly rational (he always picks up the alternative with the highest utility) and that his behavior is purely probabilistic, while recent studies on real stated preferences (see [8]) have proved that these assumptions are sometimes too restrictive, or, at least, too faraway from reality. Our proposal is to include in the random utility model a further component, aside to the stochastic one, that represent the measured rationality of the decision-maker. In this way, the inconsistencies in the behavior of the decision-maker are explained by both a stochastic and a psychometric component. The work is divided into six sections: after this brief introduction, we will introduce fuzzy revealed preference theory (Section 2), fuzzy rationality measures (Section 3) and random utility models applied to discrete choice models (Section 4). Sections 5 and 6 contain our proposals and future prospects.

2 Fuzzy Revealed Preference Theory

Revealed preference theory, in its original version by Sen, Arrow and Suzumura [1, 12, 13], has been firstly proposed in order to lay bare the connections between two different ways of representing preferences of decision-makers: choice functions and binary preference relations. Both formalizations contains information about the choices and preferences of individuals, but they express it in different ways: preference relations are represented through $n \times n$ matrices and express the binary preference between any pair of alternatives, while choice functions are operators that assign to any non-empty subsets of the choice space another set, called choice set, containing the chosen alternatives. Nevertheless, since both formalism are used to represent choice, some connection between them is expected. Revealed preference has then been defined as the binary preference relation that correspond to a given choice function. Rationality is doubtless the most intriguing property: in fact it is relatively easy to determine whenever a binary preference relation is rational (studying properties such as completeness, reflexivity, transitivity and acyclicity), while is not trivial to define an equivalent concept for choice functions. On the other hand it has been proved in [11] that when the uncertainty is represented by a convex set of probabilities (instead of a single probability distribution), then coherent choice functions does not reduce to binary comparison. A milestone in this theory is the Arrow-Sen theorem, that establishes a set of properties of the choice function that are equivalent to the rationality of the revealed preference relation. Recent works by Banerejee and Georgescu [2, 6] have renewed the research on the field by using a fuzzy approach. Their proposal is substantially based on a fuzzification of the concept of choice function, revealed preference and the related properties, like transitivity or acyclicity. Despite the fact that fuzzy preference relation are commonly used and widely accepted, fuzzy choice functions are a relatively new and unexplored concept. They first appeared in the literature in [2, 10]. Banerjee claims that fuzzy choice functions are potentially observable, as long as the decision maker is able to tell to an interviewer the degree of his inclination for one alternative (or a set of them) when faced to a choice problem. Hence, while there may be problems of estimation, fuzzy choice functions are, in theory, observable. In this contribution we only need some basic definitions, while the interested reader can find an appropriate discussion on the subject in [6]. Let $X = \{x_1, \ldots, x_n\}$ be a finite set of alternatives and $\mathcal{B} \subseteq 2^X$ a family of non-empty fuzzy subsets of X. In [6], the elements of $\mathcal{B}$ are usually denoted by capital letters like S or T and are called available sets. The value $S(x_i)$ can be considered as the degree of availability of the alternative x_i with respect to the set S. A fuzzy choice function is a function $C : \mathcal{B} \times X \to [0, 1]$, such that, for every $S \in \mathcal{B}$ there exists at least one alternative $x_j \in X$ such that $C(S)(x_j) > 0$ and for every $x_i \in X$, it hold $C(S)(x_i) \leq S(x_i)$. The value of $C(S)(x_i)$ can be interpreted as the extent to which x_i belongs to the set of chosen alternatives, when the set of available ones is S. Given a fuzzy choice function C, a fuzzy binary preference relation R can be revealed from it, using

$$R(x_i, x_j) = \sup_{S \in \mathcal{B}}(T(C(S)(x_i), S(x_j))) \qquad \forall x_i, x_j \in X \tag{1}$$

where T denotes a left-continuous t-norm. A fuzzy binary preference relation R is considered rational when it shows one of the following properties:

T-transitivity if $T(R(x_i, x_j), R(x_j, x_k)) \leq R(x_i, x_k)$, for every $x_i, x_j, x_k \in X$;

Acyclicity for all $m \geq 2$ and $(x_1, \ldots, x_m) \in X$ such that $R(x_i, x_{i+1}) > R(x_{i+1}, x_i)$, for all $i \in \{1, \ldots, m-1\}$, it holds $R(x_1, x_n) \geq R(x_n, x_1)$.

For the present work, the theory just described is useful since it allows to reconstruct preference-like information starting from choice-like information, that is usually easier and cheaper to gather. Once fuzzy revealed preference is constructed, measurements over the rationality of the decision-maker can be performed, as we will see in the next Section 3.

3 Fuzzy Rationality Measures

According to [5], when we deal with fuzzy preference relations, also notions like acyclicity, transitivity and rationality in general, should be treated as fuzzy concepts. The possibility of assigning a degree of rationality is really attractive since it allows to differentiate between decision-makers that are often irrational and other that are mainly coherent in their behavior, gaining much more insight compared to the case of a crisp differentiation. For a matter of space we will not introduce here the formalism of these fuzzy rationality measures. The interested reader can enjoy the reading of [5], in which the properties of such measures are defined and can find many examples of fuzzy rationality measures. Also Georgescu in [6, 7] proposed new measures of rationality, in the sake of clarifying the connections between the rationality concept used for fuzzy preference relations and the one that applies to fuzzy choice functions. We just recall here the main characteristics of a fuzzy rationality measure $\rho : \mathcal{P}(X) \to [0, 1]$, where $\mathcal{P}(X)$ indicates the space of all fuzzy preference relations over a fixed and finite set of alternatives X:

- $\rho(R) = 1$, for any crisp linear order. It indicates the maximum degree of rationality possible;
- $\rho(R) = 0$ indicates an absolutely incoherent decision-maker, while $\rho(R) \in (0, 1)$ indicates a decision-maker that has an intermediate degree of rationality;
- moreover, ρ is invariant over permutations of the labels of alternatives in X, it assigns the same value to opposite preference relations, it does not allow to increase the degree of rationality by just adding more alternatives and it behave well when the preference relation changes only w.r.t. a pair of alternatives x, y, while the others remain unchanged.

4 Random Utility Models Applied to Discrete Choice

Discrete choice models are statistical tools used to predict the choices of a decision-maker over a bundle of alternatives and that assign to every alternative a probability of being chosen. Since the choice of the decision-maker depends on many factors, some of which are observable and others that do not, the analyst needs an instrument that can describe the effects of both observable and unobservable factors. Hence, utility models are introduced: the idea is that the decision-maker assigns to every alternative x_i an utility U_i according to the features of x_i and to his own feelings and then he choses the one with higher utility. To deal with unobservable factors, the utility is designed as a random variable $U_i = V_i + \epsilon_i$, where U_i is the real value of the utility of alternative x_i, V_i is the deterministic (or representative, or systematic) part computed by the analyst taking into account all observable factors and ϵ_i is the disturbance or random component. Then, the probabilities for every alternatives are computed by

$$\begin{aligned} P(x_i \text{ is chosen}) &= P(U_i \geq U_j, \forall j \neq i) \\ &= P(V_i + \epsilon_i \geq V_j + \epsilon_j, \forall j \neq i) \\ &= P(V_i - V_j \geq \epsilon_j - \epsilon_i, \forall j \neq i) \,. \end{aligned} \tag{2}$$

The analyst has two main tasks: first of all, the specification of the deterministic part, i.e. he needs to create a function V_i that faithfully connects the observable variables with the real utility and he has to choose which variables to take into account. Usually, V_i is represented like $V_i(y_i, y_{DM}, \beta)$, where y_i represents a vector of measurable features of the alternative x_i, y_{DM} represents a vector of measurable features of the same decision-maker and β is a set of parameter that needs to be estimated. The functional form of V can go from a simple linear model to much more complex specification. The second task of the analyst is the specification of the random component, i.e. he has to establish the distribution of the component ϵ_i in order to capture the effect of all unobservable factors, like unobserved attributes of alternatives and decision-maker, taste variations, measurement errors, instrumental variables, etc. Choosing the distribution of ϵ_i usually face the analyst with the problem of selecting between an accurate fitting or an easy-to-compute distribution. Well known discrete choice models are logit, probit, nested logit and mixed logit, among many others (see [3, 14]).

5 A Novel Approach

According to the separation in deterministic and stochastic part just described, the main goal of the analyst will be to capture as much utility as possible into the deterministic part, letting the random component quite similar to a white noise. Our proposal goes in this direction: we first suggest an intervention into the deterministic component in order to take into account the information that can be gathered through a revealed preference study. On the other hand we propose to change the way in which uncertainty is treated, by introducing a component that measures the rationality of the decision-maker, aside the stochastic component already mentioned.

5.1 Intervention in the Deterministic Part

Let start by an example: consider one client of a supermarket that always choses the same brand of cereals, no matter its price, the possible offer on other brands, etc. This kind of information can only be captured by a preference relation study, since the utility model alone will vary the utility of that brand of cereals according to modifications of its price and other observable variables. In order to capture similar situations with a discrete choice model, it seems reasonable to introduce a new variable into the deterministic component that represents the observed preferences of the decision-maker over the set of alternatives, but how to do that? One way should be to ask directly to the decision-maker to express his preferences for every pair of alternatives in the set X, but this approach has evident limitations, such as the cardinality of X or the cost of such a survey in a market with a lot of decision-makers. Another possible approach is to reveal the preferences of the decision-maker from his past choices. Imagine the case of a supermarket that keeps track of the purchases of its clients or a web-page that tracks data of its visitors. With this kind of data, choices are easy to observe and much cheaper compared to a direct survey. Then, a fuzzy choice function C can be constructed starting from this kind of data, where the available set can be defined *a priori* by an expert or constructed with a clustering algorithm. The choice of the decision-maker over the different bundles of alternatives can be counted and the relative frequencies can be considered as the degree of choice of the decision-maker of any alternative within a bundle of alternatives. Once the choice function has been computed, a fuzzy preference relation R can be revealed from it using Eq. 1, one for every decision maker. To introduce this information into the deterministic component of the random utility model, we propose to modify the function V_i described in Section 4, in order to include a new variable y_{RP}, that indicates a value or a vector of values that supply information on the revealed preference. In the following we list some proposals: $V_i(y_i, y_{DM}, y_{RP}, \beta)$ where y_{RP} can be

- the value of $R(x_i, x_j)$, for any $x_j \neq x_i$, if Eq. 2 is used to compute the probability of $U_i \geq U_j$.
- the degree of dominance of alternative x_i in X according to revealed preference R, $D_X^R(x_i)$ (for the details on the computation of D_X^R, see Chapter 8.2 of [6]);

5.2 Intervention in the Random Component

Quoting Ben-Akiva [3] is probably the best way to introduce our proposal:"*[...]In choice experiments individuals have been observed not to select the same alternative in repetitions of the same choice situations. [...] A probabilistic choice mechanism was introduced to explain these behavioral inconsistencies. One can argue that human behavior is inherently probabilistic. It can also be argued, however, that behavior which is probabilistic amounts to an analyst's admission of a lack of knowledge about individuals' decision process. If it were possible to specify the causes of these inconsistencies, the deterministic choice theory (opposite to random choice theory, N/A) could be used. These causes, however, are usually unknown, or known but not measurable.*" Focusing on the last sentences, we glimpse that the choice of describing incoherent behavior with a probabilistic approach is just one possible solution, likely the easiest, but for sure not the only one, neither the most correct. In this sense we make our second proposal: in fact, we think that inconsistencies in human behavior can also be explained by irrationality of the decision-maker, as it has been proved, e.g., by García-Lapresta in [8]. Our idea is that the difference between expected utility and observed utility (i.e. ϵ_i) can be explained by both a rationality component and a stochastic one: $\epsilon_i = f(r_{DM}, s_i)$, $\forall i$, where r_{DM} is an indicator of the degree of rationality of the decision-maker, s_i is the stochastic component, as before, while f is a function that properly combines the two components. The value of r_{DM} can be chosen among the fuzzy rationality measures detailed in Section 3. Which choice can we make for the function f? Our intuition is that an high level of rationality shifts all the inconsistency into the stochastic part, i.e. if a person is strongly coherent, we expect that his deviation from the predicted utility is mainly due to unobservable and/or unpredictable factors, that are well-modeled by the stochastic component. On the other hand, for a strong incoherent decision-maker, the deviation from the predicted utility should be explained by both a stochastic component and another factor that accounts for that irrationality and in general, his behavior is expected to be more unpredictable. For the function f we propose the following properties:

- $f(1, s_i) = s_i$;
- $f(r_{DM}, s_i) \geq f(r'_{DM}, s_i)$, whenever $r_{DM} \leq r'_{DM}$.

5.3 Justification

To justify our intuition that rationality plays an important role in decision problems, we considered a relatively small data-set, called `Catsup`, that contains the purchases of four brands of ketchup. The sales are recorded in the database registering the price of the four available products, the chosen one for that sale and an identifier for every client. In this way the data set can be divided with respect to client's ID and their past choices can be observed. Starting from these choices, we constructed a fuzzy preference relation for every client and we computed a rationality index for all of them. Three discrete choice models have been computed using BIOGEME [4]: the first one uses the complete data-set, the second one uses only *rational* clients and the last one the remaining *non-rational* clients. The prices of the four brands are considered explicative variable. The models have been trained over the 90% of the available data and tested on the remaining 10%. The results on the correctness of the predictions are the following: full data-set, 55,0%; only no-rational clients, 57,7%; only rational clients 66,6%. The results are strongly encouraging, since they lay bare the importance that rationality assumes in choice problems.

6 Future Work

As stated in Section 1, all the proposed methods haven't been proved yet, neither in theoretical nor in practical sense. Hence, in the immediate future, we are planning to test the efficacy of the proposals explained earlier, also by using different databases in order to prove their reliability. Among the problems that we will inevitably have to face we mention:

- Construction of the choice space: revealed preference theory and discrete choice models use different, but though similar, choice spaces, so a common starting point need to be formalized;
- we have different possible interventions to apply to an already existing model: this implies a choice among several proposals, that need to be tested and justified;
- Finding several appropriate data sets, in order to test the novel approach.

Acknowledgements. The research reported in this paper has been partially supported by project MTM2010-17844 and the Foundation for the promotion in Asturias of the scientific and technologic research BP10-090.

References

1. Arrow, K.J.: Rational choice functions and orderings. Economica 26, 121–127 (1959)
2. Banerjee, A.: Fuzzy preferences and arrow-type problems in social choice. Social Choice and Welfare 11, 121–130 (1994)
3. Ben-Akiva, M., Lerman, S.R.: Discrete Choice Analysis: Theory and Application to Travel Demand. MIT Press, Cambridge (1986)
4. Bierlaire, M.: BIOGEME: A free package for the estimation of discrete choice models. In: Proc. of the 3rd Swiss Transportation Research Conference, Ascona, Switzerland (2003)
5. Cutello, V., Montero, J.: Fuzzy Rationality Measures. Fuzzy Set. Syst. 62, 39–54 (1994)
6. Georgescu, I.: Fuzzy Choice Functions, a Revealed Preference Approach. Springer, Berlin (2007)
7. Georgescu, I.: Acyclic rationality indicators of fuzzy choice functions. Fuzzy Set. Syst. 160, 2673–2685 (2009)

8. García-Lapresta, J.L., Meneses, L.C.: Individual-valued preferences and their aggregation: consistency analysis in a real case. Fuzzy Set. Syst. 151, 269–284 (2005)
9. McFadden, D.: Modeling the Choice of Residential Location. In: Karlqvist, et al. (eds.) Spatial Interaction Theory and Residential Location, Amsterdam, Netherlands, pp. 75–96 (1978)
10. Orlovsky, S.A.: Decision-making with fuzzy preference relation. Fuzzy Set. Syst. 1, 155–167 (1978)
11. Seidenfeld, T., Schervish, M.J., Kadane, J.B.: Coherent choice functions under uncertainty. In: 5th Int. Symp. Imprecise Probability: Theories and Applications, Prague, Czech Republic (2007)
12. Sen, A.: Choice function and revealed preference. Rev. Econ. Stud. 38, 307–317 (1971)
13. Suzumura, K.: Rational choice and revealed preference. Rev. Econ. Stud. 43, 149–159 (1976)
14. Train, K.E.: Discrete Choice Methods with Simulation. Cambridge University Press, Cambridge (2003)

8. García-Lapresta, J.L., Meneses, L.C.: Individual-valued preferences and their aggregation: consistency analysis in a real case. Fuzzy Sets Syst. 151, 269–284 (2005)
9. McFadden, D.: Modeling the Choice of Residential Location. In: Karlqvist, A., et al. (eds.) Spatial Interaction Theory and Residential Location, Amsterdam, Netherlands, pp. 75–96 (1978)
10. Orlovsky, S.A.: Decision-making with fuzzy preference relation. Fuzzy Sets Syst. 1, 155–167 (1978)
11. Seidenfeld, T., Schervish, M.J., Kadane, J.B.: Coherent choice functions under uncertainty. In: 5th International Symposium on Imprecise Probability: Theories and Applications, Prague, Czech Republic (2007)
12. Sen, A.: Choice functions and revealed preference. Rev. Econ. Stud. 38, 307–317 (1971)
13. Suzumura, K.: Rational choice and revealed preference. Rev. Econ. Stud. 43, 149–158 (1976)
14. Train, K.E.: Discrete Choice Methods with Simulation. Cambridge University Press, Cambridge (2003)

Fuzzy Rule-Based Ensemble Forecasting: Introductory Study

David Sikora, Martin Štěpnička, and Lenka Vavříčková

Abstract. There is no individual forecasting method that is generally for any given time series better than any other method. Thus, no matter the efficiency of a chosen method, there always exists a danger that for a given time series the chosen method is inappropriate. To overcome such a problem and avoid the above mentioned danger, distinct ensemble techniques that combine more individual forecasting methods are designed. These techniques basically construct a forecast as a linear combination of forecasts by individual methods. In this contribution, we construct a novel ensemble technique that determines the weights based on time series features. The protocol that carries a knowledge how to combine the individual forecasts is a fuzzy rule base (linguistic description). An exhaustive experimental justification is provided. The suggested ensemble approach based on fuzzy rules demonstrates both, lower forecasting error and higher robustness.

Keywords: Ensembles, fuzzy rules, time series.

1 Introduction

In time series forecasting we are given a finite sequence $y_1, y_2, \ldots, y_t$ of reals that is called a time series and our task is to determine future values $y_{t+1}, y_{t+2}, \ldots, y_{t+h}$, where h is a *forecasting horizon.*

David Sikora
Department of Informatics and Computers, Faculty of Science,
University of Ostrava, Ostrava, Czech Republic
e-mail: david.sikora@osu.cz

Martin Štěpnička · Lenka Vavříčková
Centre of Excellence, IT4Innovations, Institute for
Research and Applications of Fuzzy Modeling, Ostrava, Czech Republic
e-mail: {martin.stepnicka,lenka.vavrickova}@osu.cz

R. Kruse et al. (Eds.): Synergies of Soft Computing and Statistics, AISC 190, pp. 379–387.
springerlink.com © Springer-Verlag Berlin Heidelberg 2013

Distinct mainly statistical time series forecasting methods have been designed and are nowadays widely used [24]. However, it is a known fact there is no method that would be superior to any other. Thus, relying on a single method is a highly risky strategy which may lead to a choice of an inappropriate method for a given series. We stress that even searching for methods that outperform any other for narrower specific subsets of time series have not been successful yet, see [5]:

"Although forecasting expertise can be found in the literature, these sources often fail to adequately describe conditions under which a method is expected to be successful".

Based on the above observations, so called *ensemble techniques* have started to be designed and successfully applied. The main idea of *"ensembles"* consists in an appropriate combination of more forecasting methods in order to avoid the risk of choosing a single inappropriate one. Typically, ensemble techniques are constructed as a linear combination of the individual ones. So, if we are given a set of M individual methods and j-th individual method prediction is denoted by

$$\hat{y}^{(j)}_{t+1}, \hat{y}^{(j)}_{t+2}, \ldots, \hat{y}^{(j)}_{t+h}, \quad j = 1, \ldots, M$$

the ensemble forecast is given as follows

$$\hat{y}_{t+i} = \frac{1}{\sum_{j=1}^{M} w_j} \cdot \sum_{j=1}^{M} w_j \cdot \hat{y}^{(j)}_{t+i}, \quad i = 1, \ldots, h \tag{1}$$

where $w_j \in \mathbb{R}$ is a weight of the j-th individual method. Usually, the weights are normalized (their sum equals to unity).

Let us recall works that firstly showed gains in accuracy through ensemble techniques [7] and also lower error variance then of individual forecasts [25] which may be interpreted as a higher robustness (decreasing the danger of inaccurate forecasts). For other sources, we refer to [6]. Nevertheless, how to determine appropriate weights in (1) is still to large extent an open question. One would expect sophisticated approaches to dominate but taking a simple average – the so called "equal weights approach" – usually outperforms taking a weighted average [10]. In other words, equal weights that is an arithmetic mean, is a benchmark that is hard to beat[1] and finding appropriate non-equal weights rather leads to a random damage of the main averaging idea that is behind the robustness and accuracy improvements [11].

[1] This does not mean that equal weights or ensembles generally provide better results than any individual methods for any given time series. Vice-versa, usually, there are, say, one or two individual methods that outperform a given ensemble. However, so far there is no way how to determine which methods are these desirable ones in advance. Thus, one often chooses a wrong one. This is mirrored on results measured on sets of time series where the ensemble approach outperform the individual ones.

Although the equal weights performs as accurately as mentioned above, there are works that promisingly show the potential of more sophisticated approaches. We recall [21] that is based on distinct features of time series such as: measure of the strength of trend, measure of the strength of seasonality, skewness or kurtosis. Given time series as elements of the feature space were clustered by the k-means algorithm and individual methods were ranked according to their performance on each cluster. Three best methods for each cluster were combined using convex linear weights. For a given new time series, first the closest cluster is determined and then the given combination of the three best methods determined the ensemble forecast. This approach performed very well on a sufficiently big set of time series. The main motivation is that [21] demonstrates a dependence between time series features and accuracy of forecasting methods that allows to efficiently determine weights.

The second major motivation comes from the so called *Rule-Based Forecasting* [5, 11]. These authors came up with linguistically given (IF-THEN) rules that encode an expert knowledge "how to predict a given time series". Only some of them rules from [5, 11] do set up weights however, mostly they set up rather a specific model parameter, e.g. the smoothing factors of the Brown's exponential smoothing with trend. The rules very often use properties that are not crisp but rather vague, e.g. "unstable recent trend; recent trend is long", in antecedents. For such cases, using crisp rules, as in [11], that are either fired or not and nothing between seems to be less natural than using fuzzy rules. Similarly, the use of crisp consequents such as: "add 10% to the weight", seems to be less intuitive than using vague expression that are typical for fuzzy rules.

By following the main ideas of rule-based forecasting [11] and of using time series features (meta-learning) [21], we aim at obtaining an interpretable and understandable model that besides providing an unquestionable forecasting power helps to understand "when what works", nonchalantly said.

2 Implementation

To develop and validate the model we have used data from the known M3 data-set repository that contains 3000 time series from the M3 forecasting competition [23] and that serves as a generally accepted benchmark database. We have chosen 99 time series from fields such as Microeconomics, Industry, Macroeconomics, Finance and Demography for the model identification and different 99 time series constitute the testing set used in order to test whether the determined knowledge encoded in the fuzzy rules works generally also for other time series.

The global performance of a forecasting model is evaluated by an error measure. It should be noted that very popular measures such as *Mean Absolute Error* or *(Root) Mean Squared Error* are inappropriate for comparison across more time series because they are scale-dependent and therefore, we use scale independent *Symmetric Mean Absolute Percentage Error* [19]:

$$\text{SMAPE} = \frac{1}{h} \sum_{t=T+1}^{T+h} \frac{|e_t|}{(|y_t| + |\hat{y}_t|)/2} \times 100\%, \tag{2}$$

where $e_t = y_t - \hat{y}_t$ for $t = T+1, \ldots, T+h$.

We have chosen the most often used forecasting methods that are at disposal to the widest community: Decomposition Techniques (DT), Exponential Smoothing (ES), seasonal Autoregressive Integrated Moving Average (ARIMA), Generalized Autoregressive Conditional Heteroscedasticity (GARCH) models, Moving Averages (MA) and finally, Random Walk process (RW) and Random Walk process with a drift (RWd). For details about the methods, we only refer to the relevant literature [9, 16, 24].

In order to avoid any bias from a naive implementation of the above listed methods, we adopted implementations of these methods by professional software package: ForecastPro® for ES, ARIMA and MA; Gretl® for GARCH and RWd and NCSS® for DT. These tools executed fully automatic parameter selection and optimization which made possible to concentrate the investigation purely on the combination technique. Furthermore, their arithmetic mean (AM), that represents the equal weights, was also used as a benchmark.

We are fully aware of the potential of recent computational intelligence methods, e.g. neural networks [13], evolutionary computation [12], fuzzy techniques [30] or their combinations [29], for time series forecasting. However, the implementation of such techniques is not sufficiently standardized yet and at disposal to the widest community. In order to keep the main direction of the paper that is the ensemble technique and not the individual method setting and optimization, we leave this direction for future research.

From the given time series, the following features were extracted: *trend, seasonality, length of the time series, skewness, kurtosis, coefficient of variation, stationarity, forecasting horizon and frequency.*

Most of them are standard and we only briefly describe some of them. Stationarity assumes that the mean and the autocovariances of a time series do not change in time and it may be determined by many tests. We employ the Augmented Dickey-Fuller test [14] that under the null hypothesis assumes that a given time series is stationary, under the alternative hypothesis that time series is non-stationary. Trend and seasonality are similarly determined on such tests. Forecasting horizons are adopted from the M3 competition instructions. Since the range of each feature can significantly vary, it is crucial to normalize the range of features to the interval [0,1].

3 Fuzzy Rule-Based Ensemble

In this section, we briefly describe how we reach the goal – an identification of a fuzzy rule base that serves as a flexible model determining ensemble weights purely based on the chosen time series features and thus, describing the dependence of forecasting efficiency of individual time series on the features.

The desired fuzzy rule base may be identified by distinct approaches. Because of the missing reliable expert knowledge mentioned above, from the very beginning we omit the identification by an expert and we focus on data-driven approaches that vice-versa, may bring us the interpretable knowledge that is hidden in the data.

At this first stage, we employ the multiple linear regression for determining relationships among two or more variables. We used the linear regression on the training set of time series to estimate relationship between features of the time series and normalized SMAPE errors for each individual forecasting method. In this way we obtain seven regression models – one for each forecasting method – that model the relationships between features and forecasting errors of each forecasting method.

For the sake of clarity and simplicity, only significant features were considered. For the choice of statistically significant features we applied forward stepwise regression. This algorithm begins with no variables (features) in a model and then it selects a feature that has the largest R-squared value. This step is repeated until all remaining features are not significant. Only the statistically significant features were used in the multiple regression which estimated the unknown parameters by ordinary least squares. Furthermore, only in the case that a p-value of an parameter estimated in the multiple regression was smaller than a certain threshold of significance we rejected the null hypothesis and the estimated parameter was found statistically significant. For each method, different features played the significant role.

Because the ensemble weights should be (proportionally) higher if a given method is supposed to provide lower SMAPE error, SMAPE values were replaced by $(1 - \text{SMAPE})$ values and the obtained models were sampled. This way we obtained nodes in the reduced features spaces with only significant features and $(1-\text{SMAPE})$ errors. These nodes served as learning data for the so called *linguistic learning algorithm* [8] that automatically generates linguistic descriptions (fuzzy rule bases with linguistic evaluating expressions) that jointly with a specific fuzzy inference method *Perception-based Logical Deduction*[2] [27] may derive conclusions based on imprecise observations. Thus, we obtained seven linguistic descriptions – each of them determining weights of a single individual method based on transparent and interpretable rules, such as:

[2] Perception-based Logical deduction models fuzzy rules with use of the Łukasiewicz implication but it does aggregate them with the conjunction (as in the standard fuzzy relational approach) but views them as a list of independent rules. The implemented "perception" chooses the most appropriate rule from such a list. This approach performs well only if antecedents and consequents are *evaluative linguistic expressions*[26] which underlines the linguistic nature of the whole model. This is the reason why we have chosen this approach. Of course, it does not mean that fuzzy relational approach would not perform well.

IF *Trend* is *Big* AND *Variation* is *Small* THEN $Weight_{ARIMA}$ is *ExBig*.

At this stage of investigation, the above described construction of fuzzy rule bases is the only one that has been already experimentally justified however, other approaches that are supposed to be combined with the above one are starting to be developed and tested as well.

Naturally, we focus on fuzzy clustering. If we consider each time series as a point in a feature space extended by another feature: (1 – SMAPE) of a single individual method on the given time series fuzzy cluster analysis may then provide us with clusters of time series that have similar features and a similar forecasting performance by the chosen method. By a projection of clusters on individual feature axes we may obtain antecedent fuzzy sets; by a projection on (1 – SMAPE) axis we may obtain consequent fuzzy sets determining the weight of the chosen method.

Fuzzy clustering is a very natural approach however, it brings some complications (choosing the particular cluster algorithm; determination of the number of clusters). So, we also employ linguistic associations mining [2]. This approach firstly introduced as GUHA method [17, 18] finds distinct statistically approved associations between attributes of given objects. Particularly, we employ the fuzzy variant of this method [20, 28] and search for implicative associations that may be directly interpreted as fuzzy rules.

No matter the origin of fuzzy rules, such an ensemble technique will be naturally called *"Fuzzy Rule-Based Ensemble"* (FRBE).

4 Results, Conclusions and Future Work

In order to judge its performance, the fuzzy rule-based ensemble was applied on the 99 time series from a testing set. Table 1 shows that in both the average and the standard deviation of SMAPE values, the suggested ensemble outperforms all the individual methods and moreover, even highly suggested equal weights method (AM) has been outperformed as well.

One might consider the improvement to be rather low. To some extent it is true but it is evident that fuzzy rule-based approach performed very well. The fact that the victory has been reached not only in the accuracy but also in the robustness (standard deviation of the SMAPE error) definitely is worth noticing. In order to support our claim about the potential of FRBE approach, we proceeded the *One-Sample t-test* in order to test the null hypothesis that the mean of ($\mathrm{SMAPE}_{\mathrm{AM}} - \mathrm{SMAPE}_{\mathrm{FRBE}}$) is equal to 0. This hypothesis has not been rejected on the significance level 0.05 however, it has been rejected on the level 0.1 and we accepted the alternative hypothesis that the mean is higher than 0, which demonstrates, that on this level of significance, AM is not as good as FRBE which was found more accurate.

Table 1 Average of the SMAPE forecasting errors and standard deviation of the SMAPE forecasting errors over the testing set

Methods	Average Error	Methods	Error Std. Deviation
DT	21.59	DT	24.52
GARCH	17.27	GARCH	21.22
RWd	15.95	RWd	20.62
RW	15.26	RW	19.43
MA	15.11	MA	19.27
ARIMA	14.44	ARIMA	20.31
ES	14.43	ES	18.39
AM	14.40	AM	18.42
FRBE	**14.18**	FRBE	**18.03**

One should also note that this is an introductory study that opens this topic for further steps. The other techniques for fuzzy rule base identifications (fuzzy cluster analysis, linguistic associations mining etc.) are supposed to be experimentally evaluated in order to improve the results shortly. Furthermore, deep redundancy [15] and/or consistency analysis of obtained fuzzy rule bases is highly desirable in order to improve the knowledge that is linguistically encoded in them, is also highly desirable.

Thus, it may be concluded that in this introductory study, we have clearly stated the motivations and main ideas and mainly we have demonstrated the promising potential of the fuzzy rule-based forecasting that entitles us to continue in the foreshadowed future work.

Acknowledgements. This work was supported by the European Regional Development Fund in the IT4Innovations Centre of Excellence project (CZ.1.05/1.1.00/02.0070). Furthermore, we gratefully acknowledge partial support of projects KONTAKT II - LH12229 of MŠMT ČR and SGS11/PřF/2012 of the University of Ostrava.

References

1. Adya, M., Armstrong, J.S., Collopy, F., Kennedy, M.: An application of rule-based forecasting to a situation lacking domain knowledge. Int. J. Forecasting 16, 477–484 (2000)
2. Agrawal, R., Srikant, R.: Fast algorithms for mining association rules. In: Proc. of 20th Int. Conf. on Very Large Databases, pp. 487–499. AAAI Press (1994)
3. Armstrong, J.S.: Evaluating methods. In: Principles of Forecasting: A Handbook for Reasearchers and Practitioners. Kluwer, Dordrecht (2001)
4. Armstrong, J.S., Collopy, F.: Error measures for generalizing about forecasting methods: Empirical comparisons. Int. J. Forecasting 8, 69–80 (1992)

5. Armstrong, J.S., Adya, M., Collopy, F.: Rule-Based Forecasting Using Judgment in Time Series Extrapolation. In: Principles of Forecasting: A Handbook for Reasearchers and Practitioners. Kluwer, Dordrecht (2001)
6. Barrow, D.K., Crone, S.F., Kourentzes, N.: An Evaluation of Neural Network Ensembles and Model Selection for Time Series Prediction. In: Proc. of 2010 IEEE IJCNN. IEEE Press, Barcelona (2010)
7. Bates, J.M., Granger, C.W.J.: Combination of Forecasts. Oper. Res. Q 20, 451–468 (1969)
8. Bělohlávek, R., Novák, V.: Learning Rule Base of the Linguistic Expert Systems. Soft Comput. 7, 79–88 (2002)
9. Box, G., Jenkins, G.: Time Series Analysis: Forecasting and Control. Holden-Day, San Francisco (1976)
10. Makridakis, S., Andersen, A., Carbone, R., Fildes, R., Hibon, M., Lewandowski, R., Newton, J., Parzen, E., Winkler, R.: The Accuracy of Extrapolation (Time-Series) Methods — Results of a Forecasting Competition. J. Forecasting 1, 111–153 (1982)
11. Collopy, F., Armstrong, J.S.: Rule-Based Forecasting: Development and Validation of an Expert Systems Approach to Combining Time Series Extrapolations. Manage. Sci. 38, 1394–1414 (1992)
12. Cortez, P., Rocha, M., Neves, J.: Evolving Time Series Forecasting ARMA Models. J. Heuristics 10, 415–429 (2004)
13. Crone, S.F., Hibon, M., Nikolopoulos, K.: Advances in forecasting with neural networks? Empirical evidence from the NN3 competition on time series prediction. Int. J. Forecasting 27, 635–660 (2011)
14. Dickey, D.A., Fuller, W.A.: Distribution of the estimators for autoregressive time series with a unit root. J. Am. Stat. Assoc. 74, 427–431 (1979)
15. Dvořák, A., Štěpnička, M., Vavříčková, L.: Redundancies in systems of fuzzy/linguistic IF-THEN rules. In: Proc. EUSFLAT 2011, pp. 1022–1029 (2011)
16. Hamilton, J.D.: Time Series Analysis. Princeton University Press (1994)
17. Hájek, P.: The question of a general concept of the GUHA method. Kybernetika 4, 505–515 (1968)
18. Hájek, P., Havránek, T.: Mechanizing hypothesis formation: Mathematical foundations for a general theory. Springer, Berlin (1978)
19. Hyndman, R., Koehler, A.: Another look at measures of forecast accuracy. Int. J. Forecasting 22, 679–688 (2006)
20. Kupka, J., Tomanová, I.: Some extensions of mining of linguistic associations. Neural Netw. World 20, 27–44 (2010)
21. Lemke, C., Gabrys, B.: Meta-learning for time series forecasting in the NN GC1 competition. In: Proc. of 2010 FUZZ-IEEE, Barcelona, Spain. IEEE Press (2010)
22. MacKinnon, J.G.: Numerical Distribution Functions for Unit Root and Cointegration Tests. J. Appl. Econom. 11, 601–618 (1996)
23. Makridakis, S., Hibon, M.: The M3-Competition: Results, Conclusions and Implications. Int. J. Forecasting 16, 451–476 (2000)
24. Makridakis, S., Wheelwright, S., Hyndman, R.: Forecasting methods and applications, 3rd edn. John Wiley & Sons (2008)
25. Newbold, P., Granger, C.W.J.: Experience with forecasting univariate time series and combination of forecasts. J. Roy. Stat. Soc. A Sta. 137, 131–165 (1974)
26. Novák, V.: A comprehensive theory of trichotomous evaluative linguistic expressions. Fuzzy Set. Syst. 159, 2939–2969 (2008)
27. Novák, V.: Perception-based logical deduction. In: Computational Intelligence, Theory and Applications. ASC, pp. 237–250. Springer, Berlin (2005)

28. Novák, V., Perfilieva, I., Dvořák, A., Chen, Q., Wei, Q., Yan, P.: Mining pure linguistic associations from numerical data. Int. J. Approx. Reason. 48, 4–22 (2008)
29. Štěpnička, M., Donate, J.P., Cortez, P., Vavříčková, L., Gutierrez, G.: Forecasting seasonal time series with computational intelligence: contribution of a combination of distinct methods. In: Proc. of EUSFLAT 2011, pp. 464–471 (2011)
30. Štěpnička, M., Dvořák, A., Pavliska, V., Vavříčková, L.: A linguistic approach to time series modeling with the help of the F-transform. Fuzzy Set. Syst. 180, 164–184 (2011)

Hybrid Models: Probabilistic and Fuzzy Information

Giulianella Coletti and Barbara Vantaggi

Abstract. Under the interpretation of fuzzy set as coherent conditional probability, we study inferential processes starting from a probability distribution (on a random variable) and a coherent conditional probability on "fuzzy conditional events". We characterize the coherent extensions and we analyze an example proposed by Zadeh.

Keywords: Coherent conditional probability, fuzzy sets, inference.

1 Introduction

Randomness and fuzziness may act jointly, this fact generate new problems in combining probabilistic and fuzzy information (see e.g. [15, 13, 14]).

We refer to the interpretation of a fuzzy set E^*_φ as a pair $(E_\varphi, \mu_\varphi(x))$ with $\mu_\varphi(x) = P(E_\varphi|X = x)$, where X is a variable with range $\mathcal{C}_X$, φ any property related to X, E_φ the event "You claim that X is φ" and P a coherent conditional probability (see Section 3, for details, see, e.g., [4, 5, 6], for a similar semantic see [12, 13]). In this context it has been proved in [4] that, under the hypothesis of logical independence between E_φ and E_ψ, the membership functions $\mu_{\varphi\vee\psi}$ and $\mu_{\varphi\wedge\psi}$ of the fuzzy sets $E^*_\varphi \cup E^*_\psi$ and $E^*_\varphi \cap E^*_\psi$ are extensions of the coherent conditional probabilities μ_ψ and μ_φ. Among the coherent extensions there are the extensions computed by a t-conorm and its dual t-norm of the Frank class. In this context logical independence

Giulianella Coletti
Dip. Matematica e Informatica, University of Perugia, Italy
e-mail: coletti@dmi.unipg.it

Barbara Vantaggi
Dip. S.B.A.I. University "La Sapienza", Rome, Italy
e-mail: barbara.vantaggi@sbai.uniroma1.it

R. Kruse et al. (Eds.): Synergies of Soft Computing and Statistics, AISC 190, pp. 389–397.
springerlink.com © Springer-Verlag Berlin Heidelberg 2013

is not a strong condition: in fact, for instance E_φ and $E_{\neg\varphi}$ (which differs from $\neg E_\varphi$) are logical independent, in fact You can claim both X is φ and X is $\neg\varphi$ or claim only one of them or none. Starting from a family of fuzzy sets $\mathcal{C} = \{E^*_{\varphi_i}\}$ related to a variable X, with E_{φ_i} logical independent, let us consider the closure $\langle\mathcal{C}\rangle$ of $\mathcal{C}$ with respect to the union and intersection of fuzzy sets ruled by Frank t-norms and t-conorms. Given $\mathcal{C}$ and a probability distribution on X, the main aim is to study coherent conditional probability on the "conditional fuzzy events" $A|B$, where A and B are the events related to the elements of $\langle\mathcal{C}\rangle$. The possibility of defining this conditional probability assessment in the class of interest is assured by coherence.

In previous papers [2, 4, 5, 6] we studied the above problem in a finite ambit, in this paper we consider also variables with infinite range. The framework of reference is the general one of finitely additive conditional probabilities [9, 10, 1]. We show trough an example proposed by Zadeh, the simple applicability of the above interpretation of fuzzy sets.

2 Coherent Conditional Probability

The framework of reference is coherent conditional probability, that is a function defined on an arbitrary set $\mathcal{C} = \{E_i|H_i\}_{i\in I}$ of conditional events, consistent (or coherent) with a conditional probability defined on $\mathcal{C}' = \mathcal{E} \times \mathcal{H}$, where $\mathcal{E}$ is the algebra spanned by the events E_i, H_i and $\mathcal{H}$ the additive set spanned by events H_i (see for instance [4] and [8] for a seminal work).

In the literature many characterizations of a coherent conditional probability assessment are present, we recall the following (see e.g. [3]) working for arbitrary (possibly infinite) families of conditional events: coherence for an infinite set of conditional events can be reduced to check the coherence on any finite subset. In the infinite case it is not possible to give a representation of a coherent conditional probability in terms of a class of probabilities, but it is necessary to involve charges (a finite additive measure, which is not necessarily bounded, see [1]), so in the following the integrals are in the sense of Stieltjes or Daniel (since we consider bounded positive functions T2-measurable) with respect to finitely additive measures.

Theorem 1. *Let $\mathcal{C}$ be an arbitrary family of conditional events and consider the relevant set $\mathcal{E}(\mathcal{C}) = \{E \wedge H, H : E|H \in \mathcal{C}\}$. For a real function P on $\mathcal{C}$ the following statements are equivalent:*

(a) P is a coherent conditional probability on $\mathcal{C}$;

(b) there exists a class $\{m_\alpha\}$ of functions, such that

- each $\{m_\alpha\}$ is defined on $\mathcal{B}_\alpha \subseteq \mathcal{E}(\mathcal{C})$, and is the restriction of a positive charge defined on the algebra generated by $\mathcal{B}_\alpha$; $\mathcal{B}_\alpha \subset \mathcal{B}_\beta$ for $\alpha > \beta$;

- for any conditional event $E|H \in \mathcal{C}$ there exists a unique m_α with $0 < m_\alpha(H) = \int_H dm_\alpha < \infty$ and, for every $\beta \leq \alpha$, $P(E|H)$ is solution of all the equations

$$m_\beta(E \wedge H) = x \cdot m_\beta(H)$$

(c) for any finite subset $\mathcal{F} = \{E_1|H_1, \ldots, E_n|H_n\}$ *of* $\mathcal{C}$*, denoting by* $\mathcal{A}_o$ *the set of atoms* A_r *generated by the events* $E_1, H_1, \ldots, E_n, H_n$*, there exists a class of probabilities* $\P = \{P_0, P_1, \ldots P_k\}$*,* $(k \leq n)$ *and a relevant class of sets* $\mathcal{A}_\alpha \subseteq \mathcal{A}_0$ *such that:*

- each probability P_α *is defined on* $\mathcal{A}_\alpha$*;* $\mathcal{A}_\alpha \subset \mathcal{A}_\beta$ *for* $\alpha > \beta$*;*

- for any $E_i|H_i \in \mathcal{C}$ *there exists a unique* α *such that* $P_\alpha(H_i) > 0$ *and, for every* $\beta \leq \alpha$*,* $P(E_i|H_i)$ *is solution of all the equations*

$$P_\beta(E_i \wedge H_i) = x \cdot P_\beta(H_i).$$

Any class $\{m_\alpha\}$ satisfying condition (b) is said *to agree* with the coherent conditional probability P.

The above characterization gives rise to a sufficient conditions [2]. For this aim we recall that, given a partition $\mathcal{H} = \{H_i\}_{j\in J}$, the events E_i $(i = 1, ..., n)$, are said *logically independent with respect to* $\mathcal{H}$ if, for any $H_j \in \mathcal{H}$, $\bigwedge_{i=1}^n E_i^* \wedge H_j = \emptyset$, implies $E_i^* \wedge H_j = \emptyset$ for some $i = 1, ..., n$, where E^* stands either for E or for its contrary E^c. Note that logical independence with respect to a partition is stronger than logical independence.

Corollary 1. *Let* $\mathcal{C} = \{E_i|H_j : E_i \in \mathcal{E}, H_j \in \mathcal{H}\}$ *where* $\mathcal{E} = \{E_i\}_{i=1}^n$ *and* $\mathcal{H} = \{H_j\}_{j\in J}$ *is a (not necessarily countable) partition of* Ω*, moreover the events* E_i *are logically independent with respect to* $\mathcal{H}$*. Let* $p_o(\cdot)$ *be a coherent probability on* $\mathcal{H}$*. Then for every function* $P : \mathcal{C} \to [0,1]$ *with* $P(E_i|H_j) = 0$ *if* $E_i \wedge H_j = \emptyset$ *and* $P(E_i|H_j) = 1$ *if* $H_j \subseteq E_i$*, the global assessment*

$$\P = \{P(E_i|H_j), p_o(H_j)\}_{1=i,\ldots,n;j\in J}$$

is a coherent conditional probability assessment.

Concerning coherence, we recall moreover the following fundamental result for conditional probability (essentially due to de Finetti [8]):

Theorem 2. *Let* $\mathcal{C} = \{E_i|H_i\}$ *be any family of conditional events, and* $\mathcal{C}' \supseteq \mathcal{C}$*. Let* P *be an assessment on* $\mathcal{C}$*; then there exists a (possibly not unique) coherent conditional probability assessment extending* P *to* $\mathcal{C}'$ *if and only if* P *is a coherent conditional probability assessment on* $\mathcal{C}$*.*

When $\mathcal{C}' = \mathcal{C} \cup \{E|H\}$ the possible coherent values $p = P(E|H)$ are all the values of a suitable closed interval $[\underline{p}, \overline{p}] \subseteq [0,1]$, with $\underline{p} \leq \overline{p}$.

By Theorems 1 and 2 it is possible to prove the following theorem [7]:

Theorem 3. *Let* $\mathcal{E} = \{E|H_i\}_{i\in I}$ *be an arbitrary set of conditional events such that the set* $\mathcal{H}_0 = \{H_i\}_{i\in J}$ *of conditioning events is a partition of* Ω*. Denote by* $\mathcal{F}_0$ *the* σ*-field spanned by* $\mathcal{H}_0$*,* $\mathcal{F}_0^0 = \mathcal{F}_0 \setminus \{\emptyset\}$ *and* $\mathcal{K} = \{E|H : H \in \mathcal{F}_0^0\}$*. Let* $p : \mathcal{E} \to [0,1]$ *be any function such that* $p(E|H_i) = 0$ *if* $E \wedge H_i = 0$ *and* $p(E|H_i) = 1$ *if* $H_i \subseteq E$*.*

The following statements are equivalent:

- *there exists a coherent conditional probability P extending p on $\mathcal{K}$;*
- *there exists a class $\{m_\alpha : \alpha \in W\}$ (with $|W| \le |J|$) of positive (not necessarily bounded) charges defined on suitable σ-fields $\{\mathcal{F}_\alpha\}_{\alpha \in W}$ defined by suitable families $\mathcal{B}_\alpha$ with $\mathcal{B}_\alpha \subset \mathcal{B}_\beta$ for $\alpha > \beta$ and $m_\beta(H) = 0$ iff $H \in \mathcal{B}_\alpha$, and for any conditional event $E|H \in \mathcal{K}$ there is a unique α such that $H \in \mathcal{B}_\alpha \setminus \mathcal{B}_{\alpha+1}$, with $0 < m_\alpha(H) < \infty$, and $P(E|H) = x$ is solution of the equation*

$$x \int_H d(m_\alpha(y)) = \int_H p(E|y) d(m_\alpha(y)). \tag{7}$$

3 Fuzzy Sets as Coherent Conditional Probabilities

We adopt the interpretation of fuzzy sets in terms of coherent conditional probabilities, introduced in [4, 5, 6]. We briefly recall here the main concepts.

Let X be a (not necessarily numerical) variable, with range $\mathcal{C}_X$, and, for any $x \in \mathcal{C}_X$, let us denote by x the event $\{X = x\}$, for every $x \in \mathcal{C}_X$.

We consider any *property* φ related to the variable X and refer to the state of information of a real (or fictitious) person that will be denoted by "You". Then E_φ is the Boolean event {You claim that X is φ}.

By Corollary 1, it follows that You may assign to each of these conditional events a degree of belief (subjective probability) $P(E_\varphi|x)$, without any syntactical restriction. So we can define a fuzzy subset E^*_φ of $\mathcal{C}_X$ as a pair $(E_\varphi\, , \mu_{E_\varphi})$ with $\mu_{E_\varphi}(\cdot) = P(E_\varphi|\cdot)$.

By referring to [5] we recall the operations between fuzzy subsets: under the hypothesis of logical independence between E_φ and E_ψ with respect to X (i.e. with respect the partition $\{x\}_{x \in \mathcal{C}_X}$), the binary operations of union and intersection and that of complementation can be obtained directly by using the rules of coherent conditional probability. By defining $\mu_{\varphi\wedge\psi}(x) = P(E_\varphi \wedge E_\psi|x)$ and $\mu_{\varphi\vee\psi}(x) = P(E_\varphi \vee E_\psi|x)$, it has been proved that the coherent values for $\mu_{\varphi\wedge\psi}(x)$ and $\mu_{\varphi\vee\psi}(x)$, for any given x of X, can be obtained by any Frank t-norm and its dual t-conorm.

Given a family of logically independent events E_{φ_i}, consider the algebra $\mathcal{B}$ spanned by them, any Frank's t-norm and its dual t-conorm are apt to compute any union and intersection between the relevant fuzzy sets (E_{φ_i}, μ_i). Coherence rules the extension of the conditional probability $P(\cdot|x)$ to the other events of the algebra (for instance to the events E^c_φ), which do not support a fuzzy set.

Note furthermore that two events E_φ, E_ψ can be logically independent and have the relevant μ_φ, μ_ψ with disjoint support.

We finally recall that in this context the complement of a fuzzy set is defined as

$$(E^*_\varphi)' = (E_{\neg\varphi}\,, \mu_{\neg\varphi}) = (E_{\neg\varphi}\,, 1 - \mu_\varphi). \tag{1}$$

Obviously $E_{\neg\varphi} \neq (E_\varphi)^c$, moreover $E_{\neg\varphi}$ and E_φ are logically independent with respect to X (see [5]). Then, while $E_\varphi \vee (E_\varphi)^c = \Omega$, one has $E_\varphi \vee E_{\neg\varphi} \subsetneq \Omega$ and so $\mu_{\varphi\vee\neg\varphi}(x) = \mu_\varphi(x) + \mu_{\neg\varphi}(x) - \mu_{\varphi\wedge\neg\varphi}(x)$.

The case of two fuzzy subsets E^*_φ, E^*_ψ, corresponding to the random variables Z_1 and Z_2, respectively, the following choice for the membership of conjunction and disjunction is coherent:

$$\mu_{\varphi\vee\psi}(z,z') = P(E_\varphi \vee E_\psi | A_z \wedge A_{z'}),\ \mu_{\varphi\wedge\psi}(z,z') = P(E_\varphi \wedge E_\psi | A_z \wedge A_{z'}) \tag{2}$$

with the only constraints

$$\max\{\mu_\varphi(z) + \mu_\psi(z') - 1\,, 0\} \leq \mu_{\varphi\wedge\psi}(z,z') \leq \min\{\mu_\varphi(z) + \mu_\psi(z')\} \tag{3}$$

$$\mu_{\varphi\vee\psi}(z,z') = \mu_\varphi(z) + \mu_\psi(z') - \mu_{\varphi\wedge\psi}(z,z')\,. \tag{4}$$

4 Combining Probabilistic and Fuzzy Information

First of all notice that in this context the concept of fuzzy event, as introduced by Zadeh, is nothing else than an ordinary event of the kind

$$E_\varphi = \text{“You claim that } X \text{ is } \varphi\text{”}.$$

When a probability p_o on the variable X is given, according to Corollary 1 the assessment $\{\mu_\varphi(x), p_o(x)\}_x$ is coherent and according to condition b of Theorem 3 (p_o coincides with m_o), the only coherent value for the probability of E_φ is

$$P(E_\varphi) = \int \mu_{\varphi_i}(x) dp_o(x)\,.$$

This formally coincides with the definition proposed by Zadeh in [15], but actually differs from it since p_o is in general a finitely additive (non necessarily σ-additive) probability.

Now let $\mathcal{C}$ be a finite family of fuzzy subsets $E^*_{\varphi_i} = (E_{\varphi_i}, \mu_{\varphi_i})$ of X (with the events E_{φ_i} logically independent with respect to X), denote by $\langle \mathcal{C}_\odot \rangle$ its closure with respect to intersection and union, moreover $\mathcal{F}_{\langle \mathcal{C}_\odot \rangle}$ stands for the sets of events E_{φ_i} related to the elements of $\langle \mathcal{C}_\odot \rangle$.

From the aforementioned results (see also [5]) the next result follows:

Theorem 4. *Let $\mathcal{C}$ be a finite family of fuzzy subsets $E^*_{\varphi_i} = (E_{\varphi_i}, \mu_{\varphi_i})$ of X, with the events E_{φ_i} logically independent with respect to X. For every t-norm $\odot$ in the class of Frank, and for any probability distribution p_o on X, the assessment $\{P_\odot(E_{\varphi_i}), P_\odot(E_{\varphi_i} \wedge E_{\varphi_j})\}$ with*

$$P_\odot(E_{\varphi_i}) = \int \mu_{\varphi_i}(x) dp_o(x)\ ;\ P_\odot(E_{\varphi_i} \wedge E_{\varphi_j}) = \int (\mu_{\varphi_i} \odot \mu_{\varphi_j})(x) dp_o(x) \tag{5}$$

is a coherent probability.

Moreover, the only coherent extension to the event $E_{\varphi_i} \vee E_{\varphi_j}$ *is such that*

$$P_{\odot}(E_{\varphi_i} \vee E_{\varphi_j}) = \int (\mu_{\varphi_i} \oplus \mu_{\varphi_j})(x) dp_o(x) = P_{\odot}(E_{\varphi_i}) + P_{\odot}(E_{\varphi_j}) - P_{\odot}(E_{\varphi_i} \wedge E_{\varphi_i}),$$

where $\oplus$ *is the dual t-conorm of* $\odot$.

Proof. From results given in [5], for any t-norm $\odot$ in the Frank class ([11]), the assessment $\{P_{\odot}(E_{\varphi_i} \wedge E_{\varphi_j}|x)\}_x$ is coherent with $\mathcal{C}$ since the events E_{φ_i} are logically independent with respect to X. From Corollary 1 it follows that any probability p_o on X is coherent with the above assessment and from Theorem 3 the equation (5) follows. From additivity $P_{\odot}(E_{\varphi_i} \vee E\varphi_j) = P_{\odot}(E_{\varphi_i}) + P_{\odot}(E_{\varphi_j}) - P_{\odot}(E_{\varphi_i} \wedge E_{\varphi_i})$ and furthermore the only coherent value on $\{P_{\odot}(E_{\varphi_i} \vee E_{\varphi_j}|x)\}_{x \in \mathcal{C}_X}$ is given through the dual t-conorm $\oplus$ of $\odot$ (see again [5]), so by Theorem we have 3 $P_{\odot}(E_{\varphi_i} \vee E_{\varphi_j}) = \int (\mu_{\varphi_i} \oplus \mu_{\varphi_j})(x) dp_o(x)$.

The above assessment $P_{\odot}$ is a coherent conditional probability, so from Theorem 2 it can be furthermore extended to any conditional event $A|B$ (with $B \neq \emptyset$), where A, B are events of the algebra $\mathcal{B}$ generated by $\mathcal{F}_{\langle \mathcal{C}_{\odot} \rangle} \cup \{x : x \in \mathcal{C}_X\}$. In general this extension is not unique while for the events $A = E_{\varphi_i}$ and $B = E_{\varphi_j}$ $(i \neq j)$, with $P_{\odot}(E_{\varphi_j}) = \int \mu_j(x) dp_o(x) > 0$ the only coherent extension (called in the following coherent $\odot$-extension) is:

$$P_{\odot}(E_{\varphi_i}|E_{\varphi_j}) = \frac{\int (\mu_{\varphi_i} \odot \mu_{\varphi_j})(x) dp_o(x)}{\int \mu_{\varphi_j}(x) dp_o(x)}. \tag{6}$$

When $P_{\odot}(E_{\varphi_j}) = 0$ to obtain a unique extension for $E_{\varphi_i}|E_{\varphi_j}$ we need either a charge m_β belonging to a class $\{m_\alpha\}$ of charges on the algebra $\mathcal{A}_X$ generated by X, agreeing with p_o and such that $\int \mu_{\varphi_j}(x) dm_\beta(x) > 0$ or equivalently a conditional probability $P(\cdot|C)$, with C a suitable event of the algebra $\mathcal{A}_X$ such that $\int \mu_{\varphi_j}(x) dP(x|C) > 0$. The coherence of the assessments $\{P(\cdot|\cdot), P(E_{\varphi_j}|\cdot)\}$ and $\{\{m_\alpha(\cdot)\}, P(E_{\varphi_j}|\cdot)\}$ is assured by a simple extension of Corollary 1. In this case $P_{\odot}(E_{\varphi_i}|E_{\varphi_j})$ is obtained through equation (5), by replacing p_o either with $P(\cdot|C)$ or with the m_β.

Remark 1. *The values* $P_{\odot}(E_{\varphi_i}|E_{\varphi_j})$ *through equation (6) are coherent only when the events* E_{φ_i} *and* E_{φ_j} *are logically independent with respect to* X. *For instance, the same formula cannot be used for the coherent extension of* $P_{\odot}$ *to* $E_{\varphi_j}|E_{\varphi_j} \wedge E_{\varphi_i}$ *(or* $E_{\varphi_j}|E_{\varphi_j}$*), which is necessarily 1, independently of the Frank t-norm used for computing the coherent values of* $P_{\odot}(E_{\varphi_i} \wedge E_{\varphi_j})$.

Theorem 5. *Let* X *be a variable with countable range* $\mathcal{C}_X$ *and let* $E^*_{\varphi_i}, E^*_{\varphi_j}$ *two fuzzy subsets of* $\mathcal{C}_X$. *Given a strictly positive probability distribution* p_o *on* X *and a strict t-norm* $\odot$, *if* $P_{\odot}(E_{\varphi_i}|E_{\varphi_j})$ *coherently extending the assessment* $\{P_{\odot}(E_{\varphi_i}|\cdot), P_{\odot}(E_{\varphi_j}|\cdot), p_o(\cdot)\}$ *is equal to 1, then* E_{φ_i} *and* E_{φ_j} *are not logically independent with respect to* X.

Proof. If E_{φ_i} and E_{φ_j} were logically independent with respect to X, the relevant coherent conditional probability would be computed by equation (6). Then, $P_{\odot}(E_{\varphi_i}|E_{\varphi_j}) = 1$ implies, for any $x \in \mathcal{C}_X$, $\mu_{\varphi_i} \odot \mu_{\varphi_j}(x) = \mu_{\varphi_j}(x)$, since p_o is strictly positive, and this contradicts the fact that $\odot$ is strict. Then, the two events cannot be logically independent with respect to X.

For every t-norm $\odot$, by construction of $\langle \mathcal{C}_{\odot} \rangle$, any pair of not logically independent events E_{φ}, E_{ψ} in $\mathcal{F}_{\langle \mathcal{C}_{\odot} \rangle}$ is such that either $E_{\varphi} \subseteq E_{\psi}$ or $E_{\psi} \subseteq E_{\varphi}$. Thus, the coherent values of the relevant (conditional) events are univocally determined through the min t-norm (as shown in the next result).

Theorem 6. *Let $\mathcal{C} = \{E^*_{\varphi_i}\}_i$ be a finite family of fuzzy subsets on $\mathcal{C}_X$, with E_{φ_i} logical independent with respect to X and let p_o be any probability distribution on X. Then, the assessment*

$$\{p_o(x), P_{\odot}(A|x), P_{\odot}(A|B) \,:\, x \in \mathcal{C}_X,\ A, B \in \mathcal{F}_{\langle \mathcal{C}_{\odot} \rangle} \text{ with } P_{\odot}(B) > 0\},$$

where $\odot$ is the minimum t-norm and $P_{\odot}(A|B)$ is computed by (6), is coherent.

Proof. By construction $\langle \mathcal{C} \rangle$ involves events E_{φ_i} related to the elements of $\mathcal{C}$ and the union and intersection of them. Then, by Corollary 1 the assessment $\P_1 = \{P(E_{\varphi_i}|x), p_o\}_{x \in \mathcal{C}_X}$, is coherent since the events E_{φ_i} are logically independent with respect to X.

Moreover, from Theorem 4 $P_{\odot}(B|x), P_{\odot}(B)$ are coherent with $\P_1$ (coherence holds for any Frank t-norm and so also for minimum t-norm).

In order to prove that $P_{\odot}(A|B)$ is coherent with the previous assessment, recall that B and A are logically dependent on $\mathcal{C}$ (i.e. they are union of some atoms generated by $\mathcal{C}$), so their coherent values are obtained through minimum t-norm and t-conorm as well as $P_{\odot}(A \wedge B)$, so the following cases can occur:

- A and B logically independent,
- $A \subseteq B$ (or $B \subseteq A$).

In all these cases the extension $P_{\odot}(A \wedge B) = \min\{P_{\odot}(A), P_{\odot}(B)\}$ is coherent and then $P_{\odot}(A|B)$ is given by equation (6) and so $P_{\odot}(A|B) = 1$.

5 Example

Consider the following problem posed on [16] by Zadeh: "Usually it takes Robert about an hour to get home from work. Usually Robert leaves office at about 5pm. What is the probability that Robert is home at 6:15pm."

Let us consider the variable X= "time taken by Robert to get home from work" and Y=" time that Robert leaves office", and for X the property φ = "about an hour" and for Y the property ψ = "about at 5pm". Now we are able to define the fuzzy subsets E^*_{φ} of $\mathcal{C}_X$ and E^*_{ψ} of $\mathcal{C}_Y$ with $\mu_{\varphi}(x) = P(E_{\varphi}|x)$, triangular function centered in $60'$ and support $A = [45', 90']$ and

$\mu_\psi(y) = P(E_\psi|y)$, again triangular function centered in 5pm and support $B = [4:45pm, 5:15pm]$, respectively.

The problem now is to compute

$$P(X+Y \leq 6.15|E_\varphi \wedge E_\psi),$$

which is solvable by supposing any (finitely additive) probability on the algebra spanned by (X, Y). Nevertheless in this case we could suppose independent the two variables and having X normal distribution with mean in 5pm and Y exponential distribution. Then, chosen any Frank t-norm $\odot$, we can compute

$$P_\odot(E_\varphi \wedge E_\psi) = \int_{A\times B} \mu_{\varphi\wedge\psi}(x,y) f_X(x) f_Y(y) dxdy,$$

where $\mu_{\varphi\wedge\psi}(x,y) = \mu_\varphi(x) \odot \mu_\psi(y)$ and the integral is the Riemann's one. Similarly we can compute $P_\odot(E_\varphi \wedge E_\psi|X+Y \leq 6.15)$ and then the result through Bayes formula, since $P_\odot(E_\varphi \wedge E_\psi) > 0$.

We would note that, even if X and Y are supposed independent, the probability $P_\odot(E_\varphi \wedge E_\psi)$ generally differs from the product of $P_\odot(E_\varphi)$ and $P_\odot(E_\psi)$. They coincide when the product t-norm is considered.

References

1. Bhaskara Rao, K.P.S., Bhaskara Rao, M.: Theory of charges: a study of finitely additive measures. Academic Press, London (1983)
2. Coletti, G., Gervasi, O., Tasso, S., Vantaggi, B.: Generalized Bayesian inference in a fuzzy context: From theory to a virtual reality application. Comput. Stat. Data Anal. 56(4), 967–980 (2012)
3. Coletti, G., Scozzafava, R.: Characterization of coherent conditional probabilities as a tool for their assessment and extension. Int. J. Uncertain. Fuzz. 4, 103–127 (1996)
4. Coletti, G., Scozzafava, R.: Probabilistic logic in a coherent setting. Trends in Logic, vol. 15. Kluwer, Dordrecht (2002)
5. Coletti, G., Scozzafava, R.: Conditional probability, fuzzy sets, and possibility: a unifying view. Fuzzy Set. Syst. 144, 227–249 (2004)
6. Coletti, G., Scozzafava, R.: Conditional probability and fuzzy information. Comput. Stat. Data Anal. 51, 115–132 (2006)
7. Coletti, G., Scozzafava, R., Vantaggi, B.: Integrated Likelihood in a Finitely Additive Setting. In: Sossai, C., Chemello, G. (eds.) ECSQARU 2009. LNCS (LNAI), vol. 5590, pp. 554–565. Springer, Heidelberg (2009)
8. de Finetti, B.: Sull'impostazione assiomatica del calcolo delle probabilità. Annali Univ. Trieste 19, 3–55 (1949)
9. de Finetti, B.: Teoria della probabilitá, Einaudi, Torino (1970)
10. Dubins, L.E.: Finitely additive conditional probabilities, conglomerability and disintegration. Ann. Probab. 3, 89–99 (1975)
11. Frank, M.J.: On the simultaneous associativity of $F(x, y)$ and $x + y - F(x, y)$. Aequationes Math. 19, 194–226 (1979)

12. Lawry, J.: Appropriateness measures: An uncertainty model for vague concepts. Synthese 161, 255–269 (2008)
13. Nguyen, H.T., Wu, B.: Fundamental of Statistics with Fuzzy Data. STUDFUZZ. Springer (2010)
14. Viertl, R.: Statistical Methods for Non-precise Data. CRC Press, Boca Raton (1996)
15. Zadeh, L.A.: Probability measures of fuzzy events. J. Math. Anal. Appl. 23, 421–427 (1968)
16. Zadeh, L.A.: Toward a perception-based theory of probabilistic reasoning with imprecise probabilities. J. Stat. Plan. Infer. 105, 233–264 (2002)

12. Lawry, J.: Appropriateness measures: An uncertainty model for vague concepts. Synthese 161, 255–269 (2008)
13. Nguyen, H.T., Wu, B.: Fundamentals of Statistics with Fuzzy Data. STUDFUZZ, Springer (2006)
14. Viertl, R.: Statistical Methods for Non-Precise Data. CRC Press, Boca Raton (1996)
15. Zadeh, L.A.: Probability measures of fuzzy events. J. Math. Anal. Appl. 23, 421–427 (1968)
16. Zadeh, L.A.: Toward a perception-based theory of probabilistic reasoning with imprecise probabilities. J. Stat. Plan. Infer. 105, 233–264 (2002)

Conjunction, Disjunction and Iterated Conditioning of Conditional Events

Angelo Gilio and Giuseppe Sanfilippo

Abstract. Starting from a recent paper by S. Kaufmann, we introduce a notion of conjunction of two conditional events and then we analyze it in the setting of coherence. We give a representation of the conjoined conditional and we show that this new object is a conditional random quantity, whose set of possible values normally contains the probabilities assessed for the two conditional events. We examine some cases of logical dependencies, where the conjunction is a conditional event; moreover, we give the lower and upper bounds on the conjunction. We also examine an apparent paradox concerning stochastic independence which can actually be explained in terms of uncorrelation. We briefly introduce the notions of disjunction and iterated conditioning and we show that the usual probabilistic properties still hold.

Keywords: Conditional events, conditional random quantities, conjunction, disjunction, iterated conditionals.

1 Introduction

In probability theory and in probability logic a relevant problem, largely discussed by many authors (see, e.g., [2, 3, 7]), is that of suitably defining logical operations among conditional events. In a recent paper by Kaufmann ([8]) a

---

Angelo Gilio
Dipartimento di Scienze di Base e Applicate per l'Ingegneria,
University of Rome "La Sapienza", Italy
e-mail: angelo.gilio@sbai.uniroma1.it

Giuseppe Sanfilippo
Dipartimento di Scienze Statistiche e Matematiche "S. Vianelli",
University of Palermo, Italy
e-mail: giuseppe.sanfilippo@unipa.it

R. Kruse et al. (Eds.): Synergies of Soft Computing and Statistics, AISC 190, pp. 399–407.
springerlink.com © Springer-Verlag Berlin Heidelberg 2013

theory for the compounds of conditionals has been proposed. In this paper, based on the work of Kaufmann, we develop a similar theory in the framework of coherence. We show that conjunction and disjunction of conditional events in general are not conditional events but *conditional random quantities.* We give representations for such compounds and we study the coherent extensions of a probability assessment (x, y) on two conditional events $\{A|H, B|K\}$ to their conjunction $(A|H) \wedge (B|K)$ and their disjunction $(A|H) \vee (B|K)$. In particular, by considering the conjunction, we show cases of logical dependencies in which the combination reduces to a conditional event. For reason of space we only give a short introduction to disjunction and iterated conditioning. We give the lower and upper bounds for conjunction, disjunction and iterated conditional and we show that the usual probabilistic properties still hold in terms of previsions. We also discuss an apparent paradox where $A|H, B|K$ seem to be stochastically independent, by giving an explanation in terms of uncorrelation between random quantities.

2 Preliminary Notions

We denote events and their indicators by the same symbol and we write r.q. (resp., c.r.q.) for random quantity (resp., conditional random quantity). We recall that n events are said logically independent when there are no logical dependencies among them, which amounts to say that the number of constituents is 2^n. A conditional event $A|H$, where $H \neq \emptyset$, is a three-valued logical entity which is true, or false, or void, according to whether AH is true, or A^cH is true, or H^c is true. In the setting of coherence, given an event $H \neq \emptyset$ and a finite r.q. $X \in \{x_1, x_2, \ldots, x_n\}$, agreeing to the betting metaphor the prevision $\mathbb{P}(X|H)$ of $X|H$ is defined as the amount μ you agree to pay, by knowing that you will receive the amount X if H is true, or you will receive back the amount μ if H is false (bet called off). Then, still denoting by $X|H$ the amount that you receive, it holds that $X|H = XH + \mu H^c$ and, in what follows, based on the assessment $\mathbb{P}(X|H) = \mu$, we will look at the c.r.q. $X|H$ as the *unconditional* r.q. $XH + \mu H^c$. Operatively, what you pay is *your prevision* for $X|H$; then by linearity $\mathbb{P}(X|H) = \mu = \mathbb{P}(XH + \mu H^c) = \mathbb{P}(XH) + \mu P(H^c)$, from which it follows $\mathbb{P}(XH) = P(H)\mu = P(H)\mathbb{P}(X|H)$. In particular, when X is an event A, the prevision of $X|H$ is the probability of $A|H$ and, if you assess $P(A|H) = p$, then for the indicator of $A|H$, denoted by the same symbol, we have $A|H = AH + pH^c \in \{1, 0, p\}$. Therefore $A|H \neq A$, but "conditionally on H being true", i.e. for $H = 1$, we have $A|H = A \in \{1, 0\}$, while $A|H = p$ for $H = 0$.

Some authors look at the conditional "if A then C", denoted $A \to C$, as the event $A^c \vee C$ (*material conditional*), but since some years it is becoming standard to look at $A \to C$ as the conditional event $C|A$ (see e.g. [4, 9]).

In [8], based on a complex procedure (which exploits the notion of *Stalnaker Bernoulli space*), by assuming $P(A)$ positive it is proved that $P(A \to C) = \frac{P(AC)}{P(A)} = P(C|A)$. Then, by defining truth values of $A \to C$ (like conditional events) as: $V(A \to C) = 1$, or 0, or $P(C|A)$, according to whether AC is true, or AC^c is true, or A^c, is true, it is verified that the expectation of $V(A \to C)$ is $P(C|A)$. Moreover, assuming $P(A \vee C) > 0$, Kaufmann obtains for the conjunction of $A \to B$ and $C \to D$ the formula

$$P[(A \to B) \wedge (C \to D)] = \frac{P(ABCD)+P(B|A)P(A^cCD)+P(D|C)P(ABC^c)}{P(A \vee C)}\,.$$

Based on this result, Kaufmann suggests a natural way of defining the values for the conjunction of conditionals. We will generalize the approach of Kaufmann in the setting of coherence in a direct and simpler way. Notice that in our paper the conjoined conditional will explicitly appear as a *conditional random quantity*; hence, we will speak of *previsions* (and not of probabilities).

3 Conjunction of Conditional Events

We preliminarily deepen an aspect of coherence and we exploit linearity of prevision to directly obtain the general compound prevision theorem.

Given any event $H \neq \emptyset$ and any random quantities X and Y, if $XH = YH$ (that is, for $H = 1$ it holds that $X|H = Y|H$), then coherence requires that $\mathbb{P}(X|H) = \mathbb{P}(XH|H) = \mathbb{P}(YH|H) = \mathbb{P}(Y|H)$. In other words

$$XH = YH \implies X|H = XH + \mathbb{P}(X|H)H^c = YH + \mathbb{P}(Y|H)H^c = Y|H\,. \quad (1)$$

Theorem 1. Given two events $H \neq \emptyset, K \neq \emptyset$ and a r.q. X, if the assessment (x, y, z) on $\{H|K, X|HK, XH|K\}$ is coherent, then $z = xy$.

Proof. We have

$$X|HK = XHK + y(HK)^c = XHK + yK^c + yH^cK = (XH + yH^c)K + yK^c\,;$$

moreover, by setting $\mathbb{P}[(XH + yH^c)|K] = \mu$, we have

$$(XH + yH^c)|K = (XH + yH^c)K + \mu K^c = XHK + yH^cK + \mu K^c\,.$$

As we can see: (i) $(XH + yH^c)|K = XH + yH^c$ when $K = 1$; (ii) $(XH + yH^c)|K = \mu$ when $K = 0$; moreover: (a) $X|HK = XH + yH^c$ when $K = 1$; (b) $X|HK = y$ when $K = 0$; that is, for $K = 1$, both $X|HK$ and $(XH + yH^c)|K$ coincide with $XH + yH^c$. Then, by the same reasoning as in (1), by coherence μ must coincide with y and by linearity of prevision:

$$\mu = y = \mathbb{P}(XH|K) + yP(H^c|K) = z + y(1 - x)\,;$$

hence: $z = xy$; that is: $\mathbb{P}(XH|K) = P(H|K)\mathbb{P}(X|HK)$. □

We now introduce the notion of conjunction, by first giving some logical and probabilistic remarks. Given any events A, B, H, with $H \neq \emptyset$, let us consider the conjunction AB, or the conjunction $(A|H) \wedge (B|H) = AB|H$. We have: $AB = min\{A, B\} = A{\cdot}B \in \{0,1\}$; moreover, if we assess $P(A|H) = x, P(B|H) = y$, then $A|H = AH + xH^c, B|H = BH + yH^c$ and for $H = 1$, i.e. *conditionally on H being true*, we have:

$$A|H = AH + xH^c = A \in \{0,1\}\,,\; B|H = BH + yH^c = B \in \{0,1\}\,,$$

$$AB|H = min\{A|H, B|H\}|H = min\{AH + xH^c, BH + yH^c\}|H \in \{0,1\}\,.$$

By defining $X = min\{A|H, B|H\} = min\{AH + xH^c, BH + yH^c\}$, we have $X \in \{1, 0, x, y\}$ and, for $H = 1$, $X|H = AB|H \in \{0,1\}$. Then, defining $\mathbb{P}(X|H) = \mu$, $P(AB|H) = z$, as in (1) by coherence $\mu = z$, so that for $H = 0$ we have $X|H = AB|H = z$. In other words, $min\{A|H, B|H\}|H$ and $AB|H$ are the same conditional random quantity. Then

$$(A|H) \wedge (B|H) = min\{A|H, B|H\}\,|\,H = min\{A|H, B|H\}\,|\,(H \vee H)\,. \quad (2)$$

In particular, for $B = A$, we have $A|H = (A|H)|H$, where $(A|H)|H$ is looked at as the c.r.q. $(AH + xH^c)|H$; this equality still holds from the viewpoint of iterated conditionals introduced in Section 6. Based on formula (2), we introduce below the notion of conjunction among conditional events.

Definition 1 (Conjunction). Given any pair of conditional events $A|H$ and $B|K$, with $P(A|H) = x, P(B|K) = y$, we define their conjunction as

$$(A|H) \wedge (B|K) = min\{A|H, B|K\}\,|\,(H \vee K) = (A|H) \cdot (B|K)\,|\,(H \vee K)\,.$$

Notice that, defining $Z = min\{A|H, B|K\} = (A|H) \cdot (B|K)$, the conjunction $(A|H) \wedge (B|K)$ is the c.r.q. $Z\,|\,(H \vee K)$.

Interpretation with the betting scheme. If you assess $\mathbb{P}[(A|H) \wedge (B|K)] = z$, then you agree to pay the amount z by receiving the amount $min\{A|H, B|K\}$ if $H \vee K$ is *true*, or the amount z if the bet is *called off* $(H \vee K\ false)$. That is, you pay z and you receive the amount

$$(A|H) \wedge (B|K) = \begin{cases} 1, & AHBK\ true \\ 0, & A^cH \vee B^cK\ true \\ x, & H^cBK\ true \\ y, & AHK^c\ true \\ z, & H^cK^c\ true\,; \end{cases}$$

therefore, *operatively*, $(A|H) \wedge (B|K)$ can be *represented* as:

$$(A|H) \wedge (B|K) \;=\; 1 \cdot AHBK + x \cdot H^cBK + y \cdot AHK^c + z \cdot H^cK^c\,.$$

Then, by *linearity* of prevision, it follows

$$\mathbb{P}[(A|H)\wedge(B|K)] = z = P(AHBK)+xP(H^cBK)+yP(AHK^c)+zP(H^cK^c)\,,$$

and we obtain $zP(H \vee K) = P(AHBK) + xP(H^cBK) + yP(AHK^c)$. In particular, if $P(H \vee K) > 0$, we obtain the result of Kaufmann

$$\mathbb{P}[(A|H) \wedge (B|K)] = \frac{P(AHBK) + P(A|H)P(H^cBK) + P(B|K)P(AHK^c)}{P(H \vee K)}\,.$$

Some particular cases. We examine below the conjunction of $A|H$ and $B|K$ when there are some logical dependencies among A, B, H, K and/or for special assessments (x, y) on $\{A|H, B|K\}$. We set $P(A|H) = x, P(B|K) = y, \mathbb{P}[(A|H) \wedge (B|K)] = z$.

1. If $x = y = 1$, then $(A|H) \wedge (B|K) = 1 \cdot AHBK + 1 \cdot H^cBK + 1 \cdot AHK^c + z \cdot H^cK^c = (AH \vee H^c) \wedge (BK \vee K^c)|(H \vee K) = \mathcal{C}(A|H, B|K)$, where $\mathcal{C}(A|H, B|K)$ is the *quasi conjunction* (see, e.g., [3, 5]) of $A|H$ and $B|K$.
2. $K = AH$. As "conditionally on H being true we have $(A|H) \wedge (B|AH) = AB|H$", from (1) it follows

$$(A|H) \wedge (B|AH) = 1 \cdot ABH + z \cdot H^c = AB|H = \mathcal{C}(A|H, B|AH)\,.$$

Then, by applying Theorem 1 to the family $\{A|H, B|AH, AB|H\}$, we have

$$\mathbb{P}[(A|H)\wedge(B|AH)] = P(AB|H) = P(A|H)P(B|AH) = \mathbb{P}(A|H)\mathbb{P}(B|AH)\,,$$

which means, as will see in Sec. 5, that $A|H$ and $B|AH$ are uncorrelated.
3. $A|H \subseteq B|K$, where $\subseteq$ denotes the inclusion relation of Goodman and Nguyen. In this case, coherence requires $x \leq y$; moreover, $A|H \leq B|K$, so that $min\,\{A|H, B|K\} = A|H$. Then $AHK^c = \emptyset$ and we have

$$(A|H) \wedge (B|K) = AH + xH^cBK + zH^cK^c\,, \qquad (3)$$

from which it follows $zP(H \vee K) = x[P(H) + P(H^cBK)]$. By observing that $HB^cK = \emptyset$, we have $H \vee K = H \vee H^cK = H \vee H^cBK$; then $zP(H \vee K) = xP(H \vee K)$, from which it follows $z = x$ if $P(H \vee K) > 0$. Then, by the continuity property of coherence with respect to passages to the limits, the evaluation $z = x$ is coherent also for $P(H \vee K) = 0$. By the methods of coherence, it can be shown that the extension $\mathbb{P}[(A|H) \wedge (B|K)] = z$ of the assessment $(x, 0)$ on $\{A|H, H \vee K\}$, where $A|H \subseteq B|K$, is coherent if and only if $z = x$. Then, from (3), as $H^cB^cK = \emptyset$ we obtain

$$(A|H)\wedge(B|K) = AH+x(H^cBK+H^cB^cK+H^cK^c) = A|H+xH^c = A|H\,.$$

4 Lower and Upper Bounds for $(A|H) \wedge (B|K)$

We will now determine the coherent extensions of the assessment (x, y) on $\{A|H, B|K\}$ to their conjunction $(A|H)\wedge(B|K)$. We recall that the extension $z = P(AB|H)$ of the assessment (x, y) on $\{A|H, B|H\}$, with A, B, H logically independent, is coherent if and only if: $max\{x + y - 1, 0\} \leq z \leq min\{x, y\}$. *The same results holds for* $(A|H) \wedge (B|K)$! We have

Theorem 2. Given any coherent assessment (x, y) on $\{A|H, B|K\}$, with A, H, B, K logically independent, and with $H \neq \emptyset, K \neq \emptyset$, the extension $z = \mathbb{P}[(A|H) \wedge (B|K)]$ is coherent if and only if the Fréchet-Hoeffding bounds are satisfied, that is

$$max\{x + y - 1, 0\} = z' \ \leq \ z \ \leq \ z'' = min\{x, y\}\,. \tag{4}$$

For reasons of space we give the proof in the appendix; here we only give a sketch of the proof by the following steps:

1) by the logical independence of the events A, H, B, K, it can be verified that the assessment (x, y) is coherent for every $(x, y) \in [0, 1]^2$;

2) the values z', z'' are determined by studying the coherence of the assessment $\mathcal{P} = (x, y, z)$ on $\mathcal{F} = \{A|H, B|K, (A|H) \wedge (B|K)\}$, by means of a geometrical approach (see, e.g., [5]);

3) the points associated with the constituents generated by $\mathcal{F}$ and contained in $H \vee K$ are: $Q_1 = (1, 1, 1), Q_2 = (1, 0, 0), Q_3 = (0, 1, 0), Q_4 = (0, 0, 0), Q_5 = (1, y, y), Q_6 = (0, y, 0), Q_7 = (x, 1, x), Q_8 = (x, 0, 0)$;

3) we consider the convex hull $\mathcal{I}$ of $Q_1, \ldots, Q_8$; then, we study the solvability of the linear system representing the condition $\mathcal{P} \in \mathcal{I}$, which is necessary, and in our case also sufficient, for the coherence of $\mathcal{P}$;

4) finally we obtain that, for any given pair $(x, y) \in [0, 1]^2$, the assessment $\mathcal{P}$ is coherent if and only if $max\{x + y - 1, 0\} = z' \leq z \leq z'' = min\{x, y\}$; i.e.

$$max\{P(A|H) + P(B|K) - 1, 0\} \leq \mathbb{P}[(A|H) \wedge (B|K)] \leq min\{P(A|H), P(B|K)\}\,.$$

We remark that for quasi conjunction the inequalities (4) do not hold; indeed, the extension $\gamma = P[\mathcal{C}(A|H, B|K)]$ of the assessment (x, y) is coherent if and only if $\gamma' \leq \gamma \leq \gamma''$, where $\gamma' = z' = \max\{x + y - 1, 0\}$ and $\gamma'' = S_0^H(x, y)$, where $S_0^H(x, y) = \frac{x+y-2xy}{1-xy}$ if $(x, y) \neq (1, 1)$, $S_0^H(x, y) = 1$ if $(x, y) = (1, 1)$ (Hamacher t-conorm). We observe that: $\gamma'' \geq max\{x, y\} \geq min\{x, y\} = z''$.

5 An Apparent Paradox on $(A|H) \wedge (B|K)$

In this section[1] we consider the case $HK = \emptyset$, where it seems that $A|H$ and $B|K$ are *stochastically independent*; this appears *unreasonable*; is it?

[1] The study of this case was stimulated by a discussion between D Edgington and A Gilio.

Actually, assuming $HK = \emptyset$, the constituents contained in $H \vee K$ are $C_1 = AHK^c, C_2 = A^cHK^c, C_3 = H^cBK, C_4 = H^cB^cK$ and, given the assessment $P(A|H) = x$, $P(B|K) = y$, $\mathbb{P}[(A|H) \wedge (B|K)] = z$, the associated vectors of numerical values for $A|H, B|K, (A|H) \wedge (B|K)$ are $Q_1 = (1, y, y), Q_2 = (0, y, 0), Q_3 = (x, 1, x), Q_4 = (x, 0, 0)$. Let $\mathcal{I}$ be the convex hull of $Q_1, \ldots, Q_4$. In our case, the condition $\mathcal{P} \in \mathcal{I}$, which is necessary for the coherence of $\mathcal{P}$, is also sufficient and after some computation on the associated linear system it can be verified that $\mathcal{P}$ is coherent if and only if $z = xy$ and $(x, y) \in [0, 1]^2$. Therefore, coherence requires that

$$\mathbb{P}[(A|H) \wedge (B|K)] = P(A|H)P(B|K) = \mathbb{P}(A|H)\mathbb{P}(B|K). \tag{5}$$

Does (5) mean that $A|H$ and $B|K$ are stochastically independent? The answer, as shown below, is *negative.* Indeed, we observe that:

(i) by Definition 1, $(A|H) \wedge (B|K) = (A|H) \cdot (B|K) \mid (H \vee K)$ is a *conditional random quantity,* not a conditional event; then the correct framework for giving a meaning to equality (5) is that of random quantities; moreover, in our case we have: $(A|H) \wedge (B|K) = xH^cBK + yAHK^c + zH^cK^c$;

(ii) $(A|H) \cdot (B|K) = (AH + xH^c)(BK + yK^c) = xH^cBK + yAHK^c + xyH^cK^c$;

(iii) as $z = xy$, we have $(A|H) \wedge (B|K) = (A|H) \cdot (B|K)$; that is, the conjunction is the *product* of the *conditional random quantities* $A|H, B|K$.

Then, (5) only means that $A|H$ and $B|K$ are *uncorrelated,* and does not mean that they are independent. Hence, by the previous reasoning we have proved

Theorem 3. Given any events A, B, H, K, with $H \neq \emptyset, K \neq \emptyset, HK = \emptyset$, it holds that $\mathbb{P}[(A|H) \cdot (B|K)] = \mathbb{P}(A|H)\mathbb{P}(B|K)$; that is, the random quantities $A|H$ and $B|K$ are uncorrelated.

We remark that, as shown in case 2, Section 3, where $B = AH$, $A|H$ and $B|K$ could be uncorrelated even if $HK \neq \emptyset$. Indeed, by formula (1), we have

$$(A|H) \cdot (B|AH) = ABH + xy \cdot H^c = ABH + z \cdot H^c = (A|H) \wedge (B|AH),$$

and then $\mathbb{P}[(A|H) \cdot (B|AH)] = \mathbb{P}[(A|H) \wedge (B|AH)] = \mathbb{P}(A|H)\mathbb{P}(B|AH)$.

6 Disjunction and Iterated Conditioning

We define below the notions of disjunction and of iterated conditioning; in [6] we are working on an expanded version of this paper. A notion of conditioning among random quantities has been studied in [1].

Definition 2 (Disjunction). Given any pair of conditional events $A|H$ and $B|K$, we define $(A|H) \vee (B|K) = max\,\{A|H, B|K\} \mid (H \vee K)$.

By assessing $P(A|H) = x, P(B|K) = y, \mathbb{P}[(A|H) \vee (B|K)] = \gamma$, we have $(A|H) \vee (B|K) = 1 \cdot (AH \vee BK) + x \cdot H^cB^cK + y \cdot A^cHK^c + \gamma \cdot H^cK^c$. By coherence, it can be proved that the *prevision sum rule* holds, that is $\mathbb{P}[(A|H) \vee (B|K)] = \mathbb{P}(A|H) + \mathbb{P}(B|K) - \mathbb{P}[(A|H) \wedge (B|K)]$, and from (4)

$$max\{P(A|H), P(B|K)\} \leq \mathbb{P}[(A|H) \vee (B|K)] \leq min\{P(A|H) + P(B|K) - 1, 1\}$$

Definition 3 (Iterated conditioning). Given any pair of conditional events $A|H$ and $B|K$ we define the iterated conditional $(B|K)|(A|H)$ as

$$(B|K)|(A|H) = (B|K) \wedge (A|H) + \mu A^c|H\,,$$

where μ is the prevision of $(B|K)|(A|H)$ and represents the amount you agree to pay, with the proviso that you will receive the quantity $(B|K)|(A|H)$.

If $P(A|H) = x, P(B|K) = y, \mathbb{P}[(A|H) \wedge (B|K)] = z$, the values of $(B|K)|(A|H)$ are $1, 0, y, \mu, x + \mu(1-x), \mu(1-x), z + \mu(1-x)$, respectively associated with the constituents $AHBK, AHB^cK, AHK^c, A^cH, H^cBK, H^cB^cK, H^cK^c$. By linearity of prevision: $\mathbb{P}[(B|K)|(A|H)] = \mu = \mathbb{P}[(B|K) \wedge (A|H)] + \mu P(A^c|H)$; that is: $\mu = z + \mu(1-x)$, from which it follows

$$\mathbb{P}[(B|K) \wedge (A|H)] = \mathbb{P}[(B|K)|(A|H)]P(A|H)\,. \qquad (6)$$

Then, assuming $x = P(A|H) > 0, P(H \vee K) > 0$, one has: $\mathbb{P}[(B|K)|(A|H)] = \mu = \frac{\mathbb{P}[(B|K) \wedge (A|H)]}{P(A|H)} = \frac{z}{x} = \frac{P(AHBK) + P(A|H)P(H^cBK) + P(B|K)P(AHK^c)}{P(A|H)P(H \vee K)}$, which coincides with the result of Kaufmann. If we only assign x and y, then $max\{0, x+y-1\} \leq z \leq min\{x, y\}$, and it follows $\mu \in [\mu', \mu'']$, with $\mu' = 0$, $\mu'' = 1$ for $x = 0$, and with $\mu' = max\{0, \frac{x+y-1}{x}\}$, $\mu'' = min\{1, \frac{y}{x}\}$ for $x > 0$.

Acknowledgements. We thank the anonymous referees for their useful comments.

References

1. Biazzo, V., Gilio, A., Sanfilippo, G.: On general conditional prevision assessments. In: Proc. WUPES 2009, Liblice, Czech Rep., September 19-23, pp. 23–34 (2009)
2. Capotorti, A., Vantaggi, B.: A general interpretation of conditioning and its implication on coherence. Soft Comput. 3, 148–153 (1999)
3. Dubois, D., Prade, H.: Conditional objects as nonmonotonic consequence relationships. IEEE Trans. Syst. Man Cybern. 24, 1724–1740 (1994)
4. Gilio, A., Over, D.: The psychology of inferring conditionals from disjunctions: A probabilistic study. J. Math. Psych. 56(2), 118–131 (2012)
5. Gilio, A., Sanfilippo, G.: Quasi Conjunction and Inclusion Relation in Probabilistic Default Reasoning. In: Liu, W. (ed.) ECSQARU 2011. LNCS, vol. 6717, pp. 497–508. Springer, Heidelberg (2011)

6. Gilio, A., Sanfilippo, G.: A Coherence-Based Probabilistic Theory of Compounds of Conditionals. Working Paper; Presented at the Workshop Conditionals, Counterfactuals and Causes in Uncertain Environments, Düsseldorf, May 19-22, 2011 (2012), Slides available at `http://www.phil-fak.uni-duesseldorf.de/conditionals/`
7. Goodman, I.R., Nguyen, H.T., Walker, E.A.: Conditional Inference and Logic for Intelligent Systems: A Theory of Measure-free Conditioning. North-Holland (1991)
8. Kaufmann, S.: Conditionals right and left: Probabilities for the whole family. J. Philos. Logic. 38, 1–53 (2009)
9. Pfeifer, N., Kleiter, G.D.: The conditional in mental probability logic. In: Oaksford, M., Chater, N. (eds.) Cognition and Conditionals: Probability and Logic in Human Thought, pp. 153–173. Oxford University Press (2010)

Ockham's Razor in Probability Logic

Gernot D. Kleiter

Abstract. The paper investigates the generalization, the composition, and the chaining of argument forms in conditional probability logic. Adding premises to probabilistic argument forms does not necessarily improve the information transmitted to the conclusions, but quite contrary, usually leads to probabilistic less informative conclusions. Selecting the one or two most relevant premises results in a good Ockham razor in probability logic. The consequences for modeling human uncertain reasoning and human judgment and decision making are discussed.

Keywords: Generalized inference rules, imprecise probabilities, probability logic.

1 Introduction

Assume your orthopedist recommends a replacement surgery of one of your hips. You feel very uncertain and remember the advice to get a second opinion. Will the second opinion reduce your uncertainty? Consider two probabilistic arguments about the same decision. Is the combination of the two arguments better than the better one, worse than the worse one, or a compromise of the two arguments? Often simple models are doing better than more complex ones—less can be more. The principles guiding an Ockham razor to select and use information efficiently in judgment under uncertainty and decision making were extensively investigated in psychology [4, 7].

Probability logic investigates the propagation of probability assessments from premises to conclusions. This includes the study of the probabilistic

Gernot D. Kleiter
Department of Psychology, University of Salzburg, Austria
e-mail: gernot.kleiter@sbg.ac.at

R. Kruse et al. (Eds.): Synergies of Soft Computing and Statistics, AISC 190, pp. 409–417.
springerlink.com © Springer-Verlag Berlin Heidelberg 2013

properties and the logical properties of the arguments. Typical argument forms are the MODUS PONENS or the MODUS TOLLENS, a well-known probabilistic property is p-validity [2], and a logical property is nonmonotonicity. The probability assessments of the premises are assumed to be coherent and usually assumed to be precise (point probabilities). The probability of the conclusion is usually imprecise (an interval probability). The export of typical argument forms from classical to probability logic leads, with a few exceptions, to interval probabilities of the conclusions.

The present paper is based on the work of Gilio [6]. It investigates probabilistic argument forms like MODUS PONENS or MODUS TOLLENS containing more than the usual two premises. Such arguments have, say, n premises and involve m propositional variables or events. If the basic events in the inference rules are logically independent, then $2^m - 1$ probabilities are required to infer the point probability of the conclusions (which are logically dependent on the basic events). Adding a new event (variable) requires to double the number of given probabilities attached to the premises to infer a precise probability for the conclusion again. If not $2^m - 1$ but only $n = m$ probabilities are specified, as in many generalized argument forms, then the degree of incompleteness increases rapidly as m increases. In this case $2^m - (m+1)$ constraints remain unspecified. As a consequence, conclusions become more and more imprecise. After the inclusion of only a few additional events the conclusions may obtain any probability in the interval between zero and one. An attractive property of probability logic is that it may be interpreted as a nonmonotonic inference system [5]. It allows to retract old conclusions in the light of new evidence. The rapidly growing incompleteness in the light of new evidence, however, signals a strong nonmonotonicity.

2 Pseudodiagnosticity

The pseudodiagnosticity task [3, 11] belongs to the standard repertoire in the psychology of uncertain reasoning. Imagine a medical scenario with a disease H and two symptoms E_1 and E_2. You know the prevalence of the disease, $P(H)$, and one likelihood, $P(E_1|H)$. These two values describe the *status quo*. You may acquire the value of *one* more probability. The options are (i) the likelihood of the first symptom under the alternative hypothesis, $P(E_1|H^c)$, (ii) the likelihood of the second symptom given the disease H, $P(E_2|H)$, or (iii) the likelihood of the second symptom given the alternative disease H^c, $P(E_2|H^c)$. Most subjects prefer $P(E_2|H)$ to $P(E_1|H^c)$. This is not a good choice since it does not provide the probabilities for the application of the standard form of Bayes' Theorem but only for an incomplete version of it. The incomplete version leads to an interval posterior probability where the lower probability is lower than in the *status quo*. Here additional information leads to a less informative conclusion. The fact that more information does not improve the inference but, quite to the contrary, make it worse seems highly counter-intuitive.

For the first symptom in the scenario Bayes' Theorem has the form

$$\frac{P(H)=\alpha, P(E_1|H)=\beta_1, P(E_1|H^c)=\lambda_1}{P(H|E_1)=\gamma=\frac{\alpha\beta_1}{\alpha\beta_1+(1-\alpha)\lambda_1}}.$$

If λ_1 is not given it may have any value between 0 and 1 and $P(H|E_1)$ is in the interval $\gamma \in [\gamma', 1]$, where

$$\gamma' = \frac{\alpha\beta_1}{\alpha\beta_1+1-\alpha} \quad .$$

If there are n symptoms with the likelihoods $P(E_1|H)=\beta_1,\ldots,P(E_n|H)=\beta_n$ and the likelihoods $P(E_1|H^c),\ldots,P(E_n|H^c)$ under the alternative H^c are *not* given, and if conditional independence is assumed, then the lower probability $P(H|E_1,\ldots,E_n)$ decreases monotonically with increasing n,

$$\gamma' = \frac{\alpha\prod_{i=1}^n \beta_i}{\alpha\prod_{i=1}^n \beta_i + 1-\alpha} .$$

If the conditional independence assumption is dropped, the lower probability γ' is obtained from the minimum numerator, $P(H)P(E_1,\ldots,E_n|H)$, so that

$$\gamma' = \frac{\alpha\tau'}{\alpha\tau'+1-\alpha}, \quad \text{where} \quad \tau' = \max\left\{0, \sum_{i=1}^n \beta_i + 1 - n\right\}$$

for the case of n symptoms. The upper probability is $\gamma''=1$. Here τ' is the lower probability of a conjunction when only the probabilities of its elements are given, $P(A_1,\ldots,A_n|H) \in [\max\{0,\sum_{i=1}^n P(A_i|H)+1-n\}, \min\{P(A_1|H),\ldots,P(A_n|H)\}]$.

Example 1. If $P(H)=.5$ and $P(E_1|H)=.7$ are available only, then $P(H|E_1)\in[.41,1]$. If in addition $P(E_2|H)=.7$ is included as a third premise, then $\gamma'=\frac{.5(.7+.7-1)}{.5(.7+.7-1)+.5}=.29$ and $\gamma\in[.29,1]$. We see that the interval of $P(H|E_1,E_2)$ is wider than the interval of $P(H|E_1)$.

With the inclusion of additional "confirmatory" likelihoods the posterior probability gets worse and approaches the completely non-informative $[0,1]$ interval. This is a "less is more" situation. Even if $\beta_i=1$ the result does not improve but only stay at its previous level.

3 Nonmonotonicity in Generalized Arguments

Classical logic is monotone. Adding premises to a valid argument preserves its conclusion. If p follows from X, then p follows from the union of X and an arbitrary q. Commonsense inference, however, can be nonmonotone. Often conclusions are withdrawn in the light of new evidence. Several logical systems were proposed to account for nonmonotone reasoning. Well known is SYSTEM P. It admits a probabilistic interpretation [8, 5]. The rules of SYSTEM P replace valid inference rules of classical logic by *p-valid* rules of conditional probability logic [2]. Gilio [6] investigated the propagation of probabilities in generalized SYSTEM P rules. In these generalized rules the interval probabilities of the conclusions get wider and wider approaching the non-informative $[0, 1]$ interval as more and more premises enter the premise set. This property is the same as in the pseudodiagnosticity task.

Disjunction. For the OR rule it is sufficient to consider the case of only two premises and compare the probability of the conclusion with the case of one premise. For two premises we have [6]

$$\frac{P(B|A_1) = \alpha_1, P(B|A_2) = \alpha_2}{P(B|A_1 \vee A_2) \in [\gamma', \gamma'']} \quad ,$$

where $\gamma' = \frac{1}{1+u}$ with $u = \frac{1-\alpha_1}{\alpha_1} + \frac{1-\alpha_2}{\alpha_2}$ and $\gamma'' = \frac{v}{1+v}$ with $v = \frac{\alpha_1}{1-\alpha_1} + \frac{\alpha_2}{1-\alpha_2}$. We see that α_1 and α_2 are both inside $[\gamma', \gamma'']$. If the number of events in the disjunction increases, the interval approaches $[0, 1]$. We lose information.

Predictive probabilities. The probability of a success in trial $n + 1$ after having observed r successes and s failures in the preceding n trials, is called a *predictive probability* (corresponding to CAUTIOUS MONOTONICITY in SYSTEM P). In the special case of identical probabilities of the events, $P(E_i) = \alpha, i = 1, \ldots, n + 1$, the conditional probability for the next success, $P(E_{n+1}|E_1, \ldots, E_r, E_{r+1}, \ldots, E_n)$, is in the interval $[\gamma', 1]$, where for $r = n$

$$\gamma' = \max\left\{0, \frac{(n+1)\alpha - n}{n\alpha + 1 - n}\right\}, \quad \text{if} \quad n\alpha + 1 - n > 0$$

and $\gamma' = 0$ if $n\alpha + 1 - n \leq 0$. If $r < n$, i.e., if there is at least one failure, then the predictive interval is non-informative, $\gamma \in [0, 1]$ [13].

Example 2. If $\alpha = .9$, then *after* n successes in n trials the lower probabilities that the *next* observation will be a success are for $n = 0, 1, \ldots, 8$ equal to $.9, .875, .857, .833, .8, .75, .667, .5, 0$. The upper probability is always 1.

If $\alpha = .5$, then before any observations are made the predictive probability is .5 by definition, but after one or more observations the predictive probability may have any value between 0 and 1. If no information about dependence or independence of the events is available, predictions are impossible. Even "gambler's fallacy" is coherent in this case, that is, for example, favoring failure after a series of successes.

MODUS PONENS. The probabilistic form of the MODUS PONENS (a special case of the CUT rule of SYSTEM P) is:

$$\frac{P(A) = \alpha, P(B|A) = \beta}{P(B) \in [\gamma', \gamma'']} ,$$

where $\gamma' = \alpha\beta$ and $\gamma'' = 1 - \alpha + \alpha\beta$. The following generalization of the MODUS PONENS is given by Gilio [6]:

$$\frac{P(E_1) = \alpha_1, P(E_2) = \alpha_2, \ldots, P(E_n) = \alpha_n, P(H|E_1, E_2, \ldots, E_n) = \beta}{P(H) \in [\gamma', \gamma'']} ,$$

where $\gamma' = \max\{0, \beta[1 - \sum_{i=1}^{n}(1 - \alpha_i)]\}$ and $\gamma'' = \min\{1, 1 - (1 - \beta)[1 - \sum_{i=1}^{n}(1 - \alpha_i)]\}$. This and all other results in the paper are obtained by the elementary rules of probability (as best illustrated by Gilio [6]) or by solving linear equations.

If the conditional probability β in the premises is kept constant and if the number of categorical premises increases, then the interval of the conclusion gets wider and wider. The reason for this degradation is the import of events without simultaneously importing information about their interdependencies.

Example 3. For $\alpha = .7$ and $\beta = .9$ we obtain $\gamma \in [.63, .93]$. If we have two categorical premises with identical $\alpha_1 = \alpha_2 = .7$, and assume the same $\beta = .9$, then we obtain the interval $\gamma \in [.36, .96]$. With three categorical premises and $\alpha_1 = \alpha_2 = \alpha_3 = .7$, and as before, $\beta = .9$, we obtain $\gamma \in [.09, .99]$.

MODUS TOLLENS. The elementary probabilistic MODUS TOLLENS is:

$$\frac{P(B^c) = \alpha, P(B|A) = \beta}{P(A^c) \in [\gamma', 1]}$$

$$\gamma' = \begin{cases} \max\{0, 1 - \frac{\alpha}{1-\beta}\} & \text{if } \alpha + \beta \leq 1 \text{ and } (\alpha, \beta) \neq (0, 1) \\ 1 - \frac{1-\alpha}{\beta} & \text{if } \alpha + \beta > 1 \text{ or } (\alpha, \beta) = (0, 1) \end{cases}$$

(Thanks to one of the reviewers for the compact representation!) The generalized form has n categorical premises, $P(E_i^c) = \alpha_i$, $i = 1, \ldots, n$, and the conditional premise $P(E_1, \ldots, E_n|H) = \beta$. Let $\alpha^* = max\{\alpha_1, \ldots, \alpha_n\}$. The probability of the conclusion $P(H^c)$ is in the interval $[\gamma', 1]$, where

$$\gamma' = \begin{cases} \max\{0, 1 - \frac{\sum_{i=1}^{n} \alpha_i}{1-\beta}\} & \text{if } \alpha^* + \beta \leq 1 \text{ and } (\alpha^*, \beta) \neq (0, 1) \\ 1 - \frac{1-\alpha^*}{\beta} & \text{if } \alpha^* + \beta > 1 \text{ or } (\alpha^*, \beta) = (0, 1) . \end{cases}$$

If $\alpha^* + \beta > 1$, then of all categorical premises only the one with the maximum probability is relevant.

Example 4. If there are 4 premises with $\alpha_1 = .2, \alpha_2 = .4, \alpha_3 = .6$, and $\alpha_4 = .8$, respectively, and if $\beta = .4$, then $\gamma \in [.5, 1]$. Here $\alpha^* = .8$ so that $1 - (1 - .8)/.4 = .5$. The premises with the probabilities .2, .4, and .6 are irrelevant. If $\alpha_1 = .02, \alpha_2 = .04, \alpha_3 = .06$, and $\alpha_4 = .08$ the sum is $\sum_{i=1}^{n} \alpha_i = .2$ and $\gamma \in [.667, 1]$. The interval is wider than for any of the single categorical premises alone; they are $[.966, 1]$, $[.933, 1]$, $[.9, 1]$, and $[.866, 1]$, respectively.

4 Composition of Arguments

Classical logic and commonsense reasoning accept the AND rule of the consequence relation: If p follows from X and p follows from Y, then p follows from the union of X and Y. Does an analog rule hold in probabilistic inference?

Modus ponens. The composition of two MODI PONENTES results in taking two times the best, one time for the lower and one time for the upper probability:

$$\text{If} \quad \frac{P(A_1) = \alpha_1, P(B|A_1) = \beta_1}{P(B) \in [z_1', z_1'']} \quad \text{and} \quad \frac{P(A_2) = \alpha_2, P(B|A_2) = \beta_2}{P(B) \in [z_2', z_2'']},$$

$$\text{then} \quad \frac{P(A_1) = \alpha_1, P(A_2) = \alpha_2, P(B|A_1) = \beta_1, P(B|A_2) = \beta_2}{P(B) \in [z_3', z_3'']},$$

where $z_3' = \max\{z_1', z_2'\}$ and $z_3'' = \min\{z_1'', z_2''\}$ holds. The interval of the composition is tighter than that any of the single arguments.

Modus Tollens. We have

$$\text{If} \quad \frac{P(B^c) = \alpha_1, P(B|A) = \beta_1}{P(A^c) \in [\gamma_1', 1]} \quad \text{and} \quad \frac{P(C^c) = \alpha_2, P(C|A) = \beta_2}{P(A^c) \in [\gamma_2', 1]}$$

$$\text{then} \quad \frac{P(B^c) = \alpha_1, P(C^c) = \alpha_2, P(B|A) = \beta_1, P(C|A) = \beta_2}{P(A^c) \in [\gamma_3', 1]},$$

where $\gamma_3' = \max\{\gamma_1', \gamma_2'\}$. The composition of the two arguments is obtained by the conjunction of B^cC^c and of $BC|A$, respectively. The result corresponds to the take-the-best strategy in the psychology of decision making [4].

Example 5. If $\alpha_1 = .6, \beta_1 = .8, \alpha_2 = .4, \beta_2 = .9$, for the first argument we have $\gamma_1 \in [.5, 1]$, for the second one we have $\gamma_2 \in [.33, 1]$; the composition results in $\gamma_3 \in [.5, 1]$. The best argument wins.

5 Chaining Arguments

Forward chaining. Consider three events A, B, and C. $P(A)$, $P(B|A)$, and $P(C|B)$ are known and we want to determine $P(C)$. We apply the MODUS PONENS two times in succession, first to infer $P(B)$ from $P(A)$ and $P(B|A)$ and second to infer $P(C)$ from $P(B)$ and $P(C|B)$. The result of the first step is an interval, so that the second step applies the MODUS PONENS for premises with interval assessments:

$$\frac{P(A) \in [\alpha', \alpha''], P(B|A) \in [\beta', \beta'']}{P(B) \in [\alpha'\beta', 1 - \alpha' + \alpha'\beta'']}$$

Combining the MODUS PONENS result for point probabilities and for interval probabilities leads to the result for forward chaining:

$$\frac{P(A) = \alpha, P(B|A) = \beta_1, P(C|B) = \beta_2}{P(C) \in [\alpha\beta_1\beta_2, 1 - \beta_2 + \alpha_1\beta_1\beta_2]\,.}$$

Backward chaining. Consider three events A, B, and C. Assume $P(A) = \alpha$, $P(B|A) = \beta_1$, and $P(C|B) = \beta_2$ are given. We determine the probability $P(A|BC) = \gamma$. Using Bayes' Theorem we have $P(A|BC) = \frac{P(ABC)}{P(ABC)+P(A^cBC)}$. The probability of ABC is factorized, $P(ABC) = P(A)P(B|A)P(C|AB)$. If A and C are conditionally independent given B, $P(C|AB)$ may be replaced by $P(C|B)$, so that the factorization simplifies to $P(A)P(B|A)P(C|B) = \alpha\beta_1\beta_2$. As no information about the constituents involving A^c is given, $P(A^cBC)$ may have any value between zero and one and we obtain $\gamma \in \left[\frac{\alpha\beta_1\beta_2}{\alpha\beta_1\beta_2+1-\alpha}, 1\right]$.

Now we drop the conditional independence assumption. As a consequence, the probability of the conjunction becomes an interval probability $[\omega', \omega'']$. The values of ω' and ω'' are obtained by solving a system of linear equations involving the eight constituents $ABC, A^cBC, \ldots, A^cB^cC^c$ and the constraints α, β_1, β_2. In the present context only ω', the value of the lower probability of the conjunction, is relevant. We obtain $\omega' = \max\{0, \beta_2(1 - \alpha(1 - \beta_1)) - (1 - \alpha)\}$. The upper probability ω'' is not needed as we already know that the upper value of the ratio $\frac{P(ABC)}{P(ABC)+P(A^cBC)}$ is 1 (if $P(A^cBC) = 0$, assuming $P(ABC) > 0$). The lower value of the ratio is obtained when the probability of ABC obtains its minimum.

To sum up, for backward chaining we obtain:

$$\frac{P(A) = \alpha, P(B|A) = \beta_1, P(C|B) = \beta_2}{P(A|BC) \in [\frac{\omega'}{\omega'+1-\alpha}, 1]}\,,$$

where $\omega' = \max\{0, \beta_2(1-\alpha(1-\beta_1))-(1-\alpha)\}$. When the length of the chain increases, the lower probability is quickly approaching zero.

Example 6. If $\alpha = .5, \beta_1 = .9, \beta_2 = .7$, $\omega' = .165$, then $P(A|BC) \in [.248, 1]$. If $\alpha = .5, \beta_1 = .7, \beta_2 = .7, \omega' = .095$, then $P(A|BC) \in [.16, 1]$. In the first step of the pseudodiagnosticity example (Section 2) we had, for comparison, $\alpha = .5, \beta_1 = .7$, $P(AB) = \omega' = .35, P(A|B) \in [.412, 1]$.

6 Discussion

Probability logic is a weak inference system. This is the price of nonmonotonicity. The informativeness of conclusions decrease as more premises are added. This was also observed by Adams [1, p. 9]: "... the more premisses there are, each with its 'quantum of uncertainty', the less sure we can be of the conclusions arrived at, and these quanta can 'accumulate' as the number of premisses increases, even to the point of absurdity ... ". The probabilities of the aggregation of several arguments need not be located in the middle between the values of the original arguments but may be at the maximum/minimum values or even outside the range of the original arguments.

Probability logic has been used as a framework for modeling human reasoning [10]. In human reasoning not only elementary arguments with just two premises, but the more complex argument forms that we have discussed here, are relevant. Often people reject the idea that more information can make inferences worse. Noisy information is considered to be irrelevant and discarded *before* it enters inferences. This may lead to a *take-the-best* strategy. Although such a strategy conflicts with the principle of *total evidence* there are claims saying that take-the-best strategies are often efficient and often actually employed [4]. Ockham's razor can be a rational principle in probability logic, a principle where probability logic and human judgment meet. Unfortunately, though, probability logic is dumb about correlation, dependence or independence. Conditional probability logic (such as SYSTEM P [8] or the work of Adams [2]) consists of two parts, a probabilistic part and a logical part. The probabilistic part investigates the propagation of probabilities from premises to conclusions. Some of these inferences are considered to be "acceptable", some not. p-valid inferences are, for example, considered to be acceptable because they preserve high probabilities. While dependence and independence have extensively been studied in probability theory, they have not been investigated in standard first order logic. Thus probability logic inherits this weakness from classical logic. As dependence and independence are highly important in everyday reasoning probability logic can at best be half of a frame of reference of artificial or human reasoning. Only very recently logicians started to study independence and dependence systematically [12, 9].

Acknowledgements. Supported by the Austrian Science Foundation (FWF, I 141-G15) within the LogICCC Programme of the European Science Foundation. Many thanks to Christian Wallmann and three reviewers to help to improve the paper.

References

1. Adams, E.W.: Four probability-preserving properties of inferences. J. Philos. Logic. 25, 1–25 (1996)
2. Adams, E.W.: A primer of probability logic. Center for the Study of Language and Information Publications, Stanford (1998)
3. Doherty, M.E., Mynatt, C.R., Tweney, R.D., Schiavo, M.D.: Pseudodiagnosticity. Acta Psychologica 43, 111–121 (1979)
4. Gigerenzer, G., Todd, P.M., thee ABC Research Group: Simple Heuristics that Make us Smart. Oxford University Press, New York (1999)
5. Gilio, A.: Probabilistic reasoning under coherence in System P. Ann. Math. Artif. Intel. 34, 5–34 (2002)
6. Gilio, A.: Generalization of inference rules in coherence-based probabilistic default reasoning. Int. J. Approx. Reason. 53, 413–434 (2012)
7. Goldstein, D.G., Gigerenzer, G.: The beauty of simple models: Themes in recognition heuristic research. Judgm. Decis. Mak. 6, 392–395 (2012)
8. Kraus, S., Lehmann, D., Magidor, M.: Nonmonotonic reasoning, preferential models and cumulative logics. Artif. Intel. 44, 167–207 (1990)
9. Mann, A.L., Sandu, G., Sevenster, M.: Independence-Friendly Logic. Cambridge University Press, Cambridge (2011)
10. Pfeifer, N., Kleiter, G.D.: Uncertainty in deductive reasoning. In: Manktelow, K., Over, D., Elqayam, S. (eds.) The Science of Reason: A Festschrift for Jonathan St B. T. Evans, pp. 145–166. Psychology Press, Hove (2011)
11. Tweney, R.D., Doherty, M.E., Kleiter, G.D.: The pseudodiagnosticity trap. Should subjects consider alternative hypotheses? Think Reasoning 16, 332–345 (2010)
12. Väänänen, J.: Dependence Logic. Cambridge University Press, Cambridge (2007)
13. Wallmann, C., Kleiter, G.D.: Exchangeability in Probability Logic. In: Greco, S., Bouchon-Meunier, B., Coletti, G., Fedrizzi, M., Matarazzo, B., Yager, R.R. (eds.) IPMU 2012, Part IV. CCIS, vol. 300, pp. 157–167. Springer, Heidelberg (2012)

Conglomerable Coherent Lower Previsions

Enrique Miranda and Marco Zaffalon

Abstract. Walley's theory of coherent lower previsions builds upon the former theory by Williams with the explicit aim to make it deal with conglomerability. We show that such a construction has been only partly successful because Walley's founding axiom of joint coherence does not entirely capture the implications of conglomerability. As a way to fully achieve Walley's original aim, we propose then the new theory of *conglomerable coherent lower previsions*. We show that Walley's theory coincides with ours when all conditioning events have positive lower probability, or when conditioning partitions are nested.

Keywords: Coherent lower previsions, conglomerability, sets of desirable gambles, Williams' coherence.

1 Introduction

There are two main behavioural theories of *coherent lower previsions* (these are lower expectation functionals): Walley's [3] and Williams' [4]. The main difference between them lies in the notion of *conglomerability*. This is the property that allows us to write an expectation as a mixture of conditional expectations. De Finetti discovered in 1930 that conglomerability can fail when finitely additive probabilities, as well as infinitely many conditioning events,

Enrique Miranda
University of Oviedo, Department of Statistics and Operations Research, 33007 Oviedo, Spain
e-mail: mirandaenrique@uniovi.es

Marco Zaffalon
IDSIA, 6928 Manno (Lugano), Switzerland
e-mail: zaffalon@idsia.ch

R. Kruse et al. (Eds.): Synergies of Soft Computing and Statistics, AISC 190, pp. 419–427.
springerlink.com © Springer-Verlag Berlin Heidelberg 2013

enter the picture [1]. Walley developed his theory by modifying Williams' so as to account for conglomerability in such non-finitary setting. It is controversial that conglomerability should always be imposed; however, we have argued elsewhere [5] that this should be the case when one establishes right from the start that conditional probabilities will be used to determine future behaviour.

From this discussion, it may seem that Walley's theory should be the one to use when conglomerability is required. However, some recent research has shown that a basic procedure to construct rational models in Walley's theory does not fully consider the implications of conglomerability [2]. In this paper we take a closer look at this problem by analysing the core of Walley's theory: his notion of self-consistency for coherent lower previsions, which is called *joint coherence*. This can be regarded as the single axiom of Walley's theory.

To this end, we need to work with the theory of *coherent sets of desirable gambles*, which generalise coherent lower previsions. We review these theories and present some preliminary results in Section 2. We start our actual investigation in Section 3. We define the new theory of *conglomerable coherent lower previsions* based on desirable gambles and conglomerability. The founding axiom of this theory is called *conglomerable coherence*. We argue that this is the axiom one should use whenever conglomerability is required. Then we give the relationships about the several consistency notions in Walley's, Williams', and our new theory, with special regard to conglomerability. Most importantly, we show in Example 1 that Walley's joint coherence is not equivalent to conglomerable coherence, which is stronger. In our view, this implies that Walley's theory should be regarded as an approximation to the actual theory to use under conglomerability.

This approximation becomes exact in some important cases, which we discuss in Section 4: when either (i) every conditioning event has positive lower probability or (ii) we consider nested partitions, that is, partitions that are finer and finer, then joint coherence coincides with conglomerable coherence. These two outcomes are important because working with conglomerable coherence can be much more difficult than with Walley's joint coherence.

2 Coherent Lower Previsions and Sets of Desirable Gambles

Given a possibility space Ω, a *gamble* f is a bounded real-valued function on Ω. $\mathcal{L}(\Omega)$ (or $\mathcal{L}$) denotes the set of all gambles on Ω, and $\mathcal{L}^+(\Omega)$ (or just $\mathcal{L}^+$) the set of so-called positive gambles: $\{f \in \mathcal{L} : f \gneq 0\}$ (where $f \gneq 0$ is a shorthand for $f \geq 0$ and $f \neq 0$). A *lower prevision* $\underline{P}$ is a real functional defined on $\mathcal{L}$. From any lower prevision $\underline{P}$ we can define an upper prevision

$\overline{P}$ using conjugacy: $\overline{P}(f) := -\underline{P}(-f)$. Precise previsions, which are those for which $\underline{P}(f) = \overline{P}(f)$, are denoted by $P(f)$.

Given $B \subseteq \Omega$, the real value $\underline{P}(f|B)$ denotes the lower prevision of f conditional on B. Given a partition $\mathcal{B}$ of Ω, then we shall represent by $\underline{P}(f|\mathcal{B})$ the gamble on Ω that takes the value $\underline{P}(f|\mathcal{B})(\omega) = \underline{P}(f|B)$ iff $\omega \in B$. The functional $\underline{P}(\cdot|\mathcal{B})$ is called a *conditional lower prevision*. We say that it is *separately coherent* (or just a *linear prevision*, in the precise case) when for all $B \in \mathcal{B}$, $\underline{P}(f|B)$ is the lower envelope of the expectations obtained from a set of finitely additive probabilities. We shall also use the notations $G(f|B) := B(f - \underline{P}(f|B))$ and $G(f|\mathcal{B}) := \sum_{B\in\mathcal{B}} G(f|B) = f - \underline{P}(f|\mathcal{B})$ for all $f \in \mathcal{L}$ and all $B \in \mathcal{B}$ (note how B is used also as the indicator function of event B). In the case of an unconditional lower prevision $\underline{P}$, we shall let $G(f) := f - \underline{P}(f)$ for any gamble f in its domain.

Definition 1. Let $\underline{P}(\cdot|\mathcal{B}_1), \dots, \underline{P}(\cdot|\mathcal{B}_m)$ be separately coherent conditional lower previsions. They are called *(jointly) coherent* if for every $f_i \in \mathcal{L}, i = 0, \dots, m$, $j_0 \in \{1, \dots, m\}, B_0 \in \mathcal{B}_{j_0}$ and given $H_1 := \sum_{i=1}^m G(f_i|\mathcal{B}_i)$ and $H_2 := G(f_0|B_0)$, it holds that $\sup_{\omega\in B}[H_1 - H_2](\omega) \geq 0$ for some $B \in \cup_{i=1}^m S_i(f_i) \cup \{B_0\}$, where $S_i(f_i) := \{B_i \in \mathcal{B}_i : B_i f_i \neq 0\}$.

A number of weaker conditions are of interest for this paper.

Definition 2. Under the above conditions, $\underline{P}(\cdot|\mathcal{B}_1), \dots, \underline{P}(\cdot|\mathcal{B}_m)$ are said to:

- *avoid partial loss* when $\sup_{\omega\in B} H_1(\omega) \geq 0$ for some $B \in \cup_{i=1}^m S_i(f_i)$;
- be *weakly coherent* when $\sup_{\omega\in\Omega}[H_1 - H_2](\omega) \geq 0$;
- be *Williams-coherent* when the coherence condition holds for the particular case when $S_i(f_i)$ is finite for $i = 1, \dots, m$.

Weak and strong coherence are equivalent in the particular case of two conditional lower previsions, if we assume in addition a positivity condition:

Lemma 1. *If $\underline{P}(\cdot|\mathcal{B}_1)$ and $\underline{P}(\cdot|\mathcal{B}_2)$ are weakly coherent with some coherent lower prevision $\underline{P}$ such that $\underline{P}(B_2) > 0\ \forall B_2 \in \mathcal{B}_2$ and $\underline{P}(B_1) > 0\ \forall B_1 \in \mathcal{B}_1$ different from a given $B_1' \in \mathcal{B}_1$, then $\underline{P}(\cdot|\mathcal{B}_1)$ and $\underline{P}(\cdot|\mathcal{B}_2)$ are coherent.*

We also have the following characterisation of weak coherence:

Lemma 2. $\underline{P}(\cdot|\mathcal{B}_1), \dots, \underline{P}(\cdot|\mathcal{B}_m)$ *are weakly coherent if and only if there is some coherent lower prevision $\underline{P}$ such that for all $j = 1, \dots, m$, it holds that*

$$\underline{P}(G(f|B_j)) = 0 \text{ and } \underline{P}(G(f|\mathcal{B}_j)) \geq 0\ \forall f \in \mathcal{L}, B_j \in \mathcal{B}_j.$$

The equality $\underline{P}(G(f|B_j)) = 0$ is called the *Generalised Bayes Rule* (GBR). Condition $\underline{P}(G(f|\mathcal{B}_j)) \geq 0$ represents a condition of conglomerability of $\underline{P}$ with respect to the conditional lower prevision $\underline{P}(\cdot|\mathcal{B}_j)$. More precisely, we have the following:

Definition 3. Let $\underline{P}$ be a coherent lower prevision, and $\mathcal{B}$ a partition of Ω. We say that $\underline{P}$ is *$\mathcal{B}$-conglomerable* if whenever $f \in \mathcal{L}$ and $B_1, B_2, \ldots,$ are different sets in $\mathcal{B}$ such that $\underline{P}(B_n) > 0$ and $\underline{P}(B_n f) \geq 0$ for all $n \geq 1$, it holds that $\underline{P}(\sum_{n=1}^{\infty} B_n f) \geq 0$.

A coherent lower prevision $\underline{P}$ is $\mathcal{B}$-conglomerable if and only if there is a separately coherent conditional lower prevision $\underline{P}(\cdot|\mathcal{B})$ such that $\underline{P}, \underline{P}(\cdot|\mathcal{B})$ are (jointly) coherent (see [3, Theorem 6.8.2(a)]).

The above theory of coherent lower previsions is generalised by the theory of *coherent sets of desirable gambles*, which we summarise next. Given $\mathcal{R} \subseteq \mathcal{L}$, let us denote $\text{posi}(\mathcal{R}) := \{\sum_{k=1}^{n} \lambda_k f_k \colon f_k \in \mathcal{R}, \lambda_k > 0, n \geq 1\}$.

Definition 4. A set $\mathcal{R} \subseteq \mathcal{L}$ is called *coherent* when $\mathcal{R} = \text{posi}(\mathcal{R} \cup \mathcal{L}^+)$ and $0 \notin \mathcal{R}$. It is said to *avoid partial loss* when it is included in a coherent set, and given a partition $\mathcal{B}$ of Ω, the set $\mathcal{R}$ is said to be $\mathcal{B}$-conglomerable when for any gamble f, $Bf \in \mathcal{R} \cup \{0\}$ for all $B \in \mathcal{B}$ implies that $f \in \mathcal{R} \cup \{0\}$.

Given $\mathcal{R} \subseteq \mathcal{L}$ that avoids partial loss, its smallest coherent superset is called its *natural extension*, and its smallest coherent and $\mathcal{B}$-conglomerable superset (provided it exists), its *$\mathcal{B}$-conglomerable natural extension.* The interior $\underline{\mathcal{R}}$ of a coherent set $\mathcal{R}$ (in the topology of uniform convergence) is called a set of *strictly desirable gambles.* A set of gambles induces a conditional lower prevision by

$$\underline{P}(f|B) := \sup\{\mu : B(f - \mu) \in \mathcal{R}\}. \tag{1}$$

The set of strictly desirable gambles induced by $\underline{P}(f|B)$ is the smallest coherent set of gambles on B that induces $\underline{P}(f|B)$. This allows us to establish the following characterisation.

Theorem 1 ([2, Theorem 3]). *Let $\mathcal{R}$ be a coherent set of desirable gambles, and let $\underline{P}$ be the coherent lower prevision it induces by means of Eq.* (1). *Then $\underline{P}$ is $\mathcal{B}$-conglomerable if and only if $\underline{\mathcal{R}}$ is $\mathcal{B}$-conglomerable.*

3 Conglomerable Coherent Lower Previsions

A separately coherent conditional lower prevision $\underline{P}(\cdot|\mathcal{B}_i)$ induces the following sets of gambles:

$$\mathcal{R}_i^{B_i} := \{G(f|B_i) + \varepsilon B_i : f \in \mathcal{L}(\Omega), \varepsilon > 0\} \cup \{f \in \mathcal{L}(\Omega) : f = B_i f \gneq 0\}, \tag{2}$$

where $B_i \in \mathcal{B}_i$. Similarly, a collection of separately coherent conditional lower previsions $\underline{P}(\cdot|\mathcal{B}_1), \ldots, \underline{P}(\cdot|\mathcal{B}_m)$ induces the set of desirable gambles $\cup_{i=1}^{m} \cup_{B_i \in \mathcal{B}_i} \mathcal{R}_i^{B_i}$. Using this set, we can re-formulate one of Williams' basic results [4, Section 1.1] in our language, where lower previsions are conditional on partitions:

Theorem 2. *Let $\underline{P}(\cdot|\mathcal{B}_1), \dots, \underline{P}(\cdot|\mathcal{B}_m)$ be separately coherent conditional lower previsions. Consider $\mathcal{E} := \text{posi}(\mathcal{L}^+(\Omega) \cup (\cup_{i=1}^m \cup_{B_i \in \mathcal{B}_i} \mathcal{R}_i^{B_i}))$.*

1. *If $\underline{P}(\cdot|\mathcal{B}_1), \dots, \underline{P}(\cdot|\mathcal{B}_m)$ are Williams-coherent, then $\mathcal{E}$ is coherent.*
2. *$\mathcal{E}$ induces $\underline{P}(\cdot|\mathcal{B}_1), \dots, \underline{P}(\cdot|\mathcal{B}_m)$ by means of* (1).

In contrast to Williams', here we are concerned with the additional requirement of conglomerability. This is the motivation behind the following notions, which modify some of the consistency conditions in [3, Chapter 7].

Definition 5. Let $\mathcal{R}$ be a set of desirable gambles and $\mathcal{B}$ a partition of Ω. We say that it *avoids $\mathcal{B}$-conglomerable partial loss* if it has a $\mathcal{B}$-conglomerable coherent superset.

Definition 6. Let $\mathcal{B}_1, \dots, \mathcal{B}_m$ be partitions of Ω. A set of desirable gambles that is conglomerable with respect to all the partitions $\mathcal{B}_1, \dots, \mathcal{B}_m$, shall be called *$\mathcal{B}_{1:m}$-conglomerable*.

Definition 7. Conditional lower previsions $\underline{P}(\cdot|\mathcal{B}_1), \dots, \underline{P}(\cdot|\mathcal{B}_m)$ are called *conglomerable coherent* if there is a $\mathcal{B}_{1:m}$-conglomerable coherent set of desirable gambles that induces them. They are said to *avoid conglomerable partial loss* if they have dominating conglomerable coherent extensions.

Let us illustrate the relationships between the notions of avoiding (conglomerable) partial loss for desirable gambles and coherent conditional lower previsions.

Theorem 3. *Let $\underline{P}(\cdot|\mathcal{B}_1), \dots, \underline{P}(\cdot|\mathcal{B}_m)$ be separately coherent conditional lower previsions. Let $\mathcal{R} := \cup_{i=1}^m \cup_{B_i \in \mathcal{B}_i} \mathcal{R}_i^{B_i}$, where the sets of gambles $\mathcal{R}_i^{B_i}$ are determined by Eq.* (2).

1. *If $\underline{P}(\cdot|\mathcal{B}_1), \dots, \underline{P}(\cdot|\mathcal{B}_m)$ avoid partial loss, then $\mathcal{R}$ avoids partial loss.*
2. *$\underline{P}(\cdot|\mathcal{B}_1), \dots, \underline{P}(\cdot|\mathcal{B}_m)$ avoid conglomerable partial loss if and only if the conglomerable natural extension $\mathcal{F}$ of $\mathcal{R}$ exists. Moreover, the smallest dominating conglomerable coherent extensions are induced by the conglomerable natural extension $\mathcal{F}$ of $\mathcal{R}$.*

Now we move on to characterise the different forms of coherence. We start by a preliminary result: we detail how the coherence properties of a set of desirable gambles affect those of the conditional lower previsions it induces.

Theorem 4. *Let $\mathcal{R}$ be a coherent set of desirable gambles, and for every $i = 1, \dots, m$, $B_i \in \mathcal{B}_i$, let $\underline{P}(\cdot|\mathcal{B}_i)$ denote the conditional lower prevision it induces by* (1).

1. *$\underline{P}(\cdot|\mathcal{B}_i)$ is separately coherent for all $i = 1, \dots, m$.*
2. *$\underline{P}(\cdot|\mathcal{B}_1), \dots, \underline{P}(\cdot|\mathcal{B}_m)$ are Williams-coherent.*
3. *[3, Appendix F3] If $\mathcal{R}$ is in addition $\mathcal{B}_{1:m}$-conglomerable, then the conditional lower previsions $\underline{P}(\cdot|\mathcal{B}_1), \dots, \underline{P}(\cdot|\mathcal{B}_m)$ are coherent.*

Let us take now the inverse path, where we start from separately coherent conditional lower previsions $\underline{P}(\cdot|\mathcal{B}_1), \ldots, \underline{P}(\cdot|\mathcal{B}_m)$. For every $i = 1, \ldots, m, B_i \in \mathcal{B}_i$, let $\mathcal{R}_i^{B_i}$ be given by Eq. (2). The $\mathcal{B}_i$-conglomerable natural extension of the sets $\mathcal{R}_i^{B_i}$ $(B_i \in \mathcal{B}_i)$ is the smallest $\mathcal{B}_i$-conglomerable coherent set of desirable gambles that extends the originating sets, and is given by $\mathcal{F}_i := \mathcal{L} \cap \{\sum_{B_i \in \mathcal{B}_i} B_i f_i : B_i f_i \in \mathcal{R}_i^{B_i} \cup \{0\}\} \setminus \{0\}$ (see [5, Proposition 4]). Obviously, it need not be $\mathcal{B}_j$-conglomerable for another partition $\mathcal{B}_j$, and we can show the following:

Lemma 3. *$\mathcal{F}_i$ is $\mathcal{B}_j$-conglomerable iff $\mathcal{R}_i^{B_i}$ is $\mathcal{B}_j$-conglomerable $\forall B_i \in \mathcal{B}_i$.*

The natural extension of the union of the related sets $\mathcal{F}_1, \ldots, \mathcal{F}_m$ is equal to $\mathcal{F}_1 \oplus \cdots \oplus \mathcal{F}_m := \{\sum_{i=1}^m f_i : f_i \in \mathcal{F}_i, i = 1, \ldots, m\}$, taking into account that all of these sets are coherent. We shall denote by $\mathcal{F}$ the $\mathcal{B}_{1:m}$-conglomerable natural extension of $\cup_{i=1}^m \cup_{B_i \in \mathcal{B}_i} \mathcal{R}_i^{B_i}$, provided that it exists. It can be checked that this set $\mathcal{F}$ is also the $\mathcal{B}_{1:m}$-conglomerable natural extension of $\cup_{i=1}^m \mathcal{F}_i$.

Theorem 5. *1. $\underline{P}(\cdot|\mathcal{B}_1), \ldots, \underline{P}(\cdot|\mathcal{B}_m)$ are conglomerable coherent if and only if the $\mathcal{B}_{1:m}$-conglomerable natural extension $\mathcal{F}$ of $\cup_{i=1}^m \mathcal{F}_i$ exists and it induces them by means of Eq.* (1).
2. If $\underline{P}(\cdot|\mathcal{B}_1), \ldots, \underline{P}(\cdot|\mathcal{B}_m)$ are conglomerable coherent, then $\mathcal{F}_i$ is $\mathcal{B}_j$-conglomerable for all i, j in $\{1, \ldots, m\}$, and $\underline{P}(\cdot|\mathcal{B}_1), \ldots, \underline{P}(\cdot|\mathcal{B}_m)$ are coherent.

At this point we have characterised some important relationships between coherence and conglomerable coherence. Yet, we have not addressed the most important issue: whether or not these two notions are equivalent. The next example settles the problem showing that they are not, and hence—using Theorem 5—that conglomerable coherence is indeed stronger than coherence.

Example 1. Consider $\Omega := \mathbb{N}$, and a coherent lower prevision $\underline{P}$ which is not $\mathcal{B}$-conglomerable for some partition $\mathcal{B}$ of Ω but such that there exists a dominating $\mathcal{B}$-conglomerable linear prevision with $P(B) > 0$ for all $B \in \mathcal{B}$ (one such $\underline{P}$ is given in [2, Example 5]).

Let us define $\Omega_1 := \Omega \cup -\Omega$, and the partitions of Ω_1 $\mathcal{B}_1 := \{\Omega, -\Omega\}$ and $\mathcal{B}_2 := \{B \cup -B : B \in \mathcal{B}\}$. Define $\underline{P}(\cdot|\mathcal{B}_1)$ on $\mathcal{L}(\Omega_1)$ by $\underline{P}(f|\Omega) := \underline{P}(f_1)$ and $\underline{P}(f|-\Omega) := \underline{P}(f_2)$, where

$$\begin{array}{ccc} f_1 : \Omega \to \mathbb{R} & \text{and} & f_2 : \Omega \to \mathbb{R} \\ \omega \hookrightarrow f(\omega) & & \omega \hookrightarrow f(-\omega). \end{array} \tag{3}$$

It follows from the coherence of $\underline{P}$ that $\underline{P}(\cdot|\mathcal{B}_1)$ is separately coherent.

From the linear prevision P on $\mathcal{L}$ considered above we can derive a linear prevision P_1 on $\mathcal{L}(\Omega_1)$ by $P_1(f) := P(f_1)$, where f_1 is given by Eq. (3). Then P_1 is a linear prevision satisfying $P_1(B \cup -B) = P(B) > 0$ for any $B \in \mathcal{B}$, and moreover $P_1(\Omega) = 1$. Define $P_1(\cdot|\mathcal{B}_2)$ by GBR. Then it can be checked that $P_1, P_1(\cdot|\mathcal{B}_2)$ are coherent. On the other hand, consider $\underline{P}_1(\cdot|-\Omega) := \underline{P}(\cdot|-\Omega)$

and define $P_1(\cdot|\Omega)$ from P_1 by GBR. Then $P_1, \underline{P}_1(\cdot|\mathcal{B}_1)$ are coherent, and applying Lemma 1 we deduce that $\underline{P}_1(\cdot|\mathcal{B}_1), P_1(\cdot|\mathcal{B}_2)$ are coherent.

Similarly, if we consider the linear prevision P_2 on $\mathcal{L}(\Omega_1)$ given by $P_2(f) := P(f_2)$, we can repeat the above reasoning and define $P_2(\cdot|\mathcal{B}_2)$ and $P_2(\cdot|-\Omega)$ by GBR, and let $\underline{P}_2(\cdot|\Omega)$ be equal to $\underline{P}(\cdot|\Omega)$ and we conclude that $\underline{P}_2(\cdot|\mathcal{B}_1), P_2(\cdot|\mathcal{B}_2)$ are coherent. By taking lower envelopes, we obtain coherent $\underline{Q}(\cdot|\mathcal{B}_1), \underline{Q}(\cdot|\mathcal{B}_2)$ (see [3, Theorem 7.1.6]), and the above construction implies that $\underline{Q}(\cdot|\mathcal{B}_1) = \underline{P}(\cdot|\mathcal{B}_1)$.

Now, assume ex-absurdo that $\underline{P}(\cdot|\mathcal{B}_1), \underline{Q}(\cdot|\mathcal{B}_2)$ are conglomerable coherent. Then Theorem 5(2) implies that the set $\mathcal{F}_1$ induced by $\underline{P}(\cdot|\mathcal{B}_1)$ is $\mathcal{B}_2$-conglomerable, and Lemma 3 implies then that $\mathcal{R}_1^{\Omega}$ is $\mathcal{B}_2$-conglomerable. But

$$\mathcal{R}_1^{\Omega} = \{G(f|\Omega) + \varepsilon\Omega : f \in \mathcal{L}(\Omega_1), \varepsilon > 0\} \cup \{f \in \mathcal{L}(\Omega_1) : f = \Omega f \gneq 0\}$$

is in a one-to-one correspondence with the set of strictly desirable gambles induced by $\underline{P}$. From Theorem 1, since $\underline{P}$ is not $\mathcal{B}$-conglomerable its associated set of strictly desirable gambles is not $\mathcal{B}$-conglomerable; from this we can deduce that $\mathcal{R}_1^{\Omega}$ is not $\mathcal{B}_2$-conglomerable, whence neither is $\mathcal{F}_1$ and as a consequence $\underline{P}(\cdot|\mathcal{B}_1), \underline{Q}(\cdot|\mathcal{B}_2)$ cannot be conglomerable coherent. ♦

This finding is important because it tells us that Walley's notion of coherence does not entirely capture the implications of conglomerability (as they would follow, for example, from the axioms in [3, Appendix F1]), and in this sense it is an approximation to the theory of conglomerable coherent lower previsions. In the next section we show that such an approximation becomes exact in some important special cases.

4 Particular Cases

From Lemma 2, if $\underline{P}(\cdot|\mathcal{B}_1), \dots, \underline{P}(\cdot|\mathcal{B}_m)$ are weakly coherent, then there is an unconditional lower prevision $\underline{P}$ that is pairwise coherent with them. This connects weak coherence and conglomerability:

Theorem 6. *$\underline{P}(\cdot|\mathcal{B}_1), \dots, \underline{P}(\cdot|\mathcal{B}_m)$ are weakly coherent if and only if there are coherent sets $\mathcal{R}, \mathcal{F}_1, \dots, \mathcal{F}_m$ and a coherent lower prevision $\underline{P}$ such that for all $i = 1, \dots, m$ $\mathcal{R} \cup \mathcal{F}_i$ is $\mathcal{B}_i$-conglomerable coherent and it induces $\underline{P}, \underline{P}(\cdot|\mathcal{B}_i)$.*

However, since the coherence of $\underline{P}(\cdot|\mathcal{B}_1), \dots, \underline{P}(\cdot|\mathcal{B}_m)$ is a stronger notion that their weak coherence, we have that conglomerable coherence implies weak coherence by Theorem 5(2) and that the converse is not true by Example 1.

Now, if the lower prevision $\underline{P}$ satisfies $\underline{P}(B) > 0$ for all $B \in \mathcal{B}_1 \cup \dots \cup \mathcal{B}_m$ (whence $\mathcal{B}_1, \dots, \mathcal{B}_m$ can at most be countable), we deduce that $\underline{P}(\cdot|\mathcal{B}_1), \dots, \underline{P}(\cdot|\mathcal{B}_m)$ are conglomerable coherent:

Theorem 7. *Let $\underline{P}(\cdot|\mathcal{B}_1),\ldots,\underline{P}(\cdot|\mathcal{B}_m)$ be separately coherent conditional lower previsions which are weakly coherent with some coherent lower prevision $\underline{P}$ satisfying that $\underline{P}(B)>0$ for all $B\in\mathcal{B}_1\cup\cdots\cup\mathcal{B}_m$. Then:*

1. *$\mathcal{F}_i$ is $\mathcal{B}_j$-conglomerable for $i,j=1,\ldots,m$.*
2. *$\mathcal{F}_1\oplus\cdots\oplus\mathcal{F}_m$ is $\mathcal{B}_{1:m}$-conglomerable.*
3. *$\underline{P}(\cdot|\mathcal{B}_1),\ldots,\underline{P}(\cdot|\mathcal{B}_m)$ are conglomerable coherent.*

Another interesting case where coherence and conglomerable coherence are equivalent is when we condition on partitions that are nested. Let $\mathcal{B}_1,\ldots,\mathcal{B}_m$ be partitions of Ω such that $\mathcal{B}_j$ is finer than $\mathcal{B}_{j-1}$ for all $j=2,\ldots,m$, and let $\underline{P}(\cdot|\mathcal{B}_1),\ldots,\underline{P}(\cdot|\mathcal{B}_m)$ be separately coherent conditional lower previsions.

Theorem 8. *$\underline{P}(\cdot|\mathcal{B}_1),\ldots,\underline{P}(\cdot|\mathcal{B}_m)$ are coherent if and only if for all gambles $f\in\mathcal{L}$, $B_{j-1}\in\mathcal{B}_{j-1}$, $B_j\in\mathcal{B}_j$, it holds that $\underline{P}(G(f|\mathcal{B}_j)|B_{j-1})=0$ and $\underline{P}(G(f|\mathcal{B}_j)|B_{j-1})\geq 0$. Moreover, if $\underline{P}(\cdot|\mathcal{B}_1),\ldots,\underline{P}(\cdot|\mathcal{B}_m)$ are coherent, then $\mathcal{F}_i$ and $\mathcal{F}_1\oplus\cdots\oplus\mathcal{F}_m$ are $\mathcal{B}_{1:m}$-coherent for all $i=1,\ldots,m$, and $\underline{P}(\cdot|\mathcal{B}_1),\ldots,\underline{P}(\cdot|\mathcal{B}_m)$ are conglomerable coherent.*

5 Conclusions

This paper allows us to say a few conclusive words about the quest for a general behavioural theory of coherent (conditional) lower previsions. Now we know that Walley's theory does not consider all the implications of conglomerability and should better be understood as an approximation to the theory of conglomerable coherent lower previsions we have proposed here.

On the other hand, we have shown that in some special cases we can use Walley's theory in order to obtain the same outcomes as with conglomerable coherence: when the conditioning events have positive lower probability, or when the conditioning partitions are nested. Both cases are important in the applications of probability.

In our view, the most important next step to do is to try to make the new theory of practical use in general, not only in the cases already addressed in this paper. To this end, there is a main obstacle to overcome: the computation of the conglomerable natural extension of a set of desirable gambles. We know from [2] that we can approximate it by a sequence of sets, but we neither know whether it is attained in the limit nor whether the sequence is finite. This is the main challenge that has to be faced in future work.

Acknowledgements. Work supported MTM2010-17844 and by the Swiss NSF grants nos. 200020_134759 / 1, 200020_137680 / 1, and the Hasler foundation grant n. 10030.

References

1. de Finetti, B.: Sulla proprietà conglomerativa della probabilità subordinate. Rendiconti del Reale Instituto Lombardo 63, 414–418 (1930)
2. Miranda, E., Zaffalon, M., de Cooman, G.: Conglomerable natural extension. Int. J. Approx. Reason. (accepted)
3. Walley, P.: Statistical Reasoning with Imprecise Probabilities. Chapman & Hall, London (1991)
4. Williams, P.M.: Notes on conditional previsions. Tech. rep., School of Mathematical and Physical Science, University of Sussex, UK (1975); Reprinted in Int. J. Approx. Reason. 44, 366–383 (2007)
5. Zaffalon, M., Miranda, E.: Probability and time (submitted)

Evidential Networks from a Different Perspective

Jiřina Vejnarová

Abstract. Bayesian networks are, at present, probably the most popular representative of so-called graphical Markov models. Naturally, several attempts to construct an analogy of Bayesian networks have also been made in other frameworks as e.g. in possibility theory, evidence theory or in more general frameworks of valuation-based systems and credal sets. We collect previously obtained results concerning conditioning, conditional independence and irrelevance allowing to define a new type of evidential networks, based on conditional basic assignments. These networks can be seen as a generalization of Bayesian networks, however, they are less powerful than e.g. so-called compositional models, as we demonstrate by a simple example.

Keywords: Conditional independence, conditioning, evidence theory, evidential networks, multidimensional models.

1 Introduction

Bayesian networks are, at present, probably the most popular representative of so-called graphical Markov models. Naturally, several attempts to construct an analogy of Bayesian networks have also been made in other frameworks as e.g. in possibility theory [5], evidence theory [4] or in the more general frameworks of valuation-based systems [11] and credal sets [7].

In this paper we bring an alternative to [4], which does not seem to us to be satisfactory, as graphical tools well-known from Bayesian networks are used in different sense. An attempt, using the technique of the operator of

Jiřina Vejnarová
Institute of Information Theory and Automation of the ASCR,
182 08 Prague, Czech Republic
e-mail: vejnar@utia.cas.cz

R. Kruse et al. (Eds.): Synergies of Soft Computing and Statistics, AISC 190, pp. 429–436.
springerlink.com © Springer-Verlag Berlin Heidelberg 2013

composition [9] was already presented in [13], but in that paper we concentrated ourselves only on structural properties of the network, the problem of definition of conditional basic assignments was not solved there. After solving this problem [15], in this paper we present a new concept of evidential networks, which can be seen as a generalization of Bayesian networks. However, simultaneously we show, that these evidential networks are less powerful than e.g. so-called compositional models.

The paper is organized as follows. After a brief summary of basic notions from evidence theory (Section 2), in Section 3 we recall recent concepts important for introduction of evidential networks, such as conditioning, conditional independence and irrelevance. In Section 4 we present a theorem allowing a direct generalization of Bayesian networks to evidential framework as well as a simple example demonstrating the potential weakness of these networks.

2 Basic Notions

In this section we will briefly recall basic concepts from evidence theory [10] concerning sets and set functions.

2.1 Set Projections and Joins

For an index set $N = \{1, 2, \ldots, n\}$ let $\{X_i\}_{i \in N}$ be a system of variables, each X_i having its values in a finite set $\mathbf{X}_i$. In this paper we will deal with *multidimensional frame of discernment* $\mathbf{X}_N = \mathbf{X}_1 \times \mathbf{X}_2 \times \ldots \times \mathbf{X}_n$, and its *subframes* (for $K \subseteq N$) $\mathbf{X}_K = \times_{i \in K} \mathbf{X}_i$. When dealing with groups of variables on these subframes, X_K will denote a group of variables $\{X_i\}_{i \in K}$ throughout the paper.

For $M \subset K \subseteq N$ and $A \subset \mathbf{X}_K$, $A^{\downarrow M}$ will denote a *projection* of A into $\mathbf{X}_M$:

$$A^{\downarrow M} = \{y \in \mathbf{X}_M \mid \exists x \in A : y = x^{\downarrow M}\},$$

where, for $M = \{i_1, i_2, \ldots, i_m\}$,

$$x^{\downarrow M} = (x_{i_1}, x_{i_2}, \ldots, x_{i_m}) \in \mathbf{X}_M.$$

In addition to the projection, in this text we will also need an opposite operation, which will be called a join. By a *join*[1] of two sets $A \subseteq \mathbf{X}_K$ and $B \subseteq \mathbf{X}_L$ $(K, L \subseteq N)$ we will understand a set

$$A \bowtie B = \{x \in \mathbf{X}_{K \cup L} : x^{\downarrow K} \in A \ \ \& \ \ x^{\downarrow L} \in B\}.$$

Let us note that for any $C \subseteq \mathbf{X}_{K \cup L}$ naturally $C \subseteq C^{\downarrow K} \bowtie C^{\downarrow L}$, but generally $C \neq C^{\downarrow K} \bowtie C^{\downarrow L}$.

[1] This term and notation are taken from the theory of relational databases [1].

2.2 Set Functions

In evidence theory [10] (or Dempster-Shafer theory) two dual measures are used to model the uncertainty: belief and plausibility measures. Both of them can be defined with the help of another set function called a *basic (probability or belief) assignment* m on $\mathbf{X}_N$, i.e.,

$$m : \mathcal{P}(\mathbf{X}_N) \longrightarrow [0, 1],$$

where $\mathcal{P}(\mathbf{X}_N)$ is the power set of $\mathbf{X}_N$, and $\sum_{A \subseteq \mathbf{X}_N} m(A) = 1$. Furthermore, we assume that $m(\emptyset) = 0$. A set $A \in \mathcal{P}(\mathbf{X}_N)$ is a *focal element* if $m(A) > 0$.

Belief and *plausibility measures* are defined for any $A \subseteq \mathbf{X}_N$ by the equalities

$$Bel(A) = \sum_{B \subseteq A} m(B), \qquad Pl(A) = \sum_{B \cap A \neq \emptyset} m(B),$$

respectively. It is well-known (and evident from these formulae) that for any $A \in \mathcal{P}(\mathbf{X}_N)$

$$Bel(A) \leq Pl(A), \qquad Pl(A) = 1 - Bel(A^C), \tag{1}$$

where A^C is the set complement of $A \in \mathcal{P}(\mathbf{X}_N)$. Furthermore, basic assignment can be computed from belief function via Möbius inverse:

$$m(A) = \sum_{B \subseteq A} (-1)^{|A \setminus B|} Bel(B), \tag{2}$$

i.e. any of these three functions is sufficient to define values of the remaining two.

For a basic assignment m on $\mathbf{X}_K$ and $M \subset K$, a *marginal basic assignment* of m on $\mathbf{X}_M$ is defined (for each $A \subseteq \mathbf{X}_M$):

$$m^{\downarrow M}(A) = \sum_{\substack{B \subseteq \mathbf{X}_K \\ B^{\downarrow M} = A}} m(B).$$

3 Conditioning, Independence and Irrelevance

Conditioning and independence belong to the most important topics of any theory dealing with uncertainty. They are cornerstones of Bayesian-like multidimensional models.

3.1 Conditioning

In evidence theory the "classical" conditioning rule is so-called Dempster's rule of conditioning, nevertheless a lot of alternative conditioning rules for events have been proposed [8].

However, from the viewpoint of evidential networks conditioning of variables is of primary interest. In [14] we presented two definitions of conditioning by variables, based on Dempster conditioning rule and focusing, we proved that these definitions are correct, nevertheless, their usefulness for multidimensional models is rather questionable, as thoroughly discussed in the above-mentioned paper.

Therefore, in [15] we proposed a new conditioning rule defined as follows.

Definition 1. Let X_K and X_L $(K \cap L = \emptyset)$ be two groups of variables with values in $\mathbf{X}_K$ and $\mathbf{X}_L$, respectively. Then the *conditional basic assignment* of X_K given $X_L \in B \subseteq \mathbf{X}_L$ (for B such that $m^{\downarrow L}(B) > 0$) is defined as follows:

$$m_{X_K|_P X_L}(A|_P B) = \frac{\displaystyle\sum_{\substack{C \subseteq \mathbf{X}_{K \cup L}: \\ C^{\downarrow K} = A \,\&\, C^{\downarrow L} = B}} m(C)}{m^{\downarrow L}(B)} \tag{3}$$

for any $A \subseteq \mathbf{X}_K$.

It is evident that the conditioning is defined only for focal elements of the marginal basic assignment, but we do not consider it a substantial disadvantage, because all the information about a basic assignment is concentrated in focal elements. Its correctness is expressed by Theorem 1, proven in [15].

Theorem 1. *Set function $m_{X_K|_P X_L}$ defined for any fixed $B \subseteq \mathbf{X}_L$, such that $m^{\downarrow L}(B) > 0$ by Definition 1 is a basic assignment on $\mathbf{X}_K$.*

3.2 Independence and Irrelevance

In evidence theory the most common notion of independence is that of random set independence [6]. It has already been proven [12] that it is also the only sensible one.

This notion can be generalized in various ways [3, 11, 12]; the concept of conditional non-interactivity from [3], based on conjunctive combination rule, is used for construction of directed evidential networks in [4]. In this paper we will use the concept introduced in [9, 12], as we consider it more suitable (the arguments can be found in [12]).

Definition 2. Let m be a basic assignment on $\mathbf{X}_N$ and $K, L, M \subset N$ be disjoint, $K \neq \emptyset \neq L$. We say that groups of variables X_K and X_L are *conditionally independent given X_M with respect to m* (and denote it by $K \perp\!\!\!\perp L|M\ [m]$), if the equality

$$m^{\downarrow K\cup L\cup M}(A) \cdot m^{\downarrow M}(A^{\downarrow M}) = m^{\downarrow K\cup M}(A^{\downarrow K\cup M}) \cdot m^{\downarrow L\cup M}(A^{\downarrow L\cup M}) \quad (4)$$

holds for any $A \subseteq \mathbf{X}_{K\cup L\cup M}$ such that $A = A^{\downarrow K\cup M} \bowtie A^{\downarrow L\cup M}$, and $m(A) = 0$ otherwise.

It has been proven in [12] that this conditional independence concept satisfies so-called semi-graphoid properties taken as reasonable to be valid for any conditional independence concept and it has been shown in which sense this conditional independence concept is superior to previously introduced ones [3, 11].

Irrelevance is usually considered to be a weaker notion than independence [6]. It expresses the fact that a new piece of evidence concerning one variable cannot influence the evidence concerning the other variable.

More formally: group of variables X_L is *irrelevant* to X_K ($K \cap L = \emptyset$) if for any $B \subseteq \mathbf{X}_L$ such that $Pl^{\downarrow L}(B) > 0$ (or $Bel^{\downarrow L}(B) > 0$ or $m^{\downarrow L}(B) > 0$)

$$m_{X_K|X_L}(A|B) = m^{\downarrow K}(A) \quad (5)$$

for any $A \subseteq \mathbf{X}_K$.[2]

Generalization of this notion to conditional irrelevance may be done as follows. Group of variables X_L is *conditionally irrelevant* to X_K given X_M (K, L, M disjoint, $K \neq \emptyset \neq L$) if

$$m_{X_K|X_{L\cup M}}(A|B) = m_{X_K|X_M}(A|B^{\downarrow M}) \quad (6)$$

is satisfied for any $A \subseteq \mathbf{X}_K$ and $B \subseteq \mathbf{X}_{L\cup M}$ (whenever both sides are defined).

Let us note that the conditioning in equalities (5) and (6) stands for an abstract conditioning rule [8]. However, the validity of (5) and (6) may depend on the choice of conditioning rule, as we showed in [14] — more precisely irrelevance with respect to one conditioning rule need not imply irrelevance with respect to the other. Nevertheless, when studying the relationship between (conditional) independence and irrelevance based on Dempster conditioning rule and focusing we realized that they do not differ too much from each other [14].

However, the new conditioning rule introduced by Definition 1 exhibits much suitable properties as expressed by the following theorem proven in [15].

---

[2] Let us note that somewhat weaker definition of irrelevance can be found in [2], where equality is substituted by proportionality. This notion has been later generalized using conjunctive combination rule [3].

Theorem 2. *Let K, L, M be disjoint subsets of N such that $K, L \neq \emptyset$. If X_K and X_L are independent given X_M (with respect to a joint basic assignment m defined on $\mathbf{X}_{K \cup L \cup M}$), then X_L is irrelevant to X_K given X_M under the conditioning rule given by Definition 1.*

4 Evidential Networks

However, in Bayesian networks also the reverse implication plays an important role, as for the inference, the network is usually transformed into a decomposable model. Unfortunately, in the framework of evidence theory the reverse implication is not valid, in general, as was shown in [15]. Nevertheless, the following assertion holds true.

Theorem 3. *Let K, L, M be disjoint subsets of N such that $K, L \neq \emptyset$ and $m_{X_K|_P X_{L \cup M}}$ be a (given) conditional basic assignment of X_K given $X_{L \cup M}$ and $m_{X_{L \cup M}}$ be a basic assignment of $X_{L \cup M}$. If X_L is irrelevant to X_K given X_M under the conditioning rule given by Definition 1, then X_K and X_L are independent given X_M (with respect to a joint basic assignment $m = m_{X_K|_P X_{L \cup M}} \cdot m_{X_{L \cup M}}$[3] defined on $\mathbf{X}_{K \cup L \cup M}$).*

Proof. Irrelevance of X_L to X_K given X_M means that for any $A \subseteq \mathbf{X}_K$ and any $B \subseteq \mathbf{X}_{L \cup M}$ such that $m^{\downarrow L \cup M}(B) > 0$

$$m_{X_K|_P X_{L \cup M}}(A|B) = m_{X_K|_P X_M}(A|B^{\downarrow M}).$$

Multiplying both sides of this equality by $m^{\downarrow L \cup M}(B) \cdot m^{\downarrow M}(B^{\downarrow M})$ one obtains

$$\begin{aligned} m_{X_K|_P X_{L \cup M}}(A|B) \cdot m^{\downarrow L \cup M}(B) \cdot m^{\downarrow M}(B^{\downarrow M}) \\ = m_{X_K|_P X_M}(A|B^{\downarrow M}) \cdot m^{\downarrow L \cup M}(B) \cdot m^{\downarrow M}(B^{\downarrow M}), \end{aligned}$$

which is equivalent to

$$m(A \times B) \cdot m^{\downarrow M}(B^{\downarrow M}) = m^{\downarrow K \cup M}(A \times B^{\downarrow M}) \cdot m^{\downarrow L \cup M}(B).$$

Therefore, the equality (4) is satisfied for $C \subseteq \mathbf{X}_{K \cup L \cup M}$ such that $C = A \times B$, where $A \subseteq \mathbf{X}_K$ and $B \subseteq \mathbf{X}_{L \cup M}$. Due to Theorem 1 it is evident, that

$$\sum_{A \subseteq \mathbf{X}_K, B \subseteq \mathbf{X}_{L \cup M}} m(A \times B) = 1,$$

and therefore equality (4) is trivially for satisfied also for any other $C = C^{\downarrow K \cup M} \bowtie C^{\downarrow L \cup M}$, and $m(C) = 0$ otherwise as well. Therefore, X_K and X_L

[3] Let us note that due to Theorem 1 $m_{X_{L \cup M}}$ is marginal to m and $m_{X_K|_P X_{L \cup M}}$ can be re-obtained from m via Definition 1.

are independent given X_M with respect to a joint basic assignment $m = m_{X_K|X_M} \cdot m_{X_{L\cup M}}$. □

This theorem makes possible to define evidential networks in a way analogous to Bayesian networks, but simultaneously brings a question: are these networks advantageous in comparison with other multidimensional models in this framework? The following example brings, at least partial, answer to this question.

Example 1. Let X_1, X_2 and X_3 be three binary variables with values in $\mathbf{X}_i = \{a_i, \bar{a}_i\}, i = 1, 2, 3$, and m be a basic assignment on $\mathbf{X}_1 \times \mathbf{X}_2 \times \mathbf{X}_3$ defined as follows

$$m(\mathbf{X}_1 \times \mathbf{X}_2 \times \{\bar{a}_3\}) = .5,$$
$$m(\{(a_1, a_2, \bar{a}_3), (\bar{a}_1, \bar{a}_2, a_3)\}) = .5.$$

Variables X_1 and X_2 are conditionally independent given X_3 with respect to m. Therefore also X_2 is irrelevant to X_1 given X_3, i.e.

$$m_{X_1|X_{23}}(A|B) = m_{X_1|X_3}(A|B^{\downarrow\{3\}}),$$

for any focal element B of $m^{\downarrow\{23\}}$. As both $m^{\downarrow\{23\}}$ and $m^{\downarrow\{3\}}$ have only two focal elements, namely $\mathbf{X}_2 \times \{\bar{a}_3\}$ and $\{(a_2, \bar{a}_3), (\bar{a}_2, a_3)\}$ and $\{\bar{a}_3\}$ and $\mathbf{X}_3$, respectively, we have

$$m_{X_1|_P X_{23}}(\mathbf{X}_1|\mathbf{X}_2 \times \{\bar{a}_3\}) = m_{X_1|_P X_3}(\mathbf{X}_1|\{\bar{a}_3\}) = 1,$$
$$m_{X_1|_P X_{23}}(\mathbf{X}_1|\{(a_2, \bar{a}_3), (\bar{a}_2, a_3)\}) = m_{X_1|_P X_3}(\mathbf{X}_1|\mathbf{X}_3) = 1.$$

Using these conditionals and the marginal basic assignment $m^{\downarrow\{23\}}$ we get a basic assignment $\tilde{m}$ different from the original one, namely

$$\tilde{m}(\mathbf{X}_1 \times \mathbf{X}_2 \times \{\bar{a}_3\}) = .5,$$
$$\tilde{m}(\mathbf{X}_1 \times \{(a_2, \bar{a}_3), (\bar{a}_2, a_3)\}) = .5.$$

Furthermore, if we interchange X_1 and X_2 we get yet another model, namely

$$\hat{m}(\mathbf{X}_1 \times \mathbf{X}_2 \times \{\bar{a}_3\}) = .5,$$
$$\hat{m}(\mathbf{X}_2 \times \{(a_1, \bar{a}_3), (\bar{a}_1, a_3)\}) = .5.$$ ◇

From this example it is evident, that evidential networks are less powerful than e.g. compositional models [9], as any of these threedimensional basic assignments can be obtained from its marginals using the operator of composition (cf. e.g. [9]).

5 Conclusions

We presented a conditioning rule for variables which is compatible with our notion of conditional independence — in other words, if we use this conditioning rule, we obtain conditional irrelevance concept, which is implied by this conditional independence. We also proved a theorem showing that under some specific conditions conditional irrelevance implies conditional independence. However, by a simple example we revealed the weakness of conditional basic assignments in comparison with the joint ones and therefore also the fact that evidential networks are less powerful in comparison with e.g. compositional models.

Acknowledgements. The support of Grant GAČR P402/11/0378 is gratefully acknowledged.

References

1. Beeri, C., Fagin, R., Maier, D., Yannakakis, M.: On the desirability of acyclic database schemes. J. Assoc. Comput. Mach. 30, 479–513 (1983)
2. Ben Yaghlane, B., Smets, P., Mellouli, K.: Belief functions independence: I. the marginal case. Int. J. Approx. Reason. 29, 47–70 (2002)
3. Ben Yaghlane, B., Smets, P., Mellouli, K.: Belief functions independence: II. the conditional case. Int. J. Approx. Reason. 31, 31–75 (2002)
4. Yaghlane, B., Smets, P., Mellouli, K.: Directed Evidential Networks with Conditional Belief Functions. In: Nielsen, T.D., Zhang, N.L. (eds.) ECSQARU 2003. LNCS (LNAI), vol. 2711, pp. 291–305. Springer, Heidelberg (2003)
5. Benferhat, S., Dubois, D., Gracia, L., Prade, H.: Directed possibilistic graphs and possibilistic logic. In: Proc. of the 7th Int. Conf. IPMU 1998, pp. 1470–1477 (1998)
6. Couso, I., Moral, S., Walley, P.: Examples of independence for imprecise probabilities. In: Proc. of ISIPTA 1999, pp. 121–130 (1999)
7. Cozman, F.G.: Credal networks. Artif. Intel. 120, 199–233 (2000)
8. Daniel, M.: Belief conditioning rules for classic belief functions. In: Proc. of WUPES 2009, pp. 46–56 (2009)
9. Jiroušek, R., Vejnarová, J.: Compositional models and conditional independence in evidence theory. Int. J. Approx. Reason. 52, 316–334 (2011)
10. Shafer, G.: A Mathematical Theory of Evidence. Princeton University Press, Princeton (1976)
11. Shenoy, P.P.: Conditional independence in valuation-based systems. Int. J. Approx. Reason. 10, 203–234 (1994)
12. Vejnarová, J.: On conditional independence in evidence theory. In: Proc. of ISIPTA 2009, Durham, UK, pp. 431-440 (2009)
13. Vejnarová, J.: An Alternative Approach to Evidential Network Construction. In: Borgelt, C., González-Rodríguez, G., Trutschnig, W., Lubiano, M.A., Gil, M.Á., Grzegorzewski, P., Hryniewicz, O. (eds.) Combining Soft Computing and Statistical Methods in Data Analysis. AISC, vol. 77, pp. 619–626. Springer, Heidelberg (2010)
14. Vejnarová, J.: Conditioning, conditional independence and irrelevance in evidence theory. In: Proc. of ISIPTA 2011, Innsbruck, Austria, pp. 381–390 (2011)
15. Vejnarová, J.: Conditioning in Evidence Theory from the Perspective of Multidimensional Models. In: Greco, S., Bouchon-Meunier, B., Coletti, G., Fedrizzi, M., Matarazzo, B., Yager, R.R. (eds.) IPMU 2012, Part III. CCIS, vol. 299, pp. 450–459. Springer, Heidelberg (2012)

Part V
Engineering

Generalised Median Polish Based on Additive Generators

Balasubramaniam Jayaram and Frank Klawonn

Abstract. Contingency tables often arise from collecting patient data and from lab experiments. A typical question to be answered based on a contingency table is whether the rows or the columns show a significant difference. Median Polish (MP) is fast becoming a prefered way to analyse contingency tables based on a simple additive model. Often, the data need to be transformed before applying the MP algorithm to get better results. A common transformation is the logarithm which essentially changes the underlying model to a multiplicative model. In this work, we propose a novel way of applying the MP algorithm with generalised transformations that still gives reasonable results. Our approach to the underlying model leads us to transformations that are similar to additive generators of some fuzzy logic connectives. We illustrate how to choose the best transformation that give meaningful results by proposing some modified additive generators of uninorms. In this way, MP is generalied from the simple additive model to more general nonlinear connectives. The recently proposed way of identifying a suitable power transformation based on IQRoQ plots [3] also plays a central role in this work.

Keywords: Additive generators, contingency tables, IQRoQ plots, median polish algorithm, power transformations.

Balasubramaniam Jayaram
Department of Mathematics, Indian Institute
of Technology Hyderabad, Yeddumailaram - 502 205, India
e-mail: jbala@iith.ac.in

Frank Klawonn
Department of Computer Science, Ostfalia University
of Applied Sciences, 38302 Wolfenbüttel, Germany
e-mail: f.klawonn@ostfalia.de

Frank Klawonn
Bioinformatics and Statistics, Helmholtz Centre
for Infection Research, 38124 Braunschweig, Germany
e-mail: frank.klawonn@helmholtz-hzi.de

R. Kruse et al. (Eds.): Synergies of Soft Computing and Statistics, AISC 190, pp. 439–448.
springerlink.com © Springer-Verlag Berlin Heidelberg 2013

1 Introduction

Contingency tables as in Table 1 often arise from collecting patient data and from lab experiments. The rows and columns of a contingency table correspond to two different categorical attributes. A typical question to be answered based on data from a contingency table is whether the rows or the columns show a significant difference. For the example of the contingency, one would be interested in finding out whether the education of the father or the regions have an influence on the infant mortality.

Table 1 Infant Mortality vs Educational Qualification of the Parents in deaths per 1000 live births in the years 1964-1966 (Source: U.S. Dept. of Health, Education and Welfare)

	≤ 8	$9-11$	12	$13-15$	≥ 16
North-West	25.3	25.3	18.2	18.3	16.3
North-Central	32.1	29.0	18.8	24.3	19.0
South	38.8	31.0	19.3	15.7	16.8
West	25.4	21.1	20.3	24.0	17.5

Hypothesis tests with non-parametric tests like the Wilcoxon-Mann-Whitney-U test, Analysis of variance (ANOVA) and the t-test are some of the common options. However, each of them has its own drawbacks. For more on this, please refer to [3] and the references therein.

Median polish [2] – a technique from robust statistics and exploratory data analysis – is another way to analyse contingency tables based on a simple additive model. We briefly review the idea of median polish in Section 2. Although the simplicity of median polish as an additive model is appealing, it is sometimes too simple to analyse contingency table. Very often, especially in the context of gene, protein or metabolite expression profile experiments, the measurements are not taken directly, but are transformed before further analysis. In the case of expression profile, it is common to apply a logarithmic transformation. The logarithmic transformation is a member of a more general family, the power transformations which are explained in Section 3.

However, it is not clear whether the MP applied to the transformed data would still unearth the interesting characteristics of the data, since the logarithmic transformation essentially changes the underlying model to a multiplicative model. In this work, we propose a novel way of applying the MP algorithm that still gives reasonable results. Our approach to the underlying model leads us to transformations that are similar to additive generators of some fuzzy logic connectives. In fact, we illustrate how to choose the best transformation that give meaningful results by proposing some modified additive generators of uninorms. The recently proposed way of identifying a suitable power transformation based on IQRoQ plots [3] also plays a central role in this work.

2 Median Polish

The underlying additive model of median polish is that each entry x_{ij} in the contingency table can be written in the form

$$x_{ij} \;=\; g + r_i + c_j + \varepsilon_{ij}.$$

- g represents the overall or grand effect in the table. This can be interpreted as general value around which the data in the table are distributed.
- r_i is the row effect reflecting the influence of the corresponding row i on the values.
- c_j is the column effect reflecting the influence of the corresponding column j on the values.
- ε_{ij} is the residual or error in cell (i, j) that remains when the overall, the corresponding row and column effect are taken into account.

For a detailed explanation of the MP algorithm please refer to [2]. Table 2 shows the result of median polish applied to Table 1.

Table 2 Median polish for the Infant Mortality data

Overall: 20.775						
	≤ 8	$9-11$	12	$13-15$	≥ 16	*RE*
NW	-1.475	0.075	0.0125	-1.075	0.625	-1.475 -
NC	1.475	-0.075	-3.2375	1.075	-0.525	2.375
S	10.900	4.650	-0.0125	-4.800	0.000	-0.350
W	-3.200	-5.950	0.2875	2.800	0.000	0.350
CE	7.4750	5.9250	-1.1125	0.0750	-3.6250	

The result of median polish can help to better understand the contingency table. In the ideal case, the residuals are zero or at least close to zero. Close to zero means in comparison to the row or column effects. If most of the residuals are close to zero, but only a few have a large absolute value, this is an indicator for outliers that might be of interest. Most of the residuals in Table 2 are small, but there is an obvious outlier in Southern region for fathers with the least number of years of education.

3 Median Polish on Transformed Data

Transformation of data is a very common step of data preprocessing (see for instance [1]). There can be various reasons for applying transformations before other analysis steps, like normalisation, making different attribute ranges comparable, achieving certain distribution properties of the data (symmetric, normal etc.) or gaining advantage for later steps of the analysis. The logarithm is a special instance of parametric transformations, called power transformations (see for instance [2]) that are defined by

$$t_\lambda(x) = \begin{cases} \frac{x^\lambda - 1}{\lambda} & \text{if } \lambda \neq 0, \\ \ln(x) & \text{if } \lambda = 0. \end{cases}$$

It is assumed that the data values x to be transformed are positive. If this is not the case, a corresponding constant ensuring this property should be added to the data.

We restrict our considerations on power transformations that preserve the ordering of the values and therefore exclude negative values for λ. In this way, properties like rank correlation are preserved.

3.1 The Non-additive Model

When we choose $\lambda = 0$, i.e. the logarithm for the power transformation, we obtain the following model:

$$\ln(x_{ij}) = g + r_i + c_j + \varepsilon_{ij}. \tag{1}$$

Transforming back to the original data yields the model

$$x_{ij} = e^g \cdot e^{r_i} \cdot e^{c_j} \cdot e^{\varepsilon_{ij}}.$$

So it is in principle a multiplicative model (instead of an additive model as in standard median polish) as follows:

$$x_{ij} = \tilde{g} \cdot \tilde{r}_i \cdot \tilde{c}_j \cdot \tilde{\varepsilon}_{ij}$$

where $\tilde{g} = e^g$, $\tilde{r}_i = e^{r_i}$, $\tilde{c}_j = e^{c_j}$, $\tilde{\varepsilon}_{ij} = e^{\varepsilon_{ij}}$. The part of the model which is not so nice is that the residuals also enter the equation by multiplication. Normally, residuals are always additive, no matter what the underlying model for the approximation of the data is.

Towards overcoming this drawback, we propose the following approach. We apply the median polish algorithm to the log-transformed data in order to compute g (or $\tilde{g}$), r_i (or $\tilde{r}_i$) and c_j (or $\tilde{c}_j$). The residuals are then defined at the very end as

$$\varepsilon_{ij} := x_{ij} - \tilde{g} \cdot \tilde{r}_i \cdot \tilde{c}_j. \tag{2}$$

Let us now rewrite Eq. (1) in the following form:

$$\ln(x_{ij}) = \ln(\tilde{g}) + \ln(\tilde{r}_i) + \ln(\tilde{c}_j) + \ln(\tilde{\varepsilon}_{ij}).$$

Assuming that the residuals are small, we have

$$\ln(x_{ij}) \approx \ln(\tilde{g}) + \ln(\tilde{r}_i) + \ln(\tilde{c}_j).$$

Transforming this back to the original data, we obtain

$$x_{ij} \approx \exp\left(\ln(\tilde{g}) + \ln(\tilde{r}_i) + \ln(\tilde{c}_j)\right).$$

A natural question that arises is the following: *What happens with other power transformations, i.e., for $\lambda > 0$?* In principle the same, as we obtain

$$x_{ij} \approx t_\lambda^{-1}(t_\lambda(\tilde{g}) + t_\lambda(\tilde{r}_i) + t_\lambda(\tilde{c}_j)). \tag{3}$$

Let us denote by $\oplus_\lambda$ the corresponding, possibly associative, operator obtained as follows:

$$x \oplus_\lambda y = t_\lambda^{-1}\left(t_\lambda(x) + t_\lambda(y)\right) . \tag{4}$$

Now, we can interpret Eq. (3) as $x_{ij} \approx g \oplus_\lambda \tilde{r}_i \oplus_\lambda \tilde{c}_j$.

Thus the problem of determining a suitable transformation of the data before applying the median polish algorithm essentialy boils down to finding that operator $\oplus_\lambda$ which minimises the residuals in (2), viz.,

$$\varepsilon_{ij} = x_{ij} - g \oplus_\lambda \tilde{r}_i \oplus_\lambda \tilde{c}_j.$$

3.2 A Suitable Transformation Based on IQRoQ Plots

As stated earlier, power transformations are the most commonly used transformations on data. Recently Klawonn *et al.* [3] have proposed a novel way of finding the particular λ of a power transformation to be applied on the data such that applying the Median Polish on that still reveals interesting characteristics of the data. In the following we briefly detail their technique.

An ideal result for median polish would be when all residuals are zero or at least small. The residuals get smaller automatically when the values in the contingency table are smaller. This would mean that we tend to put a high preference on the logarithmic transformation ($\lambda = 0$), at least when the values in the contingency table are greater than 1.

Neither single outliers of the residuals nor of the row or column effects should have an influence on the choice of the transformation. What we are interested in is being able to distinguish between significant row or column effects and residuals. Therefore, the spread of the row or column effects should be large whereas at least most of the absolute values of the residuals should be small.

To measure the spread of the row or column effects, [3] uses the interquartile range which is a robust measure of spread and not sensitive to outliers like the variance. The interquartile range is the difference between the 75%- and the 25%-quantile, i.e. the range that contains 50% percent of the data in the middle. They use the 80% quantile of the absolute values of all residuals to judge whether most of the residuals are small. One should not expect all residuals to be small. There might still be single outliers that are of high interest.

Finally, they compute the quotient of the interquartile range of the row or column effects and divide it by the 80% quantile of the absolute values of all residuals. They call this quotient the IQRoQ value (InterQuartile Range over the 80% Quantile of the absolute residuals). The higher the IQRoQ value, the better is the result of median polish. For each value of λ, the corresponding power transformation is applied to the contingency table and calculate the IQRoQ value. In this way, we obtain an IQRoQ plot, plotting the IQRoQ value depending on λ.

4 Transformations and Additive Generators of Fuzzy Logic Connectives

It is very interesting to note the similarity between the operator $\oplus_\lambda$ and t-norms / t-conorms [4] in fuzzy logic.

On the one hand, the above family of power transformations closely resemble the Schweizer-Sklar family of additive generators of t-norms. In fact, the power transformations are nothing but the negative of the additive generator of the Schweizer-Sklar t-norms. Note that additive generators of t-norms are non-increasing, and in the case of continuous t-norms they are strictly decreasing, which explains the need for a negative sign to make the function decreasing.

On the other hand, given continuous and strict additive generators, one constructs t-norms / t-conorms precisely by using Eq. (4).

However, it should be emphasised that additive generators of t-norms or t-conorms cannot be directly used here. The additive generator of a t-norm is non-increasing while one requires a transformation to maintain the monotonicity in the arguments. In the case of the additive generator of a t-conorm, though monotonicity can be ensured, their domain is restricted to just $[0, 1]$. This can be partially overcome by normalising the data to fall in this range. However, this type of normalisation may not be reasonable always. Further, the median polish algorithm applied to the transformed data do not always remain positive and hence determining the inverse with the original generator is not possible.

The above discussion leads us to consider a suitable modification of the additive generators of t-norms / t-conorms that can accommodate a far larger range of values both in their domain and co-domain. Representable uninorms are another class of fuzzy logic connectives that are obtained by the additive generators of both a t-norm and a t-conorm. In this work, we construct new transformations by suitably modifying the underlying generators of these representable uninorms [4].

4.1 Modified Additive Generators of Uninorms

Let us assume that the data x are coming from $(-M, M)$. Consider the following modified generator of the uninorm obtained from the additive generators of the Schweizer-Sklar family of t-norms and t-conorms. Let $e \in (-M, M)$ be any arbitrary value. Then the following is a valid transformation with $h_\lambda : [-M, M] \to \left[\frac{(-M)^\lambda - e^\lambda}{\lambda}, \frac{1}{\lambda}\right]$, for all $\lambda \in [-\infty, 0[\cup]0, \infty]$.

$$h_\lambda(x) = \begin{cases} \dfrac{x^\lambda - e^\lambda}{\lambda}, & x \in [-M, e] \\ \dfrac{1 - \left(\dfrac{M - x}{M - e}\right)^\lambda}{\lambda}, & x \in [e, M] \end{cases};$$

$$(h_\lambda)^{-1}(x) = \begin{cases} (x\lambda + e^\lambda)^{\frac{1}{\lambda}}, & x \leq 0 \\ M - (M - e)\left[(1 - x\lambda)\right]^{\frac{1}{\lambda}}, & x \geq 0 \end{cases}.$$

Note that h_λ is monotonic for all $\lambda \in [-\infty, 0[\cup]0, \infty]$ and increases with decreasing λ.

That this modified generator is a reasonable transform can be seen by applying on the following data. Consider the 10×10 table generated by the following additive model. The overall effect is 0, the row effects are 10, 20,

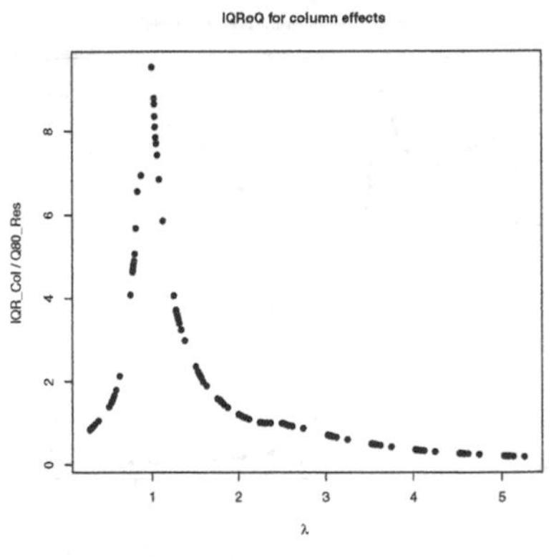

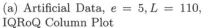

(a) Artificial Data, $e = 5, L = 110$, IQRoQ Column Plot

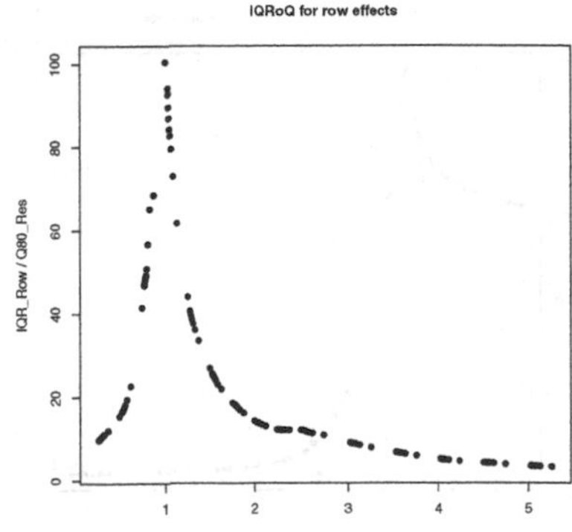

(b) Artificial Data, $e = 5, L = 110$, IQRoQ Row Plot

Fig. 1 IQRoQ plots for the column and row effects of the Artificial data with Modified Schweizer-Sklar generator

30, . . . , 100, the column effects are 1, 2, 3, . . . , 10. To each of these entries is added a noise from a uniform distribution over the interval [-0.5, 0.5]. From the IQRoQ plots for this data given in Figure 1, it can be seen that the global maxima occur at $\lambda = 1$. So the IQRoQ plots propose to apply the above transformation with $\lambda = 1$ which is a linear transformation of the data.

4.2 Finding a Suitable Transformation

In this section we present the algorithm to find a suitable transformation of the given data such that the MP algorithm performs well to elucidate the underlying structures in the data. We only consider a one parameter family of operators with the parameter denoted by λ.

The proposed algorithm is as follows. Let $\oplus_\lambda$ denote the one parameter family of operators whose domain and range allow it to be operated on the data given in the contingency table. Then for each λ the following steps are performed:

1. Apply the transformation $\oplus_\lambda$ to the contingency table.
2. Apply the median polish algorithm to find the overall, row and column effects, viz., $\tilde{g}, \tilde{r}_i, \tilde{c}_j$ for each i, j.
3. Find the residuals $\varepsilon_{ij} = x_{ij} - g \oplus_\lambda \tilde{r}_i \oplus_\lambda \tilde{c}_j$ for each i, j.
4. Determine the IQRoQ values of the above residuals.

Finally, we plot λ versus the above IQRoQ values to get the IQRoQ plots for the column and row effects.

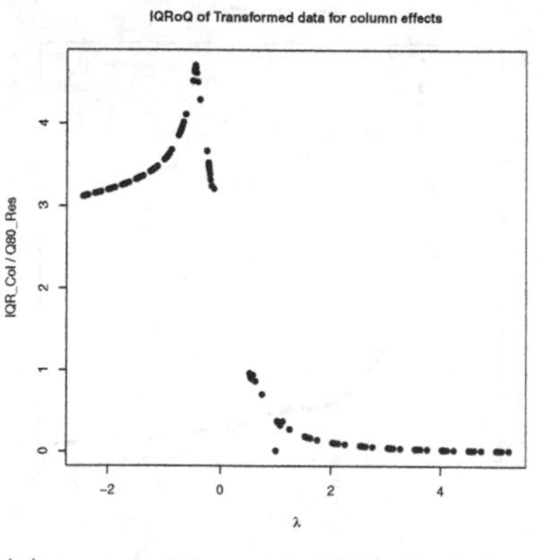

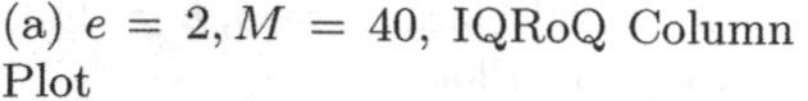
(a) $e = 2, M = 40$, IQRoQ Column Plot

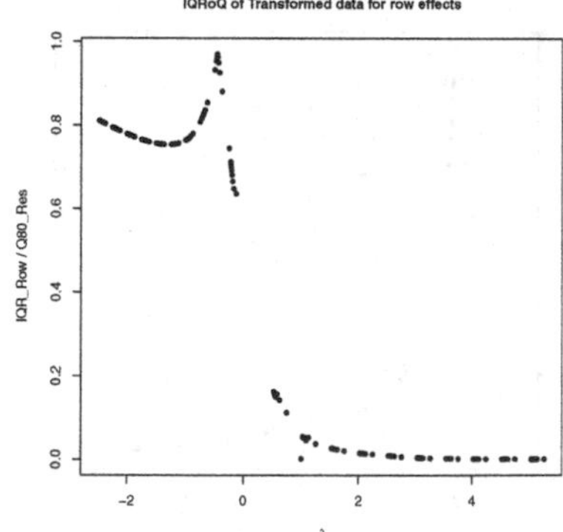

(b) $e = 2, M = 40$, IQRoQ Row Plot

Fig. 2 IQRoQ plots for the column and row effects of the Infant Mortality data

Clearly, the operator corresponding to the λ at which the above IQRoQ plots peak is a plausible transformation for the given contingency table. Though, a rigorous mathematical analysis and support for the above statement is not immediately available, an intuitive explanation is clear from the earlier work of Klawonn *et al.* [3]. Further, we illustrate the same by applying the above h_λ transformations on some real data sets and present our results in the next section.

4.3 Some Illustrative Examples

Let us consider the data given in the Contingency table Table 1. Applying the above algorithm with the transformation h_λ we obtain the IQRoQ plots - Figures 2(a) and (b) - which suggest a value of around $\lambda = -0.5$. The 'median polished' contingency table for $\lambda = -0.5$ is given in Table 3.

Table 3 Median polish on the h_λ-transformed Infant Mortality data with $\lambda = -0.5$

Overall: 0.2919985						
	≤ 8	$9-11$	12	$13-15$	≥ 16	*RE*
NW	0.00025312	0.0027983	-0.00025004	-0.010879	0.0000000	-0.010113225
NC	-0.00025312	-0.0027983	-0.01200293	0.010879	0.0078014	0.006694490
S	0.01098492	0.0091121	0.00025004	-0.044525	-0.0035433	-0.001558958
W	-0.01102793	-0.0305895	0.00456985	0.014641	0.0000000	0.001558958
CE	0.0318984143	0.0293532152	-0.0112376220	0.0002531186	-0.0294192135	

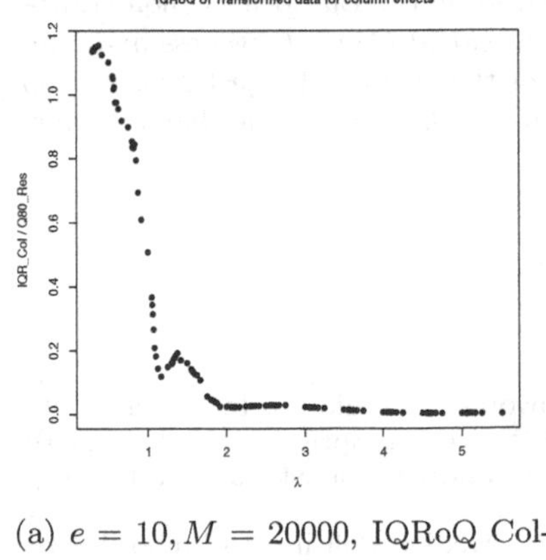

(a) $e = 10, M = 20000$, IQRoQ Column

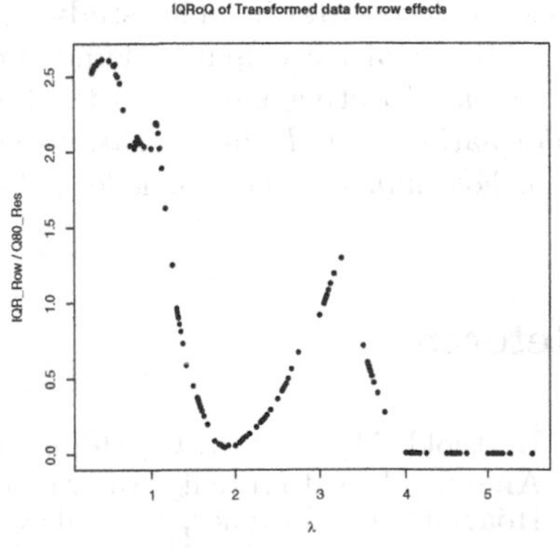

(b) $e = 10, M = 20000$, IQRoQ Row

Fig. 3 IQRoQ plots for the column and row effects of the Spleen data

We finally consider two larger contingency tables with 14 rows and 97 columns that are far too large to be included in this paper. The tables consist of a data set displaying the metabolic profile of a bacterial strain after isolation from different tissues of a mouse. The columns reflect the various substrates whereas the rows consist of repetitions for the isolates from tumor and spleen tissue. The aim of the analysis is to identify those substrates that can be utilized by active enzymes and to find differences in the metabolic profile after growth in different organs.

The corresponding IQRoQ plots shown in Figures 3(a) and (b) suggest that we choose a value of around $\lambda = 0.4$. The 'median polished' contingency table for $\lambda = 0.4$ shows that the number of residuals that are larger than the absolute value of most of the row or column effects is roughly 50%.

5 Conclusions

In this work, we have shown that that the Median Polish algorithm does not always give interpretable results when applied to raw contingency tables. This necessitates a transformation of the data. However, both the choice of the transformation and the fact that the transformation leads to changing the underlying model of the data from a simple additive to a multiplicative model become an issue. We have proposed a novel way of applying the MP algorithm even in this case that still gives reasonable results. Our approach to the underlying model leads us to transformations that are similar to additive generators of some fuzzy logic connectives. Further, we have illustrated how to choose a suitable transformation that gives meaningful results.

Acknowledgements. This study was co-financed by the European Union (European Regional Development Fund) under the *Regional Competitiveness and Employment* objective and within the framework of the Bi^{2}SON Project *Einsatz von Informations- und Kommunikationstechnologien zur Optimierung der biomedizinischen Forschung in Südost-Niedersachsen.*

References

1. Berthold, M., Borgelt, C., Höppner, F., Klawonn, F.: Guide to Intelligent Data Analysis: How to Intelligently Make Sense of Real Data. Springer, London (2010)
2. Hoaglin, D., Mosteller, F., Tukey, J.: Understanding Robust and Exploratory Data Analysis. Wiley, New York (2000)
3. Klawonn, F., Crull, K., Kukita, A., Pessler, F.: Median Polish with Power Transformations as an Alternative for the Analysis of Contingency Tables with Patient Data. In: He, J., Liu, X., Krupinski, E.A., Xu, G. (eds.) HIS 2012. LNCS, vol. 7231, pp. 25–35. Springer, Heidelberg (2012)
4. Klement, E., Mesiar, R., Pap, E.: Triangular Norms. Kluwer Academic Publishers, Dodrecht (2000)

Grasping the Content of Web Servers Logs: A Linguistic Summarization Approach

Janusz Kacprzyk and Sławomir Zadrożny

Abstract. Analyses of Web log servers are needed in many applications and can be useful to designers and analysts of computer networks, and are also an interesting research problem. Traditionally, some statistics are computed and used for analytic and design purposes. We present the use of verbalization of results of Web server log data analysis/mining through linguistic data summaries based on fuzzy logic with linguistic quantifiers. Linguistic summaries of both static and dynamic analyses are presented, with an emphasis on the latter. Examples of potentially interesting linguistic summaries are shown.

Keywords: Fuzzy logic, linguistic summarization, time series, web log.

1 Introduction

Web server logs comprise of information on accesses to resources served by a Web server considered, and the amount of such information is clearly huge. These data can be interesting and useful for many purposes, including both research and technical analyses and design of computer systems. Surveys of recent results can be found in, e.g., [17, 15, 5]. The use of various computational intelligence tools has been proposed in, e.g., [18, 16, 1, 19, 2, 3]. The use of linguistic data summaries based on fuzzy logic has been proposed

Janusz Kacprzyk · Sławomir Zadrożny
Systems Research Institute, Polish Academy of Sciences,
Warszawa, Poland
e-mail: kacprzyk,zadrozny@ibspan.waw.pl
WIT – Warsaw School of Information Technology, 01-447 Warsaw, Poland

R. Kruse et al. (Eds.): Synergies of Soft Computing and Statistics, AISC 190, pp. 449–457.
springerlink.com © Springer-Verlag Berlin Heidelberg 2013

by Zadrożny and Kacprzyk [24, 25]. Information on the available software can be found in, e.g., www.dmoz.org/Computers/Software/Internet/Site_Management/Log_Analysis/Freeware_and_Open_Source/.

This paper is an extension of our previous works in which linguistic data summaries are employed to grasp the contents of Web server log data. Semantically, its very essence is similar to the use of statistical tools and techniques, i.e. indicating what *usually* holds or happens. Our approach makes it possible to present the (huge amount of) numerical data in a more comprehensible and compressed form of short linguistic statements.

We adopt two perspectives: static and dynamic. The former views the Web server log as a database table with rows corresponding to the particular entries in the log file, and the latter views the log file data as a time series. We apply here techniques proposed by us and presented, e.g., in Kacprzyk, Yager and Zadrożny [11], Kacprzyk and Zadrożny [12, 13] for the static case, and by Kacprzyk, Wilbik and Zadrożny [7, 8, 9] for the case of dynamic analyses. We extend our previous works, cf. Zadrożny and Kacprzyk [24, 25], notably with respect to new dynamic analyses. As for a similar approach, to some extent Abraham [18, 1, 19] considers access trend analyses via fuzzy, neural, etc. tools but without a relation to natural language.

2 Log Files

Each request to a Web server is put in one or more log files. Information recorded comprises very often the fields as in Table 1 (cf. common log file format at http://www.w3.org/Daemon/User/Config/Logging.html); some extended forms are also used but not in this paper.

Table 1 Content of the web server log file

Field no.	Content
1	requesting computer name or IP address
2	username of the user triggering the request
3	user authentication data
4	the date and time of the request
5	HTTP command related to the request which includes the path to the requested file
6	status of the request
7	number of bytes transferred as a result of the request
8	software used to issue the request

A lot of software is available, both commercial and open-source (cf., e.g., AWStats at `http://awstats.sourceforge.net/`) to produce various statistics which primarily include the number of requests (or requested Web pages): per month, week, day, hour, per country or domain of the requesting computer, and often for requests from specific sources, notably search engines, statistics are generated, as well as parameters of the requesting agent, as the browser type or the operating system are analyzed. They may be computed in terms of the number of requests and/or the number of bytes transferred, and also concerns the sessions, i.e., a series of requests from the same agent which can help model agent behavior, identify navigational paths, etc.

3 Linguistic Data Summarization

We use the linguistic data summaries proposed by Yager (cf. Yager [21]) and further developed by Kacprzyk, Yager and Zadrożny [10, 11, 12]. Basically, we have: (1) $Y = \{y_1, \ldots, y_n\}$ is a set of objects (records) in a database, e.g., the set of requests to a Web server; (2) $A = \{A_1, \ldots, A_m\}$ is a set of attributes characterizing y_i's from Y, e.g., time of the request or size of the requested file, and $A_j(y_i)$ is a value of attribute A_j for object y_i.

A linguistic summary of a data set is a natural language like expression exemplified by: "Most request in the morning concern small files". The characteristic feature is here the use of *linguistic quantifiers* (e.g., *most*) and *linguistic values* (e.g., *small*). The following components of the linguistic summary should be distinguished:

- a summarizer P, i.e. an attribute A_j with, in general, a linguistic value defined on the domain of A_j (e.g. "small size"); in particular a summarizer may be composed of an attribute accompanied by a crisp value (e.g. "status code = 200"),
- a quantity in agreement Q, i.e. a linguistic quantifier (e.g. most);
- truth (validity) $\mathcal{T}$ of the summary, i.e. a number from $[0, 1]$ (e.g. 0.7);
- optionally, a qualifier R, i.e. another attribute A_k with, in general, a linguistic value defined on the domain of A_k determining a (fuzzy) subset of Y (e.g. "morning request"); again, in a special case, a qualifier may be an attribute accompanied by a crisp value.

Hence the core of a linguistic summary is a *linguistically quantified proposition* in the sense of Zadeh [22] written according to one of the following *protoforms*:

$$Qy\text{'s are } P \tag{1}$$

$$QRy\text{'s are } P \tag{2}$$

The truth (validity) $\mathcal{T}$ thus may be calculated by using, e.g., original Zadeh's calculus of linguistically quantified propositions (cf. [22]), due to (5) for (1), and (6) for (2).

The linguistic data summaries can be extended to the dynamic context, notably to the linguistic summarization of time series as proposed in a series of our papers: Kacprzyk, Wilbik and Zadrożny [8, 9], in which the analysis of how trends concerning some numerical attributes evolve over time, how long some types of behavior last, how rapid changes are, etc.

As pointed out in Zadrożny and Kacprzyk [24, 25]), various user intentions related to their needs for specific information have implied the use of different protoforms of linguistic summaries. In the dynamic context, i.e. of linguistic summaries of time series (cf. Kacprzyk, Wilbik and Zadrożny [8, 9], Kacprzyk and Wilbik [6], Wilbik and Kacprzyk [20]), we focus on trends, i.e. linear segments extracted from the time series, obtained via a piecewise linear segmentation (cf. Keogh et al. [14]), and we consider the following three features of (global) trends in time series: (1) dynamics of change (the speed of change of the consecutive values of time series), (2) duration (the length of a single trend), and (3) variability (how much scattered the times series data are); all are expressed as linguistic values from a limited dictionary using granulation (cf. Batyrshin [4]) as, e.g., increasing, slowly increasing, constant, slowly decreasing, decreasing, ... equated with fuzzy sets.

Zadeh's [23] protoforms as a convenient tool for dealing with linguistic summaries have been proposed in Kacprzyk and Zadrożny [12]. A protoform is defined as a more or less abstract prototype (template) of a linguistically quantified proposition, and in the context of linguistic summaries of time series (of trends, in fact), the first, basic protoforms used (cf. Kacprzyk, Wilbik and Zadrożny [8, 9]) were:

- a simple form (e.g., "Among all segments, *most* are *slowly increasing*"):

$$\text{Among all segments, } Q \text{ are } P \tag{3}$$

- an extended form (e.g., "Among all *short* segments, *most* are *slowly increasing*"):

$$\text{Among all } R \text{ segments, } Q \text{ are } P \tag{4}$$

Among various quality measures of a linguistic summary, the most important is its degree of truth (from $[0, 1]$) which is calculated, using the basic Zadeh's calculus of linguistically quantified propositions [22], for the simple and extended form as, respectively:

$$\mathcal{T}(\text{Among all } y\text{'s, } Q \text{ are } P) = \mu_Q\left(\frac{1}{n}\sum_{i=1}^{n}\mu_P(y_i)\right) \tag{5}$$

$$\mathcal{T}(\text{Among all } Ry\text{'s, } Q \text{ are } P) = \mu_Q\left(\frac{\sum_{i=1}^{n}\mu_R(y_i)\wedge\mu_P(y_i)}{\sum_{i=1}^{n}\mu_R(y_i)}\right) \tag{6}$$

where $\wedge$ is the minimum operation (or, for instance, a t-norm).

Kacprzyk and Wilbik [6] have etended the protoforms (3) and (4) through a temporal term E_T like: "recently", "initially", "in the spring of 2011", etc. yielding:

- a simple temporal protoform (e.g., "*Recently*, among all segments, *most* are *slowly increasing*"):

$$E_T \text{ among all segments, } Q \text{ are } P \tag{7}$$

- an extended temporal protoform (e.g., "*Initially* among all *short* segments, *most* are *slowly increasing*"):

$$E_T \text{ among all } R \text{ segments, } Q \text{ are } P \tag{8}$$

The truth value computed very similarly to the previous case. We only need to consider the temporal expression as an additional external qualifier, as it just limits the universe of interest only to the trends (segments) that occur on the time axis described by E_T. We compute the proportion of segments in which "trend is P" and occurred in E_T to those that occurred in E_T. Next we compute the degree to which this proportion is Q.

The truth value of the simple temporal protoform (7) and of the extended temporal protoform (8) are, respectively:

$$\mathcal{T}(E_T \text{ among all } y\text{'s, } Q \text{ are } P) = \mu_Q \left(\frac{\sum_{i=1}^{n} \mu_{E_T}(y_i) \wedge \mu_P(y_i)}{\sum_{i=1}^{n} \mu_{E_T}(y_i)} \right) \tag{9}$$

where $\mu_{E_T}(y_i)$ is degree to which a trend (segment) occurs during the time span described by E_T;

$$\mathcal{T}(E_T \text{among all } Ry\text{'s, } Q \text{ are } P) = \mu_Q \left(\frac{\sum_{i=1}^{n} \mu_{E_T}(y_i) \wedge \mu_R(y_i) \wedge \mu_P(y_i)}{\sum_{i=1}^{n} \mu_{E_T}(y_i) \wedge \mu_R(y_i)} \right)$$

and $\mu_{E_T}(y_i)$ can be interpreted as the average membership degree of E_T over an assumed time span $[a, b]$ (cf. Kacprzyk and Wilbik [6]).

4 Linguistic Summaries of the Content of a Web Server Log File

A Web server log file may be directly interpreted as a table of data with the columns corresponding to the fields listed in Table 1 and the rows corresponding to the requests. On the other hand the content may be obviously viewed as a time series as each request is timestamped. The timestamp, the fourth field (cf. Table 1), plays a special role as it may be used to form summaries like "Most of large files requests take place on Thursdays" which are *static*, but also summaries like "Most of decreasing trends in the number of requests are very short" which are *dynamic*. In Zadrożny and Kacprzyk [24, 25] a new approach for the static and dynamic case was proposed which will be extended here.

We denote by Y the set of all requests analyzed and assume the request attributes as in Table 2 which correspond to the fields listed in Table 1.

Table 2 Attributes of the requests used for their linguistic summarization

Attribute name	Description
domain	Internet domain extracted from the requesting computer name (if given)
hour	hour the request arrived; extracted from the date and time of the request
day of the month	as above
day of the week	as above
month	as above
filename	name of requested file (including path) extracted from HTTP command
extension	extension of the requested file extracted as above
status	the status of the request
failure	=1 if status code is of 4xx or 5xx form and =0 otherwise
success	=1 if status code is of 2xx form and =0 otherwise
size	number of bytes transferred as a result of the request
agent	name of major browser ('other" otherwise) used to issue request

We can distinguish *simple summaries*, where the qualifier R is absent. These may be exemplified by: "*Most* of requests come from the Firefox browser" or "*Almost all* requested files are *small*".

More interesting may often be *extended summaries*, e.g. "*Al*most all failures concern files with the extension ppt" or "*Most* of the requests concerning *large* files occur in the *evening*". The first summary may indicate that the maintenance of the archive of Powerpoint presentations should be carried out more carefully while the second may suggest that large reports available at the Web server should be updated in the afternoon rather than in the morning.

In the dynamic case, we deal with a numerical attribute such as the size of the requested files or the number of requests aggregated over, e.g., hours or days. The (partial) *trends* are linear segments in a piecewise linear approximation of the time series, cf., e.g., [14]. The trends are characterized by three attributes (cf. Kacprzyk, Wilbik and Zadrożny [8, 9]): dynamics of change, duration and variability. Again this attribute is treated as a linguistic variable and expressed using linguistic values (labels) such as "high", "low", etc.

We also distinguish simple and extended linguistic summaries. The dynamic linguistic summaries are classified into *frequency* and *duration* based. The former describe the partial trends using just the basic attributes (dynamics of change, duration, variability) while the latter explicitly refer to the time scale inherent in the data. For instance, a *simple frequency based summaries* is "*Most* of trends regarding the number of requests are *decreasing*", i.e. here we assume that the entity measured over time is the number of requests.

If the access data are aggregated day by day and the log file covers several months, then such a summary indicates a steady decline in the number of requests served by the Web server.

In the case of simple frequency based temporal type summaries (cf. 7), an interesting summary may be: "In *late hours* of the days, among trends regarding the number of requests *most* are *decreasing*" while an extended temporal protoform (cf. 8) may be: "In *late hours* of the days, among *long* trends regarding the number of requests *most* are *decreasing*". They indicate that in late hours in the period considered, in the case of trends, both for all and long, respectively, regarding the number of requests, most are decreasing, i.e. there is a diminishing volume of requests.

These are just some more illustrative linguistic summaries. Many more have been obtained in a numerical experiment run on the access log of one of Apache Web servers of our institute on a sample of 352 543 requests. The access log was preprocessed to obtain the data listed in Table 2. Basically, in the static case, we looked for a subclass of linguistic summaries that may be obtained using efficient algorithms for association rules mining (cf. Kacprzyk and Zadrożny [12]), with the condition and conclusion parts of an association rules corresponding to the qualifier R and summarizer S, respectively, and the truth value of the summary corresponds to the confidence measure; then we employed our FQUERY for Access software to run the experiments. For the dynamic case, we used a truth value and degree of focus based generation of linguistic summaries of time series as proposed by Kacprzyk and Wilbik [6].

We obtained many interesting linguistic summaries as, e.g.: "*All* requests with the status code 304 ("not modified") referred to *small* files" $(T = 1.0)$, "*Most* files with the `gif` extension were requested from the domain `pl`" $(T = 0.98)$. But, at the same time it does not hold that "*Most* files were requested from the domain `pl`" (true to degree 0.4 only). Another example: "*Most* files with the `gif` extension successfully fetched (with the status code equal 200) were requested from the domain `pl` $(T = 1)$"

In the dynamic case, the corresponding summaries may be: "In the *late evening*, *all* requests with the status code 304 ("not modified") referred to *small* files" $(T = 1.0)$, "In *almost all days, in evening hours, most* files with the `gif` extension were requested from the domain `pl`" $(T = 1)$, "In recent days, *almost all* files with the `gif` extension successfully fetched (with the status code equal 200) were requested from the domain `pl`" $(T = 1)$.

For lack of space we cannot quote more and provide a deeper analysis which will be given in a future paper. On can clearly seen that the addition of the simple and extended temporal protoforms of linguistic data summaries of time series data, the main extension of our previous papers (Zadrożny and Kacprzyk [24, 25]), has considerably extended the scope and usefulness of our analyses.

5 Conclusions

We have proposed an extension of the linguistic summary based analysis of Web server logs presented in our former papers to the analysis of time series of requests to a Web server by adding the simple and extended temporal protoforms of linguistic data summaries of time series. The approach proposed provides means for deeper analyses of Web logs that can be very useful both for the designers of the computer network and its extensions and persons responsible for maintenance. The resulting knowledge may help improve navigation paths, better organize paid search advertising, personalize Web site access etc. As to future work, one can mention the use of more structured information related to, e.g., session tracking, keywords present in search engine queries triggering access to given Web site etc.

References

1. Abraham, A.: Miner: A web usage mining framework using hierarchical intelligent systems. In: Proc. of the IEEE Int. Conf. on Fuzzy Systems, FUZZ-IEEE 2003, pp. 1129–1134 (2003)
2. Arotaritei, D., Mitra, S.: Web mining: a survey in the fuzzy framework. Fuzzy Set. Syst. 148, 5–19 (2004)
3. Asharaf, S., Murty, M.N.: A rough fuzzy approach to web usage categorization. Fuzzy Set. Syst. 148, 119–129 (2004)
4. Batyrshin, I.: On granular derivatives and the solution of a granular initial value problem. Int. J. Appl. Math. Comput. Sci. 12, 403–410 (2002)
5. Facca, F.M., Lanzi, P.L.: Mining interesting knowledge from weblogs: a survey. Data Knowl. Eng. 53, 225–241 (2005)
6. Kacprzyk, J., Wilbik, A.: Comparison of Time Series via Classic and Temporal Protoforms of Linguistic Summaries: An Application to Mutual Funds and Their Benchmarks. In: Borgelt, C., González-Rodríguez, G., Trutschnig, W., Lubiano, M.A., Gil, M.Á., Grzegorzewski, P., Hryniewicz, O. (eds.) Combining Soft Computing and Statistical Methods in Data Analysis. AISC, vol. 77, pp. 369–377. Springer, Heidelberg (2010)
7. Kacprzyk, J., Wilbik, A., Zadrożny, S.: A linguistic quantifier based aggregation for a human consistent summarization of time series. In: Soft Methods for Integrated Uncertainty Modelling, pp. 186–190. Springer, Berlin (2006)
8. Kacprzyk, J., Wilbik, A., Zadrożny, S.: Linguistic summarization of time series using a fuzzy quantifier driven aggregation. Fuzzy Set. Syst. 159, 1485–1499 (2008)

9. Kacprzyk, J., Wilbik, A., Zadrożny, S.: An approach to the linguistic summarization of time series using a fuzzy quantifier driven aggregation. Int. J. Intel. Syst. 25, 411–439 (2010)
10. Kacprzyk, J., Yager, R.R.: Linguistic summaries of data using fuzzy logic. Int. J. Gen. Syst. 30, 133–154 (2001)
11. Kacprzyk, J., Yager, R.R., Zadrożny, S.: A fuzzy logic based approach to linguistic summaries of databases. Int. J. Appl. Math. Comp. Sci. 10, 813–834 (2000)
12. Kacprzyk, J., Zadrożny, S.: Linguistic database summaries and their protoforms: toward natural language based knowledge discovery tools. Inf. Sci. 173, 281–304 (2005)
13. Kacprzyk, J., Zadrożny, S.: Modern data-driven decision support systems: the role of computing with words and computational linguistics. Int. J. Gen. Syst. 39, 379–393 (2010)
14. Keogh, E.J., Chu, S., Hart, D., Pazzani, M.J.: An online algorithm for segmenting time series. In: ICDM, pp. 289–296. IEEE Press (2001)
15. Kosala, R., Blockeel, H.: Web mining research: A survey. ACM SIGKDD Explor. Newslett. 2, 1–15 (2000)
16. Pal, S.K., Talwar, V., Mitra, P.: Web mining in soft computing framework: Relevance. IEEE Trans. Neural Networ. 13, 1163–1177 (2002)
17. Srivastava, J., Cooley, R., Deshpande, M., Tan, P.N.: Web usage mining: Discovery and applications of usage patterns from web data. ACM SIGKDD Explor. 1, 12–23 (2000)
18. Wang, X., Abraham, A., Smith, K.A.: Soft computing paradigms for web access pattern analysis. In: Classification and Clustering for Knowledge Discovery, pp. 233–250 (2005)
19. Wang, X., Abraham, A., Smith, K.A.: Intelligent web traffic mining and analysis. J. Netw. Comput. Appl. 28, 147–165 (2005)
20. Wilbik, A., Kacprzyk, J.: Towards a multi-criteria analysis of linguistic summaries of time series via the measure of informativeness. Int. J. Data Min. Model Manage. (forthcoming, 2012)
21. Yager, R.R.: A new approach to the summarization of data. Inf. Sci. 28, 69–86 (1982)
22. Zadeh, L.A.: A computational approach to fuzzy quantifiers in natural languages. Comput. Math. Appl. 9, 149–184 (1983)
23. Zadeh, L.A.: A prototype-centered approach to adding deduction capabilities to search engines – the concept of a protoform. In: Proc. of the Ann. Meeting of the North American Fuzzy Information Processing Society (NAFIPS 2002), pp. 523–525 (2002)
24. Zadrożny, S., Kacprzyk, J.: Summarizing the contents of web server logs: a fuzzy linguistic approach. In: 2007 IEEE Conf. on Fuzzy Systems, pp. 100–105 (2007)
25. Zadrożny, S., Kacprzyk, J.: From a static to dynamic analysis of weblogs via linguistic summaries. In: Proc. of 2011 IFSA World Congress, pp. 110–119 (2011)

Automatic Tuning of Image Segmentation Parameters by Means of Fuzzy Feature Evaluation

Arif ul Maula Khan, Ralf Mikut, Brigitte Schweitzer,
Carsten Weiss, and Markus Reischl

Abstract. Manual image segmentation performed by humans is time-intensive and inadequate for the quantification of segmentation parameters. Automatic feed-forward segmentation techniques suffer from restrictions in parameter selection and combination and are difficult in quantifying the direct parametric effect on segmentation outcome. Here, we introduce an automatic feedback-based image processing method that uses fuzzy *a priori* knowledge to adapt segmentation parameters. Therefore, a fuzzy evaluation of segment properties is performed for each parameter combination. The method was applied to biological cell imaging. An automatic tuning of the image segmentation process yields an optimal parameter set such that segments match known properties (*a priori* knowledge e.g. cell size, outline etc.).

Keywords: Automatic segment labeling, biological cell imaging, feedback, fuzzy feature evaluation, image segmentation.

1 Motivation and Overview

Image segmentation is an integral part of image analysis that divides an input image into different regions [6]. According to objective evaluation of the segmentation outcome, an image segmentation procedure can be called as *supervised* or *unsupervised*, based on the presence of reference *a priori* knowledge [14]. *A priori* knowledge can be explicit (i.e. based on known objects) or implicit (i.e. using certain set of rules or examples).

Arif ul Maula Khan · Ralf Mikut · Brigitte Schweitzer ·
Carsten Weiss · Markus Reischl
Karlsruhe Institute of Technology, 76344 Eggenstein-Leopoldshafen, Germany
e-mail: markus.reischl@kit.edu

R. Kruse et al. (Eds.): Synergies of Soft Computing and Statistics, AISC 190, pp. 459–467.
springerlink.com © Springer-Verlag Berlin Heidelberg 2013

Manual *supervised* image segmentation performed by humans delivers, in general, good results as a human brain bridges information known about the image (e.g. noise) with the information of segments (e.g. segment size and intensity etc.) to obtain a plausible outcome with respect to the posed problem. However, human image segmentation is downright time-inefficient when dealing with huge image datasets containing a variety of information.

Automatic *supervised* image segmentation techniques with manually tuned parameters (e.g. threshold values etc.) suffer from restrictions in parameters: Basically, the manual tuning seeks to optimize features like segment size, roundness etc. but the parameters only affect parameters like brightness threshold, filter size etc. Therefore, the tuning does not directly affect the features to be optimized.

The optimal parameters of an image segmentation procedure are often affected by side effects (e.g. blurriness, noise, inconsistent background illumination etc.) [12]. Computer routines additionally suffer not only from restrictions in the parameters but also from the combinatorial problem due to an increased number of parameters [1]. Thus, the optimal parameter set may not be found manually, while the tuning is very time-intensive and subject to repetitions on the arrival of a new dataset.

The parameters should be adjusted automatically in an iterative manner for improvement of segmentation results based on *a priori* knowledge. There has been an adequate work on feedback-based automatic image segmentation techniques such as [1, 2, 5, 8, 12, 13]. However, these techniques are limited in terms of well-formulated reference knowledge about object characteristics and types. Therefore, we propose a new method for an automatic feedback-driven segmentation for tuning processing parameters using fuzzy *a priori* knowledge. Since biomedical image processing is currently a quite challenging domain for image analysis, we chose two datasets of images containing living and dying cells for the evaluation of our segmentation technique. To quantify the outcome, we introduced a fuzzy evaluation criterion and built an inference machine to obtain a scalar output fed back to manipulate the segmentation routine to obtain optimal parameter set.

The paper is organized as follows: The methodology of image segmentation, subsequent feature calculation and an evaluation criterion is given in Section 2. The results of our proposed technique are given in Section 3 using two biological image datasets followed by conclusions given in Section 4.

2 Methods

The proposed scheme for feedback-based automatic image segmentation is shown in Fig. 1. It includes segmentation of a grayscale image by transforming it into a binary image containing so called binary large objects (BLOBs), which are the segments found by an image segmentation technique. An image

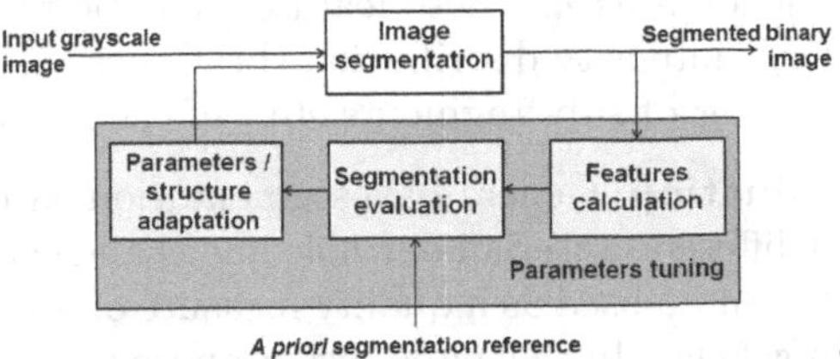

Fig. 1 Employed feedback-based automatic image segmentation scheme

pre-processing step such as image filtering using convolution etc. is included before performing binary image segmentation. The image segmentation is performed for different parameter combinations and the desired features of BLOBs are calculated. Segmentation evaluation with respect to these features is performed accordingly using given *a priori* reference features in an iterative fashion. An optimal parameter set is adopted based on a quality criterion and optimum segmentation is performed using this optimal parameter set.

Image segmentation: As an example, we used a basic sequential image processing routine consisting of a convolution filter, a thresholding and an opening routine followed by an image filling (see [7]). The convolution is done using a symmetric $r \times r$ matrix having elements equal to $\frac{1}{r^2}$, a threshold value t is set and the opening routine uses a disc of size s as a structuring element. Therefore, the image segmentation depends upon the parameter vector $\mathbf{p} = (r, s, t)^T$.

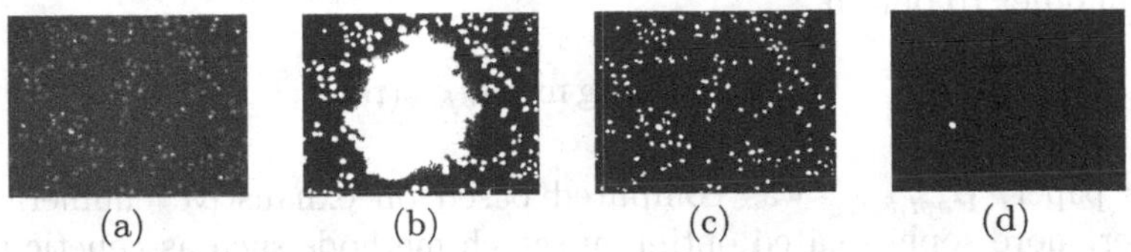

Fig. 2 Segmentation results using manual selection of $\mathbf{p}$, where Fig. 2(a) Original grayscale image and Fig. 2(b), Fig. 2(c) and Fig. 2(d) show segmentation outcome using $\mathbf{p} = (1, 1, 33700)^T$, $(3, 3, 34000)^T$ and $(5, 15, 34000)^T$ respectively.

Features calculation: With respect to $\mathbf{p}$, the objects delivered by the segmentation process not only differ in size, extent, etc. but in the underlying pixel values as well. The setting of $\mathbf{p}$[1], however, is crucial for the segmentation process (Fig. 2). To find optimal values for $\mathbf{p}$, a criterion needs to be calculated based on the feature vector $\mathbf{x}_i^T = (x_{i1}, ..., x_{im})$, considering $j = 1...m$ number of features for each segment i where $i = 1...n$, and the total number

[1] Heuristic selection of $\mathbf{p}$ in Fig. 2 was based on the image intensity histogram.

of segments found denoted as n. These features may be related to geometry (area, sphericity, etc.), intensity distribution (brightness, noise, etc.), and/or the content (e.g. number of sub-fragments etc.) of each segment.

Segmentation evaluation: Generally, a segmentation outcome is evaluated with respect to two different criteria i.e. whole image and each single segment. A variety of metrics can be used as a quality measure of features with respect to given *a priori* knowledge. In this paper, we propose fuzzy *a priori* reference features that are described by a set of membership functions to encompass a considerable level of feature variations in the reference set. Trapezoidal membership functions $\mu_j(x_j)$ with four parameters (i.e. *a*-*d* defining x-values of edges of a trapezoid) were used to formulate reference features. Fuzzy membership μ of each segment i for each feature j is denoted as μ_{ij}. It is reasonable to calculate a product of fuzzy membership μ_{ij} of all m features since it is desirable in our case to classify each segment based on the presence of each feature (i.e. $\mu_{ij} > 0 \; \forall \; j$) in its overall classification. Moreover, the total number of expected segments n in an image was also formulated as a feature of a single image segmentation process using a trapezoidal fuzzy membership function denoted as μ_c. Therefore, a criterion

$$Q_{fuzz}(\mathbf{p}) = \mu_c(\mathbf{p}) \cdot \frac{1}{n(\mathbf{p})} \sum_{i=1}^{n(\mathbf{p})} (\prod_{j=1}^{m} \mu_{ij}(\mathbf{p})), \tag{1}$$

based on aforementioned logic, is introduced to express the quality of automatic image segmentation.

Parameters/Structure adaptation: The criterion (1) needs to be maximized in order to obtain

$$\mathbf{p}_{opt,fuzzy} = \arg\max_{\mathbf{p}} Q_{fuzz}(\mathbf{p}). \tag{2}$$

In this paper, $\mathbf{p}_{opt,fuzzy}$ was computed based on exhaustive enumeration. However, more sophisticated optimum search methods such as genetic algorithms, constraint optimization etc. could be used as well.

3 Results

Benchmark dataset Human HT29 Colon Cancer 1: This benchmark dataset [3] was published in the Broad Bioimage Benchmark Collection[2]. It contains microscope images (showing cells) B_k where $k = 1...6$, shown in Fig. 3. The ground truth for B_k was only the average total number n_{ref} of cells based on two observers. For cell detection and counting, the benchmark has to be evaluated by

[2] `http://www.broadinstitute.org/bbbc/`

$$\sigma_{GD} = \frac{\|n - n_{ref}\|}{n_{ref}}. \tag{3}$$

By using $Q_{fuzz} = 1 - \min(\sigma_{GD}, 1)$, to transfer this given criterion into a fuzzy evaluation, $\mathbf{p}_{opt}$ was adopted by (3). The results for B_k are given in Fig. 1 in terms of deviation σ_{GD} from ground truth. In addition, a feed-forward automatic segmentation technique proposed by Otsu was applied [11] resulting in a threshold t. However, the results could not be directly compared based just on the consideration of the total number of segments n since in Otsu's method some segments can be considered as insignificant due to their size being considerably smaller than the normal segments. To solve this problem, an image opening (opening filter size $s = 3$ and $s = 5$) was applied in case of Otsu's method in order to remove erroneous small segments. With the addition of an image opening operation, the parameter vector for Otsu's method is described as $\mathbf{p}_{Otsu} = (s, t)^T$.

Table 1 Reference cells detection and cell count results from B_k using our method in comparison to Otsu's method using $s = 3$ and $s = 5$

	Our method		**Otsu's method ($s = 3$)**		**Otsu's method ($s = 5$)**	
Images	σ_{GD} (%)	$x_{1,mean}$	σ_{GD} (%)	$x_{1,mean}$	σ_{GD} (%)	$x_{1,mean}$
B_1	12	112	8	101	45	112
B_2	15	132	16	97	58	116
B_3	18	116	18	99	63	129
B_4	13	101	12	94	54	121
B_5	21	124	23	96	68	117
B_6	12	124	14	96	55	110
μ	**15**	**118**	**15**	**97**	**38**	**118**

The results obtained from our feedback-based parameter adaptation technique were comparable to original Otsu's feed-forward method as can be seen from values of σ_{GD} and mean cell area $x_{1,mean}$ in pixels in Tab. 1. The aggregated results using mean value μ in Tab. 1 show that our scheme was able to detect cell numbers comparable to those detected by Otsu's method but with larger mean cell area, the direct relevance or comparison of which is not stated in *a priori* reference of B_k. The μ of σ_{GD} was equal to 15 in case of our method and Otsu's method with $s = 3$. However, visual results with respect to human observation seem much more plausible when using our proposed method. This is indicated by larger value of μ of $x_{1,mean}$ equal to 118 in case of our method as opposed to Otsu's method with $s = 3$ having μ of $x_{1,mean}$ equal to 97. This effect is demonstrated in Fig. 3, where the demarcation of detected cells were seen to be inside the cell boundaries yielding smaller cell areas in case of Otsu's method with $s = 3$. Using values of s larger than 3 causes more deviation from the ground truth as can be seen from higher σ_{GD} in Tab. 1 in case of Otsu's method with $s = 5$.

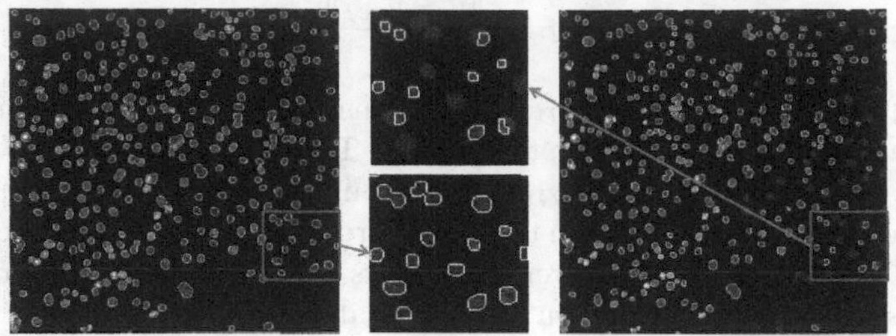

Fig. 3 Cell detection of B_2 using our feedback technique (left) in comparison to Otsu's method ($s = 3$) (right) with zoomed sections in the middle

Therefore, it can be inferred that our method not only detects and counts the cells comparable in numbers to Otsu's method that uses an image opening filter manually, but is also able to select image opening filter size automatically. Moreover, it yields much better results from subjective point of view.

Cell detection based on an heterogeneous cell dataset: A biological dataset with images P_l where $l = 1...4$ was used as shown in Fig. 2(a). This dataset consists of images showing human lung cells (A549) treated with the anticancer drug cis-platin for 24 hours and representative images were acquired as described previously in [4]. We selected three features namely, the area x_1, the roundness factor (ratio between major axis and sum of major and minor axes) x_2 and the mean of the brightness x_3 for each segment.Since we need to find specific cell classes within an image, therefore, it is recommended to define a range of confidence over which our selected features can vary. The segments for normal looking cells were labeled manually in image P_1. *A priori* knowledge was described using $\mu_j(x_j)$ for $\mathbf{x}_{ref}^T$ as shown in the Fig. 4. It can be seen from Fig. 4(a), that for a normal cell, the area lies roughly between 330 to 600 pixels. Its roundness factor (0.5 for perfectly round segments) is between 0.52 to 0.62 as shown in Fig. 4(b) and its brightness can vary from 34000 to 34300 as shown in Fig. 4(c). Moreover, the number of cells, that can be found in each image of the given dataset, can be between 60 to 190 as represented by a fuzzy function in Fig. 4(d).

Furthermore, labeling of reference cells of P_l based on parameter adaptation using fuzzy feature evaluation was also performed. Only P_1 was labeled for normal looking cells ($n_{ref} = 76$) and additional normal cells n were sought after in whole P_l. The results for P_l are given in Tab. 2 in terms of n, σ_{GD} and criterion value Q_{fuzz}. The value of σ_{GD} was given only for P_1 since only P_1 was labeled for reference cells. The results in Tab. 2 show that we were not only able to detect reference normal cells with good accuracy but additional normal cells as well in the whole P_l. P_4 with n, given as an example in Fig. 5(b), shows that our algorithm was able to label normal cells automatically based on *a priori* knowledge. P_1 in Fig. 5(a) shows demarcation of

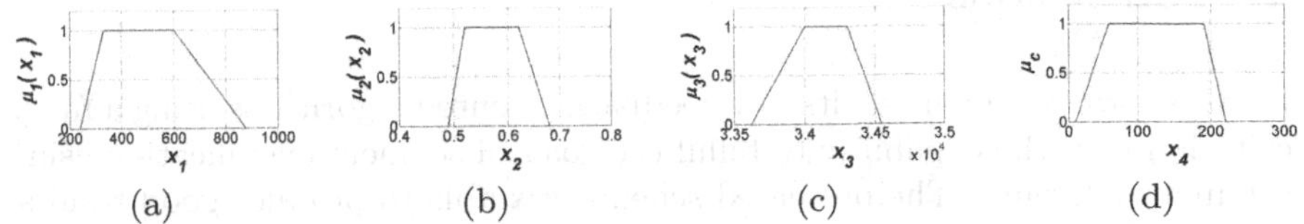

(a) (b) (c) (d)

Fig. 4 $\mu_j(x_j)$ and μ_c for reference cells based on manual inspection and labeling of P_l, where Fig. 4(a) $\mu_1(x_1)$ with (a,b,c,d) = (239, 330, 600, 878), Fig. 4(b) $\mu_2(x_2)$ with (a,b,c,d) = (0.49, 0.52, 0.62, 0.68), Fig. 4(c) $\mu_3(x_3)$ with (a,b,c,d) = (33626, 34000, 34300, 34544) and Fig. 4(d) μ_c with (a,b,c,d) = (10, 60, 190, 220).

labeled reference normal cells by an outline and n by bright stars. Among these n, 73 reference normal cells were found and are shown in Fig. 5(a) by segments having both an outline and a star. On the right side of both images in Fig. 5(a) and Fig. 5(b), a zoomed image was shown for a selected section. It can be seen from Fig. 5(b) that n are only normal cells while the dying cells with high mean brightness were not detected. Moreover, a very dull cell near the top right corner of the zoomed image, indicated by an arrow in Fig. 5(b), was also not detected. Similar trend can also be observed in Fig. 5(a).

Table 2 Reference cell detection and automatic labeling results from P_l

Images	Q_{fuzz}	σ_{GD} (%)	n
P_1	0.73	3.95	99
P_2	0.76	-	104
P_3	0.78	-	80
P_4	0.68	-	19

Hence, it was seen that the introduced fuzzy criterion in (1) was able to detect not only the labeled cells in P_l but to label additional normal looking cells based on fuzzy *a priori* knowledge.

All algorithms are implemented in MATLAB using the Image Processing Toolbox and the open source Gait-CAD Toolbox [10] for data mining.

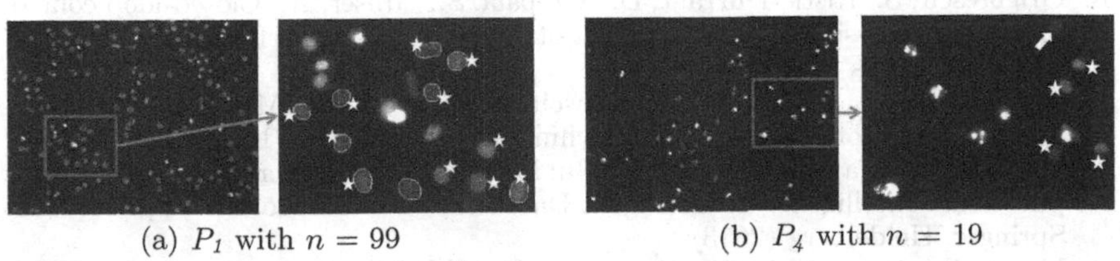

(a) P_1 with n = 99 (b) P_4 with n = 19

Fig. 5 An example of automatic cell labeling using P_l

4 Conclusions

It was shown from our results that feedback-oriented algorithms using a fuzzy criterion have the capability to fulfill the goals of segment classification using a human reference. The presented scheme was able to produce good results, using two biological image datasets, in terms of number and quality of segments found. Moreover, the automatic labeling of the whole dataset was performed. This saves a lot of time compared to manual labeling.

In the future, the techniques will be adapted for optimizing normalization in the same way as proposed in [9]. Exhaustive enumeration for finding the optimal parameter set would be replaced by nonlinear optimization. Moreover, new benchmarks would be tested even for the quality of segment features based on *a priori* knowledge.

Acknowledgements. We express our gratitude to DAAD and BioInterfaces program of the Helmholtz Association for funding this research work.

References

1. Beller, M., Stotzka, R., Müller, T.: Application of an interactive feature-driven segmentation. Biomed. Tech. 49, 210–211 (2004)
2. Bhanu, B., Lee, S., Ming, J.: Adaptive image segmentation using a genetic algorithm. IEEE Trans. Syst. Man Cyb. 25, 1543–1567 (1995)
3. Carpenter, A., et al.: CellProfiler: Image analysis software for identifying and quantifying cell phenotypes. Genome Biol. 7, R100 (2006)
4. Donauer, J., Schreck, I., Liebel, U., Weiss, C.: Role and interaction of p53, BAX and the stress-activated protein kinases p38 and JNK in benzo(a)pyrene-diolepoxide induced apoptosis in human colon carcinoma cells. Arch Toxicol 86, 329–337 (2012)
5. Farmer, M., Jain, A.: A wrapper-based approach in image segmentation and classification. IEEE Trans. Im. Proc. 14, 2060–2072 (2005)
6. Fu, K., Mui, J.: A survey on image segmentation. Pattern Recog. 13, 845–854 (1981)
7. Gonzalez, R.C., Woods, R.E., Eddins, S.L.: Digital Image Processing Using MATLAB. Prentice-Hall, Upper Saddle River (2003)
8. Grigorescu, S., Ristic-Durrant, D., Vuppala, S., Gräser, A.: Closed-loop control in image processing for improvement of object recognition. In: Proc. 17th IFAC World Congress (2008)
9. ul Maula Khan, A., Reischl, M., Schweitzer, B., Weiss, C., Mikut, R.: Feedback-Driven Design of Normalization Techniques for Biological Images Using Fuzzy Formulation of a Priori Knowledge. In: Moewes, C., Nürnberger, A. (eds.) Computational Intelligence in Intelligent Data Analysis. SCI, vol. 445, pp. 167–178. Springer, Heidelberg (2013)
10. Mikut, R., Burmeister, O., Braun, S., Reischl, M.: The open source Matlab toolbox Gait-CAD and its application to bioelectric signal processing. In: Proc. DGBMT-Workshop Biosignalverarbeitung, Potsdam, pp. 109–111 (2008)

11. Otsu, N.: A threshold selection method from gray-level histograms. IEEE Trans. Syst. Man Cyb. 9, 62–66 (1979)
12. Reischl, M., Alshut, R., Mikut, R.: On robust feature extraction and classification of inhomogeneous data sets. In: Proc. 20. Workshop Computational Intelligence, Forschungszentrum Karlsruhe, pp. 124–143 (2010)
13. Sommer, C., Straehle, C., Kothe, U., Hamprecht, F.: ilastik: Interactive learning and segmentation toolkit. In: Proc. IEEE Int. Symp. on Biomedical Imaging: From Nano to Macro, pp. 230–233. IEEE Press (2011)
14. Zhang, H., Fritts, J., Goldman, S.: Image segmentation evaluation: A survey of unsupervised methods. Comput. Vis. Image Und. 110, 260–280 (2008)

Modified Sequential Forward Selection Applied to Predicting Septic Shock Outcome in the Intensive Care Unit

Rúben Duarte Pereira, João Sousa, Susana Vieira,
Shane Reti, and Stan Finkelstein

Abstract. Medical databases often contain large amounts of missing data. This poses very strict constraints to the use of exclusively computer-based feature selection techniques. Moreover, in medical data there is usually no unique combination of features that provides the best explanation of the outcome. In this paper we propose a modified Sequential Forward Selection (SFS) approach to the problem of selecting sets of physiologic variables from septic shock patients in order to predict their outcome. We were able to achieve ten different combinations of only three physiological numerical parameters, all performing better than the best set suggested up to now. The performances of these sets are higher than 0.97 for *AUC* and up to 0.97 for *accuracy*.

Keywords: Fuzzy systems, medical data, modified sequential forward selection, septic shock.

1 Introduction

Severe sepsis (acute organ dysfunction secondary to infection) remains both an important clinical challenge and an economic burden in health care. This

Rúben Duarte Pereira · João Sousa · Susana Vieira
Technical University of Lisbon, Instituto Superior Técnico,
CIS/IDMEC-LAETA, 1049-001 Lisbon, Portugal
e-mail: rubens.dmap@gmail.com

Rúben Duarte Pereira · Stan Finkelstein
Massachusetts Institute of Technology, 02139 Cambridge, MA, USA

Shane Reti
Beth Israel Deaconess Medical Center, 02446 Boston, MA, USA

Shane Reti · Stan Finkelstein
Harvard Medical School, 02115 Boston, MA, USA

R. Kruse et al. (Eds.): Synergies of Soft Computing and Statistics, AISC 190, pp. 469–477.
springerlink.com © Springer-Verlag Berlin Heidelberg 2013

is especially so for septic shock, which is characterized by the persistence of a sepsis-induced hypotensive state, despite adequate fluid resuscitation [5]. Cost-of-illness studies in the US focusing on direct costs per sepsis patient support the premise of increasing global health care costs [4].

A review of the literature highlights previous studies in this area that have applied knowledge-based soft computing techniques to various scenarios associated with septic shock [3, 9, 10, 13].

Further than discussing the quality of the data, work developed in [3] made use of a set of 16 variables recommended by physicians to predict the outcome of septic shock patients. From the initial results it was clear that the dataset resulting from the selected features and the amount of missing data were imposing very strict restrictions to the number of patients available for study and respective samples. It also denoted the low influence of the classification method used in the prediction performance and the importance of feature selection and missing data treatment.

In [9] the 12 most frequently measured parameters were used to predict the outcome of septic shock patients. Though the performance was still similar to previous studies, the use of a different set of features and a different approach to dealing with missing values brought new insights to this area of study.

In [10], by testing different medical standard sets of variables to predict the outcome of septic shock, the group concluded that it was possible to deliver performance comparable to the use of medical scores using only 3 variables. These variables are the Arterial Systolic and Diastolic Blood Pressures and the Thrombocytes measurements.

The aim of this study is to select sets of variables related to the septic shock outcome, by making use of available medical knowledge and referring to the quality of the data available, and refine the search using computational feature selection.

The motivation for data-driven management in the health care field is that it can contribute for the prevention and cost reduction of hospital admissions, especially in the Intensive Care Units (ICU), which are the most complex and most prone to human error environments in health care.

The methodology used in this study will be described in Section 2, including a brief discussion on the modeling techniques. The results and discussion are presented in Section 3, and Section 4 concludes the paper.

2 Methodology

This study makes use of the publicly available MEDAN database [6], already partially preprocessed by the authors [6], [11].

We focused on a group of numerical physiologic variables most frequently acquired in the ICU and most commonly associated to septic shock. Only 139 patients were found to meet these inclusion criteria. All the variables comprised by these patients were used to perform the work described through the subsequent sections.

2.1 Preprocessing

From the analysis of the database we find variables present in the majority of the patients, in contrast with others that were only acquired for a much reduced number of patients. Work developed around the optimal ratio between the number of samples and the number of predicting features to use in order to achieve statistically robust results, [7], suggests using at least 3 times more patients than features for non-linear models. We used a "greedy" approach for selecting a set of features comprising at least 5 times more patients. By discarding the variable with the lowest number of patients comprising it until the ratio between the number of patients available and the number of variables is achieved we have access to at least 100 patients for the study.

Variables corresponding to medication or other external influences to the patient were left out of the study in the interest of the study for the medical staff to direct therapy towards "safe" physiological values.

First, we normalized the data from the filtered cohort basing on the mean and standard deviation calculated for all the values observed throughout each of the features. In this case the value 0 was attributed to the value corresponding to the mean minus the standard deviation and value 1 to the mean added the standard deviation. All values were then rescaled accordingly.

In [9], data from a suitable normal distribution (noise) was inserted randomly. In particular, the aim was to prevent situations where data with missing value replacements might be alternatively classified [1]. Contrastingly, the Zero-Order-Hold (ZOH) method was used in [13], proving to improve the performance of the models, and thus used in this work. This method constitutes a suitable preprocessing technique for medical data records as we are dealing with time-dependent signals, many of them with low changing frequencies, and it resembles the basis of (expert) empiric decision inference performed by physicians.

In order to model and predict the ICU survival outcome of each patient we used the mean value of the last 3 hours of measurements in each variable. This approach implicitly results in a complete dataset.

2.2 Modeling

The study presented deals with a binomial classification problem (single output) based on non-linear combinations of multiple inputs. The aim is to develop models to perform as accurately as possible in classifying test sets.

Takagi-Sugeno (TS) fuzzy models have been proven to behave as universal approximators in non-linear studies using Multiple Input Single Output (MISO) problems [16]. A TS fuzzy model [14] is a fuzzy rule-based model where the rule consequents are functions of the model input. Each rule k has a different function yielding a different value y^k for the output. The simplest consequent function is the linear affine form:

$$R^k : \text{If } \mathbf{x} \text{ is } A^k \text{ then } y^k = \left(\mathbf{a}^k\right)^T \mathbf{x} + b^k \ , \tag{1}$$

Where R^k denotes the k-th rule, $\mathbf{x}$ is the vector of antecedent variables, A^k and y^k are the (multidimensional) antecedent fuzzy sets and the one dimensional consequent variable of the k-th rule, respectively. $\mathbf{a}^k$ is a vector of parameters and b^k is a scalar offset that relate the antecedent fuzzy sets with the consequents. In this study we based on the regular Fuzzy C-means (FCM) algorithm to build the TS inference models. FCM is a clustering method which allows one piece of data to belong to two or more clusters with different degrees [2].

The `genfis3` function from the fuzzy toolbox of *MatLab* software [8] can be used to build linear TS Fuzzy Inference Systems (FIS). This algorithm initially performs FCM clustering using the output as one of the attributes, allowing the model to define the membership functions in the input features that will implicitly be related to the output. Next, the mathematic matrix division between the input, $\mathbf{X}$, and output, $\mathbf{Y}$, matrices is calculated such that if $\mathbf{X}$ is an m-by-n matrix, with $m \neq n$, and $\mathbf{Y}$ is a column vector with m components, then this division is the solution in the least squares sense to the under- or over-determined resulting system of equations. In other words, $\mathbf{a}$ minimizes the length of the vector $\mathbf{a}.\mathbf{X} - \mathbf{Y}$ (i.e.: $|\mathbf{a}.\mathbf{X} - \mathbf{Y}|$).

In order to assess and optimize the number of clusters that should be used to cluster the dataset we used the Xie-Beni index as validation measure [15]:

$$s_{XB} = \frac{\sum_{i=1}^{C} \sum_{k=1}^{N} \left(\mu_{ik}\right)^m D^2\left(\mathbf{x}_k, \mathbf{v}_i\right)}{N\left(min_{i,j,i \neq j} D^2\left(\mathbf{v}_i, \mathbf{v}_j\right)\right)} \ , \tag{2}$$

Where C is the number of clusters, N is the number of data points, μ_{ik} is the membership degree of patient k to the cluster i, $\mathbf{x}_k$ is the array of features corresponding to patient k, and $\mathbf{v}_i$ is the array of features corresponding to the center of the cluster i. D^2 was computed as the Euclidean distance. Parameter m allows different weighting of the membership degrees in the index calculation. Studies comparing several clustering validation measures [12] concluded that the Xie-Beni index provides the best response over a wide range of choices for the number of clusters (2 to 10). Calculations suggest that the best choice for m is probably in the interval [1.5, 2.5], whose mean and midpoint, $m = 2$, have often been the choice for many users of FCM [12].

The Xie-Beni index is a measure of the entropy of the system, which we want to minimize. It balances the membership degrees to each cluster relatively to the distance between clusters.

2.3 Performance Measures

Models for assessing the relation of the set of features with the outcome were iteratively built and evaluated from 100 random configurations of the dataset (100 folds), for statistical purposes. In each fold patients were randomly divided in two datasets (d_0 and d_1, 50%-50%) maintaining proportion between classes as in the whole. We then train on d_0 and test on d_1, followed by training on d_1 and testing on d_0. This is a variation on k-fold cross-validation, called $k \times 2$ cross validation.

The performance of the models was evaluated in terms of *accuracy* (correct classification rate), *sensitivity* (true positive classification rate) and *specificity* (true negative classification rate), and area under the ROC (Receiver Operating Characteristic) curve (AUC).

2.4 Modified Sequential Forward Selection

To select combinations of variables from the preprocessed dataset we used a modified Sequential Forward Selection (SFS) method, with criteria based on the predicting performance of each set evaluated. The regular SFS method sequentially adds features to the best set previously evaluated until a stopping criterion is achieved (e.g. no improvement in performance). It considers only the best set obtained in the previous step to advance to the next step. Here, we propose the addition of two criteria through which we allow more than one set to be evaluated at the next step and/or restrict the search.

At each iteration i, inside step k, we evaluate all the additions of each of the remaining features to the set used to initialize i. Thus, the number of iterations in k corresponds to the number of sets transiting from $k-1$. The algorithm is composed of the following steps:

$\mathbf{set}^{(k)} = set_{Initialization}$
While $\mathbf{set}^{(k)} \neq \emptyset$
$\mathbf{set}^{(k-1)} = \mathbf{set}^{(k)}$
$\mathbf{set}^{(k)} = \emptyset$
For $i = 1 : \text{Length}\left(\mathbf{set}^{(k-1)}\right)$
$s = p\left(\mathbf{set}_i^{(k-1)}\right)$ [1]
For $j = 1 : \text{Length}\left(\mathbf{features} \notin \mathbf{set}_i^{(k-1)}\right)$
$\mathbf{set}_i^{(k)} = \mathbf{set}_i^{(k)} \cup \left(\mathbf{set}_i^{(k-1)} \cup feature_j\right)$
end
If $p\left(best\left(\mathbf{set}_i^{(k)}\right)\right) > s$

[1] The performance associated to the initialization set is set accordingly (e.g. All performance criteria are set to 50%).

$$\mathbf{set}^{(k)} = \mathbf{set}^{(k)} \cup \left\{ \forall \mathbf{set}_{i,j}^{(k)} \in \mathbf{set}_i^{(k)} : p\left(best\left(\mathbf{set}_i^{(k)}\right)\right) - p\left(\mathbf{set}_{i,j}^{(k)}\right) < t \right\}$$

end
end
end

The step improvement criterion, s, is the minimum increment in performance, $p(\cdot)$, to be observed for any of the sets, $\mathbf{set}_i^{(k)}$, comparing to the performance of the set considered to initialize that iteration, $\mathbf{set}_i^{(k-1)}$. By increasing the step improvement we are restricting the advance of the search, and, therefore, of the number of features returned.

The parameter t is a threshold to allow as much sets as intended to advance to the next step. This parameter is set to 0 in the regular SFS method. The simplest definition for this parameter is a constant value. By increasing its value we allow more sets to be evaluated in the next step.

More restrictively we can consider the two parameters in relation to the best performance obtained for all the iterations at each step instead of only considering sets implicated in each iteration.

Usually the increment in performance tends to decrease as more variables are added and, therefore, also the minimum improvement required and the lower boundary for considering sets can be balanced in each step. These two parameters have to be selected based on initial tests to determine the differences in performance and the respective variation at each step.

The results were compared with the regular SFS and the ones obtained for models using the set of variables described in [10].

3 Results and Discussion

The selection of features based on the quantity of signals was performed directly on the MEDAN database, resulting in a total of 20 variables, all comprised by 107 patients. The consequent clustering studies and feature selection were performed with the ZOH preprocessed dataset. Assessing the optimal number of clusters through the Xie-Beni index we found that for any number of clusters higher than 2 the entropy is much higher and is not stable for several repetitions. This means the distribution of the patients along the clusters is not consistent and there might be redundancy in some clusters.

We used *accuracy* to evaluate the modified SFS since it is very representative of the overall performance. In this case, it is difficult to differentiate sets basing on *AUC* variability. Based on initial tests we defined the minimum step improvement in *accuracy* equal to 5%, resulting in the model to stop at step 2. The threshold was defined as 10% of the difference in performance between the best set evaluated in each iteration and the set considered to initialize it.

Table 1 presents the mean values for *AUC* and *accuracy* for the different datasets considered, as well as for the respective *sensitivity* and *specificity*. At each step we present the variables that were evaluated in the next step.

Working with the minimum possible number of variables (in this case we studied up to 3 variables per set) allows reaching higher statistical significance in each evaluation. We found two "base sets" in the step 1 composed by the Diastolic Arterial Blood Pressure and either the Thrombocytes ("Base Set 1") or the Urea ("Base Set 2") levels, suggesting they denote very significant information. Here we show all the combinations based on each of these sets performing above 0.95 in *accuracy* in step 3 (end of the search).

Table 1 Mean values for *AUC*, *Accuracy*, *Sensitivity* and *Specificity*

			AUC	Acc	Sens	Spec
All 20 variables:			0,962	0,938	0,932	0,943
Modified SFS:						
Step 1:	Step 2:	Step 3:				
Diastolic ABP			0,962	0,897	0,887	0,905
	Diastolic ABP Thrombocytes (BS1)		0,977	0,947	0,939	0,954
		BS1+pH	0,977	0,951	0,954	0,949
		BS1+O_2Sat	0,976	0,955	0,941	0,968
		BS1+Leucocytes	0,977	0,951	0,942	0,959
		BS1+Urea	0,979	0,956	0,955	0,956
	Diastolic BP Urea (BS2)		0,974	0,950	0,950	0,950
		BS2+Systolic ABP	0,981	0,955	0,962	0,949
		BS2+pH	0,981	0,958	0,969	0,950
		BS2+Thrombocytes	0,979	0,957	0,957	0,957
		BS2+PTT	0,980	0,971	0,969	0,972
		BS2+GOT (or ASAT)	0,977	0,964	0,970	0,958
		BS2+Total Bilirubin	0,976	0,958	0,944	0,970
		BS2+Blood Sugar	0,977	0,953	0,955	0,951
Regular SFS:						
BS2+PTT+GOT			0,980	0,972	0,965	0,977
Ref. [10]:						
Systolic ABP Diastolic ABP Thrombocytes			0,968	0,948	0,936	0,958

Abreviations : BS1: Base Set 1; BS2: Base Set 2; ABP: Arterial Blood Pressure; O_2Sat: Oxygen Saturation; PTT: Partial Thromboplastin Time; GOT (or ASAT): Glutamate Oxaloate Transaminase (or Aspartate Aminotransferase);

The mean AUC above 0.97 obtained in step 2 is evidence that two variables alone can be used to classify several train-test configurations of the dataset. In fact, adding some variables, such as the Systolic Arterial Blood Pressure, in [10], can decrease the performance comparing to the corresponding base set.

4 Conclusion

The set obtained through the regular SFS to predict the ICU outcome of septic shock patients includes four variables from the initial available set. Though this poses a unique set for study, smaller and alternative sets are more useful for the medical practice, avoiding excessive number of measurements and providing simple explanations of the disease mechanisms. The striking advantage of the methodology used here is it spans the forward search space, increasing the chances to achieve better performances using fewer predictors without performing backward selection techniques. Further studies are focusing on applying the techniques shown here to build and improve real-time monitoring systems in health care, demonstrating the opportunities to increase prevention and risk prediction in clinical settings.

References

1. Berthold, M.R.: Fuzzy-models and potential outliers. In: Proc. of the 18th Int. Conf. of the North America Fuzzy Information Processing Society. IEEE Press (1999)
2. Bezdek, J.C., Ehrlich, R., Full, W.: FCM: the fuzzy c-means clustering algorithm. Comput. Geosci. 10, 191–203 (1984)
3. Brause, R., Hamker, F., Paetz, J.: Septic Shock Diagnosis by Neural Networks and Rule Based Systems. In: Computational Intelligence Techniques in Medical Diagnosis and Prognosis. Springer US (2001)
4. Burchardi, H., Schneider, H.: Economic Aspects of Severe Sepsis: A Review of Intensive Care Unit Costs, Cost of Illness and Cost Effectiveness of Therapy. Pharmacoeconomics 22, 793–813 (2004)
5. Dellinger, R.P., et al.: Surviving Sepsis Campaign: International guidelines for management of severe sepsis and septic shock: 2008. Crit. Care Med. 36, 1394–1396 (2008)
6. Hanisch, E., Brause, R., Arlt, B., Paetz, J., Holzer, K.: The MEDAN Database (2003), `http://www.medan.de`
7. Hua, J., Xiong, Z., Lowey, J., Suh, E., Dougherty, E.R.: Optimal number of features as a function of sample size for various classification rules. Bioinform. 21, 1509–1515 (2005)
8. MathWorks: MatLab R2012a Documentation for Fuzzy Logic Toolbox (2012), `http://www.mathworks.com`
9. Paetz, J.: Knowledge-based approach to septic shock patient data using a neural network with trapezoidal activation functions. Artif. Intel. Med. 28, 207–230 (2003)

10. Paetz, J., Arlt, B.: A Neuro-fuzzy Based Alarm System for Septic Shock Patients with a Comparison to Medical Scores. In: Colosimo, A., Giuliani, A., Sirabella, P. (eds.) ISMDA 2002. LNCS, vol. 2526, pp. 42–52. Springer, Heidelberg (2002)
11. Paetz, J., Arlt, B., Erz, K., Holzer, K., Brause, R., Hanisch, E.: Data quality aspects of a database for abdominal septic shock patients. Comput. Meth. Prog. Biomed. 75, 23–30 (2004)
12. Pal, N.R., Bezdek, J.C.: On cluster validity for fuzzy c-means model. IEEE Trans. Fuzzy Syst. 3, 370–379 (1995)
13. Pereira, R.D., et al.: Predicting Septic Shock Outcomes in a Database with Missing Data using Fuzzy Modeling. In: Proc. of the IEEE Int. Conf. on Fuzzy Systems (2011)
14. Takagi, T., Sugeno, M.: Fuzzy identification of system and its applications to modelling and control. IEEE Trans. Syst. Man Cyb. 15, 116–132 (1985)
15. Xie, X.L., Beni, G.: A Validity Measure for Fuzzy Clustering. IEEE Trans. Pattern Anal. Mach. Intel. 13, 841–847 (1991)
16. Yan, S.Y., Sun, Z.Q.: Universal Approximation for Takagi-Sugeno Fuzzy Systems Combining Statically and Dynamically Constructive Methods-MISO Cases. In: Proc. of the Int. Conf. on Information Management and Engineering (2009)

Possibilistic Local Structure for Compiling Min-Based Networks

Raouia Ayachi, Nahla Ben Amor, and Salem Benferhat

Abstract. Compiling graphical models has recently been triggered much research. First investigations were established in the probabilistic framework. This paper studies compilation-based inference in min-based possibilistic networks. We first take advantage of the idempotency property of the min operator to enhance an existing compilation-based inference method in the possibilistic framework. Then, we propose a new CNF encoding which fits well with the particular case of binary networks.

Keywords: Compilation, inference, possibilistic reasoning.

1 Introduction

Knowledge compilation e.g., [2] is an important topic in many on-line applications that involve hard tasks. It transforms knowledge bases into new structures, with the intent being to improve problem-solving efficiency. One of the most prominent successful applications of knowledge compilation is in the context of *inference* in Bayesian networks [3, 5]. In [1], a direct possibilistic adaptation of the basic compilation-based inference method [5] was proposed. The main idea consists in encoding the network into a *Conjunctive Normal Form (CNF)* base using the so-called *local structure* strategy.

Raouia Ayachi · Nahla Ben Amor
LARODEC, Institut Supérieur de Gestion,
University of Tunis, Le Bardo 2000, Tunis, Tunisia
e-mail: raouia.ayachi@gmail.com,nahla.benamor@gmx.fr

Salem Benferhat
CRIL-CNRS, Université d'Artois, 62307 Lens Cedex, France
e-mail: benferhat@cril.univ-artois.fr

R. Kruse et al. (Eds.): Synergies of Soft Computing and Statistics, AISC 190, pp. 479–487.
springerlink.com © Springer-Verlag Berlin Heidelberg 2013

The resulting encoding is then compiled to efficiently compute the effect of an evidence on a set of variables. In the present paper, we propose a new possibilistic compilation-based inference method strictly more compact than the possibilistic adaptation. In fact, we will first refine the encoding phase by analyzing parameters values using a new encoding strategy: *possibilistic local structure* which goes beyond the so-called *local structure* used in probabilistic networks [3]. We also show that such strategy is exclusively useful in qualitative possibilistic framework. Then, we will improve CNF encodings, for the particular case of binary networks. Experimental results show that exploring both of possibilistic local structure and binary variables in the encoding phase has a significant improvement on inference time.

The remaining paper is organized as follows: Section 2 presents a brief refresher on possibility theory and knowledge compilation. Section 3 presents the compilation-based inference method that exploits possibilistic local structure. Section 4 describes the inference approach for the particular case of binary networks. Section 5 is dedicated to the experimental study.

2 Basic Background

Let $V = \{X_1, ..., X_N\}$ be a set of variables. We denote by D_{X_i} the domain associated with the variable X_i. By x_i (resp. x_{ij}), we denote any of the instances of X_i (resp. the jth instance of X_i). When there is no confusion we use x_i to mean any instance of X_i. In the n-ary case, $D_{X_i} = \{x_{i1}, x_{i2}, \ldots, x_{in}\}$. In the binary case, we will simply write D_{X_i} as $\{x_i, \neg x_i\}$. Ω denotes the universe of discourse, which is the cartesian product of all variable domains in V. Each element $\omega \in \Omega$ is called an *interpretation* of Ω.

2.1 Possibility Theory

This subsection briefly recalls some elements of possibility theory; for more details we refer the reader to [7, 8]. One of the basic concepts in possibility theory is the concept of a possibility distribution, denoted by π, which is a mapping from the universe of discourse to the unit interval [0, 1]. This scale can be interpreted in two ways: a *quantitative* one when values have a real sense and a *qualitative* one when values only reflect a total pre-order between the different states of the world. This paper focuses on the qualitative interpretation of possibility theory which uses the min as a conjunction operator. Given a possibility distribution π, we can define a mapping grading the possibility measure of an event $\phi \subseteq \Omega$ by $\Pi(\phi) = max_{\omega \in \phi} \pi(\omega)$. This measure evaluates the extent to which ϕ is consistent with the available beliefs.

Conditioning is a crucial notion in possibility theory. It consists in revising our initial knowledge π by the arrival of a new certain piece of information $\phi \subseteq \Omega$. The qualitative interpretation of the possibilistic scale leads to the well known definition of min-conditioning [7, 10], expressed by:

$$\Pi(\psi \mid \phi) = \begin{cases} \Pi(\psi \cap \phi) & \text{if } \Pi(\psi \cap \phi) < \Pi(\phi) \\ 1 & \text{otherwise} \end{cases} \tag{1}$$

2.2 Min-Based Possibilistic Networks

Min-based possibilistic networks can be viewed as the possibilistic counterpart of Bayesian networks [11] when we consider the qualitative interpretation of the possibilistic scale. A min-based possibilistic network over a set of N variables $V = \{X_1, X_2, ..., X_N\}$, denoted by ΠG_{min}, is composed of:

- A *graphical component* composed of a Directed Acyclic Graph (DAG) where nodes represent variables and edges encode links between variables. The parent set of any variable X_i is denoted by $U_i = \{U_{i1}, U_{i2}, ..., U_{im}\}$ where U_{ij} is the jth parent of U_i and m is the number of parents of X_i. In what follows, we use x_i, u_i, u_{ij} to denote, respectively, possible instances of X_i, U_i and U_{ij}.
- A *numerical component* that quantifies different links. Uncertainty of each node X_i is represented by a local normalized conditional possibility table (denoted by CPT_i) in the context of its parents. The set of all CPT_i is denoted by CPT. Conditional possibility tables should respect the normalization constraint for each variable $X_i \in V$ expressed by: $\forall u_i, \max_{x_i} \ \Pi(x_i|u_i) = 1$.

2.3 Compilation Concepts

Knowledge compilation is an artificial intelligence area related to a mapping problem from intractable logical theories (typically, from propositional knowledge bases in a CNF form) into suitable target compilation languages from which *transformations* and *queries* can be handled in a polynomial time with respect to the size of compiled bases [2]. There are several compilation languages as it has been studied in the knowledge map of [6]. The *Negation Normal Form (NNF)* language represents the pivotal language from which a variety of target compilation languages give rise by imposing some conditions on it. In this paper, we are in particular interested in *Decomposable Negation Normal Form (DNNF)* [4] since it is considered as one of the most compact target languages. DNNF is a subset of NNF by satisfying the *decomposability* property stating that conjuncts of any conjunction share no variables [4].

DNNF supports a rich set of polynomial-time operations which can be performed simply and efficiently. We restrict our attention to conditioning and forgetting operations:

- *Conditioning*: Let α be a propositional formula and let ρ be a consistent term, then conditioning α on ρ, denoted by $\alpha|\rho$ generates a new formula in which each propositional variable $P_i \in \alpha$ is set to $\top$ if P_i is consistent with ρ and $\bot$ otherwise.

- *Forgetting*: The forgetting of P_i from α is equivalent to a formula that do not mention P_i. Formally: $\exists P_i.\alpha = \alpha|P_i \vee \alpha|\neg P_i$.

In [9], authors have generalized the set of NNF languages by the *Valued Negation Normal Form (VNNF)* which offers an enriched representation of functions. Within VNNF's operations, we cite *max-variable elimination* which consists in forgetting variables using the max operator.

3 Compilation-Based Inference in Possibilistic Networks

Emphasis has been recently placed on compilation-based inference in graphical models, in particular Bayesian networks. This topic has been the focus of several researches [3, 5]. In [1], we proposed a direct (and naive) adaptation of the idea of compiling Bayesian networks [5] into the possibilitic framework, in particular to compile min-based possibilistic networks. This adaptation did not take into account specific features of possibility theory such as the ordinal nature of uncertainty scale. As a consequence, the size of possibilistic compiled knowledge base is the same as the one obtained in probability theory. In this section, We propose an enhancement of this adaptation, particularly for encoding min-based possibilistic networks in a more compact manner using a new strategy dedicated for qualitative approaches.

The principle of encoding is to first transform the initial network into a *Conjunctive Normal Form (CNF)*. To this end, we need to represent instances of variables and also parameters using a set of propositional variables. More precisely, *instances indicators* are associated to different instances of the network variables and *parameter variables* are relative to possibility degrees. In the probabilistic framework, [3] proposed the so-called *local structure* to decrease the number of parameter variables by associating a unique propositional variable per equal parameters per CPT_i. In this section, we propose to go one step further by taking advantage of the idempotency property of the min operator (i.e., $min(a, a) = a$) by associating a unique propositional variable per equal parameters per all conditional possibility tables (CPT). This encoding strategy, that we call *possibilistic local structure*, reduces the number of propositional variables required to encode the possibilistic network. We denote the method improved by possibilistic local structure as Π-DNNF$_{CPT}$.

The new refined CNF encoding of ΠG_{min}, denoted by C_{min}^{PLS}, needs two types of propositional variables, namely:

- $\forall X_i \in V, \forall x_{ij} \in D_{X_i}$, we associate an *instance indicator* $\lambda_{x_{ij}}$ (when there is no ambiguity, we use λ_{x_i}).
- $\forall X_i \in V$, $\forall \Pi(x_i|u_i)$, we associate a parameter variable s.t.:

$$\begin{cases} \theta_j & \text{if } occ(\Pi(x_i|u_i), CPT) > 1 \\ \\ \theta_{x_i|u_i} & \text{if } occ(\Pi(x_i|u_i), CPT) = 1. \end{cases} \quad (2)$$

where $occ(\Pi(x_i|u_i), CPT)$ is the occurrence number of $\Pi(x_i|u_i)$ in CPT.

Definition 1 outlines the new CNF encoding of ΠG_{min}.

Definition 1. *Using the set of instance indicators and parameter variables, the CNF encoding C_{min}^{PLS} contains:*

- ***Mutual exclusive clauses:*** $\forall X_i \in V$, *we have:*

$$\lambda_{x_{i1}} \vee \lambda_{x_{i2}} \vee \cdots \lambda_{x_{in}} \quad (3)$$

$$\neg\lambda_{x_{ij}} \vee \neg\lambda_{x_{ik}}, j \neq k \quad (4)$$

- ***Parameter clauses:*** $\forall X_i \in V$*:*
  - $\forall\ \theta_{x_i|u_i}$, *we have:*

$$\lambda_{x_i} \wedge \lambda_{u_{i1}} \wedge \ldots \wedge \lambda_{u_{im}} \rightarrow \theta_{x_i|u_i} \quad (5)$$

$$\theta_{x_i|u_i} \rightarrow \lambda_{x_i} \quad (6)$$

$$\theta_{x_i|u_i} \rightarrow \lambda_{u_{i1}}, \cdots, \theta_{x_i|u_i} \rightarrow \lambda_{u_{im}} \quad (7)$$

  - $\forall\ \theta_j$, *we have:*

$$\lambda_{x_i} \wedge \lambda_{u_{i1}} \wedge \ldots \wedge \lambda_{u_{im}} \rightarrow \theta_j \quad (8)$$

An inconsistent theory can be involved by applying possibilistic local structure and keeping clauses (6) and (7) for redundant values per CPT. For instance, if we encode both of these equal degrees $\Pi(a_2|b_1) = \Pi(a_1) = 0.3$ by θ_1, then we will obtain from clauses (6) and (7), $\theta_1 \rightarrow \lambda_{a_1}$ and $\theta_1 \rightarrow \lambda_{a_2}$ which is inconsistent. To avoid this problem, redundant values per CPT should be encoded using only one clause (8), however each value appearing once per CPT can be encoded using clauses (5) ,(6) and (7). Aside from equal parameters, each $\Pi(x_i|u_i)$ equal to 0, can be encoded by a shorter clause involving only indicator variables, namely: $\neg\lambda_{x_i} \vee \neg\lambda_{u_{i1}} \vee \cdots \vee \neg\lambda_{u_{im}}$, without the need for a propositional variable. Once ΠG_{min} is encoded into C_{min}^{PLS}, this latter is then compiled into the most succinct target compilation language DNNF. The resulting compiled base is finally transformed into a valued representation corresponding to a *min-max circuit*, denoted by C_{MinMax}^{PLS}. Formally,

Definition 2. *A min-max circuit C_{MinMax}^{PLS} of a DNNF sentence C_{DNNF} is a valued sentence where $\wedge$ and $\vee$ are substituted by min and max, respectively. Leaf nodes correspond to circuit inputs (i.e., indicator and parameter variables), internal nodes correspond to max and min operators, and the root corresponds to the circuit output.*

Evaluating C_{MinMax}^{PLS} consists in applying min and max operators in a bottom-up way. Inference is guaranteed to be established in polytime since it corresponds to a simple propagation from leaves to root. Note that such computation corresponds to *max-variable elimination.*

We show now that in the particular case of binary networks, we can propose a more refined encoding.

4 Inference in Binary Networks

The idea of this new compilation-based inference method, called Bin-Π-DNNF, is to reduce CNF parameters using the fact that a binary variable X_i can be encoded by a unique propositional variable. Moreover, we show that we can provide a further improvement by analyzing parameters values. Bin-Π-DNNF follows the same reasoning: from encoding to inference by going through compilation.

4.1 Encoding and Compilation Phase

Encoding a binary possibilistic network ΠG_{min} using the CNF encoding of Definition 1 yields supplementary propositional variables and clauses since the opportunity presented by binary variables is not taken into consideration. For this reason, we propose to associate one instance indicator (i.e., λ_{x_i}) instead of two (i.e., $\lambda_{x_{i1}}$ and $\lambda_{x_{i2}}$) where the positive (resp. negative) instance is represented by λ_{x_i} (resp. $\neg\lambda_{x_i}$). Moreover, each parameter $\Pi(x_i|u_i)$ (resp. $\Pi(\neg x_i|u_i)$) will be also encoded using $\theta_{x_i|u_i}$ (resp. $\neg\theta_{x_i|u_i}$). It is worthwhile to point out that by taking advantage of the opportunity presented by binary variables, we *halve the number of instance indicators and network parameters* but also *release the need for mutual exclusive clauses* and some *network parameters clauses.* The new auxiliary encoding specific to binary networks is outlined by Definition 3.

Definition 3. *Let ΠG_{min} be a binary possibilistic network, λ_{x_i} (resp. $\neg\lambda_{x_i}$) $(i = 1, \ldots, N)$ be the set of instance indicators and $\theta_{x_i|u_i}$ (resp. $\neg\theta_{x_i|u_i}$) be the*

set of parameter variables. Then $\forall X_i \in V$, its binary encoding C^b_{min} contains the following clauses:

$$\bigwedge \begin{cases} \lambda_{x_i} \wedge \lambda_{u_{i1}} \wedge \ldots \wedge \lambda_{u_{im}} \rightarrow \theta_{x_i|u_{i1},\ldots,u_{im}} \\ \neg\lambda_{x_i} \wedge \lambda_{u_{i1}} \wedge \ldots \wedge \lambda_{u_{im}} \rightarrow \neg\theta_{x_i|u_{i1},\ldots,u_{im}} \\ \vdots \qquad\qquad \vdots \qquad\qquad \vdots \\ \lambda_{x_i} \wedge \neg\lambda_{u_{i1}} \wedge \ldots \wedge \neg\lambda_{u_{im}} \rightarrow \theta_{x_i|u_{\bar{i}1},\ldots,u_{\bar{i}m}} \\ \neg\lambda_{x_i} \wedge \neg\lambda_{u_{i1}} \wedge \ldots \wedge \neg\lambda_{u_{im}} \rightarrow \neg\theta_{x_i|u_{\bar{i}1},\ldots,u_{\bar{i}m}} \end{cases} \tag{9}$$

The encoding C^b_{min} is in a CNF form where each clause encodes the fact that the possibility degree of $x_i|u_i$ (resp. $\neg x_i|u_i$) represented by the propositional formula $literal(\lambda_{x_i}) \wedge literal(\lambda_{u_{i1}}) \wedge \ldots \wedge literal(\lambda_{u_{im}})$ is equal ($\Rightarrow$ in the logical setting) to $\Pi(x_i|u_i)$ (resp. $\Pi(\neg x_i|u_i)$) represented by the propositional variable $\theta_{x_i|u_i}$ (resp. $\neg\theta_{x_i|u_i}$) where $literal(\lambda_{x_j})$ is expressed by:

$$literal(\lambda_{x_j}) = \begin{cases} \lambda_{x_j} & if \;\; x_j \in \{x_i, u_i\} \\ \neg\lambda_{x_j} & if \;\; \bar{x_j} \in \{x_i, u_i\} \end{cases} \tag{10}$$

If we emphasize on C^{PLS}_{min} and C^b_{min}, we can point out that in the binary case, we only resort to clauses (9) which represent the binary counterpart of clause (5) while omitting clauses (6) and (7). The question that may arise is: *why did we drop clauses (6) and (7) in the binary encoding?* In fact, in the binary case, these clauses lead to a contradictory encoding since from a logical point of view, $\neg\theta_{x_i|u_i}$ implies $\theta_{\bar{x_i}|u_i} \vee \theta_{x_i|\bar{u_i}} \vee \theta_{\bar{x_i}|\bar{u_i}}$, however, in our case, we assume that $\neg\theta_{x_i|u_i}$ only implies $\theta_{\bar{x_i}|u_i}$. Hence, we exclude such clauses since from a logical point of view $\neg\theta_{x_i|u_i}$ implies neither $\neg\lambda_{x_i}$ nor λ_{u_i}. This means that we move from a logical equivalence $\Leftrightarrow$ to a logical implication $\Rightarrow$.

The binary CNF encoding C^b_{min} can benefit from *possibilistic local structure*, i.e., instead of using $\theta_{x_i|u_i}$ and $\neg\theta_{x_i|u_i}$ for each pair (x_i, u_i), we can increasingly reduce the number of propositional variables associated to network parameters by according one propositional variable θ_j for each set of equal parameters. Moreover, each network parameter $\theta_{x_i|u_i}$ or $\neg\theta_{x_i|u_i}$ equal to 0 can be dropped by replacing its clause by a shorter one, namely: $\neg literal(\lambda_{x_i}) \vee \neg literal(\lambda_{u_{i1}}) \vee \cdots \vee \neg literal(\lambda_{u_{im}})$. We denote the binary method refined by possibilistic local structure as Bin-Π-DNNF$_{CPT}$.

Once the encoding phase is achieved, the resulting CNF encoding (i.e., C^{br}_{min}) is then compiled into a min-max circuit, denoted by C^{br}_{MinMax}, which will be used in the inference phase.

4.2 Inference Phase

The inference consists in efficiently computing the effect of the evidence e on the set a variables $X \subseteq V$ using C^{br}_{MinMax}, i.e. $\Pi(x|e)$. Using equation (1), it is clear that we should compute both of $\Pi(x,e)$ and $\Pi(e)$. The computation process is described in the following steps in which we will compute $\Pi_c(y)$ s.t. $y = \{(x,e);(e)\}$.

1. First, C^{br}_{MinMax} should be conditioned on y by setting λ_{x_i} (resp. $\neg\lambda_{x_i}$) to 1 if $x_i \in y$ (resp. $\bar{x_i} \in y$) and λ_{x_i} (resp. $\neg\lambda_{x_i}$) to 0 if $x_i \notin y$ (resp. $\bar{x_i} \notin y$).
2. Then, the conditioned representation, denoted by $C^{br}_{MinMax}|y$ should be valued by setting for each parameter variable the possibility degree it encodes. The resulting representation is denoted by C^{v}_{MinMax}.
3. Finally, $\Pi(y)$ is efficiently computed by applying max and min of C^{v}_{MinMax} in a bottom-up way. This efficiency is due to max-variable elimination.

5 Experimental Study

Our target through the current experimental study is to emphasize on the behavior of possibilistic local structure and local structure on CNF encodings of Π-DNNF [1], Π-DNNF$_{CPT}$ and Bin-Π-DNNF$_{CPT}$. To this end, we consider randomly generated possibilistic networks by setting the number of nodes to 50, the maximum number of parents per node to 3 and the variable cardinality to 2. We also vary possibility distributions (except for the normalization value 1) using the parameter (EP_{CPT} (%)): *the percent of equal parameters per CPT*. For each experimentation, we generate 100 possibilistic networks. Our experimental results are shown in Table 1.

It is clear from Table 1 that Bin-Π-DNNF$_{CPT}$ $<_{cnf}$ Π-DNNF$_{CPT}$ $<_{cnf}$ Π-DNNF[1]. These results prove the interest of using possibilitic local structure in general and the interest of applying the binary encoding to the particular case of binary networks.

Table 1 Π-DNNF vs Π-DNNF$_{CPT}$ vs Bin-Π-DNNF$_{CPT}$ (better values are in bold)

EP_{CPT}	**Π-DNNF**		**Π-DNNF$_{CPT}$**		**Bin-Π-DNNF$_{CPT}$**	
	variables	clauses	variables	clauses	variables	clauses
0	262	821	89	356	**84**	**334**
30	220	504	87	360	**81**	**348**
50	192	422	85	356	**81**	**340**
70	190	410	84	344	**79**	**341**
100	1560	380	84	338	**76**	**300**

[1] Π-DNNF$_{CPT}$ $<_{cnf}$ Π-DNNF means that the CNF encoding of Π-DNNF$_{CPT}$ is more compact than the one of Π-DNNF w.r.t. the number of variables and clauses.

6 Conclusion

In this paper, we improved the possibilistic compilation-based inference method of [1] by proposing a new encoding strategy, namely *possibilistic local structure*, especially for qualitative networks. Such strategy goes beyond the standard *local structure* used in Bayesian networks. It takes advantage of the idempotency property of the min operator by associating a unique propositional variable per equal parameters per CPT. Moreover, we refined the CNF encoding to deal with binary min-based possibilistis networks. With regard to future work, we would like to investigate the decision aspect under compilation.

References

1. Ayachi, R., Ben Amor, N., Benferhat, S., Haenni, R.: Compiling possibilistic networks: Alternative approaches to possibilistic inference. In: UAI, pp. 40–47 (2010)
2. Cadoli, M., Donini, F.: A survey on knowledge compilation. AI Commun. 10, 137–150 (1997)
3. Chavira, M., Darwiche, A.: Compiling bayesian networks with local structure. In: IJCAI, pp. 1306–1312 (2005)
4. Darwiche, A.: Decomposable negation normal form. J. ACM 48, 608–647 (2001)
5. Darwiche, A.: A logical approach to factoring belief networks. In: KR, pp. 409–420 (2002)
6. Darwiche, A., Marquis, P.: A knowledge compilation map. J. Artif. Intel. Res. 17, 229–264 (2002)
7. Dubois, D., Prade, H.: Possibility theory: An approach to computerized processing of uncertainty. Plenium Press, New York (1988)
8. Dubois, D., Prade, H.: Possibility theory. In: Meyers, R.A. (ed.) Encyclopedia of Complexity and Systems Science, pp. 6927–6939. Springer (2009)
9. Fargier, H., Marquis, P.: On valued negation normal form formulas. In: IJCAI, pp. 360–365 (2007)
10. Hisdal, E.: Conditional possibilities independence and noninteraction. Fuzzy Set. Syst. 1, 283–297 (1978)
11. Pearl, J.: Probabilistic reasoning in intelligent systems: networks of plausible inference. Morgan Kaufmman, San Francisco (1988)

Naïve Bayes Ant Colony Optimization for Experimental Design

Matteo Borrotti and Irene Poli

Abstract. In a large number of experimental problems the high dimensionality of the search space and economical constraints can severely limit the number of experiment points that can be tested. Under this constraints, optimization techniques perform poorly in particular when little a priori knowledge is available. In this work we investigate the possibility of combining approaches from advanced statistics and optimization algorithms to effectively explore a combinatorial search space sampling a limited number of experimental points. To this purpose we propose the Naïve Bayes Ant Colony Optimization (NACO) procedure. We tested its performance in a simulation study.

Keywords: Ant colony algorithm, combinatorial cptimization, naïve Bayes classifier.

1 Introduction

In this work we address the problem of developing a novel approach for combinatorial optimization in the context of experimental design. This approach should also improve the exploration of high dimensional spaces characterized by many interactions between variables.

More precisely we are interested in development of a new approaches for effectively exploring enzyme sequence space to improve or redesign enzyme

Matteo Borrotti · Irene Poli
European Centre for Living Technology (ECLT), San Marco 2940, Venice

Matteo Borrotti · Irene Poli
Department of Enviromental Science, Informatics and Statistics,
University Cá Foscari of Venice, Dorsoduro 2137, Venice
e-mail: {matteo.borrotti,irenpoli}@unive.it

R. Kruse et al. (Eds.): Synergies of Soft Computing and Statistics, AISC 190, pp. 489–497.
springerlink.com © Springer-Verlag Berlin Heidelberg 2013

functionality. Similar problems are presented in [1, 2]. In these works we designed a library of 95 different amino-acid sequences (*i.e.* words), that are subsequently assembled to yield a full-length enzyme (*i.e.* string) of length 4. To this regard, we can consider an enzyme as a string composed by 4 words. For each position in the string we can select an element from the set of 95 words, with repetition. The words are non-ordered discrete elements. The ultimate aim is to find the "most informative" string in according with a specific function.

Starting from this application, this work endeavours to define a new approach within statistical Design of Experiments for optimization based on Evolutionary Model Based Experimental Design [4]. In our case we have developed the Naïve Bayes Ant Colony Optimization (NACO) approach. NACO is a mixed approach that combines different methods from statistics and computer science. More precisely: (i) Naïve Bayes Classifier [6]; and (ii) Ant Colony Optimization [3], specifically $\mathcal{MAX} - \mathcal{MIN}$ Ant System [7]. In the next sections a brief description of the two approaches is given.

Our strategy identifies which elements mostly affect the response of the systems and then it adopts this information to help the metaheuristc algorithm in choosing the next set of candidate solutions.

2 Naïve Bayes Classifier

The Naïve Bayes Classifier [6] is a classification procedure based on the Bayes' rule. It assumes that the set of variables $X_1, ..., X_n$, we consider for classification, are all conditionally independent of one another, given the response Y.

We describe the conditional probability of $X = \langle X_1, ..., X_n \rangle$ on Y as:

$$P(X_1, X_2, \ldots, X_n \mid Y) = \prod_{i=1}^{n} P(X_i \mid Y)$$

This equation follows directly from the definition of conditional independence. Starting from this point, it is possible to understand how the Naïve Bayes Approach works. Assuming in fact that Y is any discrete valued variable, and the attributes $X_1, \ldots, X_n$ are any discrete or real valued variables, the goal of Naïve Bayes method is to train a classifier that will output the probability distribution over possible values of Y, for each new instance X that we want to classify.

The probability that Y will take on its k-th possible value, according to the Bayes' Rule, is

$$P(Y = y_k \mid X_1, \ldots X_n) = \frac{P(Y = y_k)P(X_1, \ldots, X_n \mid Y = y_k)}{\sum_j P(Y = y_j)P(X_1, \ldots, X_n \mid Y = y_j)}$$

where the sum is taken over all possible value y_j of Y. Assuming that X_i are conditionally independent given Y, we can write

$$P(Y = y_k \mid X_1, \dots X_n) = \frac{P(Y = y_k) \prod_i P(X_1, \dots, X_n \mid Y = y_k)}{\sum_j P(Y = y_j) \prod_i P(X_i \mid Y = y_j)} \quad (1)$$

Equation (1) is the base for the Naïve Bayes Classifier. Given a new instance $X^{new} = \langle X_1, ..., X_n \rangle$, it is possible to calculate the probability that Y will take on any given value. If we want to know the most probable value of Y, we obtain the Naïve Bayes' Rule:

$$Y \leftarrow arg \max_{y_k} \frac{P(Y = y_k) \prod_i P(X_1, \dots, X_n \mid Y = y_k)}{\sum_j P(Y = y_j) \prod_i P(X_i \mid Y = y_j)}$$

This approach reduces the complexity for learning Bayesian classifier by making a conditional independence assumption that dramatically reduces the number of parameters to be estimated when modeling $P(X \mid Y)$.

3 Ant Colony Optimization (ACO)

Ant Colony Optimization (ACO) [3] is a population-based, general-purpose stochastic search technique for the solution of difficult combinatorial problems, which is inspired by the pheromone trail laying and following foraging behaviour of some real ant species.

In ACO, each "ant" builds a solution starting from an initial state selected according to some problem dependent criteria. A solution is expressed as minimum cost (shortest) path through the states of the problem in accordance with the problem's constraints. A single ant is able to build a solution but only the cooperation among all the agents of the colony, concurrently building different solutions, is able to find high quality solutions.

In the algorithm an environment is simulated by a graph composed of a set N of states, representing nodes, and a set E of arcs fully connecting the nodes N. Let d_{ij} be the length of the arc $(i, j) \in E$, that is the distance between nodes i and j, with $i, j \in N$. The aim of the optimization in ACO is to find on the graph $G = (N, E)$ the minimal length path connecting nest to the source.

ACO uses two different types of information: pheromone and *a priori* problem-specific information (heuristic values). The combination of available pheromone and heuristic values defines *ant-decision tables*, that are, probabilistic tables used by the ants' decision policy to direct their search towards the most interesting regions of the search space.

The ant-decision table $A_i = [a^i_j(t)]_{|N_i|}$ of node i is obtained by the composition of the pheromone trail values with heuristic values as follow:

$$a_{ij} = \frac{[\tau_{ij}(t)]^\alpha[\eta_{ij}]^\beta}{\sum_{l \in N_i}[\tau_{ij}(t)]^\alpha[\eta_{ij}]^\beta} \qquad \forall j \in N_i \tag{2}$$

where $\tau_{ij}(t)$ is the amount of pheromone trail on arc (i, j) at time t; $\eta_{i,j} = 1/d_{ij}$ is the heuristic value to move from node i to node j; N_i is the set of neighbours of node i; and α and β are two parameters that control the relative weight of pheromone trail and heuristic information.

The probability with which an ant k chooses to go from node i to node $j \in N^k_i$ while building its tour at the t-th algorithm iteration, is:

$$p^k_{ij}(t) = \frac{a_{ij}(t)}{\sum_{i \in N^k_i} a_{ij}(t)} \tag{3}$$

where $N^k_i \subseteq N_i$ is the set of nodes in the neighbourhood of node i that ant k has not visited yet.

After all ants have completed their tour, pheromone evaporation on all arcs is applied in according with the pheromone trail decay coefficient (or evaporation factor) ρ, $\rho \in (0, 1]$.

In our case, we apply the $\mathcal{MAX} - \mathcal{MIN}$ Ant System ($\mathcal{MM}$AS), proposed by Stützle and Hoos in [7] because it is demonstrated that $\mathcal{MM}$AS is able to reach a strong exploitation of the search space by adding pheromone only to the best solution during the pheromone trail update. Moreover they applied a simple method for limiting the strength of the pheromone trails that effectively avoids premature convergence of the search.

4 The NACO Approach

We introduce a novel approach for optimization called Naïve Bayes Ant Colony Optimization (NACO) by combining Ant Colony Optimization technique and Naïve Bayes Classifier. NACO extracts the information from the data using the Naïve Bayes Approach and explores the search space by the ACO algorithm. At the same time, the most informative variables and interactions are identified.

In our problem we are considering a specific problem caracterized by a discrete search space where a solution is a string composed by 4 words. The words are selected from a set of 95 discrete elements with repetition. We need to reformulate the problem as a path search problem to apply ACO algorithm. For this purpose, we create a graph where each node represents a specific word. A solution is a path with length 4 composed of 4 nodes connected by 3 arcs, as shown in Fig. 1. In the biological application [1], each node corresponds to an amino-acid sequence from the initial dataset

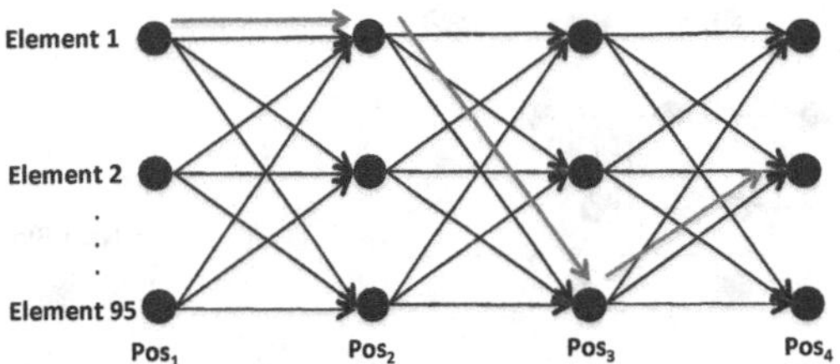

Fig. 1 A new representation of the graph where ants move. A solution is a path composed by 4 nodes and 3 arcs.

and an arc to the connection between amino-acid sequence i in position k and amino-acid sequence j in position $k+1$. A candidate solution is a path composed by 4 amino-acid sequences which represents a full-length enzyme. The total number of nodes is $95 \times 4 = 380$ with equal size intervals.

Candidate solutions' response is calculated and subsequently discretized in according with a certain constant threshold (γ, with $\gamma \in \mathbb{R}$) fixed by the experimenter. The candidate solutions (strings) are assigned to class 1 if the responses exceed the threshold otherwise to class 0. The Naïve Bayes Classifier is applied on each position of the sequence or string with class equal to 1. It calculates the maximum likelihood estimate for each word in each position given the training sequences or strings. We obtain a set of probability distributions, one for each position. These probability distributions are used to weigh the arcs on the ACO algorithm.

The following steps summarize the NACO approach:

1. Random generation and evaluation of an initial population (set of candidate points);
2. Identification of the *Iteration Best Solution* (best solution in the current iteration);
3. Calculation of the Naïve Bayes Classifier on the available evaluated solutions (N). At each iteration, it focuses on values of the response greater than a problem-specific threshold γ, with $\gamma \in \mathbb{R}$;
4. Updating the probabilities with which an ant k chooses to go from element i to element j using the information extracted in points 2 and 3;
5. Selection of the next population of candidate solutions using the principle of $\mathcal{MAX} - \mathcal{MIN}$ Ant System;
6. Evaluation of the new set of candidate solutions and inclusion of the new set in the set of solutions that has already been evaluated;
7. If stop criterion is reached, then stop. Otherwise repeat points from 2 to 6;

Fig. 2 describes point number 4. At iteration t, agents move over the graph according to the best paths identified in the previous steps (Fig. 2 (a)).

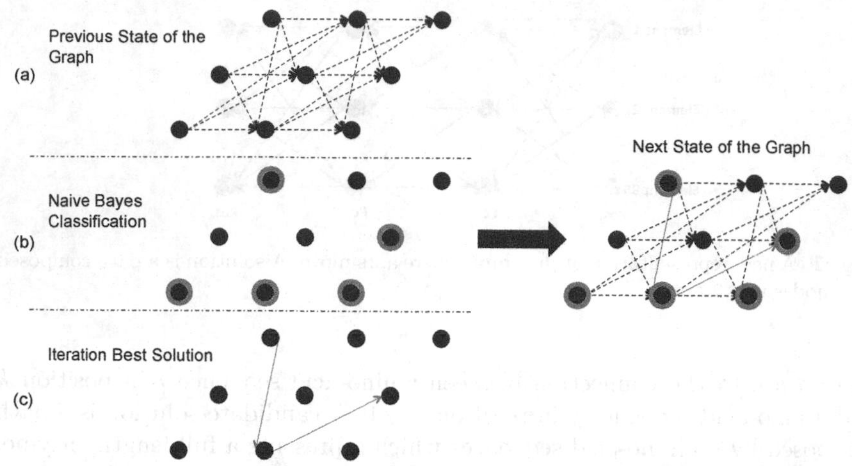

Fig. 2 Updating Phase of the Naïve Bayes Ant Colony Optimization

Following candidate solution evaluation, the *Iteration Best Solution* is identified and the corresponding pheromone path is updated (Fig. 2 (c)). At this point, using the Naïve Bayes Classifier the best variables are identified, namely those that are anticipated to yield a fitness value higher than the chosen fitness treshold γ (Fig. 2 (b)). For any given arc connecting variable (node) i with variable j, the weight λ_{ij} is changed according to the Naïve Bayes Classifier. The set of $\{\lambda_{ij}\}$ is called *Naïve Information*. Now, the ant-decision table $A_i = [a^i_j(t)]_{|N_i|}$ of node i will be obtained by the composition of the pheromone trail values with heuristic values and with *Naïve Information* as follows:

$$a_{ij} = \frac{[\tau_{ij}(t)]^\alpha [\eta_{ij}]^\beta [\lambda_{ij}]^\delta}{\sum_{l \in N_i} [\tau_{ij}(t)]^\alpha [\eta_{ij}]^\beta [\lambda_{ij}]^\delta]} \qquad \forall j \in N_i$$

where $\tau_{ij}(t)$ is the amount of pheromone trail on arc (i, j) at time t; $\eta_{i,j} = 1/d_{ij}$ is the heuristic value of moving from node i to node j; λ_{ij} is the *Naïve Information* on arc (i, j) at time t. N_i is the set of neighbors of node i; and α, β and δ are parameters that control the relative weight of pheromone trail, heuristic information and *Naïve Information*. The probability, which an ant k chooses to go from element i to element j, is then calculated according to Equation (3).

Generally, NACO will extract information from few data and it will individuate the best connection between variables.

5 Monte Carlo Simulations

In this work we develop simulative studies with the aim of testing the performance of the NACO approach. In these simulations the experiments are generated by the model $y = \varphi_h(\mathbf{x}) + \epsilon$, $\epsilon \sim N(0, \sigma_h^2)$, where $h = 1, 2$ denotes two different response surfaces, $\varphi_1(\mathbf{x})$ and $\varphi_2(\mathbf{x})$, taken from [2]. $\varphi_1(\mathbf{x})$ is called Polynomial Regression Model (PRM) and it represents a fitness landscape dominated by strong interactions, which occurs when the effect of one variables depends on the presence of another. The second formal structure, $\varphi_2(\mathbf{x})$, is called Polynomial Sparse Regression Model (PSRM) and it represents the situation where only few variables highly influence the response of the system and the others are close to 0. This kind of fitness landscapes is characterized by ruggedness and local optima. $\varphi_1(\mathbf{x})$ and $\varphi_2(\mathbf{x})$ simulate a search space of 95^4 experimental points.

In both simulations, we computed 50 Monte Carlo runs with the aim to maximize the deterministic function φ_h.

In Fig. 3, we compare the convergence of our method with the basic $\mathcal{MM}$AS. The parameter settings for $\mathcal{MM}$AS is: population size (N) 100, number of generations or experimental batches 50, evaporation factor (ρ) 0.96 and weight for the pheromone (β) 1. In this case no heuristic information is used. NACO includes two more parameters: δ that controls the weight of *Naïve Information* and γ that is the threshold considered by the Naïve Bayes Classifier. In our case δ and γ are equal to 2 and 80%, respectively. The parameter setting is chosen in accordance with preliminary studies [1].

Our empirical results (Fig. 3) show remarkable efficiency of the proposed method. The main difference between NACO and $\mathcal{MM}$AS is that the last

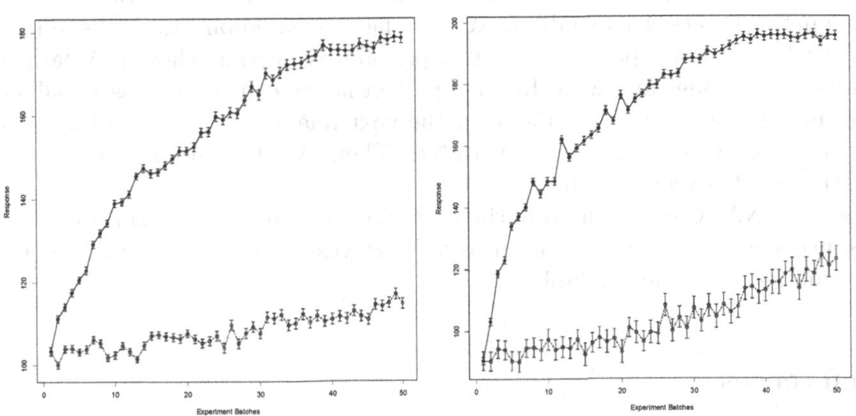

Fig. 3 Convergence behavior for Experiment 1 ($\varphi_1(\mathbf{x})$, left side) and Experiment 2 ($\varphi_2(\mathbf{x})$, right side) of the NACO (solid lines) and $\mathcal{MM}$AS (dashed lines) methods. The vertical bars are 95% Monte Carlo confidence intervals.

one does not benefit from the information obtained using the Naïve Bayes Classifier. Our results show that the inclusion of the Naïve Bayes Classifier boosts the convergence rate. This is visible after few iterations of the NACO algorithm and is due to the fact that the Naïve Bayes Classifier extracts useful information on the main variables that influence the response. The information extrapoleted from the Naïve Bayes Classifier compensates for the absence of heuristic information.

In term of computational time the two approaches do not show significante difference.

6 Conclusions

In this work, we have explored the possibility of tackling problems characterized by a very large experimental space combining bio-inspired algorithms with advanced statistical techniques. To achieve this aim, we have developed an algorithmic approach which combines some powerful features of known approaches: the Naïve Bayes Ant Colony Optimization (NACO).

We have shown that the Naïve Bayes Ant Colony Optimization (NACO) approach improves upon the limits of the individual techniques, enabling us to deal with very large experimental spaces. In fact, the Naïve Bayes Approach has a strong assumption and it assumes that the attributes $X_1, \ldots, X_n$ are all conditionally independent of one another, given the response Y. It has the advantage of simplifying the representation of the probability of X given Y but with the Naïve Bayes Classificator it is not possible to understand the relations between the attributes. This aspect can be extremely important in some experimental problems.

The combination of ACO and Naïve Bayes Classifier can describe the relative network between variables. ACO, in fact, is based on probabilistic matrices where the best path has a higher probability of being chosen. A path is composed of nodes and arcs. In our problem nodes can be seen as variables and an arc, connecting a variable to the next one, can be seen as the relation that exists between the two variables. Then, ACO implies the sequential relationship between variables.

At last, NACO can be used in the context of combinatorial optimization in absence of heuristc information since Naïve Bayes Classifier is used to extract information from the available data.

References

1. Borrotti, M.: An evolutionary approach to the design of experiments for combinatorial optimization with an application to enzyme engineering. PhD thesis, Department of Statistical Science, University of Bologna, Italy (2011)

2. Borrotti, M., De Lucrezia, D., Minervini, G., Poli, I.: A Model Based Ant Colony Design for the Protein Engineering Problem. In: Dorigo, M., Birattari, M., Di Caro, G.A., Doursat, R., Engelbrecht, A.P., Floreano, D., Gambardella, L.M., Groß, R., Şahin, E., Sayama, H., Stützle, T. (eds.) ANTS 2010. LNCS, vol. 6234, pp. 352–359. Springer, Heidelberg (2010)
3. Dorigo, M., Di Caro, G.: The ant colony optimization meta-heuristic. In: Corne, D., et al. (eds.) New Ideas Optim., pp. 11–32. McGraw-Hill, New York (1999)
4. Forlin, M., Slanzi, D., Poli, I.: Combining Probabilistic Dependency Models and Particle Swarm Optimization for Parameter Inference in Stochastic Biological Systems. In: Gaol, F.L., Nguyen, Q.V. (eds.) Proc. of the 2011 2nd International Congress CACS. AISC, vol. 145, pp. 437–444. Springer, Heidelberg (2012)
5. Kohonen, J., Talikota, S., Corander, J., Auvinen, P., Arjas, E.: A naïve bayes classifier for protein function prediction. Silico Biol. 9, 23–34 (2009)
6. Mitchell, T.M.: Machine Learning. McGraw-Hill, New York (1997)
7. Stützle, T., Hoos, H.: Max-min ant system. Future Gener. Comp. Sy. 16, 889–914 (2000)

Inferences in Binary Regression Models for Independent Data with Measurement Errors in Covariates

Vandna Jowaheer, Brajendra C. Sutradhar, and Zhaozhi Fan

Abstract. When responses along with covariates are collected from a group of independent individuals in a binary regression setup, in some practical situations the observed covariates may be subject to measurement errors differing from the true covariates values. These imprecise observed covariates, when used directly, the standard statistical methods such as naive likelihood and quasi-likelihood methods yield biased and hence inconsistent regression estimates. Because there does not exist a corrected score function for this binary measurement error model, a considerable attention is given in the literature to develop approximate unbiased estimating equation in order to obtain consistent regression estimate. In this paper, we review some of these widely used approaches and suggest a softer (approximate) quasi-likelihood approach for consistent regression parameters estimation.

Keywords: Bias correction, binary response, consistency, first order bias correction, measurement errors in covariates.

1 Introduction

In the independence set up, the estimation of the regression effects involved in binary measurement-error models with normal measurement errors in covariates has been studied extensively in the literature. For $i = 1, \ldots, K$, let y_i denote the binary response variable for the ith individual, and $x_i =$

Vandna Jowaheer
University of Mauritius, Mauritius
e-mail: vandnaj@uom.ac.mu

Brajendra C. Sutradhar · Zhaozhi Fan
Memorial University of Newfoundland, Canada
e-mail: {bsutradh,z.fan}@mun.ca

R. Kruse et al. (Eds.): Synergies of Soft Computing and Statistics, AISC 190, pp. 499–505.
springerlink.com © Springer-Verlag Berlin Heidelberg 2013

$(x_{i1}, \ldots, x_{ip})'$ be the associated $p-$dimensional covariate vector subject to normal measurement errors. Let $z_i = (z_{i1}, \ldots, z_{ip})'$ be the unobserved true covariate vector which may be fixed constant or random, and β be the regression effect of z_i on y_i. We consider the measurement error model

$$x_i = z_i + \delta v_i, \tag{1}$$

with $v_i \sim N_p(0, \Lambda = \text{diag}[\sigma_1^2, \ldots, \sigma_p^2])$.

Note that if for a sample $(y_i, x_i)(i = 1, \ldots, K)$ the covariates $\{z_i\}$ are unknown constants, a functional error-in-variables model (also known as Berkson error model) is obtained; if $\{z_i\}$ are independent and identically distributed random vectors from some unknown distribution, a structural error-in-variables model is obtained (see [2], ch. 29 and [7]). In this paper we consider the functional measurement error models. Nakamura [4] has proposed a corrected score (CS) estimation approach in functional setup, which provides closed form estimating equation for β for the Poisson regression model, but, the binary logistic regression model does not yield a corrected score function which is a limitation to this approach.

Stefanski and Carroll [7] proposed a method based on conditional scores (CNS). In this approach, unbiased score equations are obtained by conditioning on certain parameter dependent sufficient statistics for the true covariates z, and the authors have developed the approach in both functional and structural setups. The conditional score equations have a closed form for generalized linear models such as for normal, Poisson and binary logistic models. Obtaining a closed form unbiased equation for logistic regression parameter by this conditional approach is an advantage over the direct corrected score approach [4] which does not yield corrected score function.

Note that as in the absence of measurement errors, regression parameters involved in generalized linear models such as for count and binary models, may be estimated consistently and efficiently by using the first two moments based quasi-likelihood (QL) approach [10], there has been a considerable attention to modify the naive QL approach (that directly uses observed covariates ignoring measurement errors) in order to accommodate measurement errors in covariates and obtain bias corrected QL (BCQL) estimates. Some of these BCQL approaches are developed for both structural and functional models, some are developed for the functional models and others are more appropriate for structural models only. Stefanski [5] proposed a small measurement error variance based QL (SVQL) approach for structural models, the autors of [1] have used a similar small measurement error variance based QL approach which is developed to accommodate either of the structural or functional models or both. Liang and Liu [3] have discussed a BCQL approach for structural model, which was later on generalized by Wang *et al.* [9] to accommodate correlated replicates in covariates. Sutradhar and Rao [8] have used the SVQL approach from [5] for the longitudinal binary data, independence setup being a special case, under functional model only. In the next section, we provide a brief review of some of these existing simpler BCQL approaches which are suitable for functional models.

In Section 3, we provide a softer BCQL (SBCQL) approach which provides a first order approximate QL regression estimates. We give some remarks in Section 4 on future works involving this binary measurement error model.

2 Some Bias Correction Estimation Approaches

2.1 *Conditional Score (CS) Approach*

To understand the conditional score approach, consider for example, the functional version of the logistic measurement error model with scalar predictor z_i so that the measurement error v_i in (1) follows $N_1(0, \sigma_1^2)$ (see [6], Section 4.1). For convenience, consider $\delta = 1$ in (1). In this case, the density of $(y_i; x_i)$ is given by

$$f(y_i, x_i; \beta, z_i) = [\frac{\exp(z_i'\beta)}{1+\exp(z_i'\beta)}]^{y_i} [\frac{1}{1+\exp(z_i'\beta)}]^{1-y_i} \frac{1}{\sigma_1} \phi(\frac{x_i - z_i}{\sigma_1}),$$

where $\phi(.)$ is the standard normal density function. The estimation of β also requires the estimation of the nuisance parameters z_i or some functions of z_i's for $i = 1, \ldots, K$. However, Stefanski and Carroll [7] have demonstrated that the parameter dependent statistic $\lambda_i = x_i + y_i \sigma_1^2 \beta$ is sufficient for unknown z_i in the sense that the conditional distribution of (y_i, x_i) given λ_i does not depend on the nuisance parameter z_i. This fact was exploited to obtain unbiased estimating equation for β using either conditional likelihood method or mean variance function models (based on conditional density of y_i given λ_i) and quasi-likelihood methods. For the scalar regression parameter β, the unbiased estimating equation has the form ([8], Eq. (2.10))

$$\sum_{i=1}^{K} (\lambda_i - \sigma_1^2 \beta)(y_i - \tilde{p}_i) = 0, \tag{2}$$

where $\tilde{p}_i = F[\{\lambda_i - (\sigma_1^2/2)\beta\}\beta]$ with $F(t) = 1/[1+\exp(-t)]$. Let $\hat{\beta}_{\text{CNS}}$ denote the solution of (2) for β.

2.2 *Small Measurement Error Variance Based QL (SVQL) Approach*

Note that if z_i were known, then one would have obtained a consistent estimator of β by solving the so-called quasi-likelihood (QL) estimating equation

$$\sum_{i=1}^{K} z_i(y_i - \mu_{iz}) = \sum_{i=1}^{K} \psi_i(y_i, z_i, \beta) = 0, \tag{3}$$

where $\mu_{iz} = \exp(z_i'\beta)/[1 + \exp(z_i'\beta)]$ is the mean of y_i. Note that this QL estimating equation is also a likelihood estimating equation. However, because the true covariate z_i is not observed, one can not use the estimating equation (3) for the estimation of β.

Suppose that by replacing z_i with x_i in (3), one constructs a naive QL (NQL) estimating equation, namely

$$\sum_{i=1}^{K} x_i(y_i - \mu_{ix}) = \sum_{i=1}^{K} \psi_i(y_i, x_i, \beta) = 0, \tag{4}$$

where $\mu_{ix} = \exp(x_i'\beta)/[1 + \exp(x_i'\beta)]$. Let $\hat{\beta}$ be the naive estimator obtained from (4). This estimator does not converge to β, it rather converges to a different parameter say $\beta(\delta\Lambda)$. Thus, the naive estimator $\hat{\beta}$ is biased and hence inconsistent for β. As a remedy, assuming that δ is small, by expanding the expected function $E_x \sum_{i=1}^{K} \psi_i(y_i, x_i, \beta)$ about $\delta = 0$, and then equating the expanded function to zero followed by replacing z_i with x_i and β with $\hat{\beta}$, one obtains a SVQL estimator of β [5] by using the iterative equation

$$\hat{\beta}_{\text{SVQL}}(\delta) = \hat{\beta} + \frac{1}{2}\delta^2 \left[-\sum_{i=1}^{K} \hat{p}_{ix} x_i x_i' \right]^{-1} \times \left[\sum_{i=1}^{K} \hat{p}_{ix}\hat{q}_{ix}\{1 - \hat{q}_{ix}\}\hat{\beta}'\Lambda\hat{\beta}x_i - 2\hat{p}_{ix}\hat{q}_{ix}\Lambda\hat{\beta} \right], \tag{5}$$

(see also [8], Eq. (2.2), p. 181) where $\hat{q}_{ix} = 1 - \hat{p}_{ix}$.

3 A Softer BCQL Approach Using Corrected Estimating Function

We propose a bias correction approach along the lines of [4]. The difference between Nakamura's and our approach is that in [4] a corrected score function $\ell^*(\beta; y, x)$ is developed such that its expectation is the true but unknown score function, that is, $E_x[\ell^*(\beta; y, x)] = \ell(\beta; y, z)$, and then solved the corrected score equation for β obtained from $\ell^*(\beta; y, x)$, whereas in our approach we develop a corrected quasi-likelihood function, say $Q^*(y, x, \beta)$, such that

$$E_x[Q^*(y, x, \beta)] = \psi(y, z, \beta) = \sum_{i=1}^{K} z_i[y_i - \frac{\exp(z_i'\beta)}{1 + \exp(z_i'\beta)}], \tag{6}$$

and then solve the corrected QL equation, that is, $Q^*(\beta, y, x) = 0$ for β. But, because the true mean function is given by $\mu_{iz} = \exp(z_i'\beta)/[1 + \exp(z_i'\beta)]$, it is difficult to derive the corrected QL function $Q^*(y, x, \beta)$. However, a softer, that is, a first order approximate BCQL (SBCQL) estimating function may be developed as follows. We denote this SBCQL function as $\tilde{Q}_S(y, x, \beta)$ which will be approximately unbiased for $\psi(y, z, \beta)$, that is,

$$E_x[\tilde{Q}_S(y, x, \beta)] \simeq \psi(y, z, \beta).$$

Note that under the Gaussian measurement error model (1), that is when $x_i \sim N_p(z_i, \delta^2 \Lambda)$, one obtains

$$E_x[\exp(x_i'\beta - \xi)] = \exp(z_i'\beta), \tag{7}$$

$$E_x[\{x_i - \delta^2 \Lambda\beta\} \exp(x_i'\beta - \xi)] = z_i \exp(z_i'\beta), \tag{8}$$

where $\xi = \frac{\delta^2}{2}\beta'\Lambda\beta$. It then follows that

$$E_x\left[\frac{\{x_i - \delta^2\Lambda\beta\}\exp(x_i'\beta - \xi)}{1 + \exp(x_i'\beta - \xi)}\right] \simeq \frac{z_i \exp(z_i'\beta)}{1 + \exp(z_i'\beta)} = \frac{\mu_{W_{z,N}}}{\mu_{W_{z,D}}}. \tag{9}$$

Next because, the true QL function has the form

$$\psi(y, z, \beta) = \sum_{i=1}^{K} z_i y_i - \sum_{i=1}^{K}[\frac{z_i \exp(z_i'\beta)}{1 + \exp(z_i'\beta)}],$$

by using (9), one may write a softer BCQL (SBCQL) estimating equation as

$$\sum_{i=1}^{K}\left[x_i y_i - \frac{\{x_i - \delta^2\Lambda\beta\}\exp(x_i'\beta - \xi)}{1 + \exp(x_i'\beta - \xi)}\right] = 0. \tag{10}$$

We denote the solution of the SBCQL estimating equation (10) by $\hat{\beta}_{\text{SBCQL}}$.

However, because the first order approximation used in (9) may not be able to reduce the bias sufficiently, we consider an improvement by writing

$$E_x\left[\frac{\{x_i - \delta^2\Lambda\beta\}\exp(x_i'\beta - \xi)}{1 + \exp(x_i'\beta - \xi)}\right] = E_x\left[\frac{W_{x,N}}{W_{x,D}}\right]$$
$$\simeq \frac{\mu_{W_{z,N}}}{\mu_{W_{z,D}}} - \frac{\hat{\text{cov}}[W_{x,N}, W_{x,D}]}{\hat{\mu}^2_{W_{z,D}}} + \frac{\hat{\mu}_{W_{z,N}}}{\hat{\mu}^3_{W_{z,D}}}\hat{\text{var}}[W_{x,D}], \tag{11}$$

where we use

$$\hat{\mu}_{W_{x,N}} = \frac{1}{K}\sum_{i=1}^{K}[\{x_i - \delta^2 \Lambda\beta\}\exp(x_i'\beta - \xi)]$$

$$\hat{\mu}_{W_{x,D}} = \frac{1}{K}\sum_{i=1}^{K}[1 + \exp(x_i'\beta - \xi)]$$

$$\hat{\mathrm{var}}[W_{x,D}] = \frac{1}{K}\sum_{i=1}^{K}[1 + \exp(x_i'\beta - \xi)]^2 - \hat{\mu}^2_{W_{x,D}}$$

$$\hat{\mathrm{cov}}[W_{x,N}, W_{x,D}] = \frac{1}{K}\sum_{i=1}^{K}\left[\{(x_i - \delta^2\Lambda\beta)\exp(x_i'\beta - \xi)\}\{1 + \exp(x_i'\beta - \xi)\}\right] - \hat{\mu}_{W_{x,N}}\hat{\mu}_{W_{x,D}} \tag{12}$$

We now re-write (11) as

$$E_x\left[\frac{\{x_i - \delta^2\Lambda\beta\}\exp(x_i'\beta - \xi)}{1 + \exp(x_i'\beta - \xi)} + t_c\right] = \frac{\mu_{W_{x,N}}}{\mu_{W_{x,D}}}, \tag{13}$$

where

$$t_c = \frac{cov[W_{x,N}, W_{x,D}]}{\hat{\mu}^2_{W_{x,D}}} - \frac{\hat{\mu}_{W_{x,N}}}{\hat{\mu}^3_{W_{x,D}}}\hat{\mathrm{var}}[W_{x,D}].$$

Thus, instead of (10), we now solve the improved SBCQL estimating equation given by

$$\sum_{i=1}^{K}\left[x_i y_i - \frac{\{x_i - \delta^2\Lambda\beta\}\exp(x_i'\beta - \xi)}{1 + \exp(x_i'\beta - \xi)} - t_c\right] = 0. \tag{14}$$

4 Concluding Remarks

Among many alternative bias correction approaches, in this paper, we have reviewed the two widely used, namely, conditional score and small variance based QL approaches. We have provided a new softer bias corrected QL (SBCQL) approach and a further improvement as a generalization of the classical QL approach in [10]. We are currently working on the finite sample performance of the proposed approach and the results will be reported elsewhere.

Acknowledgements. The authors would like to thank two referees for their comments on the last version.

References

1. Carroll, R.J., Stefanski, L.A.: Approximate quasi-likelihood estimation in models with surrogate predictors. J. Am. Stat. Assoc. 85, 652–663 (1990)
2. Kendall, M.G., Stuart, A.: The Advanced Theory of Statistics. Griffin, London (1979)
3. Liang, K.-Y., Liu, X.: Estimating equations in generalized linear models with measurement error. In: Godambe, V.P. (ed.) Estimating Functions, pp. 47–63. Oxford University Press, New York (1991)
4. Nakamura, T.: Corrected score function for errors-in-variables models: Methodology and application to generalized linear models. Biometrka 77, 127–137 (1990)
5. Stefanski, L.A.: The Effects of measurement error on parameter estimation. Biometrika 72, 583–592 (1985)
6. Stefanski, L.A.: Measurement error models. J. Am. Stat. Assoc. 95, 1353–1358 (2000)
7. Stefanski, L.A., Carroll, R.J.: Conditional scores and optimal scores for generalized linear measurement-error models. Biometrika 74, 703–716 (1987)
8. Sutradhar, B.C., Rao, J.N.K.: Estimation of regression parameters in generalized linear models for cluster correlated data with measurement error. Can J. Stat. 24, 177–192 (1996)
9. Wang, N., Carroll, R.J., Liang, K.-Y.: Quasilikelihood estimation in measurement error models with correlated replicates. Biometrics 52, 401–411 (1996)
10. Wedderburn, R.: Quasilikelihood functions, generalized linear models, and the Gauss-Newton method. Biometrika 61, 439–447 (1974)

Multi-dimensional Failure Probability Estimation in Automotive Industry Based on Censored Warranty Data

Mark Last, Alexandra Zhmudyak, Hezi Halpert, and Sugato Chakrabarty

Abstract. The warranty datasets available for various car models are characterized by extremely imbalanced classes, where a very low amount of under-warranty vehicles have at least one matching claim ("failure") of a given type. The failure probability estimation becomes even more complex in the presence of censored warranty data, where some of the vehicles have not reached yet the upper limit of the predicted interval. The actual mileage rate of under-warranty vehicles is another source of uncertainty in warranty datasets. In this paper, we present a new, continuous-time methodology for failure probability estimation from multi-dimensional censored datasets in automotive industry.

Keywords: Automotive Industry, censored data, multi-dimensional failure prediction, warranty data.

1 Introduction

Auto manufacturers are interested in estimating the probability of a failure expected in certain systems or subsystems of an individual under-warranty vehicle within a given time and mileage range as a function of various predictive factors. For example, an automaker may be interested in estimating the

Mark Last · Alexandra Zhmudyak · Hezi Halpert
Department of Information Systems Engineering Ben-Gurion University of the Negev Beer-Sheva 84105, Israel
e-mail: {mlast,zhmudyak,halpertc}@post.bgu.ac.il

Sugato Chakrabarty
India Science Lab, General Motors Global Research and Development, GM Technical Centre India Pvt Ltd, Bangalore - 560 066, India
e-mail: sugato.chakrabarty@gm.com

R. Kruse et al. (Eds.): Synergies of Soft Computing and Statistics, AISC 190, pp. 507–515.
springerlink.com © Springer-Verlag Berlin Heidelberg 2013

probability of a tire failure in a specific vehicle during the next three months and 5,000 miles. In case of the failure probability exceeding a pre-defined threshold, the car owner may be issued a warning suggesting to visit an authorized dealer at his/her earliest convenience. Ideally, if on-board diagnostic systems were installed in every vehicle, it would be possible to record all the necessary data in a vehicle and determine its condition in real-time. Since such diagnostic systems are not present in every vehicle, due to cost and other issues, the next best source of data happens to be the warranty data of past failures. However, the warranty datasets often include "right-censored data", where some of the vehicles have not reached yet the warranty limit or even the upper limit of the prediction interval and thus can produce warranty claims after the cutoff date of the data collection process. The vehicle usage rate provides an additional aspect of uncertainty in warranty datasets. Since the vehicle mileage is recorded only at the time of a claim, the actual mileage rate of non-claimed vehicles is not known at all, whereas vehicles with past claims provide us only partial information for estimating their current usage rate. In this paper, a new, continuous-time methodology is presented for failure probability estimation from multi-dimensional censored datasets.

The rest of this paper is organized as follows. In Section 2, we provide a brief overview of previous works on failure probability estimation from warranty claims data. Our approach to failure probability estimation is presented and demonstrated on a small numeric example in Section 3. Section 4 concludes the paper with an outline of some future research directions.

2 Related Work

Assuming that the failure data are collected up to some current date (also called "database cutoff date"), Hu *et al.* [2] define the censoring time as the minimum of the vehicle's current mileage and the mileage at which it passes out of the warranty plan. The observed data includes only failures, which occurred before the censoring time of a given vehicle. The paper [2] shows that if the censoring time distribution G is known and the time-to-failures can only take discrete values (e.g., one day, two days, etc.), the failure probability in a given discrete period t (e.g., on day 30 or day 45) can be estimated by the following maximum likelihood formula:

$$\widehat{f}_{ML}(t) = \frac{n_t}{M\overline{G}(t)} \tag{1}$$

Where n_t is the number of failures observed at time t and M is the total population size of the model in question. Thus, the failure probability is estimated as a ratio between the number of failures in a given, relatively short period t and the proportion of vehicles whose censoring time is at least equal to t. This discrete-time approach totally ignores the *duration* of the prediction interval t whereas summing the failure probabilities over multiple periods may require rescaling of each estimate [2]. A similar approach is used by Majeske [3] to estimate the cumulative number of warranty claims up to time in service t. He calculates the number of vehicles at risk at the time in service t (measured in months) by removing from the population all vehicles that exceeded one of the warranty limitations (time or mileage) at time t. However, these and other previous works on failure probability estimation in automotive industry do not provide any direct method for calculating the failure probability of a specific vehicle in an arbitrary one-dimensional interval or two-dimensional cell of continuous attributes (such as mileage and/or time-in-service) given censored warranty data with partial usage observations.

To forecast the future usage of each vehicle from its last recorded claim until the end of its observable life, [1] uses the average rate of mileage accumulation over time between the sale date and the vehicle last claim. The vehicles without claims are assumed to have the same distribution of daily mileage rate as vehicles with claims. Contrary to the approach of [1], the author of [4] argues that in general, the usage (e.g., mileage) distributions of non-failed products are different from those of failed products. The actual usage distributions of such products (like vehicles without claims) may be obtained using the supplementary data approach. Customer surveys and periodic inspections are suggested as two potential sources of supplementary usage data.

3 Failure Probability Estimation from Censored Warranty Data

The sub-sections of this section present the following parts of the failure probability estimation process given censored data with partial usage observations:

- Mileage rate estimation for claimed and non-claimed vehicles
- Failure probability estimation in one continuous dimension (e.g., mileage or time in service)
- Failure probability estimation in two continuous dimensions (e.g., mileage *and* time in service)

We estimate failure probability from a dataset of vehicles and their warranty claims. Our approach is demonstrated on a small sample of simulated warranty data. Let us assume that the cutoff date of our dataset is 08/23/11 and it includes five vehicles and three claims (see Table 1). The index k represents the identification number of an individual vehicle in the warranty data. DIS_k stands for the $k-th$ vehicle days in service on the cutoff date. All three claims are related to battery failure.

Table 1 Vehicles data

k	Sale Date	DIS_k on the Cutoff Date	Claim Date	Odometer on Claim Date	DIS_k on Claim Date	Failure Type
1	02/24/11	180	05/13/11	2800	78	Battery
2	04/01/11	144				
3	05/01/11	114				
4	06/01/11	83	06/15/11	600	14	Battery
5	05/24/11	91	08/23/11	4550	91	Battery

3.1 Mileage Rate Estimation

Our method of failure probability estimation is based on the mileage M_k of a vehicle k on the cutoff date of the warranty dataset. However, the exact mileage is known only for claimed vehicles on the claim date, whereas for all other vehicles we can only estimate mileage according to the Mileage-per-Day (MPD) distribution of all vehicles or the previous claims. Our mileage estimation procedure follows a common assumption that during the warranty period, the usage rate of a given vehicle does not change significantly over time. In our example we assume that the MPD distribution table contains the following three bins only: 0 (Prob. = 0.008), 30 (Prob. = 0.6), and 45 (Prob. = 0.392). Each vehicle can belong to one of three following cases:

1. The vehicle k had exactly one claim of any type before the cutoff date
2. The vehicle k had more than one claim of any type before the cutoff date
3. The vehicle k had no claims before the cutoff date

In the first case, the mileage rate is calculated as the ratio between mileage and days in service at the first claim (denoted by M_k^1 and DIS_k^1, respectively). Consequently, the mileage at the cutoff date is found by the following formula:

$$M_k = \frac{M_k^1}{DIS_k^1} \times DIS_k \quad (2)$$

Equation 2 applies to vehicles 1,4,5 in Table 2.

In the second case, the mileage rate is calculated from the mileage and days in service at the last two claims (l and $l-1$) using the following formula:

$$M_k = M_k^l + \frac{M_k^l - M_k^{l-1}}{DIS_k^l - DIS_k^{l-1}} \times (DIS_k - DIS_k^l) \quad (3)$$

where DIS_k^l is the vehicle's days in service at claim l.

Finally, a vehicle without claims (Case 3) is represented by portions of mileage rates taken from the mileage per day (MPD) distribution of all vehicles. It is assumed here that a MPD distribution is available for use, but the proposed method does not depend on how the MPD is calculated. The cutoff date mileage of j-th portion of a vehicle k is calculated by:

$$M_k^j = MPD_j \times DIS_k \tag{4}$$

where MPD_j is the mileage per day value in the j-th bin of the MPD distribution table. The cutoff date mileage M_k^j is stored for the j-th portion of a vehicle k along with P_j - the proportion of vehicles having MPD_j. Equation 4 applies to vehicles 2,3 in Table 2.

Table 2 Mileage Estimation

k	Equation	Mileage1	Mileage2	Mileage3
1	2	6462		
2	4	0	4320	6480
3	4	0	3420	5130
4	2	3557		
5	2	4550		

3.2 *Failure Probability Estimation in One Continuous Dimension*

Let us first describe our failure probability calculation method for the Days-in-Service (DIS) dimension, which is always known exactly for every sold vehicle. We calculate the number of potential claims in the interval $[I_{x-1}; I_x)$ as follows:

- For vehicles which had at least one claim in the interval $[I_{x-1}; I_x)$, the number of potential claims is equal to the number of actual claims. This assumption eliminates the possibility of obtaining failure probabilities greater than one, especially in case of small intervals with a low number of claims.
- Vehicles which have not reached the lower bound of the interval on the cutoff date cannot produce claims in this interval and their potential is zero.
- Take relative portions of vehicles which have crossed the lower bound but have not reached the upper bound of the interval on the cutoff date. This differs from the existing discrete-time approaches that assume the potential of such vehicles to be zero as well.

- Vehicles which have crossed the upper bound of the interval on the cutoff date have a potential of one claim. This is a reasonable assumption as most failure types have a very low probability of more than one claim per vehicle during the entire warranty period.

We define the following parameters:

$N[I_{x-1}; I_x)$ - total number of actual claims inside the interval $[I_{x-1}; I_x)$
$D[I_{x-1}; I_x)$ - total number of potential claims between I_{x-1} and I_x
$c_k[I_{x-1}; I_x)$ - number of claims for a vehicle k inside the interval $[I_{x-1}; I_x)$
$portion_k$ - the potential of a vehicle k to produce a claim inside the interval $[I_{x-1}; I_x)$ based on its days in service.

The equations of failure probability estimation in the DIS interval $[I_{x-1}; I_x)$ are:

$$P[I_{x-1}; I_x) = \frac{N[I_{x-1}; I_x)}{D[I_{x-1}; I_x)} \tag{5}$$

$$D[I_{x-1}; I_x) = \sum_k ((portion_k), c_k[I_{x-1}; I_x) = 0) + N[I_{x-1}; I_x) \tag{6}$$

$$portion_k = \begin{cases} 0, & DIS_k < I_{x-1} \\ \frac{DIS_k - I_{x-1}}{I_x - I_{x-1}}, & I_{x-1} \le DIS_k < I_x \\ 1, & DIS_k \ge I_x \end{cases} \tag{7}$$

$c_k[I_{x-1}; I_x) = 0$ means that vehicle k does not have claims inside the interval $[I_{x-1}; I_x)$.

We assume the failure rate inside interval to be fixed because the interval size is relatively small. In the case of longer intervals, variable failure rate may be assumed.

Mileage is also a continuous dimension, but unlike days-in-service it can only be estimated for each vehicle. According to the approach presented in sub-section 3.1 above, we divide each vehicle k into portions with each portion j having its own mileage value M_k^j.

Thus, we define the following additional parameters:

M_k - total mileage for vehicle k on the cutoff date
j - MPD bin index
M_k^j - total mileage of vehicle k for a specific bin MPD_j
P_j - probability of the MPD bin j
$portion_k^j$ - the potential of a portion j of a vehicle k to produce a claim inside the interval $[I_{x-1}; I_x)$ based on its mileage

The number of potential claims in the mileage interval $[I_{x-1}; I_x)$ is calculated by the following formulas:

$$D[I_{x-1}; I_x) = \sum_k (\sum_j (portion_k^j), c_k[I_{x-1}; I_x) = 0) + N[I_{x-1}; I_x) \quad (8)$$

$$portion_k^j = \begin{cases} 0, & M_k^j < I_{x-1} \\ \frac{M_k^j - I_{x-1}}{I_x - I_{x-1}} \times P_j, & I_{x-1} \le M_k^j < I_x \\ P_j, & M_k^j \ge I_x \end{cases} \quad (9)$$

3.3 Failure Probability Estimation in Several Continuous Dimensions

First let us assume that our failure prediction model is based on two continuous attributes (dimensions): DIS and Mileage. This means that we are interested in failure probabilities given combinations of DIS intervals and mileage intervals of arbitrary length. Now we define the following parameters:

f -attribute index. We assume that $f = 1$ for DIS and $f = 2$ for Mileage.

$[I_{x-1}^f; I_x^f)$ - interval from I_{x-1} value to I_x value of continuous attribute f.

$comb = [[I_{x-1}^1; I_x^1), [I_{y-1}^2; I_y^2)]$ - combination of DIS and Mileage intervals.

In our example $comb = [0; 90), [0; 4500)$, where DIS interval is $[0; 90)$ and Mileage is $[0; 4500)$.

$c_k[comb]$ - number of claims for the vehicle k inside $comb$

j- MPD bin index

P_j - probability of the MPD bin j

$a_{f,k}$ - attribute f value for vehicle k on the cutoff date. We show the values of $a_{1,k}$ in the column "DIS_k on the Cutoff Date" of Table 1.

N_{comb} - number of actual claims in the combination $comb$

D_{comb} - number of potential claims in the combination $comb$

$portion_{1,k}$ - claim potential of a vehicle k based on its days in service.

$portion_{2,k}^j$ - claim potential of a j-th portion of a vehicle k based on its mileage

The failure probability in each combination is estimated using the following formulas:

$$P_{comb} = \frac{N_{comb}}{D_{comb}} \quad (10)$$

As shown below, in our example, $P_{[0;90)[0;4500)} = 2/4.816 = 0.4153$.

$$N_{comb} = \sum_k c_k[comb] \quad (11)$$

In our example, $N_{[0;90),[0;4500)} = 2$ (Vehicles 1 and 4 in Table 1).

$$D_{comb} = \sum_k ((portion_{1,k}) * (\sum_j portion_{2,k}^j), c_k[comb] = 0) + N_{comb} \quad (12)$$

In our example, $D_{comb} = 0.968 + 0.848 + 1 + 2 = 4.816$.

$$portion_{1,k} = \begin{cases} 0, \; a_{1,k} < I^1_{x-1} \\ \frac{a_{1,k} - I^1_{x-1}}{I^1_x - I^1_{x-1}}, \; I^1_{x-1} \leq a_{1,k} < I^1_x \\ 1, \; a_{1,k} \geq I^1_x \end{cases} \tag{13}$$

For vehicle number 2 with respect to dimension 1 (DIS) $portion_{1,2} = 1$.

$$portion^j_{2,k} = \begin{cases} 0, \; a^j_{2,k} < I^2_{y-1} \\ \frac{a^j_{2,k} - I^2_{y-1}}{I^2_y - I^2_{y-1}} * P_j, \; I^2_{y-1} \leq a^j_{2,k} < I^2_y \\ P_j, \; a^j_{2,k} \geq I^2_y \end{cases} \tag{14}$$

For vehicle number 2 with respect to dimension 2 (mileage) $portion^1_{2,2} = 0, portion^2_{2,2} = \frac{4320-0}{4500-0} * 0.6 = 0.576, portion^3_{2,2} = 1 * 0.392 = 0.392$.

For each combination of attribute intervals, the number of actual claims is calculated by equations [10-14]. The above procedure can be easily extended to l continuous attributes with known (like DIS) or estimated (like mileage) values. In this case, each combination will be defined by l intervals of l continuous attributes.

4 Conclusions and Future Work

In this short paper, we have presented a new, continuous-time approach to failure probability estimation from multi-dimensional censored datasets in automotive industry. The proposed approach was demonstrated on a small numeric example. Future research directions include extending this approach to censored datasets from other multi-dimensional domains (e.g., credit approval). To compare our approach with other probability estimation methods, we can use the estimated probabilities to induce failure prediction models and evaluate the accuracy of these models using ROC analysis.

Acknowledgements. This work was supported in part by the General Motors Global Research & Development – India Science Lab.

References

1. Chukova, S., Robinson, J.: Estimating mean cumulative functions from truncated automotive warranty data. In: Wilson, A., Limnios, N., Keller-McNulty, S., Armijo, Y. (eds.) Mathematical and Statistical Methods in Reliability. Series on Quality, Reliability and Engineering Statistics, pp. 121–136. World Scientific, Singapore (2005)

2. Hu, X.J., Lawless, J.F., Suzuki, K.: Nonparametric estimation of a lifetime distribution when censoring times are missing. Technometrics 40, 3–13 (1998)
3. Majeske, K.D.: A non-homogeneous Poisson process predictive model for automobile warranty claims. Reliab. Eng. Syst. Safe 92, 243–251 (2007)
4. Wu, S.: Warranty Data Analysis: A Review. Qual. Reliab. Eng. Int. (published online first, 2012)

Resolution of Inconsistent Revision Problems in Markov Networks

Aljoscha Klose, Jan Wendler, Jörg Gebhardt, and Heinz Detmer

Abstract. The item planning system at Volkswagen uses Markov networks in its core, with revision of item demands being one of the most important operations. The represented dependencies and given stipulations of the application are highly complex and make the revision problems prone to inconsistencies. We present an approach that solves inconsistent revision problems by fair adaptations of the conflicting probability assignments and fits neatly into the knowledge-based setting of Markov networks.

Keywords: Item planning, markov networks, revision inconsistencies.

1 Item Planning at Volkswagen Group

The Volkswagen Group favors a marketing policy that provides their customers with a maximum degree of freedom in choosing individual specifications of vehicles. In case of the VW Golf—being Volkswagens most popular car class—there are about 200 item families with typically 4 to 8 (but up to 150) values each, that together fully describe a car to be produced, and many of these families can directly be chosen by the customer. Although of course not all item combinations are possible, the customers utilize the given variety, as in spite of the vast number of produced cars only a diminishing fraction thereof are completely identical.

Aljoscha Klose · Jan Wendler · Jörg Gebhardt
Intelligent Systems Consulting, Celle, Germany
e-mail: klose@isc-gebhardt.de

Heinz Detmer
Volkswagen AG, Wolfsburg, Germany
e-mail: heinz.detmer@volkswagen.de

R. Kruse et al. (Eds.): Synergies of Soft Computing and Statistics, AISC 190, pp. 517–524.
springerlink.com © Springer-Verlag Berlin Heidelberg 2013

The EPL software system at Volkswagen Group (EPL for EigenschaftsPLanung) supports item planning, parts demand calculation, and capacity management for short- and medium-term forecasts. The high quality of planning results is achieved by a combination of several relevant information sources: *rules* describe buildable vehicle specifications, *production history* reflects customers preferences, and *market forecasts* lead to stipulations of modified item rates and capacity restrictions.

Since logical rule systems can be transformed into a relational setting, and rates for item combinations may be identified as (frequentistic or subjective) occurrence probabilities, it suggests itself to model the item demand as a probability distribution. The enormous complexity, however, makes it necessary to decompose this distribution using a graphical model [7, 6, 1]. Especially Markov networks turned out to be the most promising environment to satisfy the given modeling purposes.

The modeling of EPL in general and details of the *revision operation*, which is central for the application, can be found in [3, 4]. We will outline only the most important features of Markov networks and the revision operation in the next section. Section 3 describes why inconsistencies in the planning process are hardly avoidable. The original revision operation does not cope well with inconsistent revision problems. In Section 4 we present our approach to fairly resolve inconsistencies by a construct we named *partition mirrors*.

2 Markov Networks and the Revision Operation

Suppose that we are given a Markov network $M = (H, \Psi)$ which represents a joint probability distribution $P(V)$ on a set $V = \{X_1, ..., X_n\}$ of variables with finite domains $\Omega(X_i), i = 1, ..., n$. We assume that $H = (V, \{C_1, ..., C_m\})$ denotes a hypertree and $\Psi = (P(C_j))_{j=1}^m$ a family of probability distributions defined on the (maximal) cliques of H. In this setting, H and its associated undirected dependency graph $G(H)$ reflect the conditional independencies between the involved variables, and Ψ shows the resulting factorization property $P(V) = \prod_{j=1}^m P(C_j)/P(S_j)$, where S_j symbolize the separators in some representation of H as a tree of cliques.

In addition, let $\Sigma = (\sigma_s)_{s=1}^S$ be a so-called revision structure that consists of revision assignments σ_s, each of which refers to a (conditional) assignment scheme $(R_s|K_s)$ with a context scheme K_s, $K_s \subseteq V$ and a revision scheme R_s, where $\varnothing \neq R_s \subseteq V$. We assume that σ_s is specified by a set of assignment components $P^*(\rho_s^{(l)}|\kappa_s)$, where $\kappa_s \subseteq \Omega(K_s)$ is its context component and ${\rho_s}^{(l)}$ its revision component, respectively. The set $\{\rho_s^{(l)}|l = 1, ..., l^*(s)\}$ forms a partitioning of $\Omega(R_s)$. Hence, each revision assignment specifies the modifications of a probability distribution $P(R_s|\kappa_s)$. In case of the empty scheme $K_s = \varnothing$, we deal with an assignment of the (non-conditioned) probabilities $P^*(\rho_s^{(l)})$.

Finally, we suppose that for all $s = 1, ..., S$ there are cliques $C(s) \in \{C_1, ..., C_m\}$ such that $K_s \cup R_s \subseteq C_{j(s)}$. This guarantees that we do not have cross-over dependencies between cliques, which could not be expressible in the structure of the given Markov network.

Let $M = (H, \Psi)$ be a Markov network with associated joint probability distribution $P(V)$. Furthermore, let $\Sigma = (\sigma_s)_{s=1}^S$ be a revision structure. A probability distribution $P_\Sigma(V)$ is called **solution of the revision problem** $(P(V), \Sigma)$, if and only if the following conditions hold:

- **The revision assignments are satisfied:**

$$(\forall s \in \{1, ..., S\})(\forall l \in \{1, ..., l^*(s)\}) \left(P_\Sigma(\rho_s^{(l)} | \kappa_s) = P^*(\rho_s^{(l)} | \kappa_s) \right)$$

- **The interaction structure is preserved:** Except from the modifications induced by the revision assignments, $P_\Sigma(V)$ holds all probabilistic dependencies of $P(V)$.

Essentially, the required preservation of the interaction structure coincides with the decision-theoretical presupposition that the revision operator does not modify the cross product ratios of conditional events outside the influence areas of the revision assignments (*principle of minimal change* [2]).

It can be proven (cf. [4]) that in case of existence, the solution of the revision problem $(P(V), \Sigma)$ is uniquely defined. $P_\Sigma(V)$ can be calculated as the limit probability distribution when the revision procedure of *iterative proportional fitting* [8] with parameters Σ is applied to the initial distribution $P(V)$.

3 Inconsistencies

From a practical point of view, in most cases of real world applications of sufficient complexity, we have to take into account that revision problems $(P(V), \Sigma)$ specified by human experts are not solvable. The reason for that observation is the fact that revision problems $\Sigma = (\sigma_s)_{s=1}^S$ tend to contradict to some of the restrictions given by the zero values of the initial probability distribution $P(V)$. Note that assignment components $P^*(\rho_s^{(l)} | \kappa_s) > 0$ may require to change some probabilities $P(\omega) = 0$ to a strictly positive value. This kind of modification does not conform with the dependency preservation requirement of the revision operator, as zero probabilities show the absence of any interaction structure. Hence, a resulting probability $P_\Sigma(\omega) > 0$ would introduce a new interaction structure, which is the typical focus of the (in some sense complementary) *updating operations*.

Such zero-probabilities $P(\omega) = 0$ are typically generated by the rule system that introduces dependencies between variables in the form of impossible combinations.

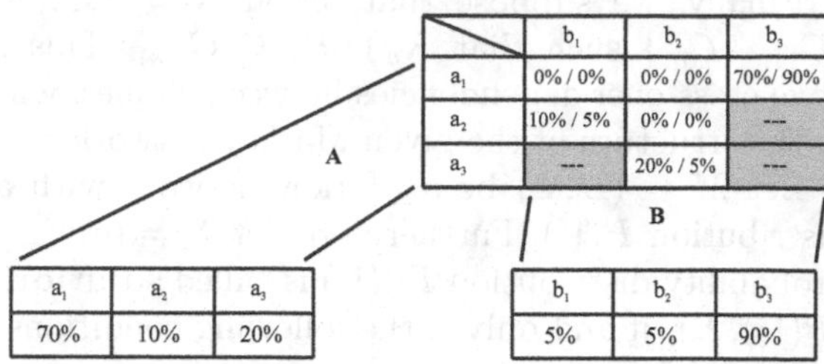

Fig. 1 Example of inconsistent stipulations. The grayed out combinations (a_3, b_1), (a_3, b_3), and (a_2, b_3) have zero probability, i.e. they are invalid. The revision generates alternating distributions as indicated by the separating slashes.

We will introduce a simple example of inconsistent stipulations that we will use to demonstrate the ideas of our approach. We consider two variables A and B with values $a_i, i = 1...3$ and $b_i, i = 1...3$. We assume that there are two rules, *"a_3 induces b_2"* and *"b_3 induces a_1"*. Therefore the combinations (a_3, b_1), (a_3, b_3), and (a_2, b_3) are not valid, i.e. they have zero probability that cannot be increased by the revision operation. The partitions $\rho_s^{(l)}$ of the revision problem correspond directly with the values of A and B, the context scheme K_s is empty. The probability assignments are shown in Fig. 1. The problem is obviously not solvable, e.g. from $(b_3 \rightarrow a_1)$ follows that $P(a_1) \geq P(b_3)$, which is in conflict with $P^*(b_3) = 0.9$ and $P^*(a_1) = 0.7$. The iterative revision algorithm will not converge to a final distribution. Independently of the initial distribution, the revision will oscillate between two limit distributions as shown in the figure, showing that the revision problem is inconsistent.

The planning experts at Volkswagen are confronted with a very high complexity. The number of families and items, and their complex—often transitively induced—dependencies make it almost impossible to assign non-conflicting stipulations. In previous work our focus was to support the users of EPL in avoiding, detecting and explaining inconsistencies.

However, it turned out that inconsistencies are not only hard to avoid, but sometimes express the concurring interests of different planning roles. The most apparent conflict exists between *market oriented* and *production oriented* planning. The former is stating what customers would like to buy; the latter describes what can be produced, considering capacities of item suppliers and production plants.

The most important tool for dealing with such inconsistencies is the introduction of priorities for stipulations. We implemented an algorithmic framework for the revision that iteratively processes the stipulations in order of their priority, decides whether the resulting planning is conflicting, and automatically adapts the stipulations to the closest consistent values. Details on the efficient implementation of this approach can be found in [5].

This algorithm guarantees that for a set of inconsistent stipulations the stipulation with the weakest priority is adapted. This paper deals with a different requirement that arose from the application. While it is quite natural to prioritize production capacities over market demands, the assignment of priorities for stipulations of one planning role is often not obvious. Moreover, it is often desirable that in case of inconsistencies several conflicting stipulations are commonly adapted. A typical example are a number of country-wise stipulations stating the market demand for an item, being in conflict with a global capacity stipulation for that item. In that case, the capacity stipulation has certainly a higher priority. The shortage of the item, however, should be fairly compensated between all countries, instead of fully meeting the demand of some higher prioritized countries whilst allocating nothing to the lower prioritized ones.

Therefore the task was to devise an approach to automatically adjust the probability assignments of a set of concurring, equally prioritized stipulations in a fair and consistent way. An obvious idea is to allow certain deviations between the probability assignments made by the user and those used in the system. These deviations should be optimized according to some measure. A measure like *least squared error* is commonly used to average deviations. However, the approach must fit into our probabilistic, knowledge-based setting. It should take the *principle of minimal change* into account. Furthermore, the algorithm needs to be efficient enough to be used with a large number of stipulations. Therefore, we propose the following approach.

4 Partition Mirrors

If planning problems contain inconsistencies, the probabilities assigned by the experts can obviously not be completely right. Therefore, the basic assumption of our approach is that stipulations are slightly distorted versions of unknown, true probability assignments. The given stipulations are only probably right.

From the perspective of knowledge representation with Markov networks, this is modeled by introducing *mirror variables* into the network structure, couple the states of these variables to their origins by suitable initial distributions, and reformulate the stipulations to assign probabilities to these new variables.

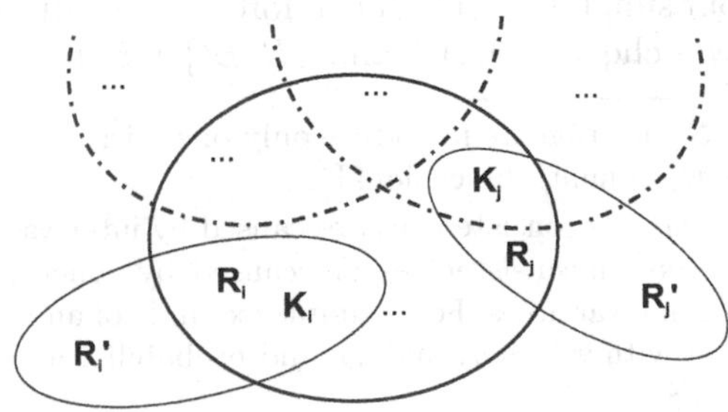

Fig. 2 Temporary modification of the clique structure: the partition mirror cliques are attached to the Markov network structure

More precisely, for each stipulation σ_s with revision scheme R_s and a context component κ_s, we introduce a mirror variable R'_s with one corresponding state ${\rho'}_s^{(l)}$ for each assignment component partition $\rho_s^{(l)}$.[1] σ_s is then replaced by a stipulation σ'_s with the same values, but using ${\rho'}_s^{(l)}$ instead of $\rho_s^{(l)}$.

We then have to define the interaction structure between R_s and R'_s. If there were no distortions, it would hold $R = {\rho'}_s^{(l)} \Leftrightarrow R' = \rho_s^{(l)}$ given $K = \kappa_s$. If distortions are considered possible, though highly improbable, and taking into account that no further information on the distortions is given (*insufficient reasoning principle*), we set the following conditional probabilities:

$$P(R'_s = {\rho'}_s^{(l_1)} | R_s = \rho_s^{(l_2)}, K = \kappa_s) := \begin{cases} 1 - \varepsilon \cdot (|\Omega(R'_s)| - 1), & \text{if } l_1 = l_2 \\ \varepsilon, & \text{else.} \end{cases}$$

For the compensation algorithm we make a (notional or temporary) modification of the Markov network structure as shown in Fig. 2. For each stipulation σ_s, there exists by definition a clique $C_{j(s)}$ with $K_s \cup R_s \subseteq C_{j(s)}$, to which we attach a new clique $C'_s = K_s \cup R_s \cup {R'}_s$. This clique's potential distribution can be calculated as $P(R_s, {R'}_s, K) = P(R'_s|R_s, K) \cdot P(R_s, K)$. The replacement stipulation σ'_s can be assigned to this clique. We call these cliques *partition mirrors*.

Our approach can cope with both, *outer* and *inner* inconsistencies[2], because an independent partition mirror is introduced for each stipulation σ_s, even if several stipulations use intersecting or equal revision schemes R_s.

The revision is performed without further modifications, and the revision problem is guaranteed to be solvable. Generally, the revision operation increases any epsilon-cells of the cliques' potentials to significant probabilities if and only if there are inconsistencies that directly force some probability mass on these combinations. For the partition mirrors this means that distortions of the original stipulations, i.e. combinations (ρ_i, ρ'_j), $i \neq j$, only occur if there is no other solution of the revision problem. The inconsistency mass of the revision problem equals the probability mass on these cells. Furthermore, if several stipulations are involved in an inconsistency, the revision distributes the inconsistency mass fairly among the corresponding partition mirrors, again according to the principle of minimal change, taking into account the initial distribution and the stipulations.

In the following we will show the results of applying the partition mirrors to our simple example. Therefore, we introduce two new variables A' and B', and two cliques $\{A, A'\}$ and $\{B, B'\}$ for the partition mirrors. The stipulations

---

1 Notice that we introduce only one mirror variable R'_s, even if the revision scheme R_s is multi-dimensional.

2 *Outer* inconsistencies are caused by inter-variable dependencies induced by rules; *inner* inconsistencies are caused by concurring probability assignments for the same variables. For a simple example of an inner inconsistency consider a variable A with values a_1 and a_2, and probability assignments $P^*(a_1) = 0.8$ and $P^*(a_2) = 0.8$.

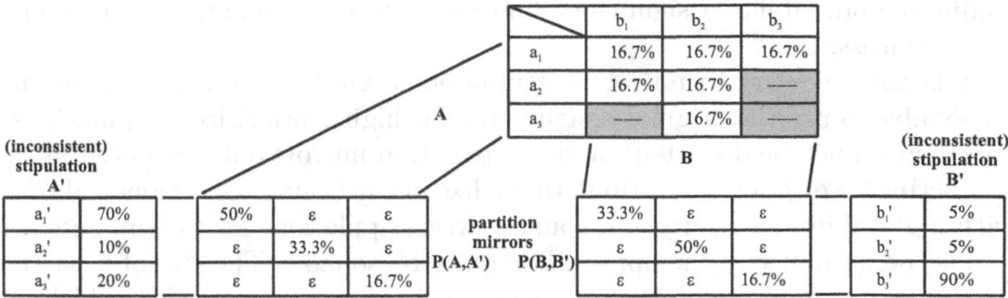

Fig. 3 Probabilities of Markov network (potential of clique $\{A, B\}$) and partition mirrors (cliques $\{A, A'\}$ and $\{B, B'\}$) before applying the revision operation

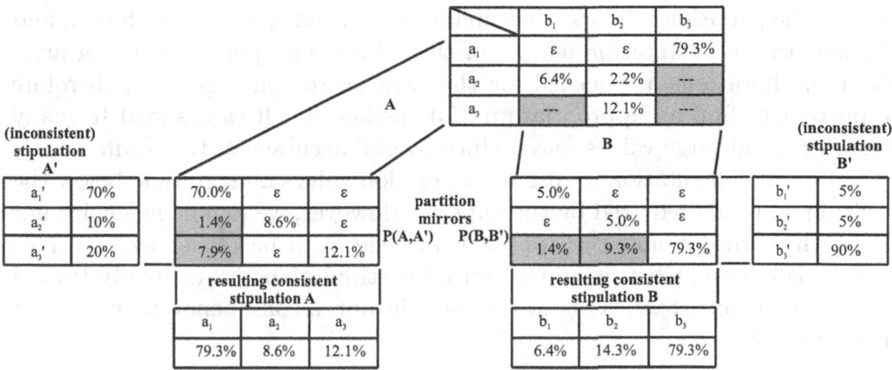

Fig. 4 Resulting probabilities of the partition mirrors and adapted stipulations. The shadowed cells of the partition mirrors hold significant inconsistency mass.

are transformed to assign probabilities $P^*(A')$ and $P^*(B')$. We assume that the allowed combinations of $P(A, B)$ are initially equally probable. With the partition mirrors initialized as described above and initial propagation from the original Markov network into the partition mirror cliques, we get the situation as shown in Fig. 3.

After application of the revision operation, the Markov network holds the probabilities as shown in Fig. 4. As can be seen, the total inconsistency mass is $20\%(= \sum_s \sum_{i \neq j} P(\rho_s^{(i)}, \rho'^{(j)}_s) = 7.9\% + 1.4\% + 1.4\% + 9.3\%)$. The marginal distributions $P(A)$ and $P(B)$ can be interpreted as the *true*, initially unknown, consistent stipulations for the "distorted" stipulations made by the expert.

4.1 Conclusions and Remarks

The presented approach of *partition mirrors* is suited to transform any inconsistent revision problem into a solvable one. The solution of the transformed revision problem yields a fair, probabilistically sound compensation of the conflicting probability assignments, and allows to determine the total inconsistency mass.

It is not necessary to mirror all stipulations. On the contrary, it is often reasonable to use the original assignments for higher prioritized stipulations (that should not be distorted), and use partition mirrors only for potentially conflicting lower prioritized stipulations, like in the example mentioned above with a global item capacity and country-wise stipulations for the same item.

The performance of the approach is quite reasonable. The size of a partition mirror is quadratic in the number of revision components $|R'_s|$, which is harmless compared to the usual cliques' potential sizes. However, the speed of convergence depends on several factors. In case of inconsistencies, the revision operation ends when the inconsistency mass is shifted to the epsilon cells of the partition mirrors. The smaller the initial epsilons are chosen, and the smaller the adaptation factors of the iterative proportional fitting are,[3] the more iterations are needed for the revision to converge. It is therefore important to find an appropriate initial epsilon. Small values lead to many iterations, and large values may reduce overall accuracy of the result.

In our implementation we use initial epsilon values a magnitude below the revision accuracy required by the experts. However, we found the choice not to be that critical, and the approach in general to be rather robust. This is especially true, when it is used only for stipulations that already turned out to be inconsistent, as it is the case in our inconsistency management framework [5].

References

1. Borgelt, C., Kruse, R.: Graphical Models—Methods for Data Analysis and Mining. J. Wiley & Sons, Chichester (2002)
2. Gärdenfors, P.: Knowledge in flux: modeling the dynamics of epistemic states. MIT Press, Cambridge (1988)
3. Gebhardt, J., Detmer, H., Madsen, A.: Predicting parts demand in the automotive industry—an application of probabilistic graphical models. In: Proc. Int. Joint Conf. on Uncertainty in Artificial Intelligence, UAI 2003 (2003)
4. Gebhardt, J., Borgelt, C., Kruse, R., Detmer, H.: Knowledge revision in markov networks. Mathware and Soft Computing, Special Issue From Modelling to Knowledge Extraction 11, 93–107 (2004)
5. Gebhardt, J., Klose, A., Wendler, J.: Markov Network Revision: On the Handling of Inconsistencies. In: Moewes, C., Nürnberger, A. (eds.) Computational Intelligence in Intelligent Data Analysis. SCI, vol. 445, pp. 153–165. Springer, Heidelberg (2013)
6. Lauritzen, S.L.: Graphical Models. Oxford University Press (1996)
7. Lauritzen, S.L., Spiegelhalter, D.J.: Local computations with probabilities on graphical structures and their application to expert systems. J. Roy. Stat. Soc. Ser. B 2, 157–224 (1988)
8. Whittaker, J.: Graphical models in applied multivariate statistics. J. Wiley & Sons, Chichester (1990)

[3] Interestingly, smaller inconsistencies yield smaller adaptation factors, and thus lead to slower speeds of convergence.

Analysis of Dynamic Brain Networks Using VAR Models

Christian Moewes, Rudolf Kruse, and Bernhard A. Sabel

Abstract. In neuroscience it became popular to represent neuroimaging data from the human brain as networks. The edges of these (weighted) graphs represent a spatio-temporal similarity between paired data channels. The temporal series of graphs is commonly averaged to a weighted graph of which edge weights are eventually thresholded. Graph measures are then applied to this network to correlate them, e.g. with clinical variables. This approach has some major drawbacks we will discuss in this paper. We identify three limitations of static graphs: selecting a similarity measure, averaging over time, choosing an (arbitrary) threshold value. The latter two procedures should not be performed due to the loss of brain activity dynamics. We propose to work on series of weighted graphs to obtain time series of graph measures. We use vector autoregressive (VAR) models to facilitate a statistical analysis of the resulting time series. Machine learning techniques are used to find dependencies between VAR parameters and clinical variables. We conclude with a discussion and possible ideas for future work.

Keywords: Dynamic networks, elctroencephalography, neuroimaging, regression, vector autoregressive model.

1 Introduction

In the last decade, a new trend in neuroscience emerged which focuses on the analysis of complex functional brain networks (see e.g. [18]). These networks

---

Christian Moewes · Rudolf Kruse
Faculty of Computer Science, University of Magdeburg,
39114 Magdeburg, Germany
e-mail: `cmoewes@ovgu.de`, `kruse@iws.cs.uni-magdeburg.de`

Bernhard A. Sabel
Medical Faculty, University of Magdeburg,
39120 Magdeburg, Germany
e-mail: `bernhard.sabel@med.ovgu.de`

R. Kruse et al. (Eds.): Synergies of Soft Computing and Statistics, AISC 190, pp. 525–532.
springerlink.com © Springer-Verlag Berlin Heidelberg 2013

are obtained from neuroimaging data by several technologies, e.g. electroencephalography (EEG), electrocorticography (ECoG), magnetoencephalography (MEG) or functional magnetic resonance imaging (fMRI). These methods record activities of brain regions (e.g. on the skull, on the brain meninges or inside of the brain). We denote these brain regions as *variables*. The sampling rates of the data highly depend on the chosen method (kHz for EEG, MEG and ECoG, Hz for fMRI).

Whenever two brain regions are *co-active*, they are *connected* to each other. This connection induces a complex brain network that gives some high-level representation of the really connected nervous cells of the brain. There is general consensus that the analysis of such brain networks will help to better understand the functionality of different brain centers and the brain as a whole [15]. Clearly, if the dynamic behavior of such networks is ignored, then valuable information will be lost pertaining to the brain networks being studied.

While it is challenging to even define nodes for brain networks [2], here this definition is explicitly given by a specific application. We discuss the analysis of dynamic brain networks obtained from patients' EEG. The patients from which we recorded EEGs have visual field deficits (visual impairments) that resulted from optic nerve damages [21]. Our goal is to find network features that correlate with clinically relevant variables. Since we deal with (partly) blind subjects, we assume that it has an effect on their functional connectivity networks. We also hypothesis that there is a correspondence between the extend of the vision loss and the dynamics of the brain connectivity.

It is well-known that brain damage leads to significant and long-lasting neurological deficits, e.g. paralysis or blindness. Therefore, it is of interest to study the relationship between the structural network damage and the functional one that might be observable by neuroimaging methods. The first study related to this problem was performed by C. Stam in 2007 [16]. His group analyzed the differences in EEG data between 15 patients with Alzheimer's disease (AD) and 13 control subjects. Functional connectivity was computed using synchronization likelihood (SL) [17] and the obtained brain network were measured by small-world network criteria [19]. Correlating these measures with clinical variables, they showed that AD is characterized by a loss of small-world network characteristics. Note that they used an averaged network with thresholded edge weights to eventually obtain an unweighted graph.

While we also describe brain networks using SL (to facilitate the choice of one similarity measure between two EEG channels), a previous analysis with averaged unweighted networks did not result in high correlations using a variety of graph measures [7]. So, in this paper we are going to extend this approach manifold: (1) we consider the complete series of EEG networks without averaging over time, (2) we study weighted networks without any edge weight threshold, (3) we use machine learning techniques to find dependencies between network measures and output variables instead of correlation coefficient analyses.

2 Functional Connectivity

To obtain a complex brain network from neuroimaging data, it is necessary to define and measure functional connectivity of two brain regions. Estimating functional connectivity does not find causal connections inside the human brain. Functional connectivity can only be interpreted as statistical relationship between brain regions without implying any causal coherence [11]. Naturally a variety of different functional connectivity methods can been found in the literature (see [20] for a good overview on EEG measures).

Here, we only introduce the concept of SL [17]. Consider a multivariate time series (e.g. a multichannel EEG recording) of length N with n variables. Let measurement $x_{i,k}$ be observed at timestamp i in channel k. Firstly, a time-delay embedding is computed by

$$X_{i,k} = \left(x_{i,k}, x_{i+L,k}, x_{i+2\cdot L,k}, \dots, x_{i+(m-1)\cdot L,k}\right)$$

where L is the lag and m the dimension of the embedding. These state vectors $X_{i,k}$ shall capture the relevant patterns of the signal. Now consider only two channels A, B. Then, the probability that $X_{i,k}$ are closer to each other than ε is

$$P^{\varepsilon}_{i,k} = \frac{1}{2(W_2 - W_1)} \sum_{\substack{j \\ W_1<|i-j|<W_2}}^{N} \theta(\varepsilon - d(X_{i,k}, X_{j,k}))$$

where d is typically the Euclidean distance. For each k and i the critical distance $\varepsilon_{i,k}$ is computed such that $P^{\varepsilon_{i,k}}_{i,k} = p_{\text{ref}}$ whereas $p_{\text{ref}} \ll 1$ is some user-defined threshold. Then, for each pair of points in time (i, j) within $W_1 < |i-j| < W_2$, the number of channels $H_{i,j}$ for which $d(X_{i,k}, X_{j,k}) < \varepsilon_{i,k}$ is computed by

$$H_{i,j} = \theta(\varepsilon_{i,A} - d(X_{i,A}, X_{j,A})) + \theta(\varepsilon_{i,B} - d(X_{i,B}, X_{j,B}))$$

where $\theta(x) = 0$ if $x \leq 0$ and $\theta(x) = 1$ for $x > 0$. The synchronization likelihood is then given by

$$SL_i = \frac{1}{2p_{\text{ref}}(W_2 - W1)} \sum_{\substack{j \\ W_1<|i-j|<W_2}}^{N} (H_{i,j} - 1) \tag{1}$$

The set of free parameters for SL can be reduced down to a size of two by prior information about the frequency range and temporal resolution of the signal [10].

3 Brain Graphs

A brain graph is created when computing functional connectivity, e.g. using SL, for each pair of variables at a given point in time. Such a network simply serves as graphical representation of pairwise statistical dependencies among all variables. Typically, these networks are described by graph measures (e.g. density, clustering coefficient, average path length). Usually, the hope is that they might correlate to clinical variables. Due to simplicity, we demand that brain graphs are simple, i.e. they do not have any loops or multiple edges. Since dealing with SL, we know that the brain graphs are symmetric.

3.1 The Meaning of Edges

An edge represents some kind of statistical dependency between two brain regions, i.e. functional connectivity. The edge weight corresponds to the strength of the functional connectivity. Most measures are normalized to $[0,1]$ or $[-1,1]$ which enables a straightforward interpretation of an edge weight. Commonly, researchers do not use weighted edges for graph analysis. Instead an arbitrarily chosen threshold is used to cancel out edge with "low" weights. Clearly, the remaining edges are unweighted. Despite the loss of information, some researchers argue that one can show different effects with a binary graph [13].

4 Critical Remark and Proposal

So far, we just reported standard techniques that convert neuroimaging data into brain networks. Let us now consider a critique we face when dealing with this approach and how to handle this problem.

We mainly argue that averaging brain graphs over time is generally not beneficial. Averaging should only be permitted if the variations of the binary time series (unweighted edge) or the numerical time series (weighted edge) were close to zero and stationary. To illustrate this, just consider a graph with two nodes, one edge and a linear trend in the evolution of the edge weight. Then, averaging would diminish this important information. In our application, a static analysis did not show very high correlations [7]. We will present much stronger correlations in Section 5 using the following approach.

Remember that the patients are at rest. Thus every EEG time series can have a different length and so does the series of networks. Now, recall that certain graph measures are applied to each network to find global relations between them and clinical variables. The series of networks is eventually transformed into a multivariate time series of real-valued graph measures.

Then, the question arises how these new time series with different lengths can be correlated to the clinical variables. We propose to use a model-based representation of these time series as it is independent from its length.

EEG data in rest do not show any trend and so do the graph measures. Thus the easiest way to fit the time series of graph measures is to use vector autoregressive (VAR) models [8]. A simple VAR model with p lags is given by

$$\mathbf{x}_t = c + \sum_{i=1}^{p} A_i \mathbf{x}_{t-i} + \epsilon_t.$$

where c is a constant, A_i is a matrix storing the interdependence between every pair of variable at point $t - i$ and ε_t is white noise. Its coefficients A_i can be simply flattened as vector which can eventually be correlated to the output variables. In the next section we will evaluate this approach using a real-world application.

5 Application and Experiments

In our experiments we used EEG data from 25 visually impaired subjects suffering from optic nerve damages [21]. Enabling the relation of EEG graph measures to clinical variables, so-called "visual field charts" were obtained from every patient. They indicate the location and size of the optic nerve damage [14]. Based on them, an expert defined the following clinical variables:

- proportion of intact/white sectors,
- proportion of relatively defected/gray sectors,
- proportion of absolutely defected/black sectors,

All of these were transformed by the cortical magnification function (CMF) [4] resulting into 3 further variables.

To preprocess the data we did the following steps in EEGLAB [5]:

- manually removal of noisy time frames at beginning/end of each recording,
- removal of uncommon EEG channels across all subjects (28 were used),
- high-pass filtering with cutoff frequency at 1 Hz to remove slow movements,
- notch filtering 50 Hz to cope with European power line frequency,
- low-pass filtering with cutoff frequency at 95 Hz,
- re-referencing by the average electrode,
- down-sampling to 150 Hz to reduce the costs of SL computation,
- removal of biological artifacts using independent component analysis [9].

Biological artifacts that stem from electromyographic (EMG) or electrocardiograph (EKG) signal appear as noise in the recorded EEG signal in all variations. For EMG/ECG removal, ICA was applied to very carefully remove noisy components.

We used FIR filters to obtain the conventional separation into frequency bands. They are typically associated with different brain states [6]. These bands are δ: $f \in (1,4]$ Hz, θ: $f \in (4,7]$ Hz, α: $f \in [8,12]$ Hz, β: $f \in [13,30]$ Hz, γ: $f \in [30,50]$ Hz and μ: $f \in [8,13]$ Hz. We expected the clinical variables to be somehow explainable by functional connectivity changes in these bands. Functional connectivity was established by SL [17] for each frequency band. We used an outer window length of $W_2 = 3\,\text{s}$ and a reference probability of $p_{\text{ref}} = 0.02$. To capture most of the dynamics, the sliding window shifted every .5 s, i.e. an overlay of $5/6$. Note that a statistical analysis of the averaged graphs has been published by these authors [7].

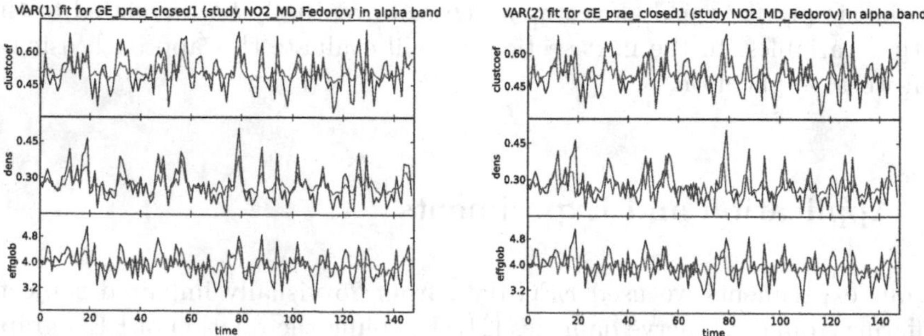

Fig. 1 Original and fitted time series of graph measures in the α band of one subject for $p = 1, 2$ on the left and right side, respectively

We applied 3 measures to every brain graph [3], i.e. *average clustering coefficient*, *density* and *global efficiency*, resulting in a multivariate time series for each subject and each frequency band. Every time series was modeled by a VAR model with $p = 1, 2$ for simplicity. Thus we obtained $p \cdot n \cdot n = 9$ and 18 parameters, respectively, describing the dynamics of the corresponding multivariate time series. Figure 1 shows an example of these time series. Every clinical variable served as variable being depended from these inputs. We used ridge regression (a penalized version of least-squares) with generalized cross validation on different penalizer $\alpha \in \{.1, .2, \ldots, .9, 1\}$ [12]. Thus we correlated brain graph dynamics with clinical variables.

Regression performance was measured using the coefficient of determination R^2 which is defined as

$$R^2 \equiv 1 - \frac{\sum_{i=1}^{N}(y_i - f_i)}{\sum_{i=1}^{N}(y_i - \bar{y})}$$

where $\bar{y} = 1/n \sum_{i=1}^{N} y_i$. Thus the closer R^2 is to 1, the better is the linear fit.

Table 1 summarizes this analysis. High scores were obtained for the proportion of intact and absolutely defected sectors (more or less independent

Table 1 Regression scores R^2 for every pair of frequency band and clinical variable fitting VAR(1) (left) and VAR(2) (right)

$p=1$	δ	θ	α	β	γ	μ
w	.198	.727	.715	.276	.207	.370
g	.193	.101	.156	.240	.273	.189
b	.226	.605	.692	.328	.269	.400
w CMF	.179	.698	.608	.288	.232	.338
g CMF	.177	.105	.183	.446	.185	.226
b CMF	.206	.630	.696	.311	.273	.364

$p=2$	δ	θ	α	β	γ	μ
w	.466	.818	.606	.517	.389	.448
g	.276	.328	.608	.376	.391	.318
b	.517	.844	.827	.519	.408	.496
w CMF	.470	.822	.590	.551	.389	.434
g CMF	.319	.331	.584	.403	.318	.341
b CMF	.516	.850	.819	.526	.406	.471

from CMF). VAR models based on δ and α seem to produce suitable features. This regression analysis clearly shows that (1) we could actually find network features describing the dynamics of the weighted graphs and (2) these features correlate with the extend of the vision loss.

6 Conclusion

Recently neuroscientists started transforming neuroimaging data into brain networks. The idea behind this approach is to use graph theory and its algorithms to produce meaningful features that can help to understand brain recordings. Using any kind of time series similarity measure, the similarity of two data channels (nodes) at some point in time produces a new time series (edge incident to the nodes). The series of graphs is typically averaged to one network of which its edge weights are thresholded resulting in an unweighted network. We do not follow this "classical" approach as it omits the dynamics of the functional connectivity. Usually several graph measures are applied to differentiate between brain networks of distinct subjects or conditions.

Keeping the series of graphs thus creates a multivariate time series of these measures that need to be analyzed. Therefore in this paper we used model-based descriptions, i.e. VAR models. Regression was applied to find linear dependencies between VAR parameters and certain variables. An EEG application of visually impaired subjects showed that this approach is useful for the generation of features.

In the future, we want to work with specialized graph measures for brain networks that capture the high fluctuations and dynamics (see e.g. [1]). Last not least, we hope to obtain more intuitive results using rule-based machine learning approaches to group or classify subjects based on learned models.

Acknowledgements. The authors thank Carolin Gall and her students from the Medical Faculty for collecting the EEG data. We give thanks to Hermann Hinrichs from the Medical Faculty for pointing out several hints to preprocess the EEG data. Finally, we thank the two anonymous reviewers for their very helpful comments to improve this paper.

References

1. Bunke, H.: On a relation between graph edit distance and maximum common subgraph. Pattern Recognit. Lett. 18(8), 689–694 (1997)
2. Butts, C.T.: Revisiting the foundations of network analysis. Science 325(5939), 414–416 (2009)
3. Csárdi, G., Nepusz, T.: The igraph software package for complex network research. InterJournal Complex Systems 1695 (2006)
4. Daniel, P.M., Whitteridge, D.: The representation of the visual field on the cerebral cortex in monkeys. J. Physiol. 159(2), 203–221 (1961)
5. Delorme, A., Makeig, S.: EEGLAB: an open source toolbox for analysis of single-trial EEG dynamics including independent component analysis. J. Neurosci. Methods 134(1), 9–21 (2004)
6. Edwards, E.: Electrocortical activation and human brain mapping. PhD thesis, University of California, Berkeley, CA, USA (2007)
7. Held, P., Moewes, C., Braune, C., Kruse, R., Sabel, B.A.: Advanced Analysis of Dynamic Graphs in Social and Neural Networks. In: Borgelt, C., Gil, M.Á., Sousa, J.M.C., Verleysen, M. (eds.) Towards Advanced Data Analysis. STUDFUZZ, vol. 285, pp. 205–222. Springer, Heidelberg (2012)
8. Lütkepohl, H.: New Introduction to Multiple Time Series Analysis. In: Econometrics / Statistics. Springer, Heidelberg (2005)
9. Makeig, S., Bell, A.J., Jung, T., Sejnowski, T.J.: Independent component analysis of electroencephalographic data. In: Touretzky, D.S., Mozer, M.C., Hasselmo, M.E. (eds.) Advances in Neural Information Processing Systems, vol. 8, pp. 145–151. MIT Press, Cambridge (1996)
10. Montez, T., Linkenkaer-Hansen, K., van Dijk, B.W., Stam, C.J.: Synchronization likelihood with explicit time-frequency priors. Neuroimage 33(4), 1117–1125 (2006)
11. Pearl, J.: Causal inference in statistics: An overview. Stat. Surv. 3, 96–146 (2009)
12. Pedregosa, F., Varoquaux, G., Gramfort, et al.: Scikit-learn: Machine learning in python. JMLR 12, 2825–2830 (2011)
13. Rubinov, M., Sporns, O.: Complex network measures of brain connectivity: Uses and interpretations. Neuroimage 52(3), 1059–1069 (2010)
14. Sabel, B.A., Fedorov, A.B., Naue, N., Borrmann, A., Herrmann, C., Gall, C.: Non-invasive alternating current stimulation improves vision in optic neuropathy. Restor. Neurol. Neurosci. 29(6), 493–505 (2011)
15. Sporns, O.: Networks of the Brain. MIT Press, Cambridge (2010)
16. Stam, C., Jones, B., Nolte, G., Breakspear, M., Scheltens, P.: Small-World networks and functional connectivity in alzheimer's disease. Cerebral Cortex 17(1), 92–99 (2007)
17. Stam, C.J., van Dijk, B.W.: Synchronization likelihood: an unbiased measure of generalized synchronization in multivariate data sets. J. Phys. D: Nonlinear Phenom. 163(3-4), 236–251 (2002)
18. Varela, F., Lachaux, J., Rodriguez, E., Martinerie, J.: The brainweb: phase synchronization and large-scale integration. Nat. Rev. Neurosci. 2(4), 229–239 (2001)
19. Watts, D.J., Strogatz, S.H.: Collective dynamics of 'small-world' networks. Nature 393(6684), 440–442 (1998)
20. Wendling, F., Ansari-Asl, K., Bartolomei, F., Senhadji, L.: From EEG signals to brain connectivity: A model-based evaluation of interdependence measures. J. Neurosci. Methods 183(1), 9–18 (2009)
21. Wüst, S., Kasten, E., Sabel, B.A.: Blindsight after optic nerve injury indicates functionality of spared fibers. J. Cogn. Neurosci. 14(2), 243–253 (2002)

Visualising Temporal Item Sets: Guided Drill-Down with Hierarchical Attributes

Fabian Schmidt and Martin Spott

Abstract. In the past years pattern detection has gained in importance for many companies. As the volume of collected data increases so does typically the number of found patterns. To cope with this problem different interestingness measures for patterns have been proposed. Unfortunately, their usefulness turns out to be limited in practical applications. To address this problem, we propose a novel visualisation technique that allows analysts to explore patterns interactively rather than presenting analysts with static ordered lists of patterns. Specifically, we focus on an interactive visualisation of temporal frequent item sets with hierarchical attributes.

Keywords: Exploration, frequent item sets, guided visualization, hierarchical attributes, temporal association rule mining.

1 Introduction

The market environment companies are faced with today changes faster than ever. Product life cycles are getting shorter, and prices and therefore profit margins shrink due to harder competition. At the same time costumers demand higher service levels. To tackle the problem, organisations collect vast amounts of data about customers, internal processes and external influences at increasing rates in order to make smart business decisions. Nevertheless,

Fabian Schmidt
ISC Gebhardt, Celle, Germany
e-mail: schmidt@isc-gebhardt.de

Martin Spott
BT Innovate and Design, Ipswich, UK
e-mail: martin.spott@bt.com

R. Kruse et al. (Eds.): Synergies of Soft Computing and Statistics, AISC 190, pp. 533–541.
springerlink.com © Springer-Verlag Berlin Heidelberg 2013

they still fail to unfold the full potential of the data for their decision making. The resulting lack of information often leads to suboptimal decisions.

Pattern detection is at the heart of most data analysis activities. However, analysts often only find the patterns they are looking for. Typically the analysis process consists of formulating a hypothesis based on domain knowledge and testing their validity. For instance, if the number of incoming jobs of a service provider is rising, then an analyst may test whether this was driven by an increasing demand for certain products, in certain areas, by specific types of job etc. The number of hypotheses is limited by analysts' time and imagination. To make things worse, some scientists like pharmacologists even argue that they do not know what they are looking for, that they will only know when they see it. This suggests that we must look for a way to present potentially interesting patterns to users rather than expect them to come up with hypotheses.

Unfortunately, machines still struggle to evaluate the interestingness of found patterns, to automatically make decisions and trigger actions based on them, mainly due to the lack of domain knowledge. For that reason, we are interested in exploratory data analysis, where machines focus on the mechanical part of analysing large amounts of data and only guide the analysts' exploration of the results through interactive visualisations. The analysts can then trigger further analysis by the machine and make decisions based on the results.

This paper focuses on temporal frequent item set mining [5], where frequent item sets are extracted from data in regular time intervals and their statistical properties like support values are tracked and analysed for trends over time. We present a new approach to an interactive, graph-based visualisation of the results. It utilises the hierarchical nature of items we often find in practical applications to create a drill-down functionality for exploration (see Fig. 1 for an example). The drill-down is guided by using interestingness measures, in that different parts of the graph are coloured according to their potential interestingness. For instance, if a pattern occurs in the entire UK, the visualisation will tell the user if a drill-down reveals interesting deviations in different parts of the UK (drill-down recommended) or if the same pattern occurs in all subregions of the UK (no drill-down recommended). Thereby, the interestingness measures takes the trend that the pattern exhibits into account. Compared to other visualisations this technique enables a guided exploration of patterns rather than just their presentation.

Section 2 assesses the state of the art in visualising association rules and frequent item sets against our requirements. In Section 3 the new approach is introduced. Section 4 then describes how to use the concept for temporal item set mining. Section 5 shows a scenario where this method has been successfully implemented before we conclude the paper in Section 6 with suggestions for further research.

2 Requirements and Existing Visualisation Techniques

The following requirements need to be fulfilled by a visualisation technique to be considered for our problem. First, it needs to clearly show the association of items in an item set. Secondly, given hierarchical items, it must be possible to drill down into (open up) lower layers in the hierarchy or hide them and do so simultaneously for several hierarchies. Thirdly, the technique must provide a way to show the value of interestingness measures such that the user can be guided in the exploration of patterns.

For visualising association rules and item sets, a variety of techniques have been proposed. The first group are colourised *2D and 3D bar charts* [16, 8, 9, 12]. They are effective as a simple visualisation, but they lack the ability to incorporate hierarchies. *Parallel coordinates* are discussed in [15, 10, 18, 19, 6, 13]. This technique is quite suitable for structured data stored in databases. In addition it can be easily adapted for guided exploration. However, it is again difficult to incorporate the hierarchical information of items. In [2, 1, 3, 4] a *3D world representation* of association rules is proposed for interactive exploration. Is has specifically been designed for interactive exploration, but it focuses on association rules and would lose some of its expressiveness when being adapted to item sets. Relations between items are not shown well, which is crucial in the context of item sets.

Finally, *graph-based visualisations* [11, 6, 14, 7, 17] are very good to show relations between items and item sets. Furthermore, they are useful to explore patterns, since methods for guiding the user can be incorporated by either colouring or grouping nodes. As a hierarchy tree is a graph by definition, they are also usable in that regard.

In conclusion, graph-based visualisation techniques have the highest potential to solve our problem, however existing methods still require extensions to meet our requirements, which we will discuss in the following section.

3 New Exploratory Visualisation

The new solution has two main components, the actual visualisation of patterns and the technique to guide the exploration. Regarding the visualisation, we first need to consider that result sets of frequent item set mining usually contain a large number of patterns. For that reason, the visualisation should provide possibilities to zoom into the pattern set from a rather general view to more specific details. Let us assume a supermarket analyses shopping patterns in the UK and that the regions of the UK are structured hierarchically as shown in Fig. 1. An item set returned by the analysis may be $I = \{Bread, Butter, UK\ South\}$, i.e. customers in the South of the UK are frequently buying bread and butter. The same pattern will probably occur in subregions in the South, however, we may want to start the analysis at the highest level and then drill down into subregions.

As hierarchical attributes like region can naturally be represented as trees, we propose to use trees as the visualisation technique rather than more general graphs. Thereby, we assume that every element in the hierarchy other than the root has exactly one parent. Almost all our real-world data sets fulfil this property. Fig. 2 shows the item set from the example above where one of the three items, the region, is part of a hierarchy. Assuming the supermarket would only sell one kind of bread and butter, these two items would only be displayed as singletons (leaves) in the tree, directly attached to the root node.

To better handle item sets with several hierarchies, a radial tree layout has been chosen. The root node is in the middle and each layer is represented by a circle around the root node with increasing diameters. In this way, hierarchies can more easily spread out from the root than with vertical or horizontal layouts.

Patterns can easily be explored by drilling down along the paths of the tree. Subtrees of a node can be hidden and expanded to control the visible amount of information (see triangular nodes in Fig. 2 to mark a hidden subtree).

Generally, items at lower levels in the hierarchy will only be shown, if a frequent item set contains them. However, the number of patterns might still be too high to recognise the most interesting ones. For that reason a guidance system for exploring the set of patterns is provided to further guide analysts.

Guidance in this context means to draw the analysts' attention to potentially interesting patterns. This can be achieved by hiding redundant, potentially uninteresting or obvious item sets. In addition to reducing the number of patterns by hiding certain ones, potentially interesting patterns can be highlighted, for instance by colouring nodes. The following section gives an example for interestingness measures that can be used for this purpose.

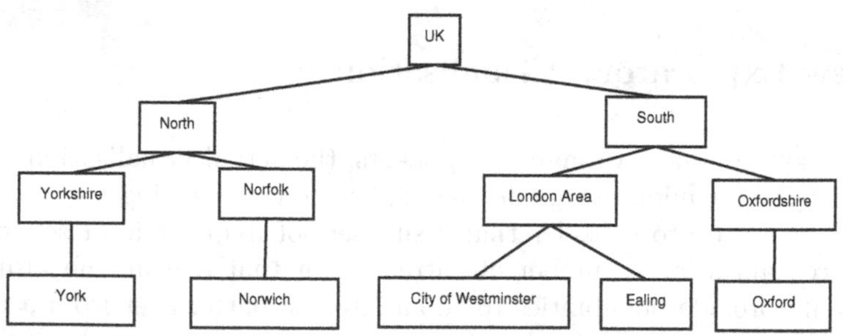

Fig. 1 Example for the (partial) hierarchy of a region. We assume that every node other than the root has exactly one parent.

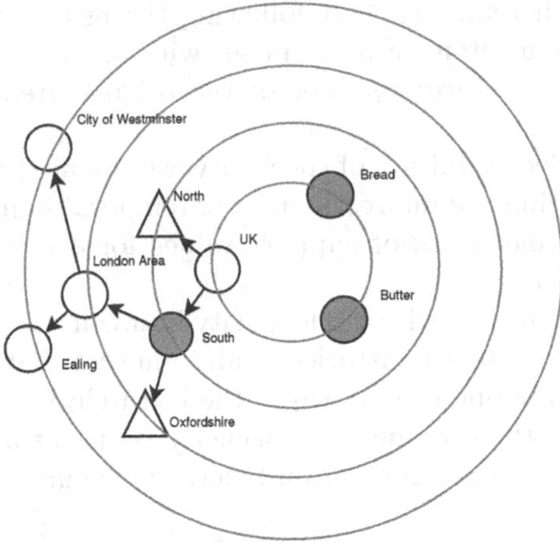

Fig. 2 Visualisation of the item set $I = \{Bread, Butter, South\}$ (green circles). The item *South* is an element of the hierarchical attribute *region* that can be drilled down into. Triangles mark nodes with a hidden hierarchy underneath that can be unfolded.

4 Interestingness Measures for Guided Exploration

To evaluate our concept, it was implemented in the context of temporal item set mining as suggested in [5]. The authors introduced techniques to measure the interestingness of temporal item sets by analysing how their support and confidence values develop over time.

In this paper the homogeneity measure from [5] is proposed as a basis for the visual guidance. Homogeneity measures how members of a group behave in relation to each other. In hierarchies it is especially interesting to reveal nodes that behave differently from their parent or siblings. Going back to the example above, let us assume the support of the item set $\{Bread, Butter, South\}$ would rise over time, which makes the pattern potentially interesting. In a subregion like London Area the support time series may show a different behaviour than in the region South. The homogeneity measure for hierarchical structures can reveal such subtrees. Typically, a homogeneous subtree is viewed as less interesting than an inhomogeneous one since drilling down does not reveal new information.

In order to make use of the homogeneity measure from [5], the concept *specificity of an item set* has to be defined in the context of hierarchies. Usually, a more specific item set is a superset of the more general one. For example, $I_1 = \{Bread, Butter, UK\}$ is more specific than $I_2 = \{Bread, Butter\}$ as $I_1 \subset I_2$. This definition cannot directly be applied to hierarchical attributes like the region. On the face of it, $I_3 = \{Bread, Butter, South\}$ is not comparable to I_1, because the items *South* and *UK* are different. However, since they are part of the same hierarchy, *South* is in fact just shorthand for

$\{UK, South\}$ and therefore $I_3 \subset I_1$ following the usual definition. In other words, if we replace an item in an item set with one at a lower level of the same hierarchy, we produce a subset of the original item set, i.e. a more specific one.

Using this extended definition of specificity, we can adapt the homogeneity measure for calculating the interestingness of temporal item sets proposed in [5]. We assume a time series of support values for every item set, further called *support history*.

Step 1 calculates the pairwise homogeneity between the support history of one item set and the support histories of all item sets that have exactly one more item (or that are one level down in the hierarchy).

Step 2 aggregates the obtained homogeneity values at a level in the hierarchy. [5] suggests different aggregation functions like min, max or averaging operators.

For Step 1 two methods for calculating homogeneity have been proposed, a heuristic one and one based on entropy. In both formulas $H(I_1) = (v_1, ..., v_n)$ and $H(I_2) = (w_1, ..., w_n)$ are the support histories of a pair of item sets I_1 and I_2 with $I_1 \supset I_2$. T_i is the time period i associated with the support values v_i and w_i. The following two formulas are used to calculate homogeneity (or difference):

$$\phi_{entropy}(H_m(I_1), H_m(I_2)) := \sum_{i=1}^{n} P(T_i \mid I_2) log_2 \frac{P(T_i \mid I_2)}{P(T_i \mid I_1)} \quad (1)$$

$$\phi_{heuristic}(H_m(I_1), H_m(I_2)) := \sum_{i=1}^{n} |v_i - w_i| \quad (2)$$

The result of Step 2 can be used if a subtree has been hidden by just showing its parent node. If the subtree shows a high level of homogeneity, drilling down is not necessary. This can be indicated by giving the node a particular colour or shape. If a subtree has already been unfolded, then the homogeneity values from Step 1 can be used to mark the differences between items at the same level in the hierarchy and the more general item one level above. Again, lower homogeneity values typically indicate a higher level of interestingness.

5 Application

In the following we give a simple example to demonstrate the usefulness of the approach. It is based on real data from a telecommunications company that describes the weekly number of tasks the mobile workforce has to work on over a period of 32 weeks. The company needs to understand, how the numbers of different types of tasks develop over time in order to plan the deployment of technicians and to spot emerging problems early. Amongst many others,

Table 1 Homogeneity of 9 areas with their parent region. Larger values mean that the area differs more from the parent region than areas with lower values.

$area_1$	$area_2$	$area_3$	$area_4$	$area_5$	$area_6$	$area_7$	$area_8$	$area_9$
0.081	0.098	0.055	0.074	0.123	0.105	0.063	0.065	0.134

typical attributes are service product, type of task, required technician skill, geographical region, type of customer and service level agreement.

The techniques described in [5] are used to produce support histories of item sets. In order to show that homogeneity is a useful measure to guide the drill-down, we picked a particular type of task represented by an item set and only vary the region, which is a hierarchical attribute, similar to the example shown in Fig. 1. The homogeneity measure is then used to compare the support histories of the higher level region with the ones of the subregions $area_1$ to $area_9$. Since the measure based on entropy gives similar results as the heuristic one, we focussed on the latter for reasons of simplicity. Table 1 illustrates that two areas, namely $area_5$ and $area_9$ have significantly higher values indicating a lower homogeneity. This means that they differ from the behaviour of the higher level region and therefore qualify for further drilling down. In the hierarchy tree shown in the visualisation these two areas would be coloured to highlight their potential interestingness.

Furthermore, the values in Tab. 1 can be aggregated into a single number representing the overall homogeneity in the parent region. Different aggregation operators are currently under investigation including combinations of measures for the mean and variance. Depending on the level of homogeneity, the node of the parent can then be coloured to indicate, if a drill-down into the underlying areas is interesting (low homogeneity) or not (high level of homogeneity).

6 Conclusion

In the past decade a lot of methods for analysing huge amounts of collected business data have been developed. Due to the increasing numbers of resulting patterns, novel ways are needed to find interesting ones in the result set.

For that reason, this paper proposes an interactive visualisation technique for a guided exploration of patterns using drill-downs in hierarchies. In particular, we have shown how the development of frequent item sets over time can be analysed in order to guide analysts towards potentially interesting patterns. With guided exploration it becomes easier to detect unexpected interesting patterns, in contrast to the traditional analysis method of just searching for expected ones.

We see a lot of potential in such methods and are working on extensions in terms of patterns other than temporal item sets and using other interestingness measures.

References

1. Blanchard, J., Guillet, F., Briand, H.: A user-driven and quality-oriented visualization for mining association rules. In: 3rd IEEE Int. Conf. on Data Mining (ICDM 2003), pp. 493–496. IEEE Press (2003)
2. Blanchard, J., Guillet, F., Briand, H.: Exploratory visualization for association rule rummaging. In: Proc. of the 4th Int. Workshop on Multimedia Data Mining (MDM/KDD 2003), pp. 107–114 (2003)
3. Blanchard, J., Guillet, F., Briand, H.: Interactive visual exploration of association rules with rule-focusing methodology. Knowl. Inf. Syst. 13, 43–75 (2006)
4. Blanchard, J., Pinaud, B., Kuntz, P., Guillet, F.: A 2D–3D visualization support for human-centered rule mining. Comput. Graph. 31, 350–360 (2007)
5. Böttcher, M.: Discovering Interesting Temporal Changes in Association Rules. Diplomarbeit, Faculty of Computer Science, University of Magdeburg, Germany (2005)
6. Bruzzese, D., Buono, P.: Combining visual techniques for Association Rules exploration. In: Proceedings of the Working Conf. on Advanced Visual Interfaces (AVI 2004), pp. 381–384 (2004)
7. Cao, N., Liu, S., Tan, L.: HiMap: Adaptive visualization of large-scale online social networks. In: 2009 IEEE Pacific Visualization Symposium, pp. 41–48 (2009)
8. Couturier, O., Hamrouni, T., Yahia, S.B., Nguifo, E.M.: A scalable association rule visualization towards displaying large amounts of knowledge. In: 2007 11th Int. Conf. Information Visualization (IV 2007), pp. 657–663 (2007)
9. Couturier, O., Rouillard, J., Chevrin, V.: An interactive approach to display large sets of association rules. In: Human Interface and the Management of Information: Methods, Techniques and Tools in Information Design, pp. 258–267 (2007)
10. Han, J., An, A., Cercone, N.J.: CViz: An Interactive Visualization System for Rule Induction. In: Hamilton, H.J. (ed.) Canadian AI 2000. LNCS (LNAI), vol. 1822, pp. 214–226. Springer, Heidelberg (2000)
11. Hao, M., Dayal, U., Hsu, M., Sprenger, T., Gross, M.: Visualization of directed associations in e-commerce transaction data. In: Data Visualization 2001: Proc. of the Joint Eurographics–IEEE TCVG Symposium on Visualization, Ascona, Switzerland, May 28-30, pp. 185–194. Springer, Wien (2001)
12. Herawan, T., Yanto, I.T.R., Deris, M.M.: SMARViz: Soft Maximal Association Rules Visualization. In: Badioze Zaman, H., Robinson, P., Petrou, M., Olivier, P., Schröder, H., Shih, T.K. (eds.) IVIC 2009. LNCS, vol. 5857, pp. 664–674. Springer, Heidelberg (2009)
13. Kopanakis, I.: Visual data mining modeling techniques for the visualization of mining outcomes. J. Visual. Lang. Comput. 14, 543–589 (2003)
14. Perer, A., Shneiderman, B.: Systematic yet flexible discovery: guiding domain experts through exploratory data analysis. In: Proc. of the 13th Int. Conf. on Intelligent User Interfaces, pp. 109–118. ACM (2008)
15. Rainsford, C.P., Roddick, J.: Visualisation of Temporal Interval Association Rules. In: Leung, K.-S., Chan, L., Meng, H. (eds.) IDEAL 2000. LNCS, vol. 1983, pp. 91–96. Springer, Heidelberg (2000)

16. Wong, P., Whitney, P., Thomas, J.: Visualizing association rules for text mining. In: Proc. of the 1999 IEEE Symp. on Information Visualization (Info Vis 1999), pp. 120–123 (1999)
17. Yamamoto, C., de Oliveira, M., Rezende, S.: Visualization to assist the generation and exploration of association rules. Post-Mining of Association Rules: Techniques for Effective Knowledge Extraction, pp. 224–245 (2009)
18. Yang, L.: Visualizing Frequent Itemsets, Association Rules, and Sequential Patterns in Parallel Coordinates. In: Kumar, V., Gavrilova, M.L., Tan, C.J.K., L'Ecuyer, P. (eds.) ICCSA 2003, Part I. LNCS, vol. 2667, pp. 21–30. Springer, Heidelberg (2003)
19. Yang, L.: Pruning and visualizing generalized association rules in parallel coordinates. IEEE Trans. Knowl. Data En. 17, 60–70 (2005)

Bayesian Block-Diagonal Predictive Classifier for Gaussian Data

Jukka Corander, Timo Koski, Tatjana Pavlenko, and Annika Tillander

Abstract. The paper presents a method for constructing Bayesian predictive classifier in a high-dimensional setting. Given that classes are represented by Gaussian distributions with block-structured covariance matrix, a closed form expression for the posterior predictive distribution of the data is established. Due to factorization of this distribution, the resulting Bayesian predictive and marginal classifier provides an efficient solution to the high-dimensional problem by splitting it into smaller tractable problems. In a simulation study we show that the suggested classifier outperforms several alternative algorithms such as linear discriminant analysis based on block-wise inverse covariance estimators and the shrunken centroids regularized discriminant analysis.

Keywords: Covariance estimators, discriminant analysis, high-dimensional data, hyperparameters.

1 Introduction

The problem of classifying high-dimensional data arizes in various applications, including gene expression arrays, different types of spectroscopy

---

Timo Koski · Tatjana Pavlenko
KTH Royal Institute of Technology, 100 44 Stockholm, Sweden
e-mail: `tjtkoski@kth.se,pavlenko@math.kth.se`

Jukka Corander
University of Helsinki, 00014 Finland
e-mail: `jukka.corander@helsinki.fi`

Annika Tillander
Karolinska Institutet, 171 77 Stockholm, Sweden
e-mail: `Annika.Tillander@ki.se`

R. Kruse et al. (Eds.): Synergies of Soft Computing and Statistics, AISC 190, pp. 543–551.
springerlink.com © Springer-Verlag Berlin Heidelberg 2013

measurments and climate studies. In this paper we establish a predictive supervised Bayesian classifier assuming that classes are represented by multivariate Gaussian distributions. The main challenge in constructing this classifier is singularity of the sample based class covariance matrix in high dimensions. To overcome this problem, we exploit the block-diagonal covariance structure and show that this approch yields en efficient classifier. Albeit our approach *assumes* the block-diagonal structure, there is a number of efficient methods of learning this structure from the data. A particulary useful scheme is based on the Lasso regularization (see e.g. [7, 2]), which makes it possible to recover structural zeros in the covariance matrix. Using this approach a class of asymptotically equivalent block-diagonal structure approximations in high-dimensional setting was derived in [11].

We assume that $N = \{1, ..., n\}$ is a set of n data items, which are to be classified into a finite set of sources using probabilistic generative models. These samples represent what may be called *test data* that lacks a *labelling* which assigns each sample to some particular source (*class*). In supervised classification there is *training* data $M = \{1, ..., m\}$, where labels are known for the samples in M and these determine exhaustively the possible sources for the test samples in N.

We let $\mathbf{x} = (x_1, \ldots, x_p)'$ be a $p \times 1$ vector in $\mathbb{R}^p$. Let $\mathbf{x}^{(N)} = \{\mathbf{x}_1, ..., \mathbf{x}_n\}$ and $\mathbf{z}^{(M)} = \{\mathbf{z}_1, ..., \mathbf{z}_m\}$ be the sample vectors for items in N and M, respectively. For any subsets $a \subset N, b \subset M$, we let $\mathbf{x}^{(a)}, \mathbf{z}^{(b)}$ denote the subsets of data for the corresponding samples.

Let now $\mathcal{C} = \{1, \ldots, k\}$ represent an ensemble of k generic classes. In the supervised classification the labelling/classification of the samples in the training set M is a partition T of $\{1, ..., m\}$ into non-empty subsets/classes $t_1, ..., t_k, 1 \leq k \leq n$, such that $t_c \cap t_{c^*} = \varnothing$, $c \neq c^*$, $c, c^* = 1, 2, ..., k$, and $\cup_{c=1}^{k} t_c = M$. Correspondingly, S is a partition of $\{1, ..., n\}$ into the classes $s_1, ..., s_k$, i.e. we have

$$\mathbf{x}^{(N)} = \cup_{c=1}^{k} \mathbf{x}^{(s_c)} = \cup_{s_c \in S} \mathbf{x}^{(s_c)}, \quad \mathbf{z}^{(M)} = \cup_{t_c \in T} \mathbf{z}^{(t_c)}.$$

Bayesian predictive inference (see e.g. [5, 6, 10]) provides the inductive tools for updating beliefs about the set of all possible classifications S by using the data observed for the items in N and M.

2 Posterior Predictive Distributions under Gaussian Models

Each test sample $\mathbf{x}_i$ is by the preceding a p-dimensional column vector $\mathbf{x}_i = (x_{i1}, ..., x_{ip})'$, and correspondingly for each training sample $i \in M$ we have $\mathbf{z}_i = (z_{i1}, ..., z_{ip})$. For given classifications S, T, each data vector assigned to a particular class c is assumed to be generated by a class-specific multivariate

Gaussian distribution with mean vector $\mu_c \in \mathbb{R}^p$ as a class centroid, and positive definite covariance matrix $\Sigma_c \in \mathbb{R}^{p \times p}$. The probability density of an observed sample $\mathbf{x}_i$ (equivalently for $\mathbf{z}_i$) can then be defined as

$$p(\mathbf{x}_i|\mu_c, \Sigma_c) = \frac{1}{(2\pi)^{p/2} |\Sigma_c|^{1/2}} \exp\left\{ -\frac{1}{2}(\mathbf{x}_i - \mu_c)' \Sigma_c^{-1}(\mathbf{x}_i - \mu_c) \right\}. \qquad (1)$$

We regard $\{\mathbf{x}_i, i \in s_c\}$ as conditionally independent outcomes of a random variable $\mathbf{X} = (\mathbf{X}_1, \ldots, \mathbf{X}_p)'$ and express (1) with $\mathbf{X} \in N(\mu_c, \Sigma_c)$.

By specifying the priors of the class parameters as a product of conjugate Gaussian-Inverse-Wishart distributions, the predictive distributions can be found explicitly. The joint prior distribution of μ_c, Σ_c is then expressed as

$$p(\mu_c, \Sigma_c) = p(\mu_c|\alpha_c, \beta^{-1}\Sigma_c)p(\Sigma_c|\delta, \Phi), \qquad (2)$$

where $p(\mu_c|\alpha_c, \beta^{-1}\Sigma_c)$ is the appropriate density with the hyperparameters $\alpha_c \in \mathbb{R}^p, \beta \in \mathbb{R}^+$, and $p(\Sigma_c|\delta, \Phi)$ is the density of inverse Wishart distribution (see [8]) with the hyperparameters $\delta \in \mathbb{R}^+, \Phi \in \mathbb{R}^{p \times p}$, where Φ is assumed to be positive definite. The density of inverse Wishart distribution can be written as

$$p(\Sigma_c|\delta, \Phi) = (2^{\delta p/2}\Gamma_p(\delta/2))^{-1}|\Phi|^{-\delta/2}|\Sigma_c|^{-(\delta+p+1)/2} \exp\left\{ -\frac{1}{2}tr(\Phi^{-1}\Sigma_c^{-1}) \right\}, \qquad (3)$$

where $\Gamma_p(\cdot)$ is the multivariate gamma function [1].

Hyperparameters β, δ and Φ of the prior are the same for all classes for any S, T and are thus not indexed with respect to the class c.

Let the sample centroid and unscaled sample covariance for class s_c be written as

$$\bar{\mathbf{x}}_c = \frac{1}{n_c} \sum_{i \in s_c} \mathbf{x}_i, \quad \mathbf{W}^{(s_c)} = \sum_{\mathbf{x}_i \in s_c} (\mathbf{x}_i - \bar{\mathbf{x}}_c)(\mathbf{x}_i - \bar{\mathbf{x}}_c)' \qquad (4)$$

respectively, where $n_c = |s_c|$.

In supervised classification, derivation of the posterior predictive distribution of test data given both training information $(\mathbf{z}^{(M)}, T)$ and the classification S can be made in the following way: the training data is first used to update the prior hyperparameters $\alpha_c, \beta, \delta, \Phi$, whereafter the posterior predictive distribution is calculated by integrating out parameters with respect to the updated prior. This corresponds to

$$p(\mathbf{x}^{(N)}|\mathbf{z}^{(M)}, S, T) = \prod_{c=1}^{k} \int_{\mathbb{R}^p} \int_{\mathbb{R}^{p \times p}} p(\mathbf{x}^{(s_c)}|\mu_c, \Sigma_c)p(\mu_c, \Sigma_c|\mathbf{z}^{(M)}, T)d\mu_c d\Sigma_c, \qquad (5)$$

where the above notation is simplified by omitting the hyperparameters from $p(\mathbf{x}^{(N)}|S, T)$ [4] and

$$p(\mu_c, \Sigma_c|\mathbf{z}^{(M)}, T) = \frac{p(\mathbf{z}^{(t_c)}|\mu_c, \Sigma_c)p(\mu_c, \Sigma_c)}{\int_{\mathcal{P}} \int_{\mathbb{R}^{p\times p}} p(\mathbf{z}^{(t_c)}|\mu_c, \Sigma_c)p(\mu_c, \Sigma_c)d\mu_c d\Sigma_c} \tag{6}$$

is the posterior distribution of μ_c, Σ_c based on the prior $p(\mu_c, \Sigma_c)$ in (2) and the data $\mathbf{z}^{(t_c)}$ in class c of T. A different class of predictive densities based on uninformative priors on Σ is found in [12, p. 51] and is originally due to [5, p. 93].

After some transforms (see [1, 3, 4]) the explicit expression for the predictive distribution becomes

$$p(\mathbf{x}^{(N)}|\mathbf{z}^{(M)}, S,T) = \tag{7}$$

$$\prod_{c=1}^{k} \frac{1}{\pi^{pn_c/2}} \left(\frac{m_c+\beta}{n_c+m_c+\beta}\right)^{\frac{p}{2}} \frac{\Gamma_p(\frac{\delta+n_c+m_c+p-1}{2})}{\Gamma_p(\frac{\delta+m_c+p-1}{2})} \frac{\det(\xi_c^1)^{\frac{\delta+m_c+p-1}{2}}}{\det(\xi_c^2)^{\frac{\delta+n_c+m_c+p-1}{2}}},$$

where $m_c = |t_c|$ is the number of training samples for class c. The determinants equal

$$\det(\xi_c^1) = \left|\Phi + \mathbf{W}^{(t_c)} + \frac{m_c\beta}{m_c+\beta}(\bar{\mathbf{z}}_c - \alpha_c)(\bar{\mathbf{z}}_c - \alpha_c)'\right|, \tag{8}$$

and

$$\det(\xi_c^2) = \left|\Phi + \mathbf{W}^{(s_c)} + \mathbf{W}^{(t_c)} + \frac{m_c\beta}{m_c+\beta}(\bar{\mathbf{z}}_c - \alpha_c)(\bar{\mathbf{z}}_c - \alpha_c)'\right. \tag{9}$$
$$+ \frac{n_c}{(n_c+m_c+\beta)(m_c+\beta)}[(m_c+\beta)\bar{\mathbf{x}}_c - m_c\bar{\mathbf{z}}_c - \beta\alpha_c]$$
$$\left.\cdot [(m_c+\beta)\bar{\mathbf{x}}_c - m_c\bar{\mathbf{z}}_c - \beta\alpha_c]'\right|,$$

where $\bar{\mathbf{z}}_c, \mathbf{W}^{(t_c)}$ are the training data counterparts of $\bar{\mathbf{x}}_c, \mathbf{W}^{(s_c)}$, respectively.

This approach to predictive classification is quite different from modeling by Gaussian processes in binary classification, as found e.g. in [9].

3 Predictive Densities with Block-Diagonal Covariance Matrices

Let us now partition the random vector $\mathbf{X} = (\mathbf{X}_1, \ldots, \mathbf{X}_p)' \in N(\mu_c, \Sigma_c)$ into b disjoint sub-vectors, i.e. $\mathbf{X} = \left(\mathbf{X}_{[1]}, \ldots, \mathbf{X}_{[b]}\right)'$, where $\mathbf{X}_{[j]} = (X_{j_1}, \ldots, X_{j_{p_j}})$, ($\mathbf{X}_{[j]}$ has values in $\mathbb{R}^{p_j}$), $j = 1, \ldots, b$ and $\sum_{j=1}^{b} p_j = p$. The block segmentation represents (in)dependencies between groups of features. The dimension p_j is not dependent of the class c.

We assume that for any $j \neq l$, $\mathbf{X}_{[j]}$ and $\mathbf{X}_{[l]}$ are conditionally independent given $c \in \mathcal{C}$. We have thus a property of block-wise interactions, and the covariance matrix Σ_c displays the structure of b blocks on the main diagonal, i.e. $\Sigma_c = \text{diag}\left[\Sigma_{c,[1]}, \ldots, \Sigma_{c,[b]}\right]$. Thus, if the sample

$\mathbf{x}_i = \left(\mathbf{x}_{i,[1]}, \ldots, \mathbf{x}_{i,[b]}\right)'$ is an outcome of $N(\mu_c, \Sigma_c)$, then $\mathbf{x}_{i,[j]}$ is a sample of $X_{[j]} \in N\left(\mu_{c,[j]}, \Sigma_{c,[j]}\right)$, $\mu_{c,[j]} \in \mathbb{R}^{p_j}$, $\Sigma_{c,[j]} \in \mathbb{R}^{p_j \times p_j}$, $j \in \{1, 2, \ldots, b\}$. Therefore

$$p(\mathbf{x}_i|\mu_c, \Sigma_c) = \prod_{j=1}^{b} p(\mathbf{x}_{i,[j]}|\mu_{c,[j]}, \Sigma_{c,[j]}). \tag{10}$$

As above, the joint prior distribution of $\mu_{c,[j]}, \Sigma_{c,[j]}$ is taken as

$$p(\mu_{c,[j]}, \Sigma_{c,[j]}) = p(\mu_{c,[j]}|\alpha_{c,[j]}, \beta^{-1}\Sigma_{c,[j]})p(\Sigma_{c,[j]}|\delta, \Phi), \tag{11}$$

where $p(\mu_{c,[j]}|\alpha_{c,[j]}, \beta^{-1}\Sigma_{c,[j]})$ is the density of multivariate Gaussian and $p(\Sigma_{c,[j]}|\delta, \Phi)$ is the conjugate Gaussian-Inverse-Wishart distribution. Hence hyperparameters β, δ and Φ are the same for all classes c and $\mu_{c,[j]}, \Sigma_{c,[j]}$ are *metaindependent*, i.e. $p(\mu_c, \Sigma_c) = \prod_{j=1}^{b} p(\mu_{c,[j]}, \Sigma_{c,[j]})$. By the preceding and the notation in section 2 we get

$$p(\mathbf{x}^{(s_c)}|\mu_c, \Sigma_c) = \prod_{i \in s_c} p(\mathbf{x}_i|\mu_c, \Sigma_c) = \prod_{i \in s_c}\prod_{j=1}^{b} p(\mathbf{x}_{i,[j]}|\mu_{c,[j]}, \Sigma_{c,[j]}), \tag{12}$$

Then the predictive density, $p(\mathbf{x}^{(N)}|S)$ with block-diagonal covariance structure given S is expressed as

$$\prod_{c=1}^{k}\prod_{j=1}^{b}\int_{\mathbb{R}^{p_j}}\int_{\mathbb{R}^{p_j \times p_j}} \prod_{i \in s_c} p(\mathbf{x}_{i,[j]}|\mu_{c,[j]}, \Sigma_{c,[j]})p(\mu_{c,[j]}, \Sigma_{c,[j]})d\mu_{c,[j]}d\Sigma_{c,[j]}.$$

In the same way as in (7) we obtain

$$p(\mathbf{x}^{(N)}|S) = \prod_{c=1}^{k}\prod_{j=1}^{b} \phi(c,j), \tag{13}$$

where

$$\phi(c,j) = \frac{1}{\pi^{p_j n_c/2}}\left(\frac{\beta}{n_c+\beta}\right)^{\frac{p_j}{2}} \frac{\Gamma_p(\frac{\delta+n_c+p_j-1}{2})}{\Gamma_p(\frac{\delta+p_j-1}{2})} \frac{|\Phi|^{\frac{\delta+p_j-1}{2}}}{\det(\eta_{c,[j]})^{\frac{\delta+n_c+p_j-1}{2}}}, \tag{14}$$

with

$$\det(\eta_{c,[j]}) = |\Phi + \mathbf{W}_{[j]}^{(s_c)} + \frac{n_c\beta}{n_c+\beta}(\bar{\mathbf{x}}_{c,[j]} - \alpha_{c,[j]})(\bar{\mathbf{x}}_{c,[j]} - \alpha_{c,[j]})'| \tag{15}$$

and

$$\bar{\mathbf{x}}_{c,[j]} = \frac{1}{n_c}\sum_{i \in s_c} \mathbf{x}_{i,[j]}, \quad \mathbf{W}_{[j]}^{(s_c)} = \sum_{\mathbf{x}_i \in s_c}(\mathbf{x}_{i,[j]} - \bar{\mathbf{x}}_{c,[j]})(\mathbf{x}_{i,[j]} - \bar{\mathbf{x}}_{c,[j]})'. \tag{16}$$

4 Bayesian Predictive Supervised and Marginal Classification with Block-Diagonal Covariance

The explicit expression for the posterior predictive distribution becomes now

$$p(\mathbf{x}^{(N)}|\mathbf{z}^{(M)},S,T) = \tag{17}$$

$$\prod_{c=1}^{k}\prod_{j=1}^{b}\frac{1}{\pi^{p_j n_c/2}}\left(\frac{m_c+\beta}{n_c+m_c+\beta}\right)^{\frac{p}{2}}\frac{\Gamma_{p_j}(\frac{\delta+n_c+m_c+p_j-1}{2})}{\Gamma_{p_j}(\frac{\delta+m_c+p_j-1}{2})}\frac{\det(\xi^1_{c,[j]})^{\frac{\delta+m_c+p_j-1}{2}}}{\det(\xi^2_{c,[j]})^{\frac{\delta+n_c+m_c+p_j-1}{2}}},$$

with $\det(\xi^1_{c,[j]})$ and $\det(\xi^2_{c,[j]})$ modified in an obvious manner from (8) and (9), respectively. In the Bayesian predictive supervised marginal classifier we have $n = 1$, i.e. $\mathbf{x}^{(N)} = \mathbf{x}$. This is an attempt to classify just one new item with sample data $\mathbf{x}$ using the information in $\mathbf{z}^{(M)}$. Then

$$p(\mathbf{x}|\mathbf{z}^{(M)},c,T) = \prod_{j=1}^{b}\phi(c,j,|\mathbf{z}^{(M)}_{[j]},T) \tag{18}$$

with $\phi(c,j,|\mathbf{z}^{(M)}_{[j]},T)$ equal to

$$\frac{1}{\pi^{p_j/2}}\left(\frac{m_c+\beta}{1+m_c+\beta}\right)^{\frac{p_j}{2}}\frac{\Gamma\left(\frac{\delta+m_c+p_j}{2}\right)}{\Gamma\left(\frac{\delta+m_c}{2}\right)}\frac{\det(\xi^1_{c,[j]})^{\frac{\delta+m_c+p_j-1}{2}}}{\det(\xi^2_{c,[j]})^{\frac{\delta+m_c+p_j}{2}}}, \tag{19}$$

since in this case we are considering the predictive distribution of $\mathbf{x}$ by conditioning on $\mathbf{z}^{(M)}$ and T under the additional condition that $\mathbf{x}$ was assigned to c. Here from (8) and (9), as this is the predictive distribution of $\mathbf{x}$,

$$\det(\xi^1_{c,[j]}) = \left|\Phi + \mathbf{W}^{(t_c)}_{[j]} + \frac{m_c\beta}{m_c+\beta}(\bar{\mathbf{z}}_{c,[j]} - \alpha_{c,[j]})(\bar{\mathbf{z}}_{c,[j]} - \alpha_{c,[j]})'\right|$$

$$\begin{aligned}\det(\xi^2_{c,[j]}) = \Big|\Phi + \mathbf{W}^{(t_c)}_{[j]} &+ \frac{m_c\beta}{m_c+\beta}(\bar{\mathbf{z}}_{c,[j]} - \alpha_{c,[j]})(\bar{\mathbf{z}}_{c,[j]} - \alpha_{c,[j]})' + \\ &+ \frac{1}{(1+m_c+\beta)(m_c+\beta)}[(m_c+\beta)\mathbf{x}_{[j]} - m_c\bar{\mathbf{z}}_{c,[j]} - \beta\alpha_{\mathbf{c},[\mathbf{j}]}]\cdot \\ &\cdot[(m_c+\beta)\mathbf{x}_{[j]} - m_c\bar{\mathbf{z}}_{c,[j]} - \beta\alpha_{c,[j]}]'\Big|.\end{aligned} \tag{20}$$

By [5, 6] we identify the sample $\mathbf{x}$ as coming from class c', if

$$c' = \text{argmax}_{c\in\mathcal{C}}\, p(\mathbf{x}|\mathbf{z}^{(M)},c,T)\pi_c.$$

where π_c is the prior probability or prevalence of c.

5 Simulation Study

To illustrate the merits of the proposed classification technique in various high-dimensional settings, we conduct a set of numerical studies. All computations are done in **R** version 2.9, `rda` and `lda` packages are used for comparative inferences.

We focus on the two-class case and generate data as $\mathbf{x}_i \in N_p(\mu_c, \Sigma)$ for $c = 1, 2$ and a range of dimensionality p with fixed block size, $p_j = 20$, where $j = 1, \ldots, p/p_j$, $i = 1, \ldots, m_c$ and $m_c = 25$. We set $\Sigma = \text{diag}[\Sigma_{[1]}, \ldots, \Sigma_{[p/p_j]}]$ where $\Sigma_{[j]} = \left[(1-\rho)I + \rho \mathbf{1}_{p_j} \mathbf{1}'_{p_j}\right]$, $0 < \rho < 1$, i.e. each $\Sigma_{[j]}$ is assumed to have equicorrelated strucure. Throughout the experiments we fix the true misclassification probability to $e = 0.1$ and recall that for Gaussian class-conditional distributions with known μ_c and Σ it holds that $e = \Phi(-D/2)$, where $\Phi(\cdot)$ is the Gaussian cumulative distribution function and $D^2 = (\mu_1 - \mu_2)'\Sigma^{-1}(\mu_1 - \mu_2)$ is the squared Mahalanobis distance between the classes. To control misclassification probability, we set the shift between class centroids to $\mu_2 - \mu_1 = d\mathbf{1}_p$, where $d = D(\mathbf{1}_p' \Sigma^{-1} \mathbf{1}_p)^{-1/2}$ and $D = -2\Phi^{-1}(e)$ is the Mahalanobis distance corresponding to $e = 0.1$. For hyperparameters in (2) and (3) we assume that $\alpha_1 = \mu_1 + \mathbf{1}_p d/6$, $\alpha_2 = \mu_2 - \mathbf{1}_p d/6$, $\Phi = I_p$, $\beta = 10$, $\delta = 30$, and compare performance of BPC with two alternative classifiers, Shrunken Centroids Regularized Discriminant Analysis (RDA) see e.g. [7] and Linear Discriminant Analysis with a block-wise estimated inverse covariance (BLDA) [11]. To optimize the strenght of regularization, λ in RDA we use 5-fold cross-validation (CV) within each training fold and select $\lambda = 0.1$ over all p. Misclassification rates were estimated using leave-one-out CV for all three classifiers since sample sizes are small. Table 1 illustrates that BPC yields more accurate classification than both RDA and BLDA for a range of high-dimensional scenarios.

Table 1 Misclassification rates estimated by CV, mean(sd) averaged over 50 runs

Classifier	$p = 40$	$p = 60$	$p = 100$	$p = 120$	$p = 140$	$p = 160$
BPC	0.11 (0.04)	0.13 (0.05)	0.13 (0.05)	0.13 (0.05)	0.13 (0.05)	0.11 (0.05)
RDA	0.25 (0.06)	0.29 (0.04)	0.31 (0.03)	0.31 (0.04)	0.31 (0.03)	0.30 (0.03)
LDA	0.17 (0.05)	0.22 (0.06)	0.24 (0.06)	0.27 (0.06)	0.25 (0.08)	0.26 (0.07)

To give an impression of how the hyperparameters effect performance accuracy we plot the misclassification profile for four choices of α with varying β and δ. We fix $p_j = 20$, $p = 200$ and $m_c = 25$. Results presented in Figure 1 indicate that shift parameter $\alpha_2 - \alpha_1$, representing prior distance between classes has the key effect whereas the influence of β and δ is less pronounced.

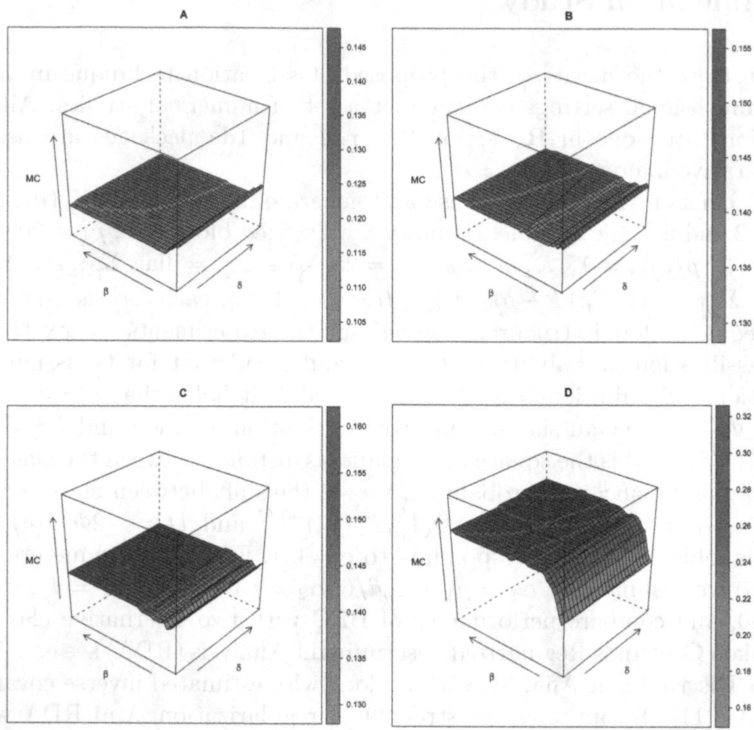

Fig. 1 Misclassification rates for BPC averaged over four runs. $\delta \in (21:30)$ and $\beta \in (1:100)$. **A**: $\alpha_2 - \alpha_1 = \mathbf{1}_p \cdot d$, **B**: $\alpha_2 - \alpha_1 = \mathbf{1}_p \cdot 2d/5$, **C**: $\alpha_2 - \alpha_1 = \mathbf{1}_p \cdot d/2$, **D**: $\alpha_2 - \alpha_1 = \mathbf{0}_p$.

References

1. Bernardo, J.M., Smith, A.F.: Bayesian Theory. John Wiley & Sons, New York (1995)
2. Bühlmann, P., van de Geer, S.: Statistics for High-Dimensional Data. Springer, New York (2011)
3. Dawid, A.P., Lauritzen, S.L.: Hyper Markov laws in the statistical analysis of decomposable graphical models. Ann. Stat. 21, 1272–1317 (1993)
4. Dawid, A.P.: Some matrix-variate distribution theory: notational considerations and a Bayesian application. Biometrika 68, 265–274 (1981)
5. Geisser, S.: Predictive Discrimination. In: Krishnaiah, P.R. (ed.) Multivariate Analysis. Academic Press, New York (1966)
6. Geisser, S.: Predictive Inference: An Introduction. Chapman and Hall, London (1993)
7. Hastie, T., Tibshirani, R., Friedman, J.: Elements of statistical learning. Springer, New York (2009)
8. Kollo, T., von Rosen, D.: Advanced multivariate statistics with matrices. Springer, New York (2005)

9. Kuss, M., Rasmussen, C.E.: Assessing approximate inference for binary gaussian process classification. J. Mach. Learn. Res. 6, 1679–1704 (2005)
10. Corander, J., Cui, Y., Koski, T., Sirén, J.: Predictive Gaussian Classifiers (under revision)
11. Pavlenko, T., Björkström, A., Tillander, A.: Covariance Structure Approximation via gLasso in High-Dimensional Supervissed Classification. J. Appl. Stat. (in press, 2012)
12. Ripley, B.D.: Pattern Recognition and Neural Networks. Cambridge University Press, Cambridge (1996)

Stochastic Measure of Informativity and Its Application to the Task of Stable Extraction of Features

Alexander Lepskiy

Abstract. In the paper we define a new notion of stochastic monotone measure. The application of this notion to solution of problem of finding of features on the noisy image is considered.

Keywords: Feature extraction, stochastic measure of informativity.

1 Introduction

As a rule in pattern recognition or, in particular, in image processing, we should identify images using sets of their features. Let $\Omega = \{\omega_i\}_{i=1}^n$ be a set of features that correspond to an image. To achieve the highest productivity and stable working of a pattern recognition system, it is necessary to choose a small subset of features in Ω with the highest information values. There are some very well-known approaches that can give us features with the highest information values based on the method of principal components, discriminant analysis and so on [2], [3]. But these methods fail to take into account structural (e.g. morphological) characteristics of object. In this situation, measure of informativity can be used [1]. By definition, an measure of informativity μ is a set function defined on the power set 2^Ω of Ω that for each $A \in 2^\Omega$ shows an information value of features in A. We assume that this function has monotone property: $\mu(A) \leq \mu(B)$ if $A \subseteq B$ for all $A, B \in 2^\Omega$ (i.e. additional information does not decrease the value of μ).

In certain tasks of image processing random nature of image features can be caused by some noisy effects. For example, if the pattern is a discrete plane curve that extracted on the image and features are some characteristics of

Alexander Lepskiy
Higher School of Economics, Moscow, Russia
e-mail: alex.lepskiy@gmail.com

R. Kruse et al. (Eds.): Synergies of Soft Computing and Statistics, AISC 190, pp. 553–561.
springerlink.com © Springer-Verlag Berlin Heidelberg 2013

curve points (e.g. feature is a estimation of curvature in given point of discrete curve [4]) then a random character of features (e.g. curvature) will be due to noise of image. In this case the expectation $\mathbf{E}[\mathrm{M}(A)]$ be characterize the level of informativeness of representation A and the variance $\sigma^2[\mathrm{M}(A)]$ be characterize the level of stability of representation to noise of pattern. Then there is the problem of finding the most stable and informative representation A of the pattern X. The complexity of solutions of this problem will be determined by the degree of dependence of random features of each other. Stochastic measure of informativity M will be additive measure if the features are independent random variables. This case was considered in [5]. In this work we will consider the case when any random feature depends from some other features. Then we get nonadditive monotone measure of informativity. More detail the problem of finding the most stable and informative representation is investigated in this paper for the most popular measure of informativity for contour image – measure of informativity by length.

2 Monotone Geometrical Measure of Informativity

Measures of informativity can be effectively used in image processing as shown in [1]. In image processing the contours of the patterns and their characteristics, for example, curvatures of smooth curves are the such features that should not depend on illumination of a scene and orthogonal transformations (such as rotation, bias, scaling). However, in reality, we have digitized curves that are given by some ordered sets of points. These curves can be corrupted by noise. This means that we can use only some statistical estimates of curvature [4] that not stable to noise. A problem of choosing an optimal polygonal representation of a contour consist in finding such a representation that preserves geometrical characteristics of contour and also that will be stable to noise. This choice can be produced by using geometrical measure of informativity that are axiomatically defined as follows.

Let X be an initial closed contour given by an ordered finite set points, i.e. $X = \{x_1, ..., x_n\}$, where $x_i \in \mathbb{R}^2$, $i = 1, ..., n$. We identify with any nonempty subset $B = \{x_{i_1}, ..., x_{i_m}\}$ a contour generated by connecting points with straight lines starting from points x_{i_1}, x_{i_2} and ending by points x_{i_m}, x_{i_1}.

Definition 1. A geometrical measure of informativity $\mu : 2^X \to [0,1]$ is a set function that has to obey the following properties: 1) $\mu(\emptyset) = 0$, $\mu(X) = 1$; 2) $A, B \in 2^X$ and $A \subseteq B$ implies $\mu(A) \leq \mu(B)$; 3) let $B = \left\{..., x_{i_{k-1}}, x_{i_k}, x_{i_{k+1}}, ...\right\} \subseteq X$ and neighbouring points $x_{i_{k-1}}, x_{i_k}, x_{i_{k+1}}$ belong to a straight line in the plane, then $\mu(B) = \mu\left(B \backslash \{x_{i_n}\}\right)$; 4) μ is invariant w.r.t. affine transformations.

Emphasize that axioms 1, 2 have been introduced by Sugeno for fuzzy measures (see [7]). Consider several ways for defining geometrical measure of informativity [1].

a) Suppose that the length of an original contour is not equal to zero and a function $L(A)$ gives us the length of subcontour $A \in 2^X$. Then a measure of informativity defined by contour length is $\mu_L(A) = \frac{L(A)}{L(X)}$.

b) Suppose that the domain limited by an original contour is convex, and a function $S(B)$ determines the area bounded by an subcontour $A \in 2^X$. Then a measure of informativity defined by contour area is $\mu_S(A) = \frac{S(A)}{S(X)}$.

c) Let $w(x, A)$ be a positive estimate of information value of the part of a contour in a neighbourhood of point $x \in A$ in a subcontour $A \in 2^X$. Then an average measure of informativity is defined by $\mu(A) = \dfrac{\sum_{x \in A} w(x, A)}{\sum_{x \in X} w(x, X)}$, where $w(x, A)$ has to be defined for any non-empty contour $A \in 2^X$ and $\mu(\emptyset) = 0$ by definition. It is easy to see that the introduced geometrical measure of informativity μ_L and μ_S can be considered as average measure of informativity. For example, for μ_L function $w(x, A) = |x - y|$, where y is a next neighbouring points in contour A; in case of μ_S function $w(x, A) = S(O, x, y)$, where O is the centroid of area, bounded by contour A, and $S(O, x, y)$ is the area of triangle with vertices in points O, x, y.

3 Stochastic Average Measure of Informativity

In real situations, values $w(x, A)$ can be considered as random values, because an original contour is corrupted by an additive probabilistic noise. To stress this, we denote these values by capital letters as $W(x, A)$. In this case we have the measure of informativity $\mathrm{M}(A) = \frac{\sum_{x \in A} W(x,A)}{\sum_{x \in X} W(x,A)}$. Then there is problem of finding the most stable and informative representation $A \in 2^X$ of the pattern X for which the expectation $\mathbf{E}[\mathrm{M}(A)]$ will be maximize and the the variance $\sigma^2[\mathrm{M}(A)]$ will be minimize. If $W(x, A) = W(x)$ and random values $W(x)$, $x \in X$, are independent random variables then the measure of informativity $\mathrm{M}(A)$ is additive and the problem of finding the most stable and informative representation was investigated in this case in [5]. We emphasize that stochastic additive measures have been already investigated in the literature (see e.g. [6]). Now we will be investigated the important case when the value $W(x, A)$ depends on two neighbouring points. For example, the geometrical information measure μ_L and μ_S are satisfied this condition.

Let $X = \{x_1, ..., x_n\}$ be an original contour and let vertices be ordered by their indices. So if we consider any subcontour $A \in 2^X$, then the order defined on A is assumed to be generated by the order on X and given by indices in the representation $A = \{x_{i_1}, ..., x_{i_m}\}$, where $i_1 < ... < i_m$. So for any $A = \{x_{i_1}, ..., x_{i_m}\} \in 2^X$ we can identify its elements by their indices and write $x_k(A) = x_{i_k}$ if $k \in \{1, ..., m\}$. We can also consider any integer index k assuming that $x_k(A) = x_l(A)$ if $l \equiv k(\mathrm{mod} m)$. To work with such indices, we use a mapping π defined by $x_k(A) = x_{\pi_A(k)}$. We suppose that $W(x_k(A), A) = W(x_k(A), x_{k+1}(A))$, $k = 1, ..., |A|$, i.e. the value $W(x_k(A), A)$ depends on two neighbouring points $x_k(A), x_{k+1}(A)$. Further, for simplicity reasons, we denote $W(x_k(A), x_{k+1}(A)) = W_{k,k+1}(A)$. Then an average monotone measure and stochastic average monotone measure have a view

$$\mu(A) = \frac{\sum_{k=1}^{|A|} w_{k,k+1}(A)}{\sum_{j=1}^{|X|} w_{k,k+1}(X)}, \mathrm{M}(A) = \frac{\sum_{k=1}^{|A|} W_{k,k+1}(A)}{\sum_{j=1}^{|X|} W_{k,k+1}(X)} \tag{1}$$

correspondingly. We call M a stochastic monotone information measure if $W_{k,k+1}(A)$, $A \in 2^X$ are random variables. In this case M has random values. In this section we find estimates of numerical characteristics of M assuming that random variables $W_{k,k+1}(A)$, $W_{l,l+1}(A)$ are independent if $|l - k| > 1$. This situation appears if we suppose that x_k, $k = 1, ..., n$, are also independent random variables.

We see that $\mathrm{M}(A) = \frac{\xi}{\eta}$, where $\xi = \sum_{k=1}^{|A|} W_{k,k+1}(A)$ and $\eta = \sum_{j=1}^{|X|} W_{k,k+1}(X)$. The following lemma is used for estimating $\mathbf{E}[\mathrm{M}(A)]$ and $\sigma^2[\mathrm{M}(A)]$.

Lemma 1. *Let ξ and η be random variables that taking values in the intervals l_ξ, l_η respectively on positive semiaxis and $l_\eta \subseteq ((1-\delta)\mathbf{E}[\eta], (1+\delta)\mathbf{E}[\eta])$, $l_\xi \subseteq (\mathbf{E}[\xi] - \delta\mathbf{E}[\eta], \mathbf{E}[\xi] + \delta\mathbf{E}[\eta])$. Then it is valid the following formulas for mean and variance of distribution of $\frac{\xi}{\eta}$ respectively*

$$\mathbf{E}\left[\frac{\xi}{\eta}\right] = \frac{\mathbf{E}[\xi]}{\mathbf{E}[\eta]} + \frac{\mathbf{E}[\xi]}{\mathbf{E}^3[\eta]}\sigma^2[\eta] + \frac{1}{\mathbf{E}^2[\eta]}\mathbf{Cov}[\xi, \eta] + r_1, \tag{2}$$

$$\sigma^2\left[\frac{\xi}{\eta}\right] = \frac{1}{\mathbf{E}^2[\eta]}\sigma^2[\xi] + \frac{\mathbf{E}^2[\xi]}{\mathbf{E}^4[\eta]}\sigma^2[\eta] - \frac{2\mathbf{E}[\xi]}{\mathbf{E}^3[\eta]}\mathbf{Cov}[\xi, \eta] + r_2, \tag{3}$$

where $\mathbf{Cov}[\xi, \eta]$ is a covariation of random variables ξ and η, i.e. $\mathbf{Cov}[\xi, \eta] = \mathbf{E}[(\xi - \mathbf{E}[\xi])(\eta - \mathbf{E}[\eta])]$; r_1, r_2 are the residuals those depends on numerical characteristics of ξ and η. It being known that $|r_1| \le \frac{\delta}{1-\delta} \cdot \frac{\mathbf{E}[\xi]+\mathbf{E}[\eta]}{\mathbf{E}^3[\eta]}\sigma^2[\eta] \le \frac{\mathbf{E}[\xi]+\mathbf{E}[\eta]}{(1-\delta)\mathbf{E}[\eta]}\delta^3$, $|r_2| \le C\delta^3$.

Proof. We prove formula (2). The formula (3) is proved by analogy. Expand the function $\phi(x, y) = \frac{x}{y}$ into a Taylor series at the point $(\mathbf{E}[\xi], \mathbf{E}[\eta])$. We get

$$\phi(x, y) = \phi(\mathbf{E}[\xi], \mathbf{E}[\eta]) + \sum_{n=1}^{\infty} \frac{1}{n!} d^n \phi(\mathbf{E}[\xi], \mathbf{E}[\eta]) =$$

$$= \phi(\mathbf{E}[\xi], \mathbf{E}[\eta]) - \frac{\mathbf{E}[\xi](y - \mathbf{E}[\eta]) - \mathbf{E}[\eta](x - \mathbf{E}[\xi])}{\mathbf{E}^2[\eta]} \sum_{n=0}^{\infty} \left(\frac{\mathbf{E}[\eta] - y}{\mathbf{E}[\eta]}\right)^n.$$

The last series converges at every point $(x, y) \in l_\xi \times l_\eta$. Then

$$\mathbf{E}\left[\frac{\xi}{\eta}\right] = \frac{\mathbf{E}[\xi]}{\mathbf{E}[\eta]} + \frac{\mathbf{E}[\xi]}{\mathbf{E}^3[\eta]}\sigma^2[\eta] - \frac{1}{\mathbf{E}^2[\eta]}\mathbf{Cov}[\xi, \eta] + r_1,$$

where

$$r_1 = -\mathbf{E}\left[\frac{\mathbf{E}[\xi](\eta - \mathbf{E}[\eta]) - \mathbf{E}[\eta](\xi - \mathbf{E}[\xi])}{\mathbf{E}^2[\eta]} \sum_{n=2}^{\infty} \left(\frac{\mathbf{E}[\eta] - \eta}{\mathbf{E}[\eta]}\right)^n\right]$$

and $|r_1| \le \frac{\delta}{1-\delta} \frac{\mathbf{E}[\xi]+\mathbf{E}[\eta]}{\mathbf{E}^3[\eta]}\sigma^2[\eta] \le \frac{\mathbf{E}[\xi]+\mathbf{E}[\eta]}{(1-\delta)\mathbf{E}[\eta]}\delta^3$. The last estimate is followed from inequality $\sigma[\eta] \le \delta\mathbf{E}[\eta]$. The lemma is proved.

We will use formulas (2) and (3) without their residuals. Respective values $\tilde{\mathbf{E}}[\mathrm{M}(A)] = \mathbf{E}[\mathrm{M}(A)] - r_1$, $\tilde{\sigma}^2[\mathrm{M}(A)] = \sigma^2[\mathrm{M}(A)] - r_2$ we will call by estimations of numerical characteristics.

Introduce the following notation: $S(A) = \sum_{i=1}^{|A|} \mathbf{E}[W_{i,i+1}(A)]$, $K(A, X) = \sum_{i=1}^{|A|} k_i^X(A)$, where $k_i^X(A) = \sum_{j=1}^{|X|} \mathbf{Cov}[W_{i,i+1}(A), W_{j,j+1}(X)]$, $A \in 2^X$. Then the formulas for $\tilde{\mathbf{E}}[\mathrm{M}(A)]$ and $\tilde{\sigma}^2[\mathrm{M}(A)]$ based on (2) and (3) can be written in the form

$$\tilde{\mathbf{E}}[\mathrm{M}(A)] = \frac{S(A)}{S(X)} + \frac{S(A)}{S^3(X)}K(X, X) - \frac{1}{S^2(X)}K(A, X), \tag{4}$$

$$\tilde{\sigma}^2[\mathrm{M}(A)] = \frac{1}{S^2(X)}K(A, A) + \frac{S^2(A)}{S^4(X)}K(X, X) - \frac{2S(A)}{S^3(X)}K(A, X). \tag{5}$$

In general, the random variable $\sum_{k=1}^{|A|} W_{k,k+1}(A)$ is not satisfied to conditions of Lemma 1. However the probability of large deviations of random length of noisy polygonal line from non-noisy length will be small if the variance of noise is small. Therefore we assume that the random length satisfied approximately to conditions of Lemma 1.

4 Stochastic Informational Measure by Contour Length

Assume that an original contour is corrupted by noise. In this case, $X = \{x_k + \mathbf{n}_k\}_{k=1}^m$, $x_k \in \mathbb{R}^2$ and $\mathbf{n}_k = (\xi_k, \eta_k)$ are random variables. Suppose also that ξ_k, η_k, $k = 1, ..., m$, are independent, normally distributed and such that $\mathbf{E}[\xi_k] = \mathbf{E}[\eta_k] = 0$, $\sigma^2[\xi_k] = \sigma^2[\eta_k] = \sigma^2$, $k = 1, ..., m$. In this section we consider a monotone measure μ and monotone stochastic measure M of view (1), where $W_{k,k+1}(A) = |x_{k+1}(A) + \mathbf{n}_{k+1}(A) - x_k(A) - \mathbf{n}_k(A)|$ and $w_{k,k+1}(A) = |x_{k+1}(A) - x_k(A)|$ correspondingly. We investigate its characteristics $\tilde{\mathbf{E}}[\mathrm{M}(A)]$ and $\tilde{\sigma}^2[\mathrm{M}(A)]$. Suppose that $W_{k,k+1}(X)$, $k = 1, ..., m$, are independent random variables. This requirement can be satisfied by the choice of some subcontour (basic contour) from the initial contour.

4.1 Numerical Characteristics of Random Variable $W_{k,k+1}(A)$

Let $\mathbf{l}_A(x) = x_+(A) - x$, where $x_+(A)$ is the next point w.r.t. x in the contour A, $l_A(x) = |\mathbf{l}_A(x)|$.

Proposition 1. *The following asymptotic equalities are valid*

$$\mathbf{E}\left[W_{k,k+1}(A)\right] = l\left(1 + \tfrac{\sigma^2}{l^2} + \tfrac{\sigma^4}{2l^4} + O\left(\tfrac{\sigma^6}{l^6}\right)\right),$$

$$\sigma^2\left[W_{k,k+1}(A)\right] = 2\sigma^2\left(1 - \tfrac{\sigma^2}{l^2} + O\left(\tfrac{\sigma^4}{l^4}\right)\right),\ l = l_A(x_k).$$

Proof. Assume that $x_{k+1}(A) - x_k(A) = (l, 0)$, $\mathbf{n}_{k+1}(A) - \mathbf{n}_k(A) = (\xi_{k+1} - \xi_k, \eta_{k+1} - \eta_k)$. Denote $\xi = \xi_{k+1} - \xi_k$ and $\eta = \eta_{k+1} - \eta_k$, $\theta = W_{k,k+1}(A) = \sqrt{(\xi + l)^2 + \eta^2}$. Then ξ, η are independent normally distributed random variables and such that $\mathbf{E}[\xi] = \mathbf{E}[\eta] = 0$ and $\sigma^2[\xi] = \sigma^2[\eta] = 2\sigma^2$. Let $u = \frac{1}{l}$. Then $\theta^2 = l^2\left(1 + 2\xi u + \ (\xi^2 + \eta^2) u^2\right)$. Let us find the representation of $\theta^* = \theta/l$ by Taylor formula at the point $u = 0$: $\theta^*(0) = 1$, $\theta^{*\prime}(0) = \xi$, $\theta^{*\prime\prime}(0) = \eta^2$, $\theta^{*\prime\prime\prime}(0) = -3\xi\eta^2$, $\theta^{*(4)}(0) = 12\xi^2\eta^2 - 3\eta^4$, $\theta^{*(5)}(0) = -60\xi^3\eta^2 + 45\xi\eta^4$, $\theta^{*(6)}(0) = 360\xi^4\eta^2 - 540\xi^2\eta^4 + 45\eta^6$. Therefore

$$\theta^*(u) = 1 + \xi u + \frac{\eta^2}{2}u^2 - \tfrac{\xi\eta^2}{2}u^3 + \tfrac{4\xi^2\eta^2 - \eta^4}{8}u^4 - \tfrac{4\xi^3\eta^2 - 3\xi\eta^4}{8}u^5 + O(u^6).$$

We compute next $\mathbf{E}[\theta^*(u)]$ taking in account that $\mathbf{E}[\xi^s] = \mathbf{E}[\eta^s] = 0$ if s is odd, $\mathbf{E}\left[\xi^2\right] = \mathbf{E}\left[\eta^2\right] = \sigma^2$, $E\left[\xi^4\right] = E\left[\eta^4\right] = 3\sigma^4$, $E\left[\xi^6\right] = E\left[\eta^6\right] = 15\sigma^6$, and that we should compute the expectation of product of independent random variables. Then we have $\mathbf{E}[\theta^*(u)] = 1 + \frac{1}{2}\sigma^2u^2 + \frac{1}{8}\sigma^4u^4 + O(u^6)$, since $E\left[\theta^{*(6)}(0)\right] = 135 \neq 0$. Compute the variance of θ: $\mathbf{E}\left[\theta^2\right] = \mathbf{E}\left[\xi^2 + 2\xi l_k + l^2 + \eta^2\right] = 2\sigma^2 + l^2$, $\mathbf{E}^2[\theta] = l^2\left(1 + \frac{\sigma^2}{l^2} + O\left(\frac{\sigma^4}{l^4}\right)\right)$, therefore, $\sigma^2[\theta] = \mathbf{E}\left[\theta^2\right] - \mathbf{E}^2[\theta] = \sigma^2\left(1 - \frac{\sigma^2}{2l^2} + O\left(\frac{\sigma^4}{l^4}\right)\right)$. The general case is also true, because values $\mathbf{E}\left[W_{k,k+1}(A)\right]$, $\sigma^2\left[W_{k,k+1}(A)\right]$ do not depend on the chosen coordinate system.

Corollary 1. *It is true the equality*

$$S(A) = \sum_{k=1}^{|A|}\mathbf{E}\left[W_{k,k+1}(A)\right] = L(A) + \sigma^2\sum_{k=1}^{|A|} l_A^{-1}(x_k) + \sigma O\left(\tfrac{\sigma^3}{\underline{l}_A^3}\right),$$

where $L(A) = \sum_{k=1}^{|A|} l_A(x_k)$ *is the length of contour* A, $\underline{l}_A = \min_k l_A(x_k)$.

By analogy with Proposition 1 and Corollary 1 we compute the covariance between random variables $W_{k-1,k}(A)$, $W_{k,k+1}(A)$.

Proposition 2. *We have*

$$\mathbf{Cov}\left[W_{k-1,k}(A), W_{k,k+1}(A)\right] =$$

$$= -\sigma^2 \cos\alpha_k \left(1 - \left(\tfrac{1}{l_{k-1}^2} + \tfrac{\cos\alpha_k}{2l_{k-1}l_k} + \tfrac{1}{l_k^2}\right)\sigma^2 + o\left(\tfrac{\sigma^2}{l^2}\right)\right),$$

where $\alpha_k = \alpha(x_k) = \left(\widehat{\mathbf{l}_{i-1}, \mathbf{l}_i}\right)$, $l_k = l_A(x_k)$, $l = \min\{l_{k-1}, l_k\}$.

Calculate the covariance $K(A,X) = \sum_i k_i^X(A)$ between the all segments of polygon A and all segments of basic polygon X with help of last proposition. Let $\alpha(x)$ ($\beta(x)$) be an inner angle of polygon A (polygon X) in vertex x, $\gamma(x)$ be an angle between the vectors $x_{+1}(A) - x$, $x_{+1}(X) - x$, where $x_{+1}(A)$ ($x_{+1}(X)$) is the next point w.r.t. x in the contour A (contour X).

Corollary 2. *To same conditions it is true equality*

$$K(A,X) = 4\sigma^2 \sum_{x\in A} \cos\tfrac{\alpha(x)}{2} \cos\tfrac{\beta(x)}{2} \cos\left(\gamma(x) + \tfrac{\alpha(x)-\beta(x)}{2}\right) + \sigma^2 o\left(\tfrac{\sigma}{\underline{l}_A}\right)$$

for $A \in 2^X$, *where* $\underline{l}_A = \min_k l_A(x_k)$.

4.2 The Numerical Characteristics of Stochastic Measure of Informativity by Length

We will find numerical characteristics of stochastic measure of informativity by length using the results of the previous item. The following theorem may be got from equality (4), Corollaries 1, 2.

Theorem 1. *The asymptotic equality*

$$\tilde{\mathbf{E}}\left[\mathrm{M}(A)\right] = \tfrac{L(A)}{L(X)} + C_1(A)\tfrac{\sigma^2}{L^2(X)} + o\left(\tfrac{\sigma^2}{\underline{l}_A^2}\right), A \in 2^X$$

is true, where

$$C_1(A) = -L(A)\sum_{x\in X} l_X^{-1}(x) + L(X)\sum_{x\in A} l_A^{-1}(x) + 4\tfrac{L(A)}{L(X)}\sum_{x\in X}\cos^2\tfrac{\beta(x)}{2} -$$

$$-4\sum_{x\in A}\cos\tfrac{\alpha(x)}{2}\cos\tfrac{\beta(x)}{2}\cos\left(\gamma(x) + \tfrac{1}{2}\alpha(x) - \tfrac{1}{2}\beta(x)\right).$$

Similarly we will find the asymptotic formula for variance of stochastic informational measure by length with help of formula (5), Corollaries 1, 2.

Theorem 2. *The asymptotic equality*

$$\tilde{\sigma}^2\left[\mathrm{M}(A)\right] = 4C_2(A)\frac{\sigma^2}{L^2(X)} + o\left(\frac{\sigma^2}{l_A^2}\right), \quad A \in 2^X$$

is true, where

$$C_2(A) = \sum_{x\in A}\cos^2\frac{\alpha(x)}{2} + \frac{L^2(A)}{L^2(X)}\sum_{x\in X}\cos^2\frac{\beta(x)}{2} -$$

$$-2\frac{L(A)}{L(X)}\sum_{x\in A}\cos\frac{\alpha(x)}{2}\cos\frac{\beta(x)}{2}\cos\left(\gamma(x)+\tfrac{1}{2}\alpha(x)-\tfrac{1}{2}\beta(x)\right).$$

The value of random error (the variance of stochastic informational measure) characterizes the degree of stability of informational measure of curve with respect to level of curve noise. We can put the task about finding of polygonal representation of fixed cardinality $A \in 2^X$, $|A| = k$, which minimized the value of variance of stochastic informational measure by length. As can be seen from Theorem 2 the polygonal representation

$$A = \arg\min_{A\in 2^X,\ |A|=k} C_2(A)$$

is a solution of indicated task for great signal-to-noise ratio $\frac{l_A^2}{\sigma^2}$.

Example 1. Let $X = \{x_1, ..., x_6\}$ be an ordered set of vertexes of regular 6-gon. Calculate the value $C_2(A)$ for various polygonal representations A of cardinality $|A| = 3$: $A_1 = \{x_1, x_3, x_5\}$, $A_2 = \{x_1, x_2, x_4\}$, $A_3 = \{x_1, x_2, x_3\}$. We have $C_2(A_1) = 1.125$, $C_2(A_2) = 1.25$, $C_2(A_3) \approx 1.66$. Thus the contour A_1 is a most stable contour to noise w.r.t. measure of informativity by length among of contours of cardinality is equal 3.

Many other tasks of finding of informative and stable representation of noisy image may be formulated and solved with help of this approach.

Acknowledgements. I would like to thank Andrew Bronevich for his helpful and stimulating comments on the manuscript of my paper. The study was implemented in the framework of The Basic Research Program of the Higher School of Economics in 2012. This work was supported by the grants 11-07-00591 and 10-07-00135 of RFBR (Russian Foundation for Basic Research).

References

1. Bronevich, A., Lepskiy, A.: Geometrical fuzzy measures in image processing and pattern recognition. In: Proc. of the 10th IFSA World Congress, Istanbul, pp. 151–154 (2003)

2. Duda, R.O., Hart, P.E., Stork, D.G.: Pattern Classification and Scene Analysis: Part I Pattern Classification. John Wiley & Sons (1998)
3. Jolliffe, I.T.: Principal Component Analysis. Springer Series in Statistics. Springer, Berlin (2002)
4. Lepskiy, A.E.: On Stability of the Center of Masses of the Vector Representation in One Probabilistic Model of Noiseness of an Image Contour. Automat. Rem. Contr. 68, 75–84 (2007)
5. Lepskiy, A.: Stable Feature Extraction with the Help of Stochastic Information Measure. In: Kuznetsov, S.O., Mandal, D.P., Kundu, M.K., Pal, S.K. (eds.) PReMI 2011. LNCS, vol. 6744, pp. 54–59. Springer, Heidelberg (2011)
6. Shiryaev, A.N.: Probability. Graduate Texts in Mathematics. Springer, New York (1995)
7. Wang, Z., Klir, G.J.: Generalized Measure Theory. IFSR International Series on Systems Science and Engineering, vol. 25. Springer, Berlin (2009)

Clustering on Dynamic Social Network Data

Pascal Held and Kai Dannies

Abstract. This paper presents a reference data set along with a labeling for graph clustering algorithms, especially for those handling dynamic graph data. We implemented a modification of Iterative Conductance Cutting and a spectral clustering. As base data set we used a filtered part of the Enron corpus. Different cluster measurements, as intra-cluster density, inter-cluster sparseness, and Q-Modularity were calculated on the results of the clustering to be able to compare results from other algorithms.

Keywords: Clustering, cluster measurements, Enron data set, graph clustering, stream data.

1 Introduction

Social network analysis has already been popular long before websites like Facebook, XING or Google+ - now commonly known as social networks - were launched. In [16] a comprehensive approach of modeling social network data as (un)directed graphs was proposed, which has become widely accepted. Over the years a lot of research has been performed on e.g. cohesiveness of groups of members in social graphs [17] or segmentation of social networks [13]. All these methods have in common that they use a static representation of the social graph underlying the respective social network.

Recent research covered also the topic of dynamic graph clustering. Kim et al.[11] tried to solve the problem of clustering dynamic social networks by

Pascal Held · Kai Dannies
Otto-von-Guericke University of Magdeburg, Faculty of Computer Science
e-mail: pascal.held@ovgu.de,kai.dannies@st.ovgu.de

R. Kruse et al. (Eds.): Synergies of Soft Computing and Statistics, AISC 190, pp. 563–571.
springerlink.com © Springer-Verlag Berlin Heidelberg 2013

using evolutionary algorithms. Goerke et al.[8] extended an algorithm based on min-cut trees [6] introducing temporal smoothness.

Attempts have been made to infer information from dynamic graphs (e.g. in [1]) but they either restrict themselves to fairly simple questions like connectivity or to path finding problems in order to cope with the changing structure of the graph. Such discretization results from some kind of binning operation performed on the data, thus leading to a loss of information, namely the exact time when an event has happened. Such an approach does not take into account the frequency with which events occur but rather lists their absolute number.

We provide a reference clustering along with a prepared data set, bases on the Enron corpus. One can download them at `http://www.ovgu.de/pheld/pub/SMPS2012`. The clustering we provide for each time step of the data set is described in Section 2.1, a divisive minimum modularity clustering, generates very well clusters with respect to cluster measurements as inter-cluster sparseness or intra-cluster density.

The paper is structured as follows: The following section gives a short summary of the Enron dataset and of the selected algorithms for clustering. Afterwards we present our experiments in Section 3 and the results in Section 4. We finish our paper with a conclusion in Section 5.

2 Related Work

2.1 Enron Dataset

We used the well-known Enron dataset (`http://www-2.cs.cmu.edu/~enron/`) as basis for our experiments. The Enron dataset is a large corpus of email messages from the Enron Corporation. This email communication is a good example for human interaction in social networks. The raw dataset contains about 620, 000 messages from 158 users [12].

For our experiments we cleaned the messages, so we removed duplicate messages, and all messages, which were not sent from one Enron employee to another. Mails from mailing lists were also dropped. The major part of the dropped messages was SPAM and duplicated messages. We interpreted mails with multiple recipients as a separate mail from the sender to every recipient. Mails with wrong addresses, like `firstnamelastname@enron.com` were matched to the correct `firstname.lastname@enron.com`.

From these messages we created an event list containing only the time stamps, the sender, and the recipient of the message. In total we got 9071 events. We used this event list to generate a dynamic graph, where every node is an Enron employee and every edge represents the communication frequency. The event list was binned to buckets of 10.000 seconds. In total we got about 10.000 of these time steps. To estimate this frequency we used a Butterworth filter with a bandpass-frequency of 0.0075. This value is the result of an optimization process, based on AIC and BIC measures. It is a good compromise between fast reaction on changing in behavior and smooth filter signal. A detailed description why we use this frequency can be found in [9]. A more detailed discussion about the resulting data set can be found at `http://www.ovgu.de/pheld/pub/SMPS2012`

The Enron dataset is not a really huge dataset, but this gives us the opportunity to cluster the resulting dynamic graph at a lot of time stamps with classical clustering methods. Later, the results of these classical algorithms can be compared with the results of dynamic clustering algorithms. Good results on this small datasets could be a indicator for good results in much larger datasets.

2.2 Clustering Algorithms

Graph clustering has obviously a close connection to the classical minimum cut problem [3], which consists in finding for a given (undirected) graph a partition of its vertices into two disjoint subsets, which minimizes the number (or the total weight) of the edges crossing between them. More precisely we try to find more than one cut here, but a "good" number of cuts for dividing the graph in the "most natural" way.

Two values of the graph are important to determine the quality of the clustering: The number of edges between the clusters and the number of edges in each cluster. In section 2.3 we will present in detail several possibilities to combine those values to a single number as a quality measure of graph clustering.

2.2.1 Divisivel Minimum Modularity Clustering (DMMC)

The basic idea of this algorithm is the same as in Iterative Conductance Cutting [10]. These algorithm aims for a cut of the graph minimizing the conductance, see section 2.3. Unfortunately finding such a cut is NP-hard. So Kannan et al.[10] used an approximation: the nodes are sorted w.r.t. their corresponding eigenvalues of the adjacancy matrix. Then calculate each possible split of this sorted set and continue with the maximal conductance.

We decided to use the recently deeply researched measurement of Modularity [2] as described in section 2.3. Besides this we decided because of the need of as good results as possible to actually solve the NP-hard problem: To try any possible cut to minimize the modularity. Even solving this problem does not guarantee an optimal solution with respect to modularity as measurement: There could be steps where a split in more than two clusters leads to a better result. So for really finding the best cluster we would have to check any possible clustering. Because this is computationally too extensive even for the given 158 nodes we hope to get a really good approximation by using the algorithm described below.

As initialization we perform a connected component analysis and then try to divide each cluster further trying each pair of nodes in the original cluster as seeds of two subsequent clusters until no further possible splitting leads to an improvement.

Algorithm 1. DMMC

```
Input:Similarity Matrix
clusters ← connectedComponents;
while Clustering changes do
    for each cluster do
        for node1,node2:cluster do          ▷ try to split clusters on each pair of nodes
            clusterCenter1 ← node1;
            clusterCenter2 ← node2;
            assign each vertex in cluster to cluster center with higher similarity;
            if clustering is better than before then
                discard cluster;
                add both new clusters;
            end if
        end for
    end for
end while
Output:Clusters A_1, ..., A_k
```

2.2.2 Other Algorithms

As second algorithm we implemented the unnormalized spectral clustering as described by Luxburg[14] and compared it with DMMC.

There are also other algorithms described in the literature. One example is Markov Clustering [5] that is simulating a random walk on the graph. Another one is Geometric MST[1] Clustering [7] which combines spectral partitioning and a geometric clustering technique. Another prominent example is a clustering based on min-cut trees developed by Flake et al.[6].

2.3 Cluster Quality Measurements

As stated above, two values of a graph clustering are especially important to measure the quality of a clustering: The sum of the weights of intra-cluster edges should be maximized and the sum of the weights of inter-cluster edges should be minimized. Because one wants only one number to compare clus-

[1] MST = minimum spanning tree.

terings, different measurements based one those two numbers were developed [3, 4]. In this section we will use following notation:

$\mid E \mid$	sum of weights of edges contained in E
$E(C)$	intra-cluster edges of Cluster C
$E(C_1, C_2)$	inter-cluster edges between C_1 and C_2
$E_{inc}(C)$	incident edges to a cluster, including inter-cluster edges from C to any other cluster as well as all intra-cluster edges of C
maxWeight	for simple, undirected graphs with $0 \leq$ *edge Weight* $\leq$ 1: $\mid V \mid \cdot \mid V - 1 \mid \cdot 0.5$

Q-Modularity

The Q-Modularity measurement was developed by Newman und Girvan [15]. One sets up a $k \times k$-Matrix $\mathcal{M}$ where k is the number of clusters. The entry (i, j) of the matrix is the sum of the edge weights between the i-th and the j-th cluster for $i \neq j$ and the sum of weights of the i-th cluster for $i = j$. The Matrix is normalized by $\mid \mathcal{M} \mid$. If this matrix has most of its weight on the principal diagonal then most of the weights are within clusters instead inbetween the clusters and therefore the clustering is good. However, only using the elements on the principal diagonal to measure the cluster quality is not sufficient, because then singleton clusters would always be optimal.

So the authors [15] choose the following quantity for taking the inter-cluster edges into account:

$$a_i = \sum_j \mathcal{M}_{i,j} = \frac{\mid E_{inc}(C_i) \mid}{maxGraphWeight} \tag{1}$$

The complete measurement is then calculated by:

$$Q\text{-}Modularity = \sum_i (\mathcal{M}_{i,i} - a_i^2) \tag{2}$$

Intra-Cluster Density and Inter-Cluster Sparseness

The basic idea of the intra-cluster density is to measure how dense the clusters are. For the intra-cluster density we used:

$$intraClusterDensity = \frac{1}{numClusters} \cdot \sum_{C \in clusters} \frac{\mid E(C) \mid}{maxWeight(C)} \tag{3}$$

The basic idea for this measurement is, that the edges inbetween the clusters should be as sparse as possible. For inter-cluster sparseness we used:

$$interClusterSparseness = 1 - \frac{\sum_{(u,v)\in E, u\in C_i, v\in C_j, i\neq j} |(u,v)|}{maxGraphWeight - \sum_{C\in clusters} |E(C)|} \quad (4)$$

3 Experiments

As stated above we used the Enron dataset for our experiments. The results achieved by processing the raw data as described in Section 2.1 lead to edge weights between zero and one, mostly much closer to zero. There are also different types of people. Some people communicate 100 times as much as other people, but they have also regularity in their communication behavior. The edge weights are approcimately lognormal distributed, so we used a logarithmic scaling:

$$scaledEdgeWeight = \frac{20 + log_{10}\, edgeWeight}{20} \quad (5)$$

Numbers smaller than zero or greater than one were set to zero or one respectively. The number 20 is used as a threshold: All values above 10^{-20} shall be considered. Any threshold you want to use can replace this threshold. The higher this number is chosen the longer clustering structure from inactive users keep alive. Isolated nodes were ignored for the clustering.

After this preprocessing we took every time step of the data and clustered it with the two algorithms described in Sections 2.2.1 and 2.2.2 where the maximal number of clusters for the second algorithm was set to 25 because more than 25 clusters on 158 nodes are hard to interpret. For the spectral clustering we stored the best result out of 100 experiments to neglect the random component as good as possible. As the target function for both algorithms we maximized the Q-Modularity. The other described measurements were used to compare the results independent of the target function, see next section.

4 Results

In Figure 1 we show the generated number of clusters over the time. The x-axis describes the timeline, where 10.000 seconds are one time step. Both algorithms generated a similar number of clusters. The K-Median algorithm is more stable than the spectral clustering algorithm in terms of number of clusters. This is caused by on the random component of the spectral clustering. This phenomenon is also reflected in the Q-Modularity. Good values for the Q-Modularity are between 0.3 and 0.7 [15]. These values were reached

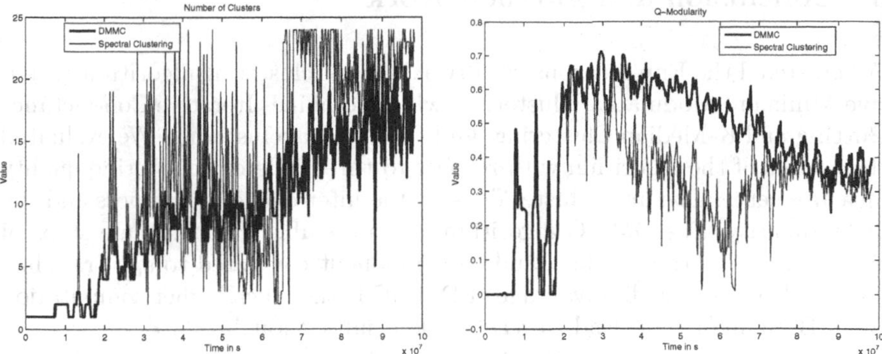

Fig. 1 left: number of clusters, right: Q-Modularity over time

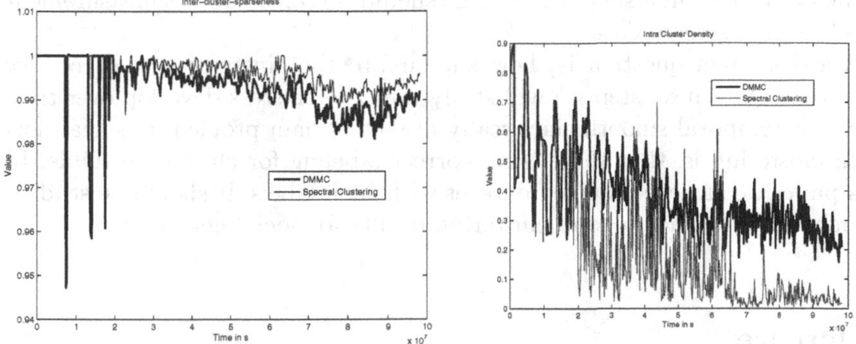

Fig. 2 inter-cluster-sparseness and intra-cluster-density over time

by the K-Median algorithm most of the time. The K-Median algorithm outperforms the spectral clustering with the Q-Modularity measurement in each time step.

The inter-cluster-sparseness, see Figure 2, is quite good for both algorithms. For this measurement the spectral clustering provides better results than the K-Median algorithm. There are some outliers at the beginning of the timeline. These results from less active elements in the graph at the first few hundred time steps. So there are no old communication structures learned and the graph is really sparse.

However the DMMC algorithm provides better results for the intra-cluster density. This leads to the conclusion that DMMC prefers the intra-cluster density, but spectral clustering prefers the inter cluster sparseness. The lower values at the end of the time series are caused by the high connectivity of the graph.

5 Conclusion and Further Work

We clustered the Enron corpus with two different clustering algorithms: Divisive Minimum Modularity Clustering, which combines Iterative Conductance Cutting and K-Median Clustering, and the spectral clustering. We evaluated the results of the algorithm with respect to three different clustering quality measurements: The intra-cluster density, the inter-cluster sparseness and the Q-Modularity. The DMMC algorithm is more stable over the time steps of the data due to the missing random component compared to spectral clustering. However, the disadvantage of DMMC is the much higher computation time. Both algorithms lead to reasonable results, though.

Despite the long computation time, the DMMC can be used as reference to test other cluster algorithms *because....* For future work one should test the described algorithms with other common stream data sets. Also the change of the clusterings in a small time range should be considered as measurement for clustering stream data.

Another open question is, how similar are the clusterings of successive timesteps. As a next step we will study how the clusters deverlop over time and how temporal smooth they really are. One main problem in social network clustering is, that there is no correct labeling for clusters availible. In this paper we enriched a given dataset with such labels. It should be studied if such measure lead to good clustering results for social networks.

References

1. Alberts, D., Cattaneo, G., Italiano, G.F.: An empirical study of dynamic graph algorithms. J. Exp. Algorithm 2 (1997)
2. Brandes, U., Delling, D., Gaertler, M., Goerke, R., Hoefer, M., Nikoloski, Z., Wagner, D.: On modularity clustering. IEEE Trans. Knowl. Data Eng. 20, 172–188 (2008)
3. Brandes, U., Gaertler, M., Wagner, D.: Experiments on Graph Clustering Algorithms. In: Di Battista, G., Zwick, U. (eds.) ESA 2003. LNCS, vol. 2832, pp. 568–579. Springer, Heidelberg (2003)
4. Delling, D., Gaertler, M., Görke, R., Nikoloski, Z., Wagner, D.: How to evaluate clustering techniques. Tech. Rep. 24, Fakultät für Informatik, Universität Karlsruhe (2006)
5. van Dongen, S.M.: Graph clustering by flow simulation. Ph.D. thesis, University Utrecht (2001)
6. Flake, G.W., Tarjan, R.E., Tsioutsiouliklis, K.: Graph clustering and minimum cut trees. Internet Math. 1, 385–408 (2004)
7. Gaertler, M.: Clustering with spectral methods. Master's thesis, University of Konstanz (2002)
8. Görke, R., Hartmann, T., Wagner, D.: Dynamic Graph Clustering Using Minimum-Cut Trees. In: Dehne, F., Gavrilova, M., Sack, J.-R., Tóth, C.D. (eds.) WADS 2009. LNCS, vol. 5664, pp. 339–350. Springer, Heidelberg (2009)

9. Held, P., Moewes, C., Braune, C., Kruse, R., Sabel, B.A.: Advanced Analysis of Dynamic Graphs in Social and Neural Networks. In: Borgelt, C., Gil, M.Á., Sousa, J.M.C., Verleysen, M. (eds.) Towards Advanced Data Analysis. STUDFUZZ, vol. 285, pp. 205–222. Springer, Heidelberg (2012)
10. Kannan, R., Vempala, S., Vetta, A.: On clustering: Good, bad and spectral. J. ACM 51, 497–515 (2004)
11. Kim, K., McKay, R., Moon, B.R.: Multiobjective evolutionary algorithms for dynamic social network clustering. In: Proc. of the 12th Ann. Conf. on Genetic and Evolutionary Computation (GECCO 2010), pp. 1179–1186. ACM, New York (2010)
12. Klimt, B., Yang, Y.: The Enron Corpus: A New Dataset for Email Classification Research. In: Boulicaut, J.-F., Esposito, F., Giannotti, F., Pedreschi, D. (eds.) ECML 2004. LNCS (LNAI), vol. 3201, pp. 217–226. Springer, Heidelberg (2004)
13. Kumar, R., Novak, J., Tomkins, A.: Structure and evolution of online social networks. In: Proc. of the 12th ACM SIGKDD Int. Conf. on Knowledge Discovery and Data Mining, pp. 611–617. ACM, New York (2006)
14. von Luxburg, U.: A tutorial on spectral clustering. Stat. Comput. 17, 395–416 (2007)
15. Newman, M.E.J., Girvan, M.: Finding and evaluating community structure in networks. Phys. Rev. E 69, 026, 113 (2004)
16. Wassermann, S., Faust, K.: Social Network Analysis: Methods and Applications. Cambridge University Press, Cambridge (1997)
17. White, D.R., Harary, F.: The cohesiveness of blocks in social networks: Node connectivity and conditional density. Sociol. Methodol. 31, 305–359 (2001)

Merging Partitions Using Similarities of Anchor Subsets

Thomas A. Runkler

Abstract. This paper addresses the problem of merging pairs of partition matrices. Such partition matrices may be produced by collaborative clustering. We assume that each subset in one partition matrix matches one of the subsets in the other partition matrix. To align the arbitrarily ordered rows in the partition matrices we use the memberships of a set of anchor points and maximize their pairwise similarities. Here, we consider various set-theoretic similarity measures. Experiments with a simplified version of the well-known BIRCH benchmark data set illustrate the effectivity of the approach and show that all considered similarity measures are well suited for partition merging.

Keywords: Fuzzy clustering, similarity measures.

1 Introduction

Fuzzy clustering partitions a data set $X = \{x_1, \ldots, x_n\} \subset \mathbb{R}^p$ into $c \in \{2, \ldots, n-1\}$ fuzzy subsets specified by a $c \times n$ membership matrix U, $u_{ik} \in [0,1]$, $i = 1, \ldots, c$, $k = 1, \ldots, n$,

$$\sum_{i=1}^{k} u_{ik} = 1, \quad \sum_{k=1}^{n} u_{ik} > 0 \tag{1}$$

A popular fuzzy clustering model is *fuzzy c-means* (FCM) [2] that minimizes

$$J(U, V, X) = \sum_{i=1}^{c} \sum_{k=1}^{n} u_{ik}^{m} \|v_i - x_k\|^2 \tag{2}$$

Thomas A. Runkler
Siemens Corporate Technology, 80200 München, Germany
e-mail: `thomas.runkler@siemens.com`

R. Kruse et al. (Eds.): Synergies of Soft Computing and Statistics, AISC 190, pp. 573–581.
springerlink.com © Springer-Verlag Berlin Heidelberg 2013

with the fuzzifier $m > 1$ and the cluster centers $V = \{v_1, \ldots, v_c\} \subset \mathbb{R}^p$. In this paper we use the Euclidean norm. Optimization of FCM can be done by alternating optimization through the conditions for local extrema of J

$$v_i = \frac{\sum\limits_{k=1}^{n} u_{ik}^m x_k}{\sum\limits_{k=1}^{n} u_{ik}}, \qquad u_{ik} = 1 \bigg/ \sum_{j=1}^{c} \left(\frac{\|v_i - x_k\|}{\|v_j - x_k\|} \right)^{\frac{2}{m-1}} \tag{3}$$

The computational complexity of this FCM optimization is asympotically linear in n and asympotically quadratic in c [18]. An approach to reduce the quadratic complexity in c is *divisive clustering* that starts with a lower value of c and subsequently divides the clusters into subclusters [5], which yields a hierarchical cluster structure. The opposite approach is *agglomerative clustering* that starts with a higher value of c and iteratively merges clusters [11, 12]. Both divisive and agglomerative clustering consider all objects of the entire data set X. Orthogonal to agglomerative clustering is *collaborative clustering* [13] where subsets of X are clustered separately and the clustering results are merged. If the data are stored in distributed and remotely located devices, then collaborative clustering reduces the communication effort, because only clustering results need to be transmitted between the devices instead of the complete data. Also data privacy issues can be solved by collaborative clustering [14]. Fig. 1 illustrates the different schemes of divisive, agglomerative, and collaborative clustering. The reverse of collaborative clustering is omitted due to triviality.

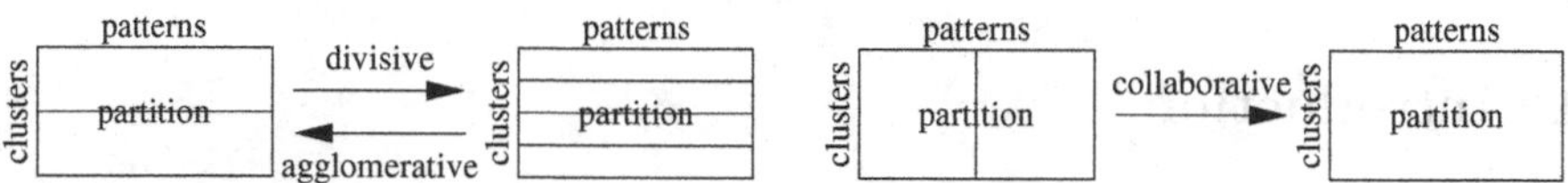

Fig. 1 Divisive, agglomerative, and collaborative clustering

This paper focusses on merging partitions resulting from collaborative clustering. Without loss of generality we consider merging a pair of partitions, as illustrated in the right view of Fig. 1. Merging more than two partitions can be done by iteratively merging pairs of partitions. Notice however that we not necessarily require pairwise merging to be commutative or associative, so merging multiple partitions may yield different results depending on the merging sequence. For simplicity we assume that both partitions have the same number of rows (clusters).

Hore and Hall suggest an approach to merge partitions U_1 and U_2 by computing the corresponding sets of cluster centers V_1 and V_2 using (3) and then applying a so-called *centroid correspondence algorithm* to merge the resulting cluster centers [7]. In this paper we do not want to merge sets of cluster centers but present an approach to explicitly merge partition matrices. This approach is based on similarity measures and requires the use of anchor subsets. The paper is structured as follows: Section 2 briefly reviews similarity measures for sets and partitions. Section 3 presents our new similarity based approach to merge partitions. Section 4 illustrates the performance of our new approach in experiments with benchmark data. Section 5 finally gives the conclusions.

2 Similarity Measures for Sets and Partitions

Measures for the similarity of fuzzy subsets can be categorized into geometric distance models, set-theoretic approaches, pattern recognition approaches, and correlation indices [21]. In this paper we restrict to set-theoretic approaches. Dubois and Prade [3] suggested a fuzzy generalization of the Jaccard index [10] or Gregson's crisp similarity [6]

$$s_1(A,B) = \frac{|A \cap B|}{|A \cup B|} \tag{4}$$

and two fuzzy generalizations of Restle's crisp similarity [16]

$$s_2(A,B) = 1 - \frac{1}{n}\big|(\neg A \cap B) \cup (A \cap \neg B)\big| \tag{5}$$

and

$$s_3(A,B) = 1 - \sup_x \big((\neg A \cap B) \cup (A \cap \neg B)\big) \tag{6}$$

where $\cap$ and $\cup$ are realized using appropriate t-norms and t-conorms [19] (in this paper we use the minimum and maximum operators), the cardinality is

$$|A| = \sum_{x \in X} \mu_A(x) \tag{7}$$

and for convenience we denote

$$\sup_x A = \sup_{x \in X} \mu_A(x) \tag{8}$$

Enta suggested a disconsistency index (or degree of separation) [4]

$$s_4(A,B) = 1 - \sup_x (A \cap B) \tag{9}$$

Fuzzy partitions are sets of fuzzy subsets, so the similarity of partitions can be defined based on the similarity of subsets. A similarity measure for partitions based on the subset similarity measure s_1 has been proposed by Runkler [17]. A fuzzy extension of the Rand index [15] has been proposed by Hüllermeyer and Rifqi [8]. Generalization of several similarity indices to fuzzy, probabilistic, and possibilistic partitions are presented in [1]. A further overview of similarity measures for partitions is going to appear in [9].

3 Similarity Based Merging of Partitions

We consider the problem of merging two partition matrices $U_1 \in [0,1]^{c\times n_1}$ and $U_2 \in [0,1]^{c\times n_2}$ specifying c fuzzy subsets of X_1 and X_2, respectively, to a joint partition matrix $U \in [0,1]^{c\times n}$ that specifies c fuzzy subsets of X, where each matrix U_1, U_2 and U holds the constraints at (1), where $X_1 \cup X_2 = X$, and where $n_1, n_2 < n$. The partition matrices U_1 and U_2 may or may not be produced by fuzzy clustering. If they come from fuzzy clustering, then they reflect a collaborative approach, where c clusters are found in each of the two subsets X_1 and X_2 of the data set X separately, and then the partitions are merged to represent c clusters of the whole data set X. We assume that U_1 and U_2 are extracted from U by a sampling process such that the c clusters in U_1 match the c clusters in U_2. Notice that this assumption does not generally hold for arbitrary partition matrices U_1 and U_2. Each row in U_1 and U_2 represents a fuzzy subset of X_1 and X_2, respectively, so merging U_1 and U_2 to U means to merge rows of U_1 and U_2 to rows of U. The order of rows in U_1 and U_2, however, is arbitrary, so we need to find out which row in U_1 matches which row in U_2. To do so, we define a set $X_A \subset X$ of anchor points that are added to X_1 and X_2 and serve as references to match the rows in U_1 and U_2. In other words, we partition X into the pairwise disjoint subsets X_1, X_2, and X_A, then find a partition matrix $[U_1, A_1]$ over $\{X_1, A\}$ and a partition matrix $[A_2, U_2]$ over $\{A, X_2\}$, and finally merge both partition matrices to $U = [U_1, A_1, U_2']$ over $X = \{X_1, A, X_2\}$, where $[\cdot]$ denotes the (horizontal) matrix concatenation. The matching of subsets in U_1 and U_2 is done by exchanging the rows in U_2 (forming a new matrix U_2') according to the matching of the anchor subsets A_1 and A_2. Figure 2 illustrates this merging scheme. Notice that this merging procedure is not symmetric but keeps A_1 and discards A_2, so merging $[U_1, A_1]$ and $[A_2, U_2]$ will usually yield a different result than merging $[U_2, A_2]$ and $[A_1, U_1]$. The matrix U_2' is generated

Fig. 2 Merging partition matrices U_1 and U_2 using anchors A_1 and A_2

from U_2 with respect to A_1 and A_2 using the following algorithm: For each row $i = 1, \ldots, c$ in A_1 find the most similar row in A_2 (using one of the subset similarity measures presented in the previous section) and use the corresponding row of U_2 as the ith row in U_2'.

4 Experiments

For our experiments we use a variant of the well-known BIRCH benchmark data set [20]. The original BIRCH data set is an artificial data set with 100 clusters in an evenly spaced 10×10 grid spaced by $4\sqrt{2}$, each consisting of 1000 points randomly generated using two-dimensional Gaussian distribution with variance $\sqrt{2}$. To simplify our experiments we consider a variant of this data set that contains only 2×2 clusters, each containing only 100 points, i.e. we have a total of $n = 400$ points with $c = 4$ clusters. The grey dots in the left view of Fig. 3 show this data set. In a first experiment we run regular FCM clustering, $c = 4$, 100 iteration steps, on this data set. The resulting cluster centers are shown as circles ($\bigcirc$). We compare the resulting partition with the ideal FCM partition computed using the ideal cluster centers $V = \{(0,0), (0, 4\sqrt{2}), (4\sqrt{2}, 0), (4\sqrt{2}, 4\sqrt{2})\}$ by (3). The left view of Fig. 4 shows the membership values for regular FCM (horizontal) versus the ideal FCM memberships (vertical). Ideally, all points should be on the unit main diagonal or on the reverse unit main diagonal (because of swapped rows in U). Here, most points are very close to the ideal case.

In our second experiment we randomly partition the data set into a set A of 40 anchor points and two subsets X_1 and X_2 with $(400 - 40)/2 = 180$ data

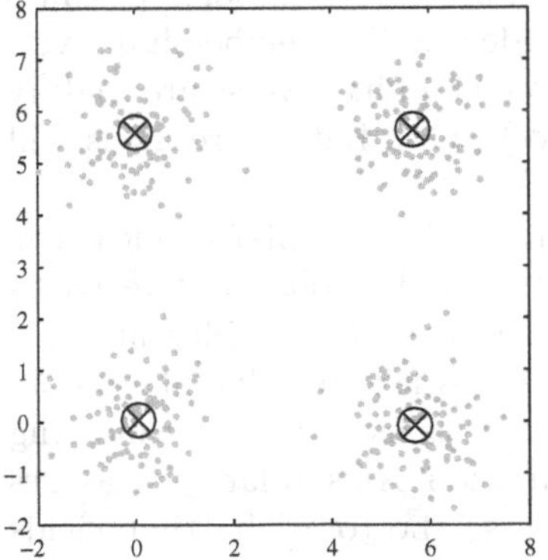

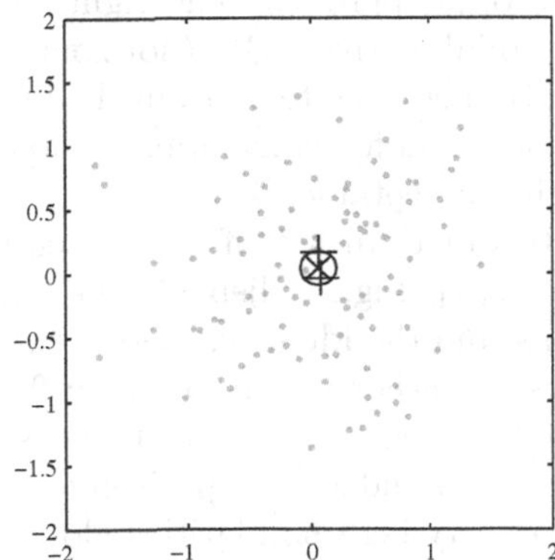

Fig. 3 Simplified BIRCH data set (grey dots), clustering results ($\bigcirc$), collaborative clustering results ($\times$). Right: Zoom of the bottom left cluster, partial clustering results ($+$).

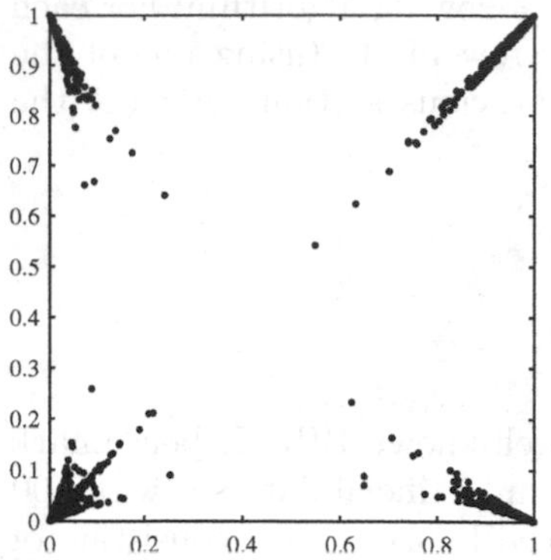

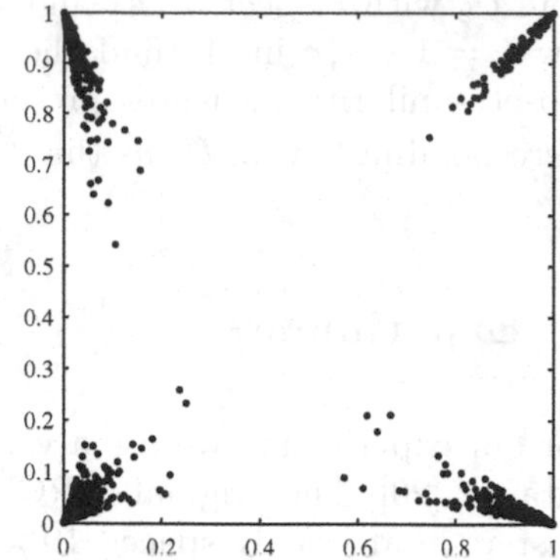

Fig. 4 Membership values of regular FCM (left, horizontal) and collaborative FCM (right, horizontal) versus ideal FCM memberships with the exact cluster centers (vertical)

points each. Then we produce the partition matrices $[U_1, A_1]$ and $[A_2, U_2]$ by running FCM, $c = 4$, 100 iteration steps, on $\{X_1, X_A\}$ and $\{X_A, X_2\}$, and merge U_1, U_2, A_1, and A_2 to $U = [U1, A_1, U_2']$ with the similarity measure s_1 (4), just as decribed in the previous section. Finally, we compute the equivalent cluster centers using (3). The ticks ($\times$) in Fig. 3 shows these cluster centers. They highly correspond with the cluster centers obtained by conventional clustering ($\bigcirc$). The right view of Fig. 3 shows a zoom of the bottom left cluster and the cluster centers (+) for the two partial results of the collaborative approach. The partial results slightly deviate from the results obtained by conventional clustering, but the merging process averages both cluster centers and finds a solution that highly coincides with the conventional approach. The right view of Fig. 4 shows the membership values for collaborative FCM (horizontal) versus the ideal FCM memberships (vertical). The deviations from the unit main diagonal and its reverse are slightly higher than for conventional (non-collaborative) FCM, but the result is still highly acceptable.

In our third set of experiments we compare all four similarity measures $s_1, \ldots, s_4$. Fig. 5 displays the similarities between the collaborative partitions and the ideal partitions (with ideal cluster centers) for different numbers of anchor points between 2 and 398 (averages for 10 different random initializations). Notice that each similarity measure is used twice: for merging partitions and for comparison with the ideal partition. All similarity measures yield very high similarities close to one except s_3 (bottom left) that yields similarities around 0.5. This is caused by the fact that s_3 is influenced by the fuzziness of the partitions. More specifically, s_3 can only achieve similarities of one for crisp partitions, and for ambiguous binary partitions with all

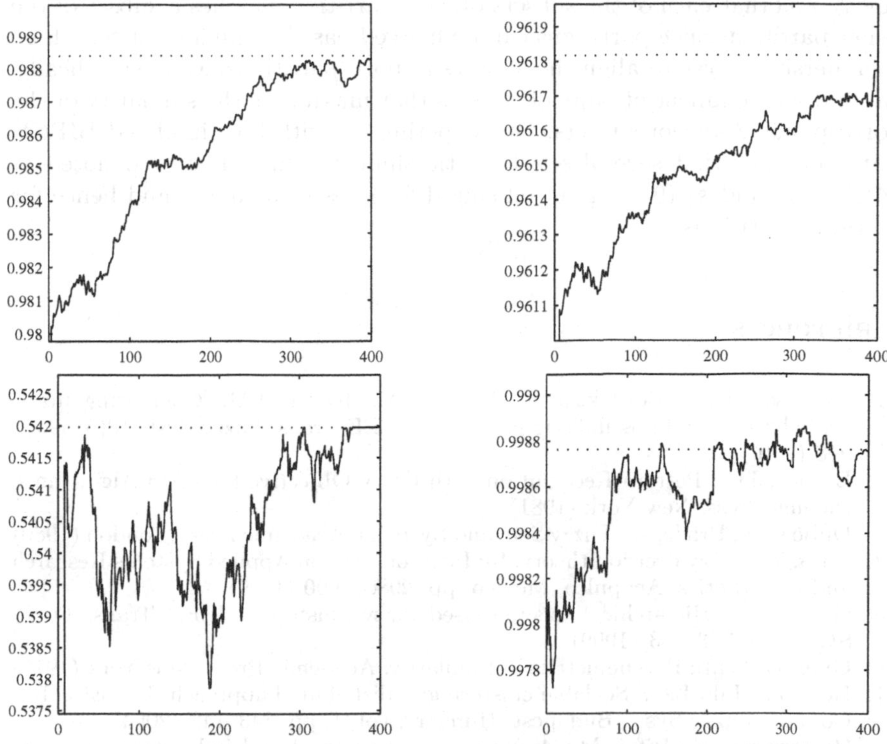

Fig. 5 Similarities between collaborative and ideal partitions for different numbers of anchor points (similarity measures $s_1, \ldots, s_4$)

$u_{ik} = 0.5$ we only get $s_3 = 0.5$. For s_1, s_2, and s_4, very high similarities are achieved, even with relatively few anchor points, and the similarity further increases with the number of anchor points which matches the intuitive expectation. For s_4 the maximum similarity is already approximately achieved for about 100 (25%) anchor points. To summarize, we obtain very good results with s_1, s_2, and s_4, but s_3 is less suitable for cluster merging.

5 Conclusions

We have introduced a novel approach to merge partitions that may or may not be generated by collaborative clustering. Without loss of generality we restricted to merge pairs of partitions; multiple merging can be realized by iterative pairwise merging. We assumed that each partition describes the same number of subsets (i.e. the partition matrices have the same number of rows), and that each of the subsets of each partition matches a subset of the other partition. Such partitions can be merged based on an anchor set whose memberships serve to align the subsets in both partitions. More specifically, we find an alignment of pairs of subsets that maximizes the similarity of the corresponding anchor subsets. Our experiments with the simplified BIRCH data set show that several set-theoretic similarity measures — denoted s_1 (4), s_2 (5), and s_4 (9) — are well suited for subset alignment and hence for merging partitions.

References

1. Anderson, D.T., Bezdek, J.C., Popescu, M., Keller, J.M.: Comparing fuzzy, probabilistic, and possibilistic partitions. IEEE Trans. Fuzzy Syst. 18(5), 906–918 (2010)
2. Bezdek, J.C.: Pattern Recognition with Fuzzy Objective Function Algorithms. Plenum Press, New York (1981)
3. Dubois, D., Prade, H.: Fuzzy Sets and Systems. Academic Press, London (1980)
4. Enta, Y.: Fuzzy decision theory. In: Int. Congress on Applied Systems Research and Cybernetics, Acapulco, Mexico, pp. 2980–2990 (1980)
5. Geva, A.B.: Hierarchical unsupervised fuzzy clustering. IEEE Trans. Fuzzy Syst. 7(6), 723–733 (1999)
6. Gregson, R.M.: Psychometrics of Similarity. Academic Press, New York (1975)
7. Hore, P., Hall, L.O.: Scalable clustering: a distributed approach. In: IEEE Int. Conf. on Fuzzy Syst., Budapest, Hungary, vol. 1, pp. 143–148 (2004)
8. Hüllermeier, E., Rifqi, M.: A fuzzy variant of the Rand index for comparing clustering structures. In: Joint IFSA World Congress and EUSFLAT Conference, Lisbon, Portugal, pp. 1294–1298 (2009)
9. Hüllermeier, E., Rifqi, M., Henzgen, S., Senge, R.: Comparing fuzzy partitions: A generalization of the Rand index and related measures. IEEE Trans. Fuzzy Syst. 20(3), 546–556 (2012)
10. Jaccard, P.: Étude comparative de la distribution florale dans une portion des alpes et des jura. Bulletin de la Société Vaudoise des Sciences Naturelles 37, 547–579 (1901)
11. Kaymak, U., Babuška, R.: Compatible cluster merging for fuzzy modelling. In: IEEE Int. Conf. on Fuzzy Syst., Yokohama, pp. 897–904 (1995)
12. Krishnapuram, R., Freg, C.P.: Fitting an unknown number of lines and planes to image data through compatible cluster merging. Pattern Recogn. 25(4), 385–400 (1992)
13. Pedrycz, W.: Collaborative fuzzy clustering. Pattern Recogn. Lett. 23, 1675–1686 (2002)
14. Pedrycz, W.: Collaborative and knowledge-based fuzzy clustering. Int. J. Innov. Comput. I 3(1), 1–12 (2007)
15. Rand, W.M.: Objective criteria for the evaluation of clustering methods. J. Am. Stat. Assoc. 66(336), 846–850 (1971)
16. Restle, F.: A metric and an ordering on sets. Psychometrica 24, 207–220 (1959)
17. Runkler, T.A.: Comparing Partitions by Subset Similarities. In: Hüllermeier, E., Kruse, R., Hoffmann, F. (eds.) IPMU 2010. LNCS, vol. 6178, pp. 29–38. Springer, Heidelberg (2010)

18. Runkler, T.A., Bezdek, J.C., Hall, L.O.: Clustering very large data sets: The complexity of the fuzzy c-means algorithm. In: European Symposium on Intelligent Technologies, Hybrid Systems and Their Implementation on Smart Adaptive Systems (eunite), Albufeira, pp. 420–425 (2002)
19. Schweizer, B., Sklar, A.: Associative functions and statistical triangle inequalities. Publ. Math–Debrecen. 8, 169–186 (1961)
20. Zhang, T., Ramakrishnan, R., Livny, M.: BIRCH: An efficient data clustering method for very large databases. In: ACM SIGMOD Int. Conf. on Management of Data, pp. 103–114 (1996)
21. Zwick, R., Carlstein, E., Budescu, D.V.: Measures of similarity among fuzzy concepts: A comparative analysis. Int. J. Approx. Reason. 1, 221–242 (1987)

Author Index